Fisheries Economics
Volume II

International Library of Environmental Economics and Policy
General Editors: Tom Tietenberg and Wendy Morrison

Titles in the Series

Fisheries Economics Volume II

Collected Essays

Edited by

Lee G. Anderson

College of Marine Studies, University of Delaware, USA

Routledge
Taylor & Francis Group

LONDON AND NEW YORK

First published 2002 by Ashgate Publishing

Reissued 2019 by Routledge
2 Park Square, Milton Park, Abingdon, Oxon OX14 4RN
52 Vanderbilt Avenue, New York, NY 10017

Routledge is an imprint of the Taylor & Francis Group, an informa business

Notice:
Product or corporate names may be trademarks or registered trademarks, and are used only for identification and explanation without intent to infringe.

Publisher's Note
The publisher has gone to great lengths to ensure the quality of this reprint but points out that some imperfections in the original copies may be apparent.

Disclaimer
The publisher has made every effort to trace copyright holders and welcomes correspondence from those they have been unable to contact.

A Library of Congress record exists under LC control number:

ISBN 13: 978-0-367-25576-3 (hbk)
ISBN 13: 978-0-367-25583-1 (pbk)
ISBN 13: 978-0-429-28850-0 (ebk)

Contents

Acknowledgements

The editor and publishers wish to thank the following for permission to use copyright material.

Academic Press Ltd for the essay: Jon M. Conrad (1992), 'A Bioeconomic Model of the Pacific Whiting', *Bulletin of Mathematical Biology*, **54**, pp. 219–39. Copyright © 1991 Society for Mathematical Biology, by permission of the publisher Academic Press.

Academic Press, USA for the essays: John R. Boyce (1996), 'An Economic Analysis of the Fisheries Bycatch Problem', *Journal of Environmental Economics and Management*, **31**, pp. 314–36. Copyright © 1996 Academic Press; Rögnvaldur Hannesson (1997), 'Fishing as a Supergame', *Journal of Environmental Economics and Management*, **32**, pp. 309–22. Copyright © 1997 Academic Press; Rögnvaldur Hannesson and Stein Ivar Steinshamn (1991), 'How to Set Catch Quotas: Constant Effort or Constant Catch?', *Journal of Environmental Economics and Management*, **20**, pp. 71–91. Copyright © 1991 Academic Press; James N. Sanchirico and James E. Wilen (1999), 'Bioeconomics of Spatial Exploitation in a Patchy Environment', *Journal of Environmental Economics and Management*, **37**, pp. 129–50. Copyright © 1999 Academic Press; Trond Bjørndal (1988), 'The Optimal Management of North Sea Herring', *Journal of Environmental Economics and Management*, **15**, pp. 9–29. Copyright © 1988 Academic Press; Kenneth E. McConnell and Jon G. Sutinen (1979), 'Bioeconomic Models of Marine Recreational Fishing', *Journal of Environmental Economics and Management*, **6**, pp. 127–39. Copyright © 1979 Academic Press; Lee G. Anderson (1993), 'Toward a Complete Economic Theory of the Utilization and Management of Recreational Fisheries', *Journal of Environmental Economics and Management*, **24**, pp. 272–95. Copyright © 1993 Academic Press; Frances R. Homans and James E. Wilen (1997), 'A Model of Regulated Open Access Resource Use', *Journal of Environmental Economics and Management*, **32**, pp 1–21. Copyright © 1997 Academic Press. By permission of the publisher Academic Press.

Blackwell Publishers & Copyright Clearance Center, Inc. for the essays: Trond Bjørndal (1987), 'Production Economics and Optimal Stock Size in a North Atlantic Fishery', *Scandinavian Journal of Economics*, **89**, pp. 145–64; Lee G. Anderson and Dwight R. Lee (1986), 'Optimal Governing Instrument, Operation Level, and Enforcement in Natural Resource Regulation: The Case of the Fishery', *American Journal of Agricultural Economics*, **68**, pp. 678–90. Copyright © 1986 American Agricultural Economics Association; Robert L. Kellogg, J.E. Easley, Jr. and Thomas Johnson (1988), 'Optimal Timing of Harvest for the North Carolina Bay Scallop Fishery', *American Agricultural Economics Association*, **70**, pp. 50–62. Copyright © 1988 American Agricultural Economics Association; Quinn Weninger (1998), 'Assessing Efficiency Gains from Individual Transferable Quotas: An Application to the Mid-Atlantic Surf Clam and Ocean Quahog Fishery', *American Journal of Agricultural Economics*, **80**, pp. 750–64. Copyright © 1998 American Agricultural Economics Association.

Series Preface

The *International Library of Environmental Economics and Policy* explores the influence of economics on the development of environmental and natural resource policy. In a series of twenty five volumes, the most significant journal essays in key areas of contemporary environmental and resource policy are collected. Scholars who are recognized for their expertise and contribution to the literature in the various research areas serve as volume editors and write an introductory essay that provides the context for the collection.

Volumes in the series reflect the broad strands of economic research including 1) Natural and Environmental Resources, 2) Policy Instruments and Institutions and 3) Methodology. The editors, in their introduction to each volume, provide a state-of-the-art overview of the topic and explain the influence and relevance of the collected papers on the development of policy. This reference series provides access to the economic literature that has made an enduring contribution to contemporary and natural resource policy.

TOM TIETENBERG
WENDY MORRISON
General Editors

Part IV
Extensions

Multispecies Models

[22]

Analysis of Open-Access Commercial Exploitation and Maximum Economic Yield in Biologically and Technologically Interdependent Fisheries

LEE G. ANDERSON[1]

College of Marine Studies and Department of Economics, University of Delaware, Newark, Del. 19711, USA

ANDERSON, L. G. 1975. Analysis of open-access commercial exploitation and maximum economic yield in biologically and technologically interdependent fisheries. J. Fish. Res. Board Can. 32: 1825–1842.

Fisheries may be interdependent because of biological relationships that exist between their stocks or because the gear of one affects mortality in the stock of the other. The problems of defining a maximum sustainable yield in these cases are discussed. A graphical analysis is used to describe the combinations of effort from both fisheries where concurrent exploitation is possible and which of these combinations will result in a simultaneous equilibrium. Finally the conditions for a combined maximum economic yield (MEY) are presented and it is shown that they will not hold if each fishery is managed to obtain an individual MEY.

ANDERSON, L. G. 1975. Analysis of open-access commercial exploitation and maximum economic yield in biologically and technologically interdependent fisheries. J. Fish. Res. Board Can. 32: 1825–1842.

Certaines pêcheries peuvent être interdépendantes par suite des relations biologiques qui existent entre les stocks ou du fait que l'engin de pêche utilisé sur l'un affecte la mortalité de l'autre. L'auteur examine les problèmes que pose, dans des cas de ce genre, la définition du maintien constant des prises maximales. A l'aide d'une analyse graphique, il décrit les diverses combinaisons d'efforts consacrés aux deux pêcheries lorsqu'une exploitation conjointe est possible, et laquelle de ces combinaisons produira un équilibre simultané. Finalement, il examine les conditions requises pour un rendement économique maximal combiné et démontre que ces conditions ne tiendront pas si chaque pêcherie est gérée en vue d'obtenir un REM individuel.

Received June 25, 1974
Accepted June 17, 1975

Reçu le 25 juin 1974
Accepté le 17 juin 1975

ALTHOUGH it is generally recognized that there are frequent interdependencies between commercially exploited fisheries, economic analysis of interdependencies has been quite limited. For example, Gordon (1953, 1954), Scott (1955), Crutchfield and Zellner (1962, p. 14), Christy and Scott (1965, p. 8) and Smith (1969) essentially assume the problem away for purposes of their analyses. Some work has been done in the general area, however. See for example Austin (1974), Quirk and Smith (1969), and C. W. Clark (unpublished data). However, there has not as yet been a complete economic analysis of it. This then is the purpose of this paper. An additional goal will be to compare it to the analysis of independent fisheries.

The format will be as follows. For each type of interdependence a basic model will be developed which will be used to describe the open-access operation and maximum economic yield (MEY) of the fisheries in a static framework.

The goal will be to provide an economic interpretation of the conditions in each. I will use a static framework because it can provide the essentials of the analysis in a straightforward manner. Also, because until recently it has been used almost exclusively in fishery economic discussions, this will allow easy comparison with earlier works. Nevertheless, a dynamic framework is necessary to obtain a complete grasp of the management problem. Therefore, in the final section the model is reformulated to consider proper control over time.

Two fisheries are biologically interdependent if their stocks engage in interspecific competition, i.e. if they compete for food, space, or some other limited facet of the environment, or if one preys upon the other. Fisheries are technologically interdependent if the gear of either affects the man-made mortality of the other. Each of these types will be discussed in turn using a general model the biological underpinnings of which are based on the works of Schaefer (1954, 1957, 1959) and Larkin (1963, 1966). Throughout, general implicit functions will be used be-

1826   J. FISH. RES. BOARD CAN., VOL. 32(10), 1975

cause interpretation of them and their derivatives is analytically more meaningful. The interested reader may derive a fairly straightforward explicit model for both of these cases by combining the above cited works.

Because I will be talking about different fisheries and different stocks, a note on definitions is in order at the outset. A fishery is composed of two components: the biological component which comprises the stock of fish that it exploits and the economic component which comprises the people and equipment that are available, in both the short and long run, to exploit the stock. It will be necessary at points in this discussion, especially with regard to technologically interdependent species, to distinguish between them.

Biological Interdependence

THE BASIC MODEL

The following is a straightforward expansion of the models used by Schaefer (1954, 1957, 1959). By equating the yield and the growth equations, it is possible to obtain a formulation for the equilibrium population size of a fish stock, which, when substituted back into the yield equation, obtains a steady-state or sustainable yield equation. This sustainable yield can then be used as the basis for the long-run revenue curve of the fishery for use in economic analysis of commercial exploitation. Essentially this procedure assumes that the fish stock obtains an instantaneous equilibrium with each change in the level of effort. This makes the analysis of economic variables much more simple.

The current model is different in that it shows the interrelationships between two fisheries, but it will basically hold to the simple steady-state analysis even though Smith (1969) has quite rightly pointed out that simultaneous biological and economic equilibrium may be difficult to obtain. The purpose here is to compare equilibrium points for different assumptions concerning the management of the fishery. Problems of possible disequilibrium will be discussed when they are relevant, however.

Consider two fisheries, Fishery 1 and Fishery 2, the fish stocks of which are biologically interdependent. This interrelationship can be expressed mathematically as:

$$P_1 = P_1(P_2) \qquad (1a)$$

$$\frac{dP_1}{dt} = g_1(P_1, P_2) \qquad (1a')$$

$$P_2 = P_2(P_1) \qquad (1b)$$

$$\frac{dP_2}{dt} = g_2(P_2, P_1) \qquad (1b')$$

P_1 and P_2 represent the size in weight of the respective populations. These equations say that population size in each stock is a function of the size in the other but that the instantaneous growth rate of each is related to the stock size of both. Equations 1a' and b' will be recognized as modified Schaefer type growth equations. Conceptually biological interdependence exists when only one of these variables, population size or growth rate, is a function of the size of the other stock. For completeness we will consider the more general case.

Let catch in each fishery be a function of the amount of effort used and the size of the populations. In this case catch in each fishery is exclusively from its own stock. Therefore, the terms catch in a particular fishery and yield from its stock can be used interchangeably. (This fortunate state of affairs does not exist in the case of technological interdependence.) These yield curves can be expressed as:

$$Y_1 = h_1[E_1, P_1(P_2)] \qquad (2a)$$

$$Y_2 = h_2[E_2, P_2(P_1)] \qquad (2b)$$

Y_1 and Y_2, and E_1 and E_2 represent the yields and the amounts of effort in the respective fisheries.

There will be a biological equilibrium in each of these fisheries when yield is equal to population growth, i.e. when:

$$h_1[E_1, P_1(P_2)] = g_1(P_1, P_2) \qquad (3a)$$

$$h_2[E_2, P_2(P_1)] = g_2(P_2, P_1) \qquad (3b)$$

Solving the above set of equations for P_1 and P_2 yields equations for the equilibrium population size in both stocks:

$$P_1 = \bar{P}_1(E_1, E_2) \qquad (4a)$$

$$P_2 = \bar{P}_2(E_2, E_1) \qquad (4b)$$

From the implicit function theorem, we know that such a solution exists if the appropriate Jacobian of 3a and b is nonvanishing. (See Lancaster 1968 p. 326ff for a discussion of this theorem.) The effect of a change in effort in one fishery on the equilibrium population size in the other can be expressed as:

$$\frac{\partial P_1}{\partial E_2} = \frac{dP_1}{dP_2}\left(\frac{\partial P_2}{\partial E_2}\right) \qquad (5a)$$

the yield from the stock of Fishery 2. Its size is dependent upon the levels of effort in both fisheries and the population size of stock 2. By the same token, catch in Fishery 2 is:

$$C_2 = Y_{22} + Y_{21}$$

Where

$$Y_{22} = H_{22}(E_2, E_1, P_2) \qquad (16a)$$

$$Y_{21} = H_{21}(E_2, E_1, P_1) \qquad (16b)$$

By definition, the catches in each fishery as described above must be such that they sum to the yield of the respective stocks; that is:

$$Y_1 = Y_{11} + Y_{21}$$

$$Y_2 = Y_{22} + Y_{12}$$

The two stocks will be in biological equilibrium when their growth is equal to their yield. Mathematically this can be expressed as:

$$G_1(P_1) = H_1(E_1, E_2, P_1) \qquad (17a)$$

$$G_2(P_2) = H_2(E_2, E_1, P_2) \qquad (17b)$$

With the same qualifications as described above, it is possible to derive equations in E_1 and E_2 for the equilibrium population size in both stocks from equations 17a and b. In their implicit form they will look like equations 4a and b, but in this case

$$\frac{\partial P_1}{\partial E_2} \text{ and } \frac{\partial P_2}{\partial E_1}$$

can be called the interfishery direct effect; a change in the level of effort in one fishery, will, through mortality, have a direct effect on the equilibrium population size in the other. Recall that in the previous case there was an indirect effect through the interrelationships of the populations (equations 5a and b).

The sustainable yield curve for each stock can be obtained by substituting the expression for the equilibrium stock size into equations 14a and b.

$$Y_1^* = H_1^* [E_1, E_2, P_1(E_1, E_2)] \qquad (18a)$$

$$Y_2^* = H_2^* [E_2, E_1, \bar{P}_2(E_2, E_1)] \qquad (18b)$$

Similarly these sustained yields can be broken down into the catch in each fishery. Sustainable catch in Fishery 1 is composed of two components:

$$Y_{11}^* = H_{11}^* [E_1, E_2, \bar{P}_1(E_1, E_2)] \qquad (15a')$$

$$Y_{12}^* = H_{12}^* [E_1, E_2, \bar{P}_2(E_2, E_1)] \qquad (15b')$$

Similarly the two parts of sustainable catch in Fishery 2 are:

$$Y_{22}^* = H_{22}^* [E_2, E_1, \bar{P}_2(E_2, E_1)] \qquad (16a')$$

$$Y_{21}^* = H_{21}^* [E_2, E_1, \bar{P}_1(E_1, E_2)] \qquad (16b')$$

Where

$$Y_1^* = Y_{11}^* + Y_{21}^*$$

$$Y_2^* = Y_{22}^* + Y_{12}^*$$

Here again, since sustained yield is a function of effort in both fisheries, the concept of maximum sustainable yield must be viewed in a different light. In this case, there will be one MSY for each stock, which can be obtained with an infinite number of combinations of effort from each fishery. Simultaneous MSY will require a unique combination of effort from each.

OPEN-ACCESS EXPLOITATION

Let α_{11} be the market price for the main catch of Fishery 1 and α_{12} be the price for its incidental catch. Similarly let α_{22} and α_{21} be the market prices for the main and incidental catches from Fishery 2. Even though the same stocks are involved, α_{11} and α_{21}, and α_{22} and α_{12} may not be equal because of differences in access to market channels, storage facilities, etc. in the two fisheries. In fact it may be possible that the market price of the incidental catch in either fishery may be zero (or even negative) if it is just returned to the sea.

The equations for the boundary curves in this case can be derived from equations 18a and b in the same manner as was described above, but they have a somewhat different meaning in this case. With biological interdependence, they represent the maximum amount of effort that one fishery can absorb for various levels of effort in the other. For example, B_1 shows how much E_1 is necessary to reduce sustainable yield in its stock to zero for the given level of effort in Fishery 2. In this case, both fisheries are obtaining catch from each stock and the boundary curves show the amount of effort in one fishery that when used in combination with the given level of effort in the other will reduce total sustainable yield for that stock to zero. That is, any combination of effort on the boundary curve for Fishery 1 will drive total sustainable yield from stock 1 to zero.

Examples of possible boundary curves are plotted in Fig. 4. The curves are both downward sloping because as effort is increased in either fishery, the amount of effort that is necessary to reduce sustainable yield in the other stock to zero decreases. Their exact slope depends upon the relative effects of effort in the two fisheries on catch and hence the population of each stock; if incidental catches are insignificant, the curves will have a very slight slope, but if they are substantial, the slopes will be quite steep.

In Fig. 4a the area bounded by the two curves

1836   J. FISH. RES. BOARD CAN., VOL. 32(10), 1975

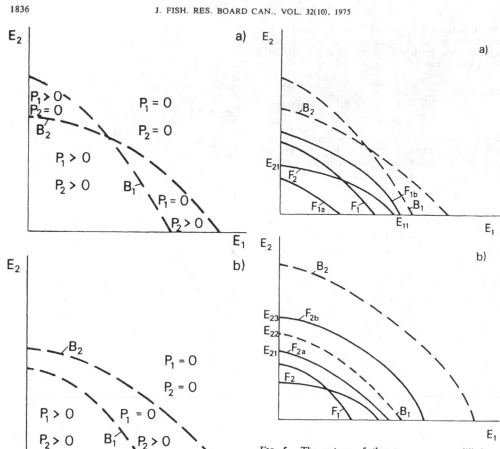

FIG. 4. The boundary curves for technologically interdependent fisheries show the combinations of effort in both fisheries that when used together will reduce sustainable yield for the stock to zero. With each stock, as the amount of effort in one fishery increases, the amount in the other that must be used with it to reduce sustainable yield to zero decreases. It is possible for one boundary curve to be completely inside the other.

represents the combinations of effort in both fisheries where both stocks can survive. In either of the areas that is inside one curve but outside the other, one of the stocks will be destroyed. In Fig. 4b, B_1 is everywhere inside B_2, which means that in this case the area bounded by B_1 represents the possible combinations of effort where both stocks will survive. In the previous case, when the boundary curves intersect, they become

FIG. 5. The nature of the open-access equilibrium depends upon the relative positions of the F curves. In Fig. 5a, note that if F_1 and F_2 are the curves, both fisheries will be commercially active. With F_{1a} and F_2 only Fishery 2 will be active and with F_{1b} and F_2 only Fishery 1 will be active. With F_1 and F_{2b} in Fig. 5b the stock from Fishery 1 will be destroyed.

horizontal or vertical straight lines because at that point both populations are equal to zero and the biological relationships no longer exist. Since the technical interrelationships are still valid when one stock is destroyed, the boundary curves have meaning beyond the intersection point in this case.

If we again let the unit cost of effort in the two fisheries be U_1 and U_2, respectively, the conditions for a simultaneous open-access equilibrium are:

$$\alpha_{11}Y_{11}{}^* + \alpha_{12}Y_{12}{}^* - U_1E_1 = 0 \qquad (19a)$$

$$\alpha_{22}Y_{22}{}^* + \alpha_{21}Y_{21}{}^* - U_2E_2 = 0 \qquad (19b)$$

A simultaneous equilibrium will occur when the sum of the revenues in both fisheries is equal to total cost. Here again the open-access equilibrium level of effort in one fishery depends upon the level of effort in the other. The equations for the open-access equilibrium curves which show the amount of effort that will be produced in one fishery for a given level of effort in the other under open-access conditions, can be derived from equations 19a and b in the manner described above. Examples of these curves are plotted in Fig. 5a and b. Because of the inter-fishery direct effect, the open-access amount of effort in each fishery will fall as the level of effort increases in the other. These F curves are subject to the same interpretation as those in biologically interdependent fisheries except for their relationship with the boundary curves. Since the F curves are related to catch from both stocks while the B curves are related to the yield of only one, the F curves are not bounded by the B curves. It sounds like a contradiction in terms, but in those areas where the F curve is beyond the B curve, the fishery in question is operating such that revenues equal costs strictly on its incidental catch; its main catch will be zero. By the same token if the F curve of one fishery is beyond the B curve of the other, its incidental catch is zero.

At the intersection of these two curves, both fisheries will be in an open-access equilibrium for the given level of effort in the other and so a simultaneous open-access equilibrium will exist. If no intersection exists in the positive quadrant, then only one or neither of the fisheries will be commercially active.

This can be demonstrated using Fig. 5a. The case where neither fishery will be active will occur when neither of the open-access curves are present in the positive quadrant. If F_1 and F_2 are the relevant curves, a simultaneous equilibrium is possible where both can operate. If F_{1a} and F_2 are the curves, the equilibrium will be on the vertical axis at E_{21}. The open-access equilibrium

level of effort in Fishery 2 will have such a high incidental catch for stock 1 that Fishery 1 will not be commercially profitable. The opposite will occur when F_{1b} and F_2 are the curves. The equilibrium will occur on the horizontal axis at E_{11}. The open-access equilibrium level of effort in Fishery 1 will obtain such a high incidental catch from stock 2 that Fishery 2 will be rendered commercially unprofitable.

The actual destruction of one of the stocks can be demonstrated in Fig. 5b where B_1 lies everywhere inside B_2. With F_1 and F_2 simultaneous commercial exploitation of both fisheries is possible at their intersection. However, if the price–cost relationship in Fishery 2 changes such that its open-access equilibrium curve shifts to F_{2a}, only that fishery will be active and it will operate at E_{21}. At this point the stock in Fishery 1 is still viable since E_{21} is less than E_{22} the point where stock 1 is destroyed when E_1 is zero. This means that Fishery 2 is taking some incidental catch of stock 1. In the case where the curve for Fishery 2 shifts out to F_{2b}, Fishery 2 will operate at E_{23} where the population of stock 1 will be destroyed. This, of course, means that not only is Fishery 1 completely out of business, but incidental catch in Fishery 2 is zero.

It should be acknowledged that the questions raised above about the actual dynamics of reaching an equilibrium point apply equally as well in this case also. In fact, because bionomic equilibrium in a stock depends upon its rate of growth and the rate at which effort will change in two fisheries, the problem may be even more complex.

MAXIMUM ECONOMIC YIELD

The analysis of individual MEYs and a combined MEY developed above applies directly to this case; only the exact makeup of the two maximizing conditions differ. For comparison purposes let us study these differences. The first order conditions for individual static MEYs are the following:

$$\alpha_{11}\left[\frac{\partial H_{11}{}^*}{\partial E_1} + \frac{\partial H_{11}{}^*}{\partial \bar{P}_1}\left(\frac{\partial \bar{P}_1}{\partial E_1}\right)\right] + \alpha_{12}\left[\frac{\partial H_{12}{}^*}{\partial E_1} + \frac{\partial H_{12}{}^*}{\partial \bar{P}_2}\left(\frac{\partial \bar{P}_2}{\partial E_1}\right)\right] - U_1 = 0 \qquad (20a)$$

$$\alpha_{22}\left[\frac{\partial H_{22}{}^*}{\partial E_2} + \frac{\partial H_{22}{}^*}{\partial \bar{P}_2}\left(\frac{\partial \bar{P}_2}{\partial E_2}\right)\right] + \alpha_{21}\left[\frac{\partial H_{21}{}^*}{\partial E_2} + \frac{\partial H_{21}{}^*}{\partial \bar{P}_1}\left(\frac{\partial \bar{P}_1}{\partial E_2}\right)\right] - U_2 = 0 \qquad (20b)$$

This says that for a given level of effort in the other fishery, each of these technically related fisheries can obtain an individual profit maximum by expanding effort until the increase in revenue from both its main and incidental catches is equal

to the unit cost of the effort. The change in sustained revenue from both types of catch resulting from an increase in effort can be divided into two components: the increase in catch due to the change in effort

$$\left(\text{i.e. } \alpha_{11} \frac{\partial H_{11}^{*}}{\partial E_1} \text{ and } \alpha_{12} \frac{\partial H_{12}^{*}}{\partial E_1} \right)$$

and the decrease due to the adverse effect on the relevant equilibrium populations

$$\left[\text{i.e. } \alpha_{11} \frac{\partial H_{11}^{*}}{\partial P_1} \left(\frac{\partial \bar{P}_1}{\partial E_1} \right) \text{ and } \alpha_{12} \frac{\partial H_{12}^{*}}{\partial P_2} \left(\frac{\partial \bar{P}_2}{\partial E_1} \right) \right].$$

This is not a true combined MEY, however, because the complete repercussions of the interfishery direct effect are not considered. A combined static MEY occurs when the sum of the profits is maximized. The first order conditions for this to occur are:

$$\alpha_{11} \left[\frac{\partial H_{11}^{*}}{\partial E_1} + \frac{\partial H_{11}^{*}}{\partial P_1} \left(\frac{\partial \bar{P}_1}{\partial E_1} \right) \right] + \alpha_{12} \left[\frac{\partial H_{12}^{*}}{\partial E_1} + \frac{\partial H_{12}^{*}}{\partial P_2} \left(\frac{\partial \bar{P}_2}{\partial E_1} \right) \right] + \alpha_{22} \left[\frac{\partial H_{22}^{*}}{\partial E_1} + \frac{\partial H_{22}^{*}}{\partial P_2} \left(\frac{\partial \bar{P}_2}{\partial E_1} \right) \right]$$
$$+ \alpha_{21} \left[\frac{\partial H_{21}^{*}}{\partial E_1} + \frac{\partial H_{21}^{*}}{\partial P_1} \left(\frac{\partial \bar{P}_1}{\partial E_1} \right) \right] - U_1 = 0 \qquad (21a)$$

$$\alpha_{22} \left[\frac{\partial H_{22}^{*}}{\partial E_2} + \frac{\partial H_{22}^{*}}{\partial P_2} \left(\frac{\partial \bar{P}_2}{\partial E_2} \right) \right] + \alpha_{21} \left[\frac{\partial H_{21}^{*}}{\partial E_2} + \frac{\partial H_{21}^{*}}{\partial P_1} \left(\frac{\partial \bar{P}_1}{\partial E_2} \right) \right] + \alpha_{11} \left[\frac{\partial H_{11}^{*}}{\partial E_2} + \frac{\partial H_{11}^{*}}{\partial P_1} \left(\frac{\partial \bar{P}_1}{\partial E_2} \right) \right]$$
$$+ \alpha_{12} \left[\frac{\partial H_{12}^{*}}{\partial E_2} + \frac{\partial H_{12}^{*}}{\partial P_2} \left(\frac{\partial \bar{P}_2}{\partial E_2} \right) \right] - U_2 = 0 \qquad (21b)$$

The difference between these and the previous conditions is that the extra terms take into account the effect of a change in effort in one fishery will have on the revenue from both main and incidental catches in the other fishery due to the direct effect of the effort itself and to the adverse effect on the equilibrium population size of both stocks. For a combined static MEY, effort should be expanded in each fishery until the change in revenue from main and incidental catches in both fisheries is equal to the unit cost of effort in that fishery.

In the case of technologically interdependent fisheries the tax per unit that will result in a combined static MEY can be conceptualized as the sum of two taxes; one to correct the misallocation that occurs in the catch from each stock in both fisheries. For example in Fishery 1, to correct the misallocation with regard to stock 1 there must be a tax equal to the difference between its average revenue from that stock and the sum of the marginal revenue of its main catch and the marginal revenue of the incidental catch in Fishery 2 both due to a change in its level of effort. This can be represented as

$$\alpha_{11} \frac{Y_{11}^{*}}{E_1} - \left(\alpha_{11} \frac{\partial Y_{11}^{*}}{\partial E_1} + \alpha_{21} \frac{\partial Y_{21}^{*}}{\partial E_1} \right).$$

To correct the misallocation in stock 2 from Fishery 1 there must be a tax equal to the difference between its average revenue for incidental catch and the sum of the marginal revenue of its incidental catch and the marginal revenue of the main catch of Fishery 2 both due to a change in

its level of effort. This can be represented as

$$\alpha_{12} \frac{Y_{12}^{*}}{E_1} - \left(\alpha_{12} \frac{\partial Y_{12}^{*}}{\partial E_1} - \alpha_{22} \frac{\partial Y_{22}^{*}}{\partial E_1} \right).$$

The sum of these two amounts is the proper tax per unit of effort. Of course, a similar tax is necessary for Fishery 2.

Dynamic Maximum Economic Yield

Static MEY considers net revenue in all periods to be strictly comparable, but dynamic MEY takes into account that net revenues in future periods must be discounted. Brown (1974), Clark (1973), Herfindahl and Kneese (1974), Neher (1974), Plourde (1970, 1971), and Quirk and Smith (1969) have studied dynamic MEY using optimal control theory and their results can be quite easily expanded to the case of interdependent fisheries as formulated here. It may be interesting to note at the outset, however, that with Schaefer type growth functions, a dynamic MEY will occur at a level of effort and an equilibrium stock size somewhere between those at the open-access equilibrium and the static maximum economic yield depending upon the size of the discount rate. The lower it is the more closely does dynamic MEY compare to static MEY.

The problem in the dynamic case is to find the proper time path of the two control variables, the amount of effort in the two fisheries, such that the discounted stream of net revenues is maximized. The problem can be solved by the application of Pontryagin's maximum principle (Intriligator

1971, page 348ff.) The proper present value Hamiltonian function for the case of biological interdependence is:

$$H = e^{-rt} [(\alpha_1 h_1 + \alpha_2 h_2 - U_1 E_1 - U_2 E_2)$$
$$+ \lambda_1 (h_1 - g_1) + \lambda_2 (h_2 - g_2)]$$

The only new parameters are λ_1 and λ_2, which are the auxiliary variables and may be interpreted as the imputed value of the respective stocks taking into account their growth potential; r, which is the discount rate; t, which represents the time

period. The term e^{-rt} is the continuous time discount factor. The first term inside the large brackets is the sum of the annual net revenues of the fisheries while the second two specify the net growth in the two stocks, i.e. the difference between yield and growth. Below these will be referred to $\dot{P}_1$, and $\dot{P}_2$. Equations 1a' and b', 2a and b are the formulations for h_1, h_2, g_1, and g_2, respectively. Note that these are not the sustainable yield curves as used above. The necessary conditions for a maximum for the Hamiltonian are:

$$\frac{\partial H}{\partial E_1} = (\alpha_1 - \lambda_1) \frac{\partial h_1}{\partial E_1} - U_1 \leqslant 0 \tag{22a}$$

$$\frac{\partial H}{\partial E_2} = (\alpha_2 - \lambda_2) \frac{\partial h_2}{\partial E_1} - U_2 \leqslant 0 \tag{22b}$$

$$e^{-rt} \left(\frac{d\lambda_1}{dt} - r\lambda_1 \right) = e^{-rt} \left[(\alpha_1 - \lambda_1) \frac{\partial h_1}{\partial P_1} + (\alpha_2 - \lambda_2) \frac{\partial h_2}{\partial P_2} \left(\frac{dP_2}{dP_1} \right) + \lambda_1 \frac{\partial g_1}{\partial P_1} + \lambda_2 \frac{\partial g_2}{\partial P_1} \right] \tag{23a}$$

$$e^{-rt} \left(\frac{d\lambda_2}{dt} - r\lambda_2 \right) = e^{-rt} \left[(\alpha_2 - \lambda_2) \frac{\partial h_2}{\partial P_2} + (\alpha_1 - \lambda_1) \frac{\partial h_1}{\partial P_1} \left(\frac{dP_1}{dP_2} \right) + \lambda_2 \frac{\partial g_2}{\partial P_2} + \lambda_1 \frac{\partial g_1}{\partial P_2} \right] \tag{23b}$$

$$\dot{P}_1 = h_1 - g_1 \tag{24a}$$

$$\dot{P}_2 = h_2 - g_2 \tag{24b}$$

To be formally precise we must also constrain the level of effort in each fishery to be nonnegative. For a steady-state interior solution (i.e. when the time paths for the levels of effort eventually settle down to nonextreme amounts each period), the above six equations must have a solution when the equality in 22a and b holds and when the time rate of change in the λ's and the net growth rates of both stocks are equal to zero. Since this is the situation most easily compared to a static MEY, it will prove useful to assume that this is the case and then interpret the meaning of the equations.

Moving in ascending order of ease of interpretation, when the net growth of the stocks are set equal to zero, equations 24a and b stipulate that natural growth must equal catch at all points in

time. Equations 22a and b state that the net value of the short-run marginal yield in both fisheries must equal its marginal cost of effort. Two things should be stressed at this point. First, we are concerned with the net value of catch, which is the difference between market price and the imputed value of a unit of stock in place. Second, we are talking about marginal catches due to a change in effort holding stock size constant. Above we are referring to the marginal catch after the stock had adjusted to the new level of effort and a new equilibrium stock size had been achieved.

Before analyzing 23a and b it will prove useful to reformulate them letting the time rate of change of the λ's equal zero and dividing both sides by the continuous time discount factor.

$$r\lambda_1 = (\alpha_1 - \lambda_1) \frac{\partial h_1}{\partial P_1} + (\alpha_2 - \lambda_2) \frac{\partial h_2}{\partial P_2} \frac{dP_2}{dP_1} + \lambda_1 \frac{\partial g_1}{\partial P_1} + \lambda_2 \frac{\partial g_2}{\partial P_1} \tag{23a'}$$

$$r\lambda_2 = (\alpha_2 - \lambda_2) \frac{\partial h_2}{\partial P_2} + (\alpha_1 - \lambda_1) \frac{\partial h_1}{\partial P_1} \frac{dP_1}{dP_2} + \lambda_2 \frac{\partial g_2}{\partial P_2} + \lambda_1 \frac{\partial g_1}{\partial P_2} \tag{23b'}$$

The first two terms on the right-hand side of each are the net values of the marginal yield in both fisheries provided by a change in the size of the stock size in the one. Remember that, just like effort, stock size is an input in the yield production function. The second two terms are the values of the marginal productivity of each stock in "providing" for natural growth in both, where the values are indicated by the λ's, the imputed value of a unit of stock in place. Note that the sign of the second and fourth term in each will depend upon the exact nature of the biological interdependence.

The left-hand side of both equations is the annual interest earned on a unit of stock in terms of the given discount rate. For a steady-state time path then, the sum of the values for the two uses of the stock (i.e. providing current catch and providing growth) in both of the interdependent stocks should be equal to the annual interest obtainable from an asset equal in value to the imputed value of the stock.

Interpreting the last two sets of equations as a unit, as indeed they must be, we come up with the same type of results as in the static analysis. Maximum economic yield will require that

changes in effort, and hence ultimately in stock size, in each fishery be evaluated by comparing the combined effects on both fisheries. This formulation, however, explicitly considers the differences between net revenue at different points in time and in doing so demonstrates that the interfishery effects on productivity of each stock in providing catch and growth must be considered with respect to the interest rates.

The analysis of technological interdependence is again quite similar and so it will only be necessary to sketch out the results. The proper Hamiltonian for the maximization of the present value of net revenues is:

$$H' = e^{-rt} [(\alpha_{11}H_{11} + \alpha_{12}H_{12} - U_1E_1 + \alpha_{22}H_{22}$$
$$- \alpha_{21}H_{21} - U_2E_2) + \lambda_1(H_{11} + H_{21} - G_1)$$
$$+ \lambda_2(H_{22} + H_{12} - G_2)]$$

where the yield and growth functions are as in equations 15a and b, 16a and b, and 13a and b, respectively. Note that here again we are not using the sustained yield functions. The first order conditions for an interior steady-state solution (other than the net growth of the stocks equaling zero) are as follows:

$$H' = e^{-rt}[(\alpha_{11}H_{11} + \alpha_{12}H_{12} - U_1E_1 + \alpha_{22}H_{22} - \alpha_{21}H_{21}$$
$$- U_2E_2) + \lambda_1(H_{11} + H_{21} - G_1) + \lambda_2(H_{22} + H_{12} - G_2)]$$

$$(\alpha_{11} - \lambda_1) \frac{\partial H_{11}}{\partial E_1} + (\alpha_{12} - \lambda_2) \frac{\partial H_{12}}{\partial E_1} + (\alpha_{22} - \lambda_2) \frac{\partial H_{22}}{\partial E_1} + (\alpha_{21} - \lambda_1) \frac{\partial H_{21}}{\partial E_1} - U_1 = 0 \quad (25a)$$

$$(\alpha_{22} - \lambda_2) \frac{\partial H_{22}}{\partial E_2} + (\alpha_{21} - \lambda_1) \frac{\partial H_{21}}{\partial E_2} + (\alpha_{11} - \lambda_1) \frac{\partial H_{11}}{\partial E_2} + (\alpha_{12} - \lambda_2) \frac{\partial H_{12}}{\partial E_2} - U_2 = 0 \quad (25b)$$

$$r\lambda_1 = (\alpha_{11} - \lambda_1) \frac{\partial H_{11}}{\partial P_1} + (\alpha_{21} - \lambda_1) \frac{\partial H_{21}}{\partial P_1} + \lambda_1 \frac{\partial G_1}{\partial P_1} \quad (26a)$$

$$r\lambda_2 = (\alpha_{22} - \lambda_2) \frac{\partial H_{22}}{\partial P_2} + (\alpha_{12} - \lambda_2) \frac{\partial H_{12}}{\partial P_2} + \lambda_2 \frac{\partial G_2}{\partial P_2} \quad (26b)$$

The first two state that the level of effort in each fishery should be expanded until the sum of the net revenues of the changes in short-run revenue in both fisheries that results are equal to the marginal cost of producing the effort. Again net value is the difference between market price of current catch and the imputed value of a unit of the respective stock in place. The second two equations can be interpreted that the net value of the increased yield in both fisheries provided by the last unit of population in each, plus the imputed value of the marginal productivity of the stock in providing growth must be equal to the

return possible from an asset equal in value to a unit of that stock.

The optimal taxes for the dynamic situation are similar to those above except that the net value of current harvest $(P_i - \lambda_i)$ is substituted for market prices and the average and marginal products of effort are evaluated using the short-run rather than the sustained yield function.

By considering only the conditions that must hold at a steady-state interior solution, we are gleaning only a small portion of the information possible concerning the path to and the makeup of the optimal point. Nevertheless, to go any further is beyond the scope of the present work. Following Clark (1973, page 956), however, the extinction of one or both species may occur at that point if the rate of interest is high relative to natural growth rates, and the market price of the last unit of stock is high relative to the cost of harvesting it. Also, following Brown (1974, page 164), both stocks may operate beyond maximum sustainable yield given the level of effort in the other fishery.

Conclusions

The purpose of this paper has been to describe the open-access exploitation of interdependent fisheries and the conditions necessary for a combined MEY. Two types of interdependent fisheries can be conceptualized: 1) biological, where the size of the stock of one fishery is dependent upon the size of the stock of the other, and 2) technological, where the gear of one fishery affects the mortality of the other. In both instances only a combined maximum sustainable yield can be defined. Although there are some fundamental differences in the nature of these interdependencies, economic analysis of the two types is substantially the same. It is possible to show the conditions under which open-access exploitation of the fisheries will result in none, one, or both of them being commercially profitable. In the case where only one of the fisheries is active it is also possible to distinguish between nonuse or only incidental use of the other stock and its actual destruction. It was also pointed out that this steady-state analysis ignores the dynamic problems of reaching a bionomic equilibrium. The basic elements of the dynamics were reviewed, however.

The analysis of maximum economic yield demonstrated that a combined MEY is different than each fishery trying to achieve an individual MEY and therefore it should be concluded that simultaneous regulation of interdependent fisheries is required. The conditions necessary for individualized and combined MEY's (both in a static and a dynamic sense) and mathematical expressions for the unit taxes on effort necessary to achieve each were provided. These spelled out in detail how the two types of interdependencies affect proper exploitation. The basic economic logic behind each was that the cost of a unit of effort in each fishery must be compared with the resulting change in revenue (in the short run and the long run) in both. It was also pointed out that at a combined MEY it is possible for one of the stocks to be unused or only incidentally used and perhaps even destroyed.

Acknowledgments

This study was begun while I was at the University of Miami and was completed at the University of Delaware. It was sponsored by their Sea Grant Institutional program which is administered by the National Oceanic and Atmospheric Administration of the United States Department of Commerce. I am grateful for comments provided by several referees and by G. Brown Jr., but accept full responsibility for the contents of the paper.

ANDERSON, L. G. 1973. Optimum economic yield of a fishery given a variable price of output. J. Fish. Res. Board Can. 30: 509–518.

AUSTIN, C. B. 1974. The economic implications of overfishing in a multi-species fishery. Ph.D. dissertation. Clark Univ., Worcester, Mass. 125 p.

BROWN, G. JR. 1974. An optimal program for managing common property resources with congestion externalities. J. Polit. Econ. 82: 163–174.

CHRISTY, F., AND A. D. SCOTT. 1965. The common wealth in ocean fisheries. The Johns Hopkins Press, Baltimore, Md. 281 p.

CLARK, C. W. 1973. Profit maximization and the extinction of animal species. J. Polit. Econ. 81: 950–961.

COPES, P. 1970. The backward-bending supply curve of the fishing industry. Scott. J. Polit. Econ. 17: 69–77.

CRUTCHFIELD, J. A., AND A. ZELLNER. 1961. Economic aspects of the Pacific halibut fishery. Fish. Ind. Res., Vol. 1. Washington, D.C.: U.S. Dep. Inter. 173 p.

GORDON, H. S. 1953. An economic approach to the optimum utilization of fisheries resources. J. Fish. Res. Board Can. 10: 442–457.

    1954. The economic theory of a common property resource. J. Polit. Econ. 62: 124–142.

GOULD, J. R. 1972. Extinction of a fishery by commercial exploitation: a note. J. Polit. Econ. 80: 1031–1038.

HERFINDAHL, O., AND A. KNEESE. 1974. Economic theory of natural resources. Merrill Publishing Co., Columbus, Ohio. 405 p.

INTRILIGATOR, M. 1971. Mathematical optimization and economic theory. Prentice-Hall, Englewood Cliffs, N.J. 508 p.

LANCASTER, K. 1968. Mathematical economics. The MacMillan Co., New York, N.Y. 411 p.

LARKIN, P. A. 1963. Interspecific competition and exploitation. J. Fish. Res. Board Can. 20: 647–678.

    1966. Exploitation in a type of predator–prey relationship. J. Fish. Res. Board Can. 23: 349–356.

NEHER, P. A. 1974. Notes on the Volterra-Quadratic fishery. J. Econ. Theory. 8: 39–49.

PLOURDE, C. G. 1970. A simple model of replenishable natural resource exploitation. Am. Econ. Rev. 60: 518–522.

    1971. Exploitation of common property replenishable natural resources. West. Econ. J. 9: 256–266.

QUIRK, J., AND V. SMITH. 1969. Dynamic economic models of fishing, p. 3–32. In A. Scott. Economics of fisheries management: a symposium. Univ. British Columbia, Vancouver, B.C.

SCHAEFER, M. B. 1954. Some aspects of the dynamics of populations important to the management of commercial marine fisheries. Inter-Am. Trop. Tuna Comm. Bull. 1: 25–56.

——— 1957. Some considerations of population dynamics and economics in relation to the management of the commercial marine fisheries. J. Fish. Res. Board Can. 14: 669–681.

——— 1959. Biological and economic aspects of the management of commercial marine fisheries. Trans. Am. Fish. Soc. 88: 100–104.

SMITH, V. L. 1969. On models of commercial fishing. J. Polit. Econ. 77: 181–98.

SOUTHEY, C. 1972. Political prescriptions in bionomic models: the case of the fishery. J. Polit. Econ. 80: 769–775.

[23]

JOURNAL OF ENVIRONMENTAL ECONOMICS AND MANAGEMENT **31**, 314–336 (1996)
ARTICLE NO. 0047

An Economic Analysis of the Fisheries Bycatch Problem*

JOHN R. BOYCE

University of Auckland, Auckland, New Zealand

Received March 13, 1995; revised November 28, 1995

Bycatch is the incidental take of a species that has value to some other group. This paper compares open access and individual transferable quota equilibria to the equilibrium in which the joint value of the fisheries is maximized. The open access induced problems can be corrected by an individual transferable quota system only if both the target species and the bycatch species have tradable quotas, and only if the bycatch species does not have existence value. There exists a range of the bycatch-to-target species harvest levels for which the total harvest of each will be exactly taken by a given technology, even under open access. However, there may not even exist a unique open access equilibrium if bycatch is allocated by "rule of capture." Prohibitions on the sale of bycatch reduce the bycatch level, but they also reduce social welfare. © 1996 Academic Press, Inc.

> It is inevitable that [halibut] will be caught in various degrees and proportions when trawling for other species.

F. H. Bell, 1981

1. INTRODUCTION

One of the most vexing problems facing managers of fishery stocks is the problem of incidental harvesting of non-targeted species. Bycatch, as the incidental catch is called, occurs with almost every fishery to some degree since the harvester does not observe exactly what he is catching until his gear is drawn to the surface.[1] However, the term "bycatch" is generally used to describe incidental catch in a fishery for which there exists another constituency with a claim on the bycatch species. Though sonar fish finders, improved technologies in trawl net design, increased use of pots, and other gear substitution may reduce bycatch, as long as the target and the non-target species intermingle it is often impossible to eliminate it entirely.[2] Public pressure (e.g., concerning dolphin bycatch in the tuna fishery in the Pacific Ocean and the Gulf of Mexico or green sea turtle bycatch in the Gulf of Mexico shrimp fishery), legal requirements such as the Endangered Species Act (e.g., concerning Columbia River chinook salmon), and political pressure from competing interest groups (e.g., concerning incidental take of halibut, crab, and salmon in the North Pacific groundfish fisheries and incidental take of chum

---

* This paper has benefited from comments made by Diane Bischak, two anonymous referees, and an associate editor. The usual disclaimer applies.

[1] Mortality rates in ocean fisheries bycatch are high because the fish being taken are pulled to the surface too quickly or in too great a mass to survive the pressure (e.g., [18]).

[2] The cod and pollock trawl fisheries in the North Pacific are moving toward nets which have square shaped spaces rather than diamond shaped spaces. The diamond shaped spaces become elongated under pressure, reducing the chance that smaller and flatter fish (e.g., halibut) can escape. The square spaces technology is an attempt to reduce this type of bycatch. However, even this method cannot exclude bycatch of different but similar sized species. (See [23, April 1993, p. 61].) An exception is the turtle excluder devices (TEDs) which have virtually eliminated turtle bycatch in the Southeast shrimp fishery [25].

ECONOMIC ANALYSIS OF THE BYCATCH PROBLEM   315

salmon in the Aleutian Islands sockeye salmon intercept fishery) all force managers to impose limits on bycatch of certain species.

Thus bycatch presents several unique problems to managers. First, the manager
is faced with the efficiency (and political) problem of determining the allocation of
the bycatch between competing interests. Second, once the manager has determined an allocation, the allocation may not be internally consistent (i.e., may not
maximize profits for individuals participating in the fishery) while at the same time
satisfying resource conservation constraints. For example, if the total allowable
catch (TAC) for the bycatch species is reached before the TAC for the target
fishery species, the TAC in the target fishery may not be harvested, escapement
may be higher than desired, and fish may be "left on the table," in the jargon of
fishermen.[3] Thus the bycatch problem presents a challenge to managers' attempts
to control harvest and escapement simultaneously in the target and bycatch
fisheries. Third, bycatch will likely change as factors such as technology and prices
change. Thus a program which is successful in one state of the world may fail in
another. This point is particularly apt for fisheries managers since most of the
current direction in regulation of bycatch is toward gear restrictions or time and
area closures. While these methods may reduce bycatch, it is the exceptional case
in which these restrictions eliminate bycatch or give fisherman an incentive to
internalize the full costs of bycatch.[4] Furthermore, while such restrictions may be
effective at reducing bycatch, their economic viability is often questionable.[5]

This paper presents a stylized model of bycatch in a fishery. The problem is
examined from the perspective of a single season. Although there are exceptions
(such as predator–prey relations between species), TAC limits on levels of harvest
for both the target and bycatch species can be treated as predetermined within a

[3] For example, in the North Pacific groundfish fishery, halibut bycatch TAC limits forced early
closure of the 1992 longline cod fleet, with approximately 27,000 metric tons of the cod TAC not taken
[23, Aug. 1992]. In 1990, the domestic flatfish fisheries in Zone 1 of the Bering Sea were closed on
February 27, 1990 because of *C. bairdi* crab bycatch, though it later resumed in March. On March 14,
1990, the domestic flatfish fishery was again closed due to halibut bycatch, and on March 19, 1990, the
entire Bering Sea and Aleutian Islands fishery was closed to domestic flatfish fisheries because of
halibut bycatch [23, May 1990]. After noting that more than 47,000 metric tons of sole ($22.8 million,
gross value) were unharvested by the domestic fleet and that about half the 127,000 metric tons of sole
would be unharvested by the joint-venture fleet, both because the 1990 halibut bycatch TAC constraint
was reached in the Bering Sea, National Marine Fishery Service Biologist Janet Smoker observed "It
looks like a lot of money will be left in the ocean this year" (quoted in [23, June 1990, p. 63]).
[4] The turtle excluder devices (TEDs) have been reported to be quite successful in the shrimp
fisheries. However, green sea turtles are only one source of bycatch. The shrimp fisheries are reported
to catch more finfish biomass than shrimp [23, May 1992, p. 25] and more uneconomič mollusks [17, pp.
18–19]. Similarly, in the North Pacific, time and area closures have protected halibut spawning grounds
from the groundfish fleets. However, halibut bycatch is almost impossible to eliminate fully given that
halibut and cod coexist in similar ecological niches (e.g., [5]). In the "Area M" salmon fishery in the
Bering Sea, which targets sockeye salmon destined for Bristol Bay, there is bycatch of chum salmon
destined for Norton Sound [9]. As the two species are similar in size and in migration patterns, simple
time and area closures will not eliminate bycatch without shutting down the target fishery.
[5] A disadvantage of technologies created by the regulatory process is that it is not clear such
technologies are economically sound. This is not true for technologies adopted under an ITQ or tax
system. If fishermen adopt new technologies, the technology is economically viable. Ward [25] has
developed a model to show the effect on stocks and allocations if a new technology is adopted, but there
are no costs of adopting the technology. Thus, while his model tells us something about what might
happen if new technology is adopted, he tells us nothing about the viability of the technology. In
addition, fishermen have complained that the process of testing new gear types is too slow under the
command and control management in the North Pacific groundfish fishery [23, Apr. 1993, p. 61.].

JOHN R. BOYCE

season due to resource conservation constraints.[6] In addition, it is assumed that the bycatch is produced incidentally by the target fishery, as pollution is produced incidentally in the production of steel or sawdust is produced incidentally in the production of lumber. The model is also determinant as there is assumed to be no uncertainty regarding harvest rates or prices. Finally, since the viability of gear restrictions is mainly an empirical question, the focus is on what effect taxes or individual transferable quotas (ITQs) have on bycatch. Looking at the problem from the context of a single season and treating the fisheries as distinct allocations, accurately represents the manner in which many fisheries are managed. Legal requirements force managers to set season limits on the catch of the target species to maintain sustainability of the stocks. Thus, even if the species are treated as part of a multi-species fishery, harvest TACs may exist for individual species. In addition, fisheries are often managed on a multi-species basis, so limits on bycatch are set in the general context of allocation of the resource among competing uses.

This paper focuses on three questions: (1) How should the bycatch species be allocated among its competing uses? (2) How does open access affect bycatch rates? (3) Can rationalization through individual transferable quotas or taxes achieve the social optimal allocation of bycatch and effort?

Of the assumptions made in this paper, the assumption regarding the bycatch technology is the most restrictive. Bycatch is treated as a function solely of the harvest rate of the target species. This is a very restrictive form of a multi-product production function. In part, it may be defended by an appeal to the fact that in many fisheries, the bycatch is actually quite small relative to the harvest level. For example, bycatch of salmon in the North Pacific groundfish fisheries is roughly one salmon per fifty tons of groundfish. This becomes economically significant only when the amount of groundfish is quite large. Bycatch of Washington and Oregon salmon by Alaska trollers targeting salmon from Alaskan rivers is even less. Clearly, however, objections can be raised to this defense. Shrimp fisheries, for example, regularly have bycatch of "trash fish", mollusks and other bottom dwellers, that is in excess of the quantity of shrimp recovered on a pound to pound basis. Other fisheries are truly multi-species fisheries. For example, the "Area M" fishery which intercepts sockeye salmon en route to Bristol Bay in Alaska also catches large proportions of chum salmon bound for the Yukon River and Norton Sound [9]. In cases such as this, depicting the fishery as a target fishery (sockeye salmon) with bycatch (chum salmon) is clearly a stretch. However, in such cases the present model may serve as a useful simplification of a difficult problem.

2. MODEL AND ASSUMPTIONS

Assume that there exist two biological species, the target and bycatch species.[7] The target stock is harvested by fisherman in Fishery One. The bycatch species

---

[6] For examples of papers where the interdependent biological relationships are considered in the context of bioeconomic models see [11] and [26].

[7] The simplification of treating the target species as a single species and the bycatch as a single species is justified on the grounds that in the North Pacific cod, pollock, and sablefish are managed as single species in the Gulf of Alaska [20]. However, in the Bering Sea, the groundfish resource (yellowfin sole, pollock, Pacific ocean perch, turbot, Atka mackerel, Pacific cod, and sablefish) is managed as a complex (i.e., as a multi-species fishery) [21]. This paper focuses on the simpler case where the target species and bycatch can be considered as separate single species.

may be the target stock in Fishery Two (e.g., in the North Pacific, crab is bycatch to the groundfish fishery), or it may be a species not targeted by any commercial fishery (e.g., dolphins to the tuna fishery). The bycatch species may have existence value (e.g., [15]) even if it has no commercial value.[8] The bycatch species is taken incidentally in Fishery One while pursuing the target species. Let S_1 denote the total allowable catch of the target species which may be removed within a season by Fishery One, and let S_2 denote the TAC of the bycatch species which may be removed by Fishery One and Two together. S_1 and S_2 are determined prior to the season, say by biological (escapement) or legal requirements. The bycatch species might be a stock which is protected by the Endangered Species Act, a stock managed independently (in which case a proportion B might be allocated to Fishery One and $S_2 - B$ allocated to Fishery Two), or it might be a stock jointly allocated between Fishery One and Fishery Two on a first come, first served, basis.

A. Technology Assumptions

Assume that fisherman within each fishery are homogeneous in terms of opportunity costs, fishing skills, and technology, although there may be differences between the technologies used in the two fisheries. In addition, assume that there are no stock or congestion externalities in either fishery. These assumptions imply that stock and aggregate effort levels do not enter into the profit function [6]. Let variable profits for harvest of the target species for vessel j in Fishery One be defined as $\pi_1(h_{1j}; P_1)$, where h_{1j} is the harvest per day of the target species by vessel j, and P_1 is the output price of the target species. (Since the output level h_{1i} is the only choice variable, input prices are ignored.) Assuming fishermen are homogeneous, the j subscript on the harvest level is omitted except where doing so will cause confusion, i.e., $h_{1j} = h_1$, $j = 1, \ldots, N_1$, where N_1 is the maximum possible number of entrants into Fishery One. Variable profits in Fishery Two are denoted by $\pi_2(h_{2j}; P_2)$. Variable profits in each fishery have the following properties (dropping the P_i arguments):

Assumption A.1. $\pi_i(0) = 0$; $\pi_i'(h_i) > 0$ for all $h_i \geq 0$; $\pi_i''(h_i) < 0$ for $h_i > 0$.

The first two parts of A.1 imply that zero harvest yields zero profits and that profits increase as the harvest rate increases. The third part of A.1 states that profits are concave in h_i.

The bycatch technology assumption is that there is a single species model with unwanted (or desired) production of bycatch being a function of the output of the target species (not necessarily in fixed proportions). Let $b(h_1)$ denote the per day removal of the bycatch species by Fishery One for a harvest level h_1 of the target species. Fishery Two is assumed to have no bycatch of the targets species for Fishery One.[9] As the rate of harvest of the target species is the only variable

---

[8] The existence value aspect of some bycatch was suggested by an anonymous referee.

[9] Bycatch is usually asymmetric in this sense. A very clear set of examples has to do with the salmon fisheries in the North Pacific. The bycatch of Columbia River sockeye salmon by the Southeast Alaska troll fishery is not symmetric; there is no corresponding Southeast Alaska salmon bycatch in the Columbia River area. Similarly, in the Area "M" intercept sockeye salmon fishery, bycatch of Norton Sound and Yukon River chum salmon occurs, but no bycatch of sockeye salmon occurs in the Norton Sound or Yukon River chum salmon fisheries. While bycatch of cod and pollock may also occur in the halibut and crab fisheries, such bycatch is trivial.

available a fisherman controls, input substitution is ignored in the analysis. The Fishery One bycatch function is assumed to have the following properties:

Assumption A.2. $b(0) = 0$; $b(h_1) > 0$, $b'(h_1) > 0$, and $b''(h_1) \geq 0$, for all $h_1 > 0$.

The first three parts of A.2 says that zero bycatch is possible only with zero output of the target species, and that bycatch is positive and increases as output of the target species increases. Thus bycatch is "essential" to the target fishery [13]. Regarding the fourth assumption, when $b'' = 0$, there is a fixed-proportion relationship between bycatch and harvest of the target species (i.e., $b(h_1) \equiv \alpha h_1$, for some non-negative constant α), and when $b'' > 0$, the ratio of bycatch to the target species increases as the harvest rate increases.[10] This functional relationship assumes that methods to reduce bycatch require greater care be taken in harvesting the target species, slowing down that harvest rate, but that it is impossible to fully eliminate bycatch with the given technology. The costs of separating, counting, and either selling or disposing of the bycatch are assumed to depend on the harvest level, so these costs are already accounted for in the π_1 function. Thus part of the reason for the decline in marginal profits ($\pi_i'' < 0$) is due to the increase in the bycatch proportion as harvest of the target species increases. Bycatch is being modeled as though it were "pollution" being generated with output (e.g., [7]), with part of the cost being internalized (sorting, counting, etc.) and part being external (the reduction in available stock to others). However, the pollution analogy is incomplete since part of the external cost is borne by others within the industry in the bycatch case due to the TAC constraint. Pollution controls generally are stated in terms of pollution allowed per firm.

To consider several cases within the context of a single model a pair of parameters (δ, γ) will be used to differentiate between the cases. Let $\delta \in \{-1, 0, 1\}$ be the weight associated with bycatch in the objective function of a vessel (or society) in Fishery One. When $\delta = 1$, the vessel is allowed to sell the bycatch (at price P_2) in addition to the selling target species harvest at price P_1. When $\delta = 0$, the vessel derives no direct value from the bycatch. When $\delta = -1$, each unit of bycatch costs society P_2, say from foregone existence value. The parameter $\gamma \in \{0, 1\}$ is used to switch between having an active commercial fishery (Fishery Two) targeting the bycatch species and not having one. When $\gamma = 0$, the bycatch species has no commercial value, although it may have existence value if $\gamma = 0$ and $\delta = -1$.

Given a season length of T_1 (e.g., the number of days the fishery is open), a market price of P_1 for the target species (dollars per fish), P_2 for the bycatch species (dollars per fish), and an identical fixed but avoidable cost k_1 (dollars per vessel), season profits to vessel j in Fishery One are[11]

$$v_{1j} = T_1 \left[\pi_{1j}(h_{1j}) + \delta P_2 b(h_{1j}) \right] - k_1, \qquad j = 1, \ldots, N_1. \tag{1}$$

[10] When $b'' = 0$, there is no way to reduce the ratio of bycatch to target species harvest ratio. However, if $b'' > 0$, the bycatch to target species harvest ratio may be reduced by slowing down the harvest rate on each vessel. Berger, *et al.* [6] have found that there is considerable variability in bycatch to harvest ratios in the Bering Sea groundfish trawl fisheries. The North Pacific Fisheries Management Council considered plans to kick individual fishermen out of the fishery if their bycatch rate was too high. This suggests that fishermen can control bycatch to some extent with the given technology.

[11] Note that when $\delta = -1$, the term in square brackets is identical in form to a pollution model with the firm paying P_2 per unit pollution, $b(h_1)$, emitted.

When $b'' > 0$, if $\delta = 1$, it is possible for the term in square brackets to not be concave. (This is not a problem if $\delta = -1$ or $\delta = 0$.) Thus, our final assumption is:

Assumption A.3. $\pi''(h_1) + \delta P_2 b''(h_1) < 0$ for $h_1 > 0$, so $\pi_1(h_1) + \delta P_2 b(h_1)$ is sufficiently concave such that second-order conditions hold.

If Fishery Two exists, then given a season length of T_2, bycatch market price of P_2, and a fixed but avoidable cost k_2 (also identical across fishermen), season profits to vessel j in Fishery Two are

$$v_{2j} = T_2 \gamma \pi_{2j} - k_2, \qquad j = 1, \dots, N_2. \tag{2}$$

Suppose that $n_1 \leq N_1$ vessels participate in the fishery targeting the target species. The TAC constraint for the target species is

$$S_1 - T_1 \sum_{j=1}^{n_1} h_{1j} = S_1 - T_1 n_1 h_1 \geq 0, \tag{3}$$

where the equality holds due to fishermen being homogeneous. Similarly, for the bycatch species, the bycatch TAC constraint is

$$S_2 - T_1 \sum_{j=1}^{n_1} b(h_{1j}) + T_2 \sum_{j=1}^{n_2} h_{2j} = S_2 - T_1 n_1 b(h_1) + T_2 n_2 h_2 \geq 0, \tag{4}$$

where the equality holds due to fishermen being homogeneous.[12] The constraint in (4) could be rewritten as two constraints, one for each fishery, if the allocation were to be divided up between Fishery One and Fishery Two as,

$$S_2 - B - T_2 n_2 h_2 \geq 0, \quad \text{and} \quad B - T_1 n_1 b(h_1) \geq 0, \quad \text{for } 0 \leq B \leq S_2. \tag{5}$$

The advantage of writing the constraint as (5) is that it shows that a positive bycatch allocation is necessary for Fishery One to exist given assumption A.2. In the event that $\gamma = 0$, and $\delta = 1$, $B \leq S_2$ denotes the share of the possible total (S_2) allowable bycatch that the fishery is allocated. Note that both (3) and (4) are quasi-convex in (T_i, n_i, h_i, B). This is used below to establish sufficiency for the Kuhn–Tucker conditions.

In addition, assume that there exists an upper bound on the length of the season for each fishery. For example, these constraints could be due to the seasonal nature of spawning of the target and bycatch species. Given the fixed-but-avoidable costs of entering, such a constraint is necessary to solve the social planner's problem (cf. Clark [8, pp. 240–43]). Thus we require[13]

$$T_1 \leq \bar{T}_1, \quad \text{and} \quad T_2 \leq \bar{T}_2. \tag{6}$$

---

[12] It is assumed that bycatch cannot be simply discarded without being counted toward one's quota. In the North Pacific groundfish fishery, this is enforced by an "observer program" where trained observers monitor what is being caught on each vessel. In the case where bycatch can be sold, this problem would not occur except for high-grading (e.g., [1], [2]).

[13] The assumption of a fixed-but-avoidable cost is necessary to obtain a determinacy for n and T (where the i subscripts have been dropped). In the simple model with no bycatch, if $k = 0$ the first-order conditions to the social planner's problem reduce to the two equations in three unknowns: $\pi' = \pi/h$ and $S = Tnh$. While h is determined exactly, T and n are not. The constraint on T is also necessary. If $k > 0$ and there is no constraint on T, then the solution to the social planner's problem involves the non-solvable equations $T\pi = k$, and $n\pi = 0$. Both assumptions are plausible for the real world. Most fisheries require some reworking of gear or travel to the fishery to participate, thus $k > 0$. Also, in many fisheries, the species is economically viable to harvest only at certain times of the year due to biological or market conditions, so the assumption of a maximum season length is also plausible.

JOHN R. BOYCE

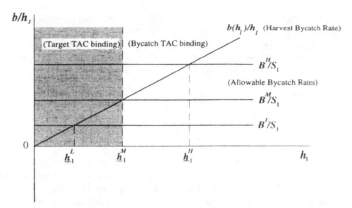

FIG. 1. Relationship between harvest level in Fishery One and which TAC constraint binds.

Finally, there are also non-negativity constraints for the h_{ij}, and n_i, and the T_i.[14]

B. Which Constraint Will Bind?

The ratio of bycatch to harvest of the target species Fishery One is defined to be $b(h_1)/h_1$. Thus if n_1 vessels fish T_1 time periods, they will take $T_1 n_1 b(h_1)$ of the bycatch and $T_1 n_1 h_1$ of the target species. Thus for a given h_1, the binding constraint will be determined by whether

$$b(h_1)/h_1 \lessgtr B/S_1. \tag{7}$$

If the bycatch ratio b/h is greater than Fishery One's bycatch species TAC to target species TAC ratio B/S_1, then the bycatch constraint will be binding; otherwise the target species TAC constraint will be binding. For future reference, define the point $\underline{h}_1$ as the value of h_1 for which (7) is an equality. When $b'' > 0$, the bycatch ratio increases as h_1 increases (i.e., $d[b/h]/dh_1 = [b'h_1 - b]/h^2 > 0$). Thus, for $h_1 > \underline{h}_1$, the bycatch species TAC constraint binds, and for $h_1 < \underline{h}_1$, the target species TAC constraint binds. Figure 1 shows three possible $\underline{h}_1$ values corresponding to three different levels of the bycatch constraint. Since $B^H > B^M > B^L$, the corresponding $\underline{h}_1$ levels are ordered as $\underline{h}_1^H > \underline{h}_1^M > \underline{h}_1^L$.[15]

3. THE SOCIAL PLANNER'S PROBLEM

A. Optimization Problem for Social Planner

Given fixed output prices, the value society obtains from the two fisheries is given by

$$V = T_1 n_1 [\pi_1(h_1) + \delta P_2 b(h_1)] - k_1 n_1 + T_2 n_2 \gamma \pi_2(h_2) - k_2 n_2, \tag{8}$$

[14] The constraints that $n_i \leq N_i$ are ignored. In open access, N_i is unlimited, and the social planner may use up to the same amount as in open access.

[15] Note that the sign of the slope of b/h and all other lines drawn in the figures can be shown to be correct, but whether the line is linear, convex, concave, or otherwise is indeterminate.

which is simply the sum of economic profits in the two fisheries. Equation (8) can be shown to be quasi-concave in (T_i, n_i, h_i, B). This plus the quasi-convexity of the constraints (3)–(6) ensure sufficiency for the Kuhn–Tucker conditions the Arrow–Enthoven sufficiency theorem, assuming the constraint qualification is met. The Lagrangian for this problem is

$$L = V + \lambda[S_1 - T_1 n_1 h_1] + \mu_1[B - T_1 n_1 b(h_1)] + \mu_2[S_2 - B - T_2 n_2 h_2]$$

$$+ \sigma_0 B + \sigma_2[S_2 - B] + \sum_{i=1}^{2} \left\{ \tau_i[\overline{T}_i - T_i] + \phi_i h_i + \theta_i n_i + \psi_i T_i \right\},$$

where the Lagrange multipliers λ, μ_i, σ_0, σ_2, τ_i, ϕ_i, θ_i, and ψ_i, $i = 1, 2$, correspond to the constraints (3), (5), (6), and the non-negativity constraints, respectively.

In fisheries such as the North Pacific groundfish fishery, bycatch of crab, halibut, and salmon are of commercial value, but they are not allowed to be sold by the groundfish fishery. (The bycatch is disposed of by dumping it back into the sea.) By the envelope theorem, it may be seen that this restriction is not based on efficiency.[16]

PROPOSITION 1. *Assuming that the bycatch species is of commercial value ($\delta \neq -1$), society would be better off if the target fishery were able to sell the incidental catch ($\delta = 1$) than not ($\delta = 0$).*

Thus the prohibition in many fisheries on Fishery One selling bycatch is based on something other than efficiency. More likely, it is based on a desire by Fishery Two to reduce the incentive for Fishery One to benefit from the bycatch. Allowing Fishery One to sell the bycatch gives them an incentive to incur higher bycatch rates. Thus it increases the competition for the bycatch species, reducing the number of vessels in Fishery Two (cf. [18]).

Assuming the social planner chooses the bycatch allocation B, effort in each fishery h_i, $i = 1, 2$, the number of entrants in each fishery n_i, $i = 1, 2$, and the season length in each fishery T_i, $i = 1, 2$, the system of first-order conditions to the social planner's problem include the constraints (3)–(6) and (with arguments of the functions suppressed)[17]

$$\partial L / \partial B = \mu_1 - \mu_2 + \sigma_0 - \sigma_2 = 0,$$

$$\sigma_0 \geq 0, \quad B\sigma_0 = 0, \quad \sigma_2 \geq 0, \quad \sigma_2[S_2 - B] = 0, \tag{9}$$

$$\partial L / \partial h_1 = T_1 n_1[\pi_1' + (\delta P_2 - \mu_1)b' - \lambda] + \phi_1 = 0,$$

$$h_1 \geq 0, \quad \phi_1 \geq 0, \quad h_1 \phi_1 = 0, \tag{10}$$

$$\partial L / \partial h_2 = T_2 n_2[\pi_2' - \mu_2]\gamma + \phi_2 = 0,$$

$$h_2 \geq 0, \quad \phi_2 \geq 0, \quad h_2 \phi_2 = 0, \tag{11}$$

$$\partial L / \partial n_1 = T_1[\pi_1 + (\delta P_2 - \mu_1)b - \lambda h_1] - k_1 + \theta_1 = 0,$$

$$n_1 \geq 0, \quad \theta_1 \geq 0, \quad n_1 \theta_1 = 0, \tag{12}$$

[16] Proofs to all propositions are available from the author.
[17] Second-order conditions can be shown to hold for all interior solutions. They are available from the author.

JOHN R. BOYCE

$$\partial L/\partial n_2 = T_2[\pi_2 - \mu_2 h_2]\gamma - k_2 + \theta_2 = 0,$$

$$n_2 \geq 0, \quad \theta_2 \geq 0, \quad n_2\theta_2 = 0, \tag{13}$$

$$\partial L/\partial T_1 = n_1[\pi_1 + (\delta P_2 - \mu_1)b - \lambda h_1] - \tau_1 + \psi_1 = 0,$$

$$T_1 \geq 0, \quad \psi_1 \geq 0, \quad T_1\psi_1 = 0, \tag{14}$$

$$\partial L/\partial T_2 = n_2[\pi_2 - \mu_2 h_2]\gamma - \tau_2 + \psi_2 = 0,$$

$$T_2 \geq 0, \quad \psi_2 \geq 0, \quad T_2\psi_2 = 0. \tag{15}$$

Equation (9) shows that if the bycatch is allocated to each fishery, then $\mu_1 = \mu_2$; i.e., the marginal value of the bycatch is equal across the fisheries. Otherwise, bycatch will be allocated entirely to the fishery with the largest μ_i. If the harvest level of the target species is positive ($h_1 > 0$), (10) says the harvest level is chosen such that the marginal profit equals the sum of the scarcity rents on the target λ and bycatch species $\mu_1 b'$. When $h_2 > 0$, (11) shows that the optimal harvest level equates marginal profit from harvesting the bycatch species with marginal scarcity rent μ_2. Equation (12) shows that when $n_1 = 0$, $\theta_1 = k_1$, and that when $n_1 > 0$, the return on the target species (and the bycatch species if $\delta = 1$) over the entire season *net* of harvesting costs and scarcity rent, λ or $\mu_1 b$ (plus the existence value cost if $\delta = -1$), just equals the cost of an additional vessel, k_1. Equation (13) has a similar interpretation as (12), with $\theta_2 = k_2$ when $n_2 = 0$, and net returns over the entire season equal the entry cost of an additional vessel if $n_2 > 0$. Equations (14) and (15) are composed of two sets of terms. The expressions in the square brackets are seen to be the net return per vessel per unit time. The value of an extension in the season lengths are thus the net return per vessel per unit time times the number of vessels n_i.

Let us now state:

PROPOSITION 2. *If the bycatch species has commercial value the social planner will allocate the bycatch and fish each fishery as follows*:

(i) *If $0 = \mu_1 < \mu_2$, $B = 0$, Fishery Two takes the entire bycatch allocation utilizing the entire season, and Fishery One does not fish (i.e., $T_1 = n_1 = h_1 = 0$, $T_2 = \bar{T}_2$, $n_2 > 0$, $h_2 > 0$, and $\lambda = 0$)*;
(ii) *If $R\mu_1 > \mu_2 = 0$, $B = S_2$, Fishery One takes the entire bycatch allocation, but not all of the target species TAC, utilizing the entire season available to it, and Fishery Two does not fish (i.e., $T_2 = n_2 = h_2 = 0$, $T_1 = \bar{T}_1$, $n_1 > 0$, $h_1 > 0$, and $\lambda > 0$)*;
(iii) *If $\mu_1 = \mu_2 > 0$, $0 < B < S_2$, and both Fishery One and Fishery Two operate for the entire season, taking the full allocation of the bycatch species (i.e., $T_1 = \bar{T}_1, T_2 = \bar{T}$)*;
(iv) *If $S_2 = 0$, neither fishery operates (i.e., $T_1 = n_1 = h_1 = T_2 = n_2 = h_2 = 0$)*.

Proposition 2 shows that if both stocks have commercial value, it is optimal to either utilize both stocks (case iii), with bycatch being utilized fully, or to fish only in the Fishery which has the highest marginal value for an additional unit of bycatch (cases i and ii). If the bycatch TAC were also a choice variable (say in a multi-season model with maximization of (8) within each season), then it is clear that the bycatch species will always be fully utilized (either as a target species, as bycatch, or as both). Thus the long run trade-off is between increasing the bycatch

TAC now and in the future. There would be tremendous short-term pressure to increase the TAC of the bycatch now, especially if one of the fisheries is shut down as a result of the low bycatch species TAC.

A corollary to Proposition 2 has to do with the case where the manager has no control over the bycatch allocation to Fishery One. In this case, B may be chosen by another agency (e.g., halibut and groundfish in the North Pacific) or as a legal limit imposed on the take of species for which there is no commercial value, but for which society has existence value (e.g., dolphins in the tuna fishery). Since there are no stock effects [6, 8],

COROLLARY 2.1. *If the allocation of bycatch to Fishery One is chosen by means other than maximizing* (8), *then Fishery One will harvest over the entire allowable season* (i.e., $T_1 = \bar{T}_1$), *for whichever allocation(s) it fully utilizes.*

B. Optimal Solution When Bycatch Has No Commercial Value

Next, let us characterize the solution under several different scenarios, beginning with the case where no Fishery Two exists (so $B = S_2$) and the target species TAC constraint binds ($\lambda > 0$).

1. *TAC for target species binding.* Let $\gamma = 0$, implying that there is no Fishery Two, and assume $\lambda > 0$, implying that the target species TAC binds. Given Corollary 2.1, let $\mu_1 = 0$. Then from (10), $\lambda^* = \pi_1' + \delta P_2 b'$, where x_1^* denotes the solution when $\gamma = 0$ and $\lambda > 0$, for $x = h, n, T, \lambda$. Using Corollary 2.1 (so $T_1 = \bar{T}_1$) and plugging λ^* into (12), the optimal harvest rate must satisfy

$$k_1/\bar{T}_1 = \pi_1 - \pi_1' h_1^* + \delta P_2[b - b' h_1^*] = \Phi_1(h_1^*) + \delta P_2 \Gamma_1(h_1^*), \qquad (16)$$

where $\Phi_i \equiv \pi_i - \pi_i' h_i$, $i = 1, 2$, and $\Gamma \equiv b_1 - b_1' h_1$. The function Φ_1 is the profit per day net of scarcity rent on the target species, ignoring bycatch. $\Phi_i' = -\pi_i'' h_i > 0$. $P_2 \Gamma' < 0$ is the reduction in net profits per day when bycatch is sold.

If $\delta = 0$ or $b'' = 0$, the bycatch does not affect profits except through the increased costs in the π_1 function for disposal. In this case (16) says that $\Phi_1 = k_1/\bar{T}_1$, or that the optimal harvest rate is set such that net profits per day equal fixed cost per day. When $\delta = 1$ and $b'' > 0$, the bycatch can be sold. Since $P_2 \Gamma' < 0$, (16) implies that a larger harvest level is optimal. This relationship is shown in Fig. 2. If each bycatch removal results in lost existence value ($\delta = -1$) the optimal harvest rate decreases relative to the case where $\delta = 0$. These results, however, are contingent upon the convexity of the bycatch function. If $b'' = 0$, then $\Gamma = 9$, and the optimal harvest level is unaffected by δ.[18] It can also be shown that (10) and A.3 are sufficient to ensure that second-order conditions are satisfied for selection of h_1^*, given $T_1 = \bar{T}_1$ and (3) determines n_1^*.

2. *TAC for bycatch binding.* Now suppose that the bycatch TAC constraint binds rather than the target species TAC constraint. Then $\lambda = 0$ and $\mu_1 > 0$. Let

---

[18] When $\pi = P_1 h_1 - c_1(h_1)$, and $b(h_1) = \alpha h_1$ (so $b'' = 0$), (16) becomes $c_1'(h_1)h_1 - c_1(h_1) = k_1/T_1$. Which shows that the optimal harvest rate is independent of both P_2 and P_1. That is, the optimal harvest rate is chosen such that costs are minimized. So long as profits are positive, neither the output price nor the bycatch price affects this decision when $b'' = 0$. When $b'' > 0$, (17) shows that the output decision is affected by P_2 since an increase in bycatch affects the cost minimizing choice.

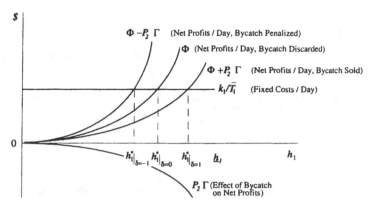

FIG. 2. Optimal harvest rate for Fishery One when target TAC is binding and no Fishery Two exists.

h_1^{**} denote the optimal level of harvest given that the bycatch constraint is binding. Using (10) to eliminate μ_1 in (12), the optimal harvest level h_1^{**} thus solves

$$k_1/\overline{T}_1 = \pi_1 - [b/b'h_1^{**}]\pi_1'h_1^{**} \equiv \Phi_1(h_1^{**}) + [1 - 1/\beta(h_1^{**})]\pi_1'(h_1^{**})h_1^{**}, \tag{17}$$

where $\beta \equiv b'h/b$ is the elasticity of bycatch with respect to harvest of the target species. A comparison of (17) and (16) shows that the difference is that in (17) the $\delta P_2\Gamma$ term from (16) is replaced by $(1 - 1/\beta)\pi_1'h_1$. However, when $b'' = 0$, $\beta = 1$ and $\Gamma = 0$, so the second term drops out in both expressions, and the optimal harvest level is the same whichever constraint is binding. When $b'' > 0$, $\beta > 1$. Thus $0 < 1 - 1/\beta < 1$. Since $\pi_1'h_1 > 0$, the optimal harvest level h_1^{**} is greater than h_1^*, the optimal harvest level when the target species TAC is binding or $b'' = 0$. However, note that $\partial h_1^{**}/\partial\delta = 0$, unlike the case where the target species TAC binds. If the bycatch constraint binds, being able to sell the bycatch has no effect on the harvest level.

A comparison of the equilibrium harvest levels is shown in Fig. 3 for the case where $\delta = 0$. For the case where $\underline{h}_1^L < h_1^{**} < h_1^*$, the target TAC cannot be binding since $h_1^* > \underline{h}_1^L$. Thus the bycatch TAC constraint is binding. When $h_1^{**} < h_1^* < \underline{h}_1^H$, the bycatch TAC cannot be binding since $h_1^{**} < \underline{h}_1^H$. Thus the target TAC constraint is binding. The next proposition shows what happens when $h_1^{**} < \underline{h}_1^M < h_1^*$:

COROLLARY 2.2. *When $h_1^{**} < \underline{h}_1^M < h_1^*$, the optimal solution is $h_1^{***} = \underline{h}_1^M$. Thus both TAC constraints bind simultaneously, so $b(h_1^{***})/h_1^{***} = B/S_1$.*

3. *The optimal of number of entrants.* Since the season length is always the maximum allowable length, the binding TAC constraint plus (16) or (17) determines the number of entering vessels. Thus for if the target species TAC is binding, $N_1^* = S_1/\overline{T}_1 h_1^{**}$ and when the bycatch TAC constraint binds, $N_1^{**} = B/\overline{T}_1 b(h_1^{**})$. In either case, an increase in h_1 means less entrants. This just show that the social planner is trading off marginal harvesting costs with marginal entry costs in (16)

Fisheries Economics II

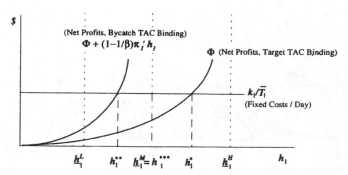

FIG. 3. Optimal harvest rate for Fishery One when bycatch TAC is binding and bycatch has no commercial value ($\delta = 0$).

and (17). In both (16) and (17), the right-hand side is an increasing function of the harvest rate. Thus, either an increase in entry costs, k_1 or a decrease in the length of the season, $\overline{T}_1$ results in an increased harvest rate per vessel. Thus with higher entry costs, the social planner uses each vessel more intensely so that the number of vessels may be reduced. Note that an increase in B or S_1 (whichever is binding) does not affect the optimal harvest per vessel, but does increase the number of vessels used.

C. Both Stocks Exploited Commercially

If both fisheries are exploited commercially, then $\gamma > 0$. Since the bycatch species has commercial value, assume that $\delta \neq -1$. First, consider the case where both stocks are fully exploited, so that $0 < B < S_2$.

1. *Both stocks fully utilized.* When both stocks are fully utilized, $T_1 = \overline{T}_1$, and $T_2 = \overline{T}_2$. In addition, the first-order conditions (9) and (11) imply,

$$\mu_1 = \mu_2 = \pi_2'(h_2), \tag{18}$$

which says that the marginal value of another unit of the bycatch stock is equal to its value in production in the Fishery Two, which targets that stock. Similarly, from (10):

$$\lambda = \pi_1'(h_1) + \delta P_2 b'(h_1) - \pi_2'(h_2) b'(h_1). \tag{19}$$

This says that the value of an additional unit of the target species stock equals the value of a marginal unit of production from that stock minus the marginal value of an additional unit of bycatch. Using (18) and (19), the optimal levels of harvest per vessel are given by

$$k_1/\overline{T}_1 = \Phi_1(\hat{h}_1^{***}) + \left[\delta P_2 - \pi_2'(\hat{h}_2^*) \right] \Gamma(\hat{h}_1^{***}), \tag{20}$$

$$k_2/\overline{T}_2 = \Phi_2(\hat{h}_2^{***}), \tag{21}$$

where Φ_i and Γ are defined as before, and where a caret $(\hat{\ })$ denotes the difference between the cases where Fishery Two exists and where it does not exist, and $\hat{h}_1^{***}$, $i = 1, 2$, indicates that *both* TAC constraints are binding. In the event that $\lambda = 0$, it can be shown that (20) collapses to (17).

Figure 4 shows that analysis. From Eq. (16), the solution is h_1^* in Fig. 4. Comparing (20) with (16) shows the difference made by taking account of the second fishery (the $-\pi_2'\Gamma$ term). When profits in both fisheries are maximized simultaneously, the harvest level in Fishery One is smaller than it would be if Fishery Two did not exist. This is because the cost of bycatch includes foregone revenues in Fishery Two when it exists. However, this result only holds if $b'' > 0$. In the event that $b'' = 0$, $\Gamma = 0$, so (20) and (16) are identical. On the basis of (16), (17), and (20), we state:

PROPOSITION 3. *If bycatch is a fixed proportion of total catch of the target species (i.e. $b'' = 0$), then the optimal harvest rate in Fishery One depends only upon entry costs k_1 relative to the marginal harvesting profits of the target species.*

Let us now compare the optimal harvest rate in Fishery One for the case where bycatch affects the optimal harvest rate in Fishery One (i.e., $b'' > 0$). As in Fig. 2, the net marginal profits when $\delta = 0(\Phi_1 - \pi_2'\Gamma)$ lies above the net marginal profits curve $[\Phi_1 + (P_2 - \pi_2')\Gamma]$ when $\delta = 1$ since $\Gamma < 0$. Thus, the harvest rate in Fishery One increases as δ increases. When the target fishery is allowed to sell the bycatch the cost of the bycatch constraint declines, and vessels increase the harvest rate, incurring a higher bycatch rate.

The optimal number of vessels are given by the TAC constraints for the target and bycatch species, respectively, i.e.,

$$\hat{n}_1^{***} = S_1/\overline{T}_1\hat{h}_1^{***}, \tag{22}$$

$$\hat{n}_2^{***} = \left[S_2 - \overline{T}_1\hat{n}_1^{***}b\left(\hat{h}_1^{***}\right)\right]/\overline{T}_2\hat{h}_2^{***}, \tag{23}$$

Note that due to the recursive nature of the solution (20)–(23), n_2^{***} is determined by what is left over after Fishery One takes its share. (I.e., $\hat{h}_2^{***}$ solves (21); $\hat{h}_1^{***}$ solves (20); $\hat{n}_1^{***}$ solves (22); so $\hat{n}_2^{***}$ solves (23).)

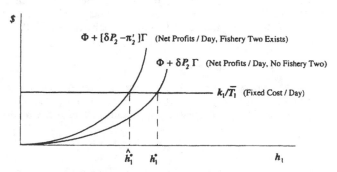

FIG. 4. Optimal harvest rate for Fishery One when bycatch is commercially harvested by Fishery Two, and both fisheries are active, compared with cost where no Fishery Two exists.

It was remarked below (20) that if $\lambda = 0$, (20) collapses to (17). It is now shown that $\lambda = 0$ is not optimal:

COROLLARY 2.3. *If the marginal value of the bycatch stock is identical across each fishery ($\mu_1 = \mu_2$), then the only unique solution occurs when both stocks are fully harvested.*

Thus each stock is exploited for the entire possible season, and over the course of that season, the entire TAC is removed for the bycatch stock, and maybe for the target stock. The optimal allocation of the bycatch species is thus $\hat{B}^{***} = \overline{T}_1 \hat{n}_1^{***} b(\hat{h}_1^{***})$ to Fishery One, and $S_2 - \hat{B}^{***} = \overline{T}_2 \hat{n}_2^{***} \hat{h}_2^{***}$ to Fishery Two.

2. *Bycatch allocated entirely to fishery two.* When the bycatch is allocated to Fishery Two, Fishery One is shut down since bycatch is essential to production in Fishery One by A.2. The harvest level in Fishery Two is given by (21) (denoted as $\hat{h}_2^*$, to distinguish from the case where both stocks are fully utilized), and the number of entrants solves

$$\hat{n}_2^* = S_2 / \overline{T}_2 \hat{h}_2^*. \tag{24}$$

3. *Bycatch allocated entirely to fishery one.* If $\mu_1 > \mu_2$, then all of the bycatch species TAC is allocated to Fishery One. Since $\mu_1 > 0$ it implies that the bycatch TAC constraint is binding for Fishery One. Thus, unless $S_2/S_1 = b/\hat{h}_1^{**}$, where $\hat{h}_1^{**}$ solves (17), $\lambda = 0$. Thus the solution is given by $\hat{h}_1^{**}$ solving (17) and the number of entrants is given by

$$\hat{n}_1^{**} = S_2 / \overline{T}_1 b(\hat{h}_2^{**}). \tag{25}$$

4. OPEN ACCESS EQUILIBRIA

Under open access each entrant chooses a harvest rate to maximize profits, but entry drives economic profits to zero. Thus

$$\pi_1(h_{1j}) + \delta P_2 b(h_{1j}) = k_1/T_1, \quad \text{and} \quad \pi_2(h_{2j}) = k_2/T_2, \forall j. \tag{26}$$

The season profits for vessel j in Fishery One depends upon which TAC constraint(s) is (are) binding. If the target species TAC is binding in Fishery One, the season length is $T_1 = S_1/\sum_{j=1}^{n_1} h_{1i}$. If the bycatch constraint is binding and $\gamma = 0$, then $T_1 = S_2/\sum_{j=1}^{n_1} b(h_{1j})$. If $\gamma = 1$ and the bycatch is allocated on a "rule of capture" basis between the target and bycatch fisheries, then $T_1 = S_2/[\sum_{j=1}^{n_1} b(h_{1j}) + \sum_{j=1}^{n_2} h_{2j}]$.

A. Only Fishery One Commercially Exploited

1. *TAC for target species binding.* If the target species harvest constraint is binding, then using the season profits (2) with the season length substituted out using (4) as above, the level of harvest which maximizes profits can be shown to satisfy[19]

$$\pi_1'(h_1) + \delta P_2 b'(h_1) = [\pi_1(h_1) + \delta P_2 b(h_1)]/n_1 h_1. \tag{27}$$

[19] Second-order conditions require: $[\pi_1'' + \delta P_2 b''][n_1 h_1]^2 - [\pi_1' + \delta P_2 b']n_i h_1 < 0$. This condition is satisfied by A.2. Similar conditions can be derived for the case where the bycatch TAC constraint is binding.

JOHN R. BOYCE

The term on the right hand side of (26) is the value placed on the stock by individual j. Note that (26) involves n_1, which is endogenous. The open access equilibrium when the target species TAC is binding are the values $\{h_1^\circ, n_1^\circ, T_1^\circ\}$ that solve (26), (27), and (4). Using (4) to eliminate T_1 in (26), and using (26) to eliminate n_1 in (27), gives an expression involving only h_1:

$$\pi_1'(h_1^\circ) + \delta P_2 b'(h_1^\circ) = k_1/S_1. \qquad (28)$$

Rewriting (27) in this fashion is convenient in that the comparative statics can be derived simply by totally differentiating (28). In particular, note that $\partial h_1^\circ/\partial \delta = -b'/[\pi_1'' + \delta P_2 b''] > 0$, by A.3. Thus allowing Fishery One to sell bycatch has the expected effect that the harvest rate (and hence, the bycatch rate) is increased. Of course, this also implies that a tax on bycatch equal to P_2 would *reduce* the harvest level. It can also be shown that $\partial h_1^\circ/\partial k_1 > 0$, and $\partial h_1^\circ/\partial S_2 > 0$, and that the equilibrium values of n_1° and T_1° are inversely related to h_1°. In contrast, in the social optimum, T_1 is independent of h_1.

2. *TAC for bycatch binding.* When the bycatch constraint is binding, from (6) the season length is $T_1 = S_2/\sum_{h=1}^{n_1} b(h_{1j})$. Thus the harvest level which maximizes profits is

$$\pi_1'(h_1) + \delta P_2 b'(h_1) = \{\beta(h_1)[\pi_1(h_1) + \delta P_2 b'(h_1)]\}/n_1 h_1. \qquad (29)$$

Thus (29) and (27) differ by the term β in the numerator of the right-hand side. Recall that $\beta \geq 1$ as $b'' \geq 0$. Thus if $b'' = 0$, then (29) and (27) are identical. That is, if bycatch is a constant proportion of catch, then the optimal harvest level under open access is unchanged by having the bycatch TAC bind instead of the target species TAC. In the social optimum condition (17), the same effect was noted. However, the solution in (17) did not depend upon δP_2. This is not the case in the open access equilibrium (29).

Note also that in (29), h_1 depends upon n_1. The equilibrium $\{h_1^{\circ\circ}, n_1^{\circ\circ}, T_1^{\circ\circ}\}$ must satisfy (29), the zero profit equation (26), and the bycatch TAC constraint (5). Using (5) and (26) to eliminate n_1 in (29), the open access harvest level $h_1^{\circ\circ}$ is given implicitly by:

$$\pi_1'(h_1^{\circ\circ}) + \delta P_2 b'(h_1^{\circ\circ}) = b'(h_1^{\circ\circ})[k_1/S_2], \qquad (30)$$

which uniquely solves for $h_1^{\circ\circ}$ by A.3. Again, comparative statics can be conducted on (30) by a total differential approach. Thus, $\partial h_1^{\circ\circ}/\partial \delta > 0$, $\partial h_1^{\circ\circ}/\partial k_1 < 0$, and $\partial h_1^{\circ\circ}/\partial S_2 > 0$. It can also be shown that both $n_1^{\circ\circ}$ and $T_1^{\circ\circ}$ are inversely related to $h_1^{\circ\circ}$.

The solutions in (28) and (30) are compared in Fig. 5. From (28) and (30), it is clear that whether $h_1^{\circ\circ}$ or h_1° holds depends upon whether $b'(h_1)/S_2 \lessgtr 1/S_1$. As in Fig. 3, if $h_1^{\circ\circ} < h_1^\circ < \underline{h}_1^H$, then the target TAC must bind. Conversely, if $\underline{h}_1^L < h_1^{\circ\circ} < h_1^\circ$, then the bycatch constraint is binding. Finally, in the event that $h_1^{\circ\circ} < h_1^M < h_1^\circ$, *both* constraints are binding. Let $h_1^{\circ\circ\circ} = \underline{h}_1^M$ denote this solution. Then the zero profit condition (26) determines $T_1^{\circ\circ\circ}$, and either the target or bycatch TAC constraint determines $n_1^{\circ\circ\circ}$. Even though each fisherman fishes at the optimal harvest level, the season length will be too short and the number of entrants too large under open access since an individual fisherman ignores the cost he imposes on other fishermen by his removals of the target and bycatch stocks.

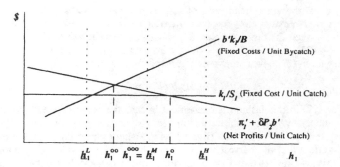

FIG. 5. Open access harvest rate in Fishery One when target TAC is binding and no Fishery Two exists.

B. Two Fisheries, One TAC Constraint on the Bycatch Species

Now suppose that the bycatch species can be used either as bycatch or as a target species, and the allocation is decided by the "rule of capture." Each fishery shuts down once the bycatch TAC is taken. Thus, $T_1 = T_2 \leq \min(\overline{T}_1, \overline{T}_2)$, which is given by:

$$T_1 = T_2 \equiv T = S_2 / [n_1 b(h_1) + n_2 h_2]. \tag{31}$$

A representative vessel in Fishery One chooses harvest level h_1 to maximize (2) given the season length is determined by (31). The harvest level which maximizes profits to a vessel in Fishery One and Fishery Two are, respectively,

$$\pi'_1(h_1) + \delta P_2 b'(h_1) = \{[\pi_1(h_1) + \delta P_2 b(h_1)]b'(h_1)\} / [n_1 b(h_1) + n_2 h_2], \tag{32}$$

$$\pi'_2(h_2) = \pi_2(h_2) / [n_1 b(h_1) + n_2 h_2]. \tag{33}$$

However, using the zero profit condition (26), and the equilibrium harvest, effort, and season length levels given by (31)–(33), the following can be shown:

PROPOSITION 4. *Management of the bycatch as a single stock is unstable if the bycatch constraint is binding for Fishery One. Either there does not exist a unique solution in terms of n_1 and n_2, or there does not exist a solution in h_1 and h_2.*

Proposition 4 suggests that an open access fishery which allocates bycatch by the rule of capture will be unstable. Thus, once a bycatch species becomes commercially viable, even if each fishery remains open access, the bycatch species is explicitly allocated between the bycatch user group (Fishery One) and the target user group (Fishery Two). To do otherwise would induce multiple equilibria, meaning that the manager would be unable to predict the economic consequences of their actions.

C. Two Fisheries, Separate TAC Constraints on the Bycatch Species

However, as we shall see in this section, the allocation may be contentious.

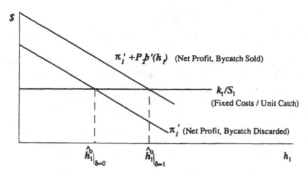

FIG. 6. Effect of allowing Fishery One to sell bycatch under open access when target TAC is binding and no Fishery Two exists.

COROLLARY 4.1. *If the bycatch constraint is binding, there does not exist an open access allocation* $(B, S_2 - B)$ *such that the season lengths are identical* $(T_1 = T_2)$ *and each fishery has an identical marginal valuation of the bycatch stock* $(\pi'_1 + \delta P_2 b' = \pi'_2)$ *at the vessel level.*

Corollary 4.1 shows that even if the manager is able to set a bycatch allocation such that a unit of the bycatch has equal value to each fishery, one of the fisheries will close before the other, creating an incentive for fishermen in the fishery with the shorter season to request a larger allocation. If the manager sets the allocation such that $T_1 = T_2$, then fishermen in one fishery or the other will have a higher value at the margin for the bycatch, creating an incentive for fishermen in the fishery with the higher marginal valuation of the bycatch to request a larger allocation. In either case, the fishery manager will face pressures to reallocate the bycatch, and ultimately, to raise TAC limits.

D. Effect of Prohibiting Sales of Bycatch by Fishery One

In the North Pacific groundfish fisheries, bycatch of halibut, salmon, and crab cannot be sold, although this is not true for all fisheries.[20] In Fig. 6, the equilibrium (28) is shown for the case where the target species TAC is binding for Fishery One. When the bycatch is able to be sold ($\delta = 1$), both (28) and (30) show that the harvest level per vessel is higher than if it cannot be sold ($\delta = 0$). The total bycatch removals in Fishery One are $H_b = T_1 n_1 b(h_1) = S_1 b(h_1)/h_1$, where the second equality is obtained by using the zero profit condition (26) and the target TAC constraint (4). Thus,

PROPOSITION 5. *If the target species TAC constraint is binding for Fishery One, prohibiting Fishery One from selling bycatch reduces the total bycatch removed by Fishery One.*

This helps to explain the prohibition on selling bycatch by Fishery One, even though it is socially inefficient (see Proposition 1). If Fishery One can sell its bycatch, it decreases their incentive to reduce bycatch. This causes a larger

[20] An anonymous referee reports that in the Mid-Atlantic scallop fishery, there is substantial bycatch of summer flounder, black sea bass, lobster, and monkfish, and that these species are all sold.

bycatch, which decreases the take available to the second fleet. Prohibiting the first fleet from selling its bycatch therefore increases the number of vessels who can participate in the second fishery (cf. [12, 14]).[21]

The prohibition on Fishery One from selling bycatch supports the position that bycatch is morally wrong. This allows Fishery Two (or whoever gets value from the bycatch) to maintain the higher moral ground in the bycatch debate, which is very useful in the political arena. A similar result has been observed by Hahn [10, p. 30] with respect to the position taken by environmentalists against marketable pollution permits. In each case, a prohibition on selling the bycatch or pollution reduces the legitimacy of the claim by the bycatch fleet or the polluter, and in both cases, a prohibition on trades reduces social welfare. Thus in each case, the prohibition has to do with one group wishing to prevent transfers to the other group.

5. RATIONALIZING BYCATCH WITH TRANSFERABLE QUOTAS

Suppose that managers rationalize the fishery using an individual transferable quota system.[22] Assume that there are *two* quota systems, one for the target species, and one for the bycatch species. Indeed, two quota systems are necessary for the system to fully rationalize the bycatch problem for all possible outcomes. Since there are no congestion or stock externalities, an ITQ system will be capable of generating the social optimum [6]. This result is extended here to the case of bycatch, but only if there exists a competitive quota market for whichever species is the binding constraint, and only if taking the bycatch imposes no lost existence value.

A. Both Species Commercially Harvested

Let m_1 and m_2 be the market clearing competitive season (rental) prices for quotas of the target and bycatch species, respectively. Assume each vessel j in Fishery One which participated in the open access fishery is given an identical quota for the target and bycatch species, q_{1j}^1 and q_{1j}^2, and that each vessel in Fishery Two which participated in the open access fishery is given a quota of the bycatch species q_{2j}^2. Assume also that the TAC for each fishery is completely allocated as quota shares. Let z_{1j}^1 and z_{1j}^2 denote purchases ($z_{ij}^k > 0$) or sales ($z_{1j}^k < 0$) of the quotas of the target ($k = 1$) and bycatch ($k = 2$) species, respectively, at the market prices m_i by vessel j in Fishery One, and let z_{2j}^2 denote the quantity of bycatch quotas bought or sold by vessel j in Fishery Two.

As each vessel is free to fish over the entire possible season $(0, \overline{T}_1)$ or $(0, \overline{T}_2)$, the season lengths are constrained by the upper bounds, $\overline{T}_1$ and $\overline{T}_2$ as in (6). In addition, for an individual vessel the harvest of the target and bycatch species is limited by his initial quota allocation net of purchases or sales

$$q_{1j}^1 + z_{1j}^1 \geq T_{1j} h_{1j}, \qquad j = 1, \ldots, N_1, \tag{34}$$

---

[21] It can also be shown that if the bycatch TAC binds for Fishery One, the prohibition on selling bycatch *increases* the total harvest of the target species.

[22] See [1, 2, 6, 14] for discussions of ITQ systems.

JOHN R. BOYCE

$$q_{1j}^2 + z_{1j}^2 \geq T_{1j} b(h_{1j}), \qquad j = 1, \ldots, N_1, \tag{35}$$

$$q_{2j}^2 + z_{2j}^2 \geq T_{2j} h_{2j}, \qquad j = 1, \ldots, N_2. \tag{36}$$

The objective of a vessel in Fishery One is to choose $\{h_{1j}, z_{1j}^1, z_{1j}^2, T_{1j}\}$ to maximize

$$v_{1j} = T_{1j}\left[\pi_{1j} + \delta P_2 b\right] - k_1 - m_1 z_{1j}^1 - m_2 z_{11j}^2, \qquad j = 1, \ldots, N_1, \tag{37}$$

subject to (6), (34), and (35). Similarly, the objective of a vessel in Fishery Two is to choose $\{h_{2j}, z_{2j}^2, T_{2j}\}$ to maximize

$$v_{2j} = T_{2j}\pi_{2j} - k_2 - m_2 z_{2j}^2, \qquad j = 1, \ldots, N_2, \tag{38}$$

subject to (6) and (36).

Finally, to participate in the fishery a vessel owner must earn at least as much from entering the fishery as from selling his quotas, i.e.,

$$T_{1j}\left[\pi_{1j} + \delta P_2 b\right] - k_1 \geq q_{1j}^1 m_1 + q_{1j}^2 m_2, \tag{39}$$

$$T_{2j}\pi_{2j} - k_{2j} \geq q_{2j}^2 m_2. \tag{40}$$

Let λ_j, μ_{1j}, τ_{1j}, μ_{2j}, and τ_j denote the multipliers for the constraints in (34)–(36) and (6), with the notation for λ and μ identical to that used in the Lagrangian for (9). Then the first-order conditions for vessel j in Fishery One include the zero profits equation (39), and the season length constraints (6), the quota constraints (34) and (35), and (dropping the vessel notation since each fishermen is identical)

$$\pi_1'(h_1) + \delta P_2 b'(h_1) - \lambda - \mu_1 b'(h_1) = 0, \tag{41}$$

$$m_1 = \lambda_1, \quad m_2 = \mu_1, \tag{42}$$

$$\pi_1(h_1) + \delta P_2 b(h_1) - \lambda_1 h_1 - \mu_1 b(h_1) = \tau_1. \tag{43}$$

Similarly, for vessel j in Fishery Two, the first-order necessary conditions include the season length constraint (6), the quota constraint (36), the zero profits condition (40), and (dropping the vessel subscripts)[23]

$$\pi_2'(h_2) - \mu_2 h_2 = 0, \tag{44}$$

$$m_2 = \mu_2, \tag{45}$$

$$\pi_2(h_2) - \mu_2 h_2 = \tau_2^2. \tag{46}$$

Equations (43) and (46) plus (39) and (40), respectively, can be rearranged to show that $T_{1i} = \overline{T}_1$ if $m_1 > 0$ and that $T_{1i} = \overline{T}_2$ if $m_2 > 0$. Furthermore, $T_{2j} = \overline{T}_2$ for all j. From (42) and (45), $\lambda_{1i} = m_1$, and $\mu_{1j} = \mu_{2j} = m_2$. Thus to each individual the shadow value of additional units of the two stocks equals the market

---

[23] The second-order conditions are satisfied by the Arrow–Enthoven sufficiency theorem for Kuhn–Tucker problems, since the objective function is quasi-concave in (T_{ij}, h_{ij}, z_{ij}), the constraints are quasi-convex in (T_{ij}, h_{ij}, z_{ij}), and the second-order partial differentials of the objective function exist at the solution.

price of those stocks. The only values of m_1 and m_2 which hold in the system of equations (34)–(36) and (39)–(46) are $m_1 = \lambda$ and $m_2 = \mu$, where λ and μ are the multipliers in the system of Eqs. (10)–(15). (Note, however that m_1 may equal zero.) Therefore:

PROPOSITION 6. *If the bycatch species has no existence value, a competitive individual transferable quota system is capable of maximizing social welfare, as defined by* (8), *but there must exist a market for both quotas.*

Both quotas are traded, since individuals in Fishery One will either be buying or selling bycatch quotas. Note also that the following corollary to Proposition 1 holds:

COROLLARY 1.1. *While the ITQ system is capable of maximizing the constrained social welfare problem where Fishery One is restricted from selling bycatch, there are gains from trade by allowing Fishery One to sell the bycatch.*

B. Bycatch Has Existence Value

Suppose that the bycatch species has no commercial value, but it does have existence value. For example, sea lions are occasionally taken as bycatch in the Bering Sea pollock fishery and the tuna fishery has bycatch of dolphins. Since the bycatch has no commercial value, it is not harvested. To the social planner, this corresponds to the case where $\delta = -1$ and $\gamma = 0$. Thus the solution involves Eqs. (10), (12), (14), plus the constraints (4) and (6). However, $\delta = 0$ to the commercial vessel in Fishery One with tradable quotas for both the pollock and the sea lions, since he cannot sell the bycatch. Thus:

PROPOSITION 7. *An ITQ system will not be sufficient to maximize social welfare* (*defined by* (9)) *unless there also exists a charge* (*e.g., a user fee or a tax*) *of P_2 for each unit of bycatch taken.*

The problem is that even if $m_2 > 0$ in this case, m_2 only reflects the scarcity of the bycatch TAC to Fishery One. It does not reflect the full social cost of taking additional units of bycatch.

C. Using Taxes Instead of ITQs

There is nothing unique about the ITQ system from a pure efficiency viewpoint. Indeed, the prices m_1 and m_2 (along with a tax of P_2 if the bycatch has existence value) could be used as taxes instead, and the same allocation would be achieved. However, a tax system would be much less politically viable since the tax would charge vessels for use of every unit of harvest and bycatch, while a tradable quota system would only charge them explicitly once they have used up their initial quota (e.g., [7]).

ITQs will not produce an efficient allocation if production externalities are present [6]. A tax system might be thought to be less sensitive to this criticism. However, for either a tax or an ITQ system to remedy the production externalities, there must be additional markets or taxes. For example, if there are congestion externalities present, then either an entry fee or a tradable limited entry permit would be necessary in addition to the taxes or tradable quotas on harvesting.

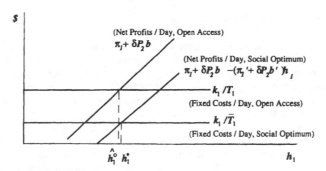

FIG. 7.   Optimal and open access harvest rates for Fishery One, bycatch not commercial harvested.

D. Will Vessels Fish "Cleaner"?

When the bycatch problem in the North Pacific groundfish fishery began to become quite serious in the late 1980s, managers were surprised at the large quantities of target species TAC left unharvested. "[We] assumed the fleet would modify its behavior when faced with bycatch caps and closures," said one management official.[24] One question that remains with rationalization is will it have discernible effects in regard to lowering the ratio of bycatch to the target species? That is, will vessels fish "cleaner"?

Figure 7 shows the relationship between the harvest rate under open access and under the social optimum. Figure 7 is based on the zero profit condition (26) for open access and the equilibrium condition for the harvest rate when only Fishery One is active and the target TAC is binding, given in (16), for the social optimum. For open access, the season length is shorter, so $k_1/T_1 > k_1/\bar{T}_1$. However, since $-(\pi_1' + \delta P_2 b')h_1 < 0$, the net profits per day for the social optimum lies below the net profits per day for the open access. Thus, the two effects are of opposite sign, and it is ambiguous whether an individual's harvest rate is higher under open access than under ITQs. Hence:

PROPOSITION 8.   *Fishery One will have lower aggregate bycatch if it is rationalized, but this will be due to the reduction in the number of vessels. Bycatch per vessel may increase or decrease.*

6. DISCUSSION AND CONCLUSIONS

This paper examined a stylized economic model of the fisheries bycatch problem. It was assumed that there are no production externalities (e.g., [6]); vessels within each fishery are homogeneous; prices are not affected by changes in harvest rates or total harvest levels; the target fishery is treated as a single species fishery [24]; there are no multiple grades or high grading problems [1, 3]; and the bycatch rate depends only on the harvest rate of the target species. All of these assumptions are

---

[24] National Marine Fishery Service regional director Steve Pennoyer (quoted in [23, June 1990, p. 64]).

over-simplifications of the environments in which managers face the bycatch problem.

Even within this simplified framework, it has been shown that open access solutions under the "rule of capture" allocation may be unstable. When bycatch is explicitly allocated between different uses, the harvest rate under open access may be optimal, but the number of entrants will be too large. These results should give policy makers an added incentive to work toward some form of rationalization. Unfortunately, this paper has given reason to be cautious about how the bycatch problem is integrated into the general problem of rationalizing fisheries. Here, the correct incentives can be given to vessels only by creating transferable quota systems for *both* the target and bycatch species which are tradable between fisheries. Even this will not be sufficient if there are externalities beyond the simple common properties externality due to lack of ownership. In particular, if the bycatch species has existence value an ITQ system will not be sufficient to eliminate the external cost of bycatch removals. Similar conclusions would be reached if there also existed production externalities.

The model in this paper has focused entirely on the in-season problem. Most of the fisheries economics literature has been concerned with optimization over an infinite planning horizon. To fully appreciate the subtleties of the problem, the model proposed here would have to be placed inside of a model such as proposed by McKelvey [16]. In this context, the present model shows that for a given technology, there may be many instances where the joint TAC constraints for the bycatch and target species are incompatible in the sense that both will not be fully harvested. This result can occur under a rationalized system as well as under an open access system. It is not surprising that in the face of such short-term pressures, managers have sought to solve the problem with a command and control focus on technology. However, it is not clear that the gear restrictions are necessarily good economic policy. Even if such restrictions were good economic policy, a command and control system is probably the least likely means of finding such a technology. Furthermore, as is shown in Figs. 3 and 5, the open access harvest rate per vessel may well equal the social optimal harvest rate (for a given technology) for a range of bycatch ratios. Just because the observed bycatch ratio equals the allowable bycatch ratio does not mean that rents are being maximized.

REFERENCES

1. L. G. Anderson, Highgrading in ITQ fisheries, *Mar. Resour. Econom.* **9**, 209–226 (1994).
2. R. Arnason, On catch discarding in fisheries, *Mar. Resour. Econom.* **9**, 189–207 (1994).
3. A. H. Barnette, The Pigouvian tax rule under monopoly, *Amer. Econom. Rev.* **70**, 1037–1040 (1980).
4. F. H. Bell, "The Pacific Halibut: The Resource and the Fishery," Alaska Northwest Publishing, Anchorage, Alaska (1981).
5. J. D. Berger, R. F. Kappenman, L.-L. Low, and R. J. Marasco, Procedures for bycatch estimation of prohibited species in the 1989 Bering Sea domestic trawl fisheries, NOAA Technical Memorandum, NMFS, F/NWC-173, U.S. Department of Commerce, Oct. 1989.
6. J. R. Boyce, Individual transferable quotas and production externalities in a fishery, *Nat. Resour. Modeling* **4**, 385–408 (1992).
7. J. M. Buchanan, and G. Tullock, Polluters' profits and political response: Direct controls versus taxes, *Amer. Econom. Rev.* **65**, 139–147 (1975).

8. C. W. Clark, "Mathematical Bioeconomics: The Optimal Management of Renewable Resources," Wiley, New York (1976).
9. K. R. Criddle, Predicting the consequences of alternative harvest regulations in a sequential fishery, unpublished manuscript, Department of Economics, University of Alaska, Fairbanks (1994).
10. R. W. Hahn, Economic prescriptions for environmental problems: How the patient has followed the doctor's orders, *J. Econom. Perspect.* 3, 95–114 (1989).
11. R. Hannesson, Optimal harvesting of ecologically interdependent fish species, *J. Environ. Econom. Management* 10, 318–345 (1983).
12. R. N. Johnson, and G. D. Libecap, Contracting problems and regulation: The case of the fishery, *Amer. Econom. Rev.* 72, 1005–1022 (1982).
13. D. W. Jorgensen, Technology and decision rules in the theory of investment behavior, *Quart. J. Econom.* 87, 523–542 (1973).
14. J. Karpoff, Suboptimal controls in a common resource management: The case of the fishery, *J. Polit. Econom.* 95, 179–194 (1987).
15. J. B. Loomis, and D. M. Larson, Total economic values of increasing gray whale populations: Results from a contingent valuation survey of visitors and households, *Mar. Resour. Econom.* 9, 275–286 (1994).
16. R. McKelvey, The fishery in a fluctuating environment: Coexistence of specialist and generalist fishing vessels in a multipurpose fleet, *J. Environ. Econom. Management* 10, 287–309 (1983).
17. *National Geographic*, Nov. 1995.
18. Natural Resource Consultants, "The Nature and Scope of Fishery Dependent Moralities in the Commercial Fisheries of the Northeast Pacific," Seattle, Washington, June (1990).
19. P. A. Neher, R. Arnason, and N. Mollett, Eds., "Rights Based Fishing," Kluwer Academic, Dordrecht (1989).
20. North Pacific Fisheries Management Council, "Draft Resource Assessment Document for the 1989 Gulf of Alaska Groundfish Fishery," Anchorage, Alaska, Sept. 1988.
21. North Pacific Fisheries Management Council, "Draft Resource Assessment Document for the 1989 Bering Sea–Aleutian Islands Groundfish Fishery," Anchorage, Alaska, Sept. 16, 1988.
22. North Pacific Fisheries Management Council, (various issues), Newsletter, Anchorage, Alaska.
23. *Pacific Fishing*, various issues.
24. D. Squires, Public regulation and the structure of production in multiproduct industries: An application to the New England otter trawl industry, *Rand J. Econom.* 18, 232–247 (1987).
25. J. M. Ward, The bioeconomic implications of a bycatch reduction device as a stock conservation management measure, *Mar. Resour. Econom.* 9, 227–240 (1994).
26. J. E. Wilen, and G. Brown, Jr., Optimal recovery paths for perturbations of trophic bioeconomic systems, *J. Environ. Econom. Management* 13, 224–234 (1986).

International Utilization

[24]

The great fish war: an example using a dynamic Cournot-Nash solution

David Levhari*

and

Leonard J. Mirman**

In recent years there have been numerous international conflicts about fishing rights. These conflicts are wider in scope than those captured by the model presented in this paper. Yet the model sheds light on the economic implications of these conflicts as well as on the implications of other duopolistic situations in which the decisions of the participants affect the evolution of an underlying population of interest. Our model has two basic features: the underlying population changes as a result of the actions of both participants, and each participant takes account of the other's actions. This strategic aspect is studied, for an example, by using the concept of a Cournot-Nash equilibrium in which each participant's reaction depends on the stock of fish and not on previous behavior. Thus, the model is a discrete-time analog of a differential game. The paper examines the dynamic and steady-state properties of the fish population that results from the participants' interactions.

1. Introduction

■ In recent years we have witnessed numerous international conflicts about fishing rights in various seas and water zones. The most well-known conflict is the "Cod War" between Iceland and the United Kingdom. These conflicts have much wider scope and motives than those which are captured by our simple model. Yet the model sheds some light on the economic implications inherent in the fishing conflicts and also in other duopolistic or oligopolistic situations in which the decisions of the participants have an effect on the evolution of the underlying population of interest, which in our model is fish.

Our model has two basic features. First, there is the strategic aspect: each of the participants must take account of the actions of the other participant. The second feature is that the underlying population is changing, so that the actions of both participants affect the future size or rate of growth of the fish population. In effect, these two features together create a dynamic externality which is the essence of the problem being studied. Although there are many

* The Hebrew University, Jerusalem.
** University of Illinois, Urbana.
Research support from National Science Foundation under grant SOC 76-11583 and U.S.-Israel BSF 1828–79 is gratefully acknowledged.

areas of economics to which this type of model is applicable, e.g., imperfect competition among firms or macroeconomic stabilization, the fishing context will be used for expository reasons.

The dynamics of the fish population is a natural phenomenon governed by the "law of growth" for fish.[1] The economics of natural resources and fishing, in particular, has been studied quite extensively during the last several years. In particular, the pioneering works of Gordon (1954), Scott (1955), and Smith (1968) have been especially important. These papers use the natural dynamics of the fish population to study the economic implications of fishing for the population of fish. While these authors develop positive models, the more recent literature, which is concerned with normative models, uses dynamic optimization techniques to study the optimal rate of extraction of fish and its long-run implications. The papers of Clark (1971), Clark and Monro (1975), Long (1977), and Levhari, Michener, and Mirman (1978) study various aspects of optimal fishing policies and their effect on the size or possible depletion of the fish population.[2] There are, then, several ways to study dynamic models of fishing.

The point of view of this paper is that the object of each country is to maximize the sum of discounted utilities.[3] The maximization technique used is essentially discrete-time dynamic programming, which in this context is a discrete-time analog of a differential game. Differential games are useful for modeling many economic problems which involve both dynamics and strategic behavior and have been recently surveyed by Clemhout and Wan (1978).[4]

To capture the strategic aspects in our model, we assume that each of the participants acts as a Cournot duopolist in a dynamic framework and takes the policy of the other participant as given, while trying to maximize his own discounted sum of utilities. The two players may have different subjective discount factors. It is quite possible, of course, that one of the major elements in the fishing conflict is the different time preferences of the participants. However, as we show, in the present model, this difference in subjective time preferences is quite consistent with the existence of a stable equilibrium.

To study the dynamic duopoly problem in its most simplified form we shall make several assumptions. In particular, we assume that only economic considerations count. Moreover, threats are not allowed, so that actions based on retaliation for behavior in the past will not be considered. Finally, each country is interested in the maximization of the welfare of its own citizens: the catch cannot be used for resale or for profit.

The structure of the dynamic Cournot equilibrium used in this paper is similar to the simple duopoly problem first studied by Cournot in the static context.

---

[1] For a discussion of the laws of biological growth see Lotka (1956).

[2] Much, although not all of this literature, is discussed in Clark (1976) or the surveys of Smith (1977) and Fisher and Peterson (1976).

[3] Discounted profit maximization is another criterion which is often used; unfortunately, in the present context, *a second externality is introduced by this criterion*. If profit maximization were the objective, then each country would have to consider the effect of the other countries' catch on the current market price, the "market" externality, as well as on the future fish population, the dynamic externality. These two externalities combined make the analysis very difficult.

[4] Moreover several recent papers have used dynamic dominant player models to study the economics of imperfect competition, see for example Kydland (1975).

Although it may be interpreted in a dynamic context, the original Cournot solution is static in nature, as is made clear by the Nash analysis in the theory of games. In the context of the fishing model, however, there is a true dynamic structure to the problem. In particular, the fish population responds in a natural way to the quantity of fish extracted by both countries. Hence, although each period's equilibrium is a Cournot-Nash solution, there may be a change in the size of the fish population. Moreover, the sequence of decisions by both countries is in fact itself a Cournot-Nash equilibrium, even though the fish population need not be in equilibrium. In this framework it is natural to ask whether or not the size of the fish population eventually settles down to an equilibrium in the dynamic sense. This equilibrium will be referred to as a steady-state equilibrium. It is desirable to keep the two notions of equilibrium separate. The system which will be studied in this paper is always in equilibrium in the Cournot-Nash sense, while it may or may not be in a steady-state equilibrium.

We derive the infinite horizon Cournot-Nash policies by finding finite horizon Cournot-Nash policies and letting the horizon tend to infinity. The size of the fish population under the influence of these infinite horizon Cournot-Nash policies is also studied. We show in the context of an example, that a positive finite steady state is eventually attained. Moreover, we derive for this example a closed-form solution for both the Cournot-Nash policy functions and the steady-state quantity of fish. These Cournot-Nash policy functions and the steady-state quantity of fish are then compared with the optimal policy functions and the optimal steady-state quantity of fish when both countries form a cooperative venture and pool their resources. We find, not surprisingly since it is utility that is maximized, that the Cournot-Nash policies imply a greater harvest of fish, and, therefore, a smaller steady state.

Finally, we apply these same techniques to several special cases. For example, we study the von Stackelberg case with a leader-follower structure. This example is especially interesting since it is applicable to areas of government policy making (Kydland and Prescott, 1977) and to imperfect competition among firms (Kydland, 1975, 1977; Brock, 1975). We also give a linear example which shows that it is possible for the stock to grow infinitely large under a cooperative solution, while under a Cournot-Nash solution the stock tends to zero.

2. The model

■ Let x_t be the quantity of fish at time t. Suppose that, if uninterrupted, the quantity of fish would grow according to the biological rule,

$$x_{t+1} = x_t^\alpha, \qquad 0 < \alpha < 1. \tag{1}$$

Thus one easily observes that $x_t = x_{t+1} = 1$,[5] is a stable steady state of the fish population. See Figure 1. Suppose, however, that two countries fish the waters. Each country has a utility for the fish it catches in each period and thus an interest in the long-run effect of its present catch. Moreover, each country must take the catch of the other country into consideration when deciding on its own catch. The former consideration is accounted for by using a dynamic programming argument, and the latter by using the concept of a Cournot-Nash

---

[5] A normalization of the fish population is employed for expositional purposes.

FIGURE 1

NATURAL "LAW OF GROWTH" FOR FISH

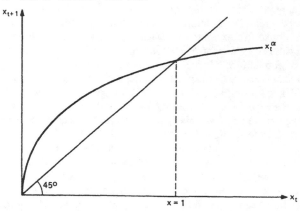

equilibrium. To be more precise, suppose that country i has a utility function u_i for present consumption. Let c_i be the present consumption of country i and suppose that the utility function of country i is logarithmic, i.e., $u_i(c_i) = \log c_i$. Let $0 < \beta_i < 1$ be the discount factor for country i. Suppose, moreover, that the objective of each country is to maximize the sum of the discounted utility of fish.

Consider the problem in the context of a finite horizon. Assume that if there were no future period, each country would get[6] an equal share (or any other prescribed share) of all the remaining fish. The initial level of fish is given by x. If there is a *one-period horizon*[7] and if country 1 takes country 2's actions as given, the optimal response for country 1 is found from the maximization problem,

$$\underset{0 \le c_1 \le x - c_2}{\text{Max}} \{\log c_1 + \beta_1 \log \tfrac{1}{2}(x - c_1 - c_2)^\alpha\}$$

$$= \underset{0 \le c_1 \le x - c_2}{\text{Max}} \{\log c_1 + \alpha\beta_1 \log (x - c_1 - c_2) + \beta_1 \log \tfrac{1}{2}\}, \quad (2)$$

where c_1 is the optimal present consumption of country 1, given the consumption c_2 of country 2. The value $x - c_1 - c_2$ is the remaining stock of fish which becomes $(x - c_1 - c_2)^\alpha$ in the next period. The first-order condition for this problem is,

$$(1 + \alpha\beta_1)c_1 + c_2 = x. \quad (3)$$

Equation (3) represents the reaction curve of country 1.

Using a similar argument for country 2, the optimal response is given by

$$c_1 + (1 + \alpha\beta_2)c_2 = x. \quad (4)$$

These policies are depicted in Figure 2, as the reaction functions for each of the countries. The Cournot-Nash $\bar{c}_1$, $\bar{c}_2$ is the intersection of these two reaction functions.

---

[6] This assumption actually plays no role in the derivation of the Cournot-Nash policies, as will become clear below.

[7] Note that here horizon refers to the number of future periods. Hence, a one-period horizon problem is in fact a two-period maximization problem.

326 / THE BELL JOURNAL OF ECONOMICS

FIGURE 2

COURNOT—NASH EQUILIBRIA

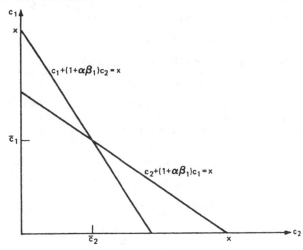

In this case, the Cournot-Nash equilibrium is given by the simultaneous solution of equations (3) and (4). The solution, as seen in Figure 2, is

$$\bar{c}_1 = \frac{\alpha\beta_2}{\alpha^2\beta_1\beta_2 + \alpha\beta_1 + \alpha\beta_2} \, x, \tag{5}$$

and

$$\bar{c}_2 = \frac{\alpha\beta_1}{\alpha^2\beta_1\beta_2 + \alpha\beta_1 + \alpha\beta_2} \, x. \tag{6}$$

The remaining stock of fish is then given by

$$x - \bar{c}_1 - \bar{c}_2 = \frac{\alpha^2\beta_1\beta_2}{\alpha^2\beta_1\beta_2 + \alpha\beta_1 + \alpha\beta_2} \, x. \tag{7}$$

Next consider a two-period horizon problem with the first country again reacting to the second country in the present under the *assumption* that in the future the one-period horizon Cournot-Nash solution, given above, will prevail. To solve this problem we must find a one-period horizon Cournot-Nash "valuation" function. To find it, consider the one-period horizon objective function of country 1 under the Cournot-Nash equilibrium given above (or the one-period "valuation" function),

$$\log \bar{c}_1 + \beta_1 \log \tfrac{1}{2} + \alpha\beta_1 \log (\bar{x} - \bar{c}_1 - \bar{c}_2) = (1 + \alpha\beta_1) \log x + A_1,$$

where

$$A_1 = \log \frac{(\alpha\beta_2)(\alpha^2\beta_1\beta_2)^{\alpha\beta_1}}{(\alpha^2\beta_1\beta_2 + \alpha\beta_1 + \alpha\beta_2)^{1+\alpha\beta_1}} + \beta_1 \log \tfrac{1}{2}.$$

Notice that A_1 is a constant (independent of x) and will have no effect on the optimal policy. Hence the objective function for the two-period horizon problem is

$$\log c_1 + \alpha\beta_1(1 + \alpha\beta_1) \log (x - c_1 - c_2) + A_1.^8 \qquad (8)$$

Again the optimal response c_1 of country 1 corresponding to the decision c_2 of country 2 satisfies the first-order condition,

$$(1 + \alpha\beta_1 + \alpha^2\beta_1^2)c_1 + c_2 = x. \qquad (9)$$

Similarly the optimal response for country 2 is given by

$$c_1 + (1 + \alpha\beta_2 + \alpha^2\beta_2^2)c_2 = x. \qquad (10)$$

Once again, by solving (9) and (10) as a simultaneous system, we can find the Cournot-Nash equilibrium for the two-period horizon case.

The process can be repeated for an n-period horizon yielding the Cournot-Nash policies,

$$\bar{c}_1 = \frac{\alpha\beta_2[\sum_{j=0}^{n-1} (\alpha\beta_2)^j]x}{(\sum_{j=0}^{n} (\alpha\beta_1)^j)(\sum_{j=0}^{n} (\alpha\beta_2)^j) - 1}, \qquad (11)$$

$$\bar{c}_2 = \frac{\alpha\beta_1[\sum_{j=0}^{n-1} (\alpha\beta_1)^j]x}{(\sum_{j=0}^{n} (\alpha\beta_1)^j)(\sum_{j=0}^{n} (\alpha\beta_2)^j) - 1}, \qquad (12)$$

and

$$x - \bar{c}_1 - \bar{c}_2 = \frac{\alpha^2\beta_1\beta_2(\sum_{j=0}^{n-1} (\alpha\beta_1)^j)(\sum_{j=0}^{n-1} (\alpha\beta_2)^j)x}{(\sum_{j=0}^{n} (\alpha\beta_1)^j)(\sum_{j=0}^{n} (\alpha\beta_2)^j) - 1}. \qquad (13)$$

As the horizon tends to infinity, the limiting values of $\bar{c}_1$, $\bar{c}_2$ and $x - \bar{c}_1 - \bar{c}_2$ are:

$$\bar{c}_1 = \frac{\alpha\beta_2(1 - \alpha\beta_1)x}{1 - (1 - \alpha\beta_1)(1 - \alpha\beta_2)}, \qquad (14)$$

$$\bar{c}_2 = \frac{\alpha\beta_1(1 - \alpha\beta_2)x}{1 - (1 - \alpha\beta_1)(1 - \alpha\beta_2)}, \qquad (15)$$

and

$$x - \bar{c}_1 - \bar{c}_2 = \frac{\alpha^2\beta_1\beta_2 x}{\alpha\beta_1 + \alpha\beta_2 - \alpha^2\beta_1\beta_2}.^9 \qquad (16)$$

---

[8] Note we have let c_i be the optimal reaction and $\bar{c}_i$ be the Cournot-Nash policy in the first period under both a one- and a two-period horizon. They will in fact be different, with the length of the horizon playing an essential role. However, the context should make clear which horizon is being used and no confusion should arise.

[9] It can be shown that for the infinite-horizon maximization problem for both countries, the optimal policies satisfy the following functional equations,

$$\begin{matrix} i, j = 1, 2 \\ i \neq j \end{matrix} \quad u_i'(g_i(x)) = \beta\mu_i'[g_i(f(x - g_1(x) - g_2(x)))]f'(x - g_1(x) - g_2(x))[1 - g_j'],$$

where u_i represents the one period utility of country i, $f(x)$ is the growth function, and $g_i(x)$ is

Equations (14) and (15) represent consumption policies for countries 1 and 2, respectively, when an infinite horizon is considered, while equation (16) represents both countries' combined investment policy for fish. Note that these policies are applicable in each period and no longer depend upon the length of the horizon. Hence these policies may be used in deriving the dynamic behavior of the stock of fish from which the steady-state solution may be found. Let $x_0 > 0$ be any initial stock of fish. Under a Cournot-Nash equilibrium the dynamic equation for fish becomes

$$x_{t+1} = [x_t - c_1(x_t) - c_2(x_t)]^\alpha. \tag{17}$$

In particular,

$$x_1 = [x_0 - c_1(x_0) - c_2(x_0)]^\alpha = \left(\frac{\alpha^2 \beta_1 \beta_2}{\alpha\beta_1 + \alpha\beta_2 - \alpha^2 \beta_1 \beta_2} \right)^\alpha x_0^\alpha,$$

and

$$x_2 = \left(\frac{\alpha^2 \beta_1 \beta_2}{\alpha\beta_1 + \alpha\beta_2 - \alpha^2 \beta_1 \beta_2} \right)^{\alpha + \alpha^2} x_0^{\alpha^2}.$$

In general,

$$x_t = \left[\frac{\alpha^2 \beta_1 \beta_2}{\alpha\beta_1 + \alpha\beta_2 - \alpha^2 \beta_1 \beta_2} \right]^{\Sigma_{j=1}^t \alpha^j} x_0^{\alpha^t}. \tag{18}$$

Hence

$$\lim_{t \to \infty} x_t = \left[\frac{\alpha^2 \beta_1 \beta_2}{\alpha\beta_1 + \alpha\beta_2 - \alpha^2 \beta_1 \beta_2} \right]^{\alpha/(1-\alpha)} = \bar{x}, \tag{19}$$

where $0 < \bar{x} < 1$ (notice that $0 < \alpha\beta_1, \alpha\beta_2 < 1$).

Thus the "volume" of fish x_t converges to a steady-state level $\bar{x}$. The steady-state level may be rewritten as

$$\bar{x} = \left(\frac{1}{\dfrac{1}{\alpha\beta_1} + \dfrac{1}{\alpha\beta_2} - 1} \right)^{\alpha/(1-\alpha)} \tag{20}$$

This dynamic behavior is illustrated in Figure 3. From (20) observe that the higher $\beta_i (i = 1, 2)$—that is, the higher the discount factor of either country—the higher the steady-state level of fish $\bar{x}$. If both countries have the same rate of time preference so that $\beta_1 = \beta_2 = \beta$, the steady state is

$$\bar{x} = \left(\frac{1}{\dfrac{2}{\alpha\beta} - 1} \right)^{\alpha/(1-\alpha)} = \left(\frac{\alpha\beta}{2 - \alpha\beta} \right)^{\alpha/(1-\alpha)} \tag{21}$$

Let us compare the Nash-Cournot steady state with the steady-state solution to the problem when the countries combine their take so as to max-

---

the Cournot-Nash policy function for country i. In our example $u_i(x) = \log x$, $f(x) = x^\alpha$. Setting $g_1(x) = \lambda_1 x$ and $g_2(x) = \lambda_2 x$, we find the set of two linear equations,

$$\lambda_1 + (1 - \alpha\beta_1)\lambda_2 = 1 - \alpha\beta_1 \quad \text{and} \quad (1 - \alpha\beta_2)\lambda_1 + \lambda_2 = 1 - \alpha\beta_2.$$

Therefore,

$$\lambda_1 = \frac{(1 - \alpha\beta_1)\alpha\beta_2}{1 - (1 - \alpha\beta_1)(1 - \alpha\beta_2)} \quad \text{and} \quad \lambda_2 = \frac{(1 - \alpha\beta_2)\alpha\beta_1}{1 - (1 - \alpha\beta_1)(1 - \alpha\beta_2)}.$$

A different approach has been followed in the paper, since to justify, even intuitively, the functional equations would be a lengthy procedure.

FIGURE 3

COURNOT–NASH DYNAMICS

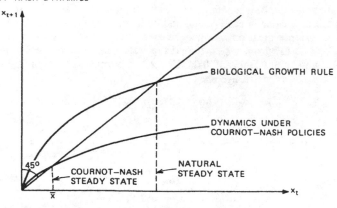

imize the discounted sum of both countries' utilities. The identical rate of time preference serves as the common discounting factor.

If the countries plan for an infinite number of periods, then the optimal policy in any period can be found from the equation

$$x - 2c = \alpha\beta x.^{10} \tag{22}$$

Here the entire catch is $2c$, while c is the part of the catch consumed by each country. Under this policy the dynamic equation is

$$x_{t+1} = (\alpha\beta x_t)^\alpha \tag{23}$$

and the steady-state quantity of fish[11] is

---

[10] The objective function for this "cooperative" solution, assuming both countries are identical, is

$$\operatorname*{Max}_{(c_0, \ldots, c_t, \ldots)} \sum_{t=0}^{\infty} \beta^t \{2u(c_t)\}$$

subject to $x_t + 2c_t = f(x_{t-1})$, $t = 0, \ldots$, and the initial output $x = f(x_{-1})$. For this problem the optimal policy must satisfy the equations

$$2u'(c_t) = \beta 2u'(c_{t+1})f'(x_t) \qquad t = 0, \ldots.$$

Letting $u(c) = \log c$ and $f(x) = x^\alpha$, it follows that

$$\frac{1}{c_t} = \frac{\alpha\beta}{c_{t+1}} x_t^{\alpha-1}. \tag{i}$$

It is clear that if $f(x_{t-1}) = x$, then $c_t = \lambda x$, $c_{t+1} = \lambda(x - \lambda x)^\alpha = \lambda x^\alpha(1-\lambda)^\alpha$, and $x_t = (1-\lambda)x$. Equation (22) follows from equation (i) by replacing c_t, c_{t+1}, x_t and solving for λ. Also compare this with the discussion in footnote 9.

[11] The *steady-state level* under joint management is identical in the present example to that achieved when there is just one country with the aim of maximizing

$$\sum_{t=0}^{\infty} \beta^t \log c_t$$

subject to $x_{t+1} = (x_t - c_t)^\alpha$. The steady-state catch of both countries (combined) is the same as that of a single country with the same utility function. This seems to stem from the logarithmic form of the utility function since the utility of twice the catch does not affect the optimal size of the catch. Moreover, doubling the utility has no effect on the optimal size of the catch. However, it is more generally true, since in the joint management case only technological consideration and the rate of discount determine the steady state.

330 / THE BELL JOURNAL OF ECONOMICS

$$\hat{x} = (\alpha\beta)^{\alpha/(1-\alpha)}. \tag{24}$$

Comparing the steady-state quantity of fish with that of the Cournot-Nash duopoly, one easily observes that

$$\hat{x} = (\alpha\beta)^{\alpha/(1-\alpha)} > \left(\frac{\alpha\beta}{2 - \alpha\beta}\right)^{\alpha/(1-\alpha)} = \bar{x}. \tag{25}$$

Hence, as expected, a Cournot-Nash duopoly implies a smaller steady-state quantity of fish.

Note that in the steady state the constant level of fish consumed by each country in the Cournot-Nash case is

$$c_N = \frac{\alpha\beta(1 - \alpha\beta)}{1 - (1 - \alpha\beta)(1 - \alpha\beta)}\left(\frac{\alpha\beta}{2 - \alpha\beta}\right)^{\alpha/(1-\alpha)} = \left(\frac{1 - \alpha\beta}{2 - \alpha\beta}\right)\left(\frac{\alpha\beta}{2 - \alpha\beta}\right)^{\alpha/(1-\alpha)} \tag{26}$$

Under joint exploitation of the seas' resources, the steady-state level of catch of each country is:

$$c_J = \left(\frac{1 - \alpha\beta}{2}\right)(\alpha\beta)^{\alpha/(1-\alpha)}. \tag{27}$$

It is not difficult to verify that $c_J > c_N$; that is, by combined management the two countries will consume, for each level of the population of fish, smaller quantities of fish, but will be able to achieve a higher "permanent" catch. Hence the conflict implicit in the duopoly problem leads both countries to overconsume, with less left for future generations.

3. Extensions

■ In Section 2 we assumed that the two countries operate either as Cournot duopolists or as "good neighbors" exploiting jointly the "fruits of the sea." Another possibility which has been raised in the duopoly literature is that one of the countries is "sophisticated" while the other is "naive." This is a dynamic version of the von Stackelberg duopoly analysis. The "sophisticated" country takes into account its ability to manipulate the other country's output, while the "naive" country follows the same Cournot type assumption as made above. For definiteness we assume that country 1 is the "leader," while country 2 is the "follower."

The aim of the follower with a one-period horizon (assuming that in the last period there is a given rule of division, possibly each receiving $x/2$) is

$$\underset{0 \leq c_2 \leq x - c_1}{\text{Max}} \{\log c_2 + \alpha\beta_2 \log (x - c_1 - c_2)\}. \tag{28}$$

Operating in Cournot fashion, taking c_1 as given, the first-order condition for country 2 is

$$c_2 = \frac{x - c_1}{1 + \alpha\beta_2}. \tag{29}$$

Country 1 takes the policy of the second country as given and thus its aim with a one-period horizon is

$$\underset{c_1}{\text{Max}} \left\{\log c_1 + \alpha\beta_1 \log \left(x - c_1 - \frac{x - c_1}{1 + \alpha\beta_2}\right)\right\}. \tag{30}$$

Notice that the aim of country 1 is similar to that of a monopolist. The first-order condition is

$$c_1 = \frac{1}{1 + \alpha\beta_1} x. \qquad (31)$$

Substituting the policy c_1 in country 2's policy, we find

$$c_2 = \frac{x - \dfrac{1}{1 + \alpha\beta_1} x}{1 + \alpha\beta_2} = \frac{\alpha\beta_1 x}{(1 + \alpha\beta_2)(1 + \alpha\beta_1)}. \qquad (32)$$

The remaining quantity of fish is

$$x - c_1 - c_2 = \frac{\alpha^2 \beta_1 \beta_2 x}{(1 + \alpha\beta_1)(1 + \alpha\beta_2)}. \qquad (33)$$

Hence, with a two-period horizon under a standard dynamic programming formulation, the aim of country 2 is

$$\max_{0 \le c_2 \le x - c_1} \{\log c_2 + \alpha\beta_2(1 + \alpha\beta_2) \log (x - c_1 - c_2) + \text{constant}\}. \qquad (34)$$

Repeating this process, we find, in the limit,

$$c_1 = (1 - \alpha\beta_1)x, \qquad c_2 = \alpha\beta_1(1 - \alpha\beta_2)x$$

and

$$x - c_1 - c_2 = (\alpha\beta_1)(\alpha\beta_2)x.$$

Thus, using these "sophisticated-naive" policies, the quantity of fish x_t approaches the steady state

$$\hat{\hat{x}} = [(\alpha\beta_1)(\alpha\beta_2)]^{\alpha/(1-\alpha)}.$$

Comparing this steady state with the Cournot-Nash steady-state solution $\bar{x}$, it is clear that $\bar{x} > \hat{\hat{x}}$ since $0 < (1 - \alpha\beta_1)(1 - \alpha\beta_2) < 1$. Thus the steady-state quantity of fish in the von Stackelberg solution is smaller than that in the Cournot-Nash solution. The "sophisticated" country enjoys the *short-run* catch while in the steady state comparisons are ambiguous.[12]

☐ The growth function used in the previous example was of the form x^α, $0 < \alpha < 1$. This is a quite reasonable growth function possessing a natural steady state at $x = 1$ when there is no external interference. However, to show in a more pronounced fashion the difference between duopoly and combined exploitation of the sea, we shall use a linear growth function. Moreover, there are situations, e.g., resource depletion models, in which this model can be applied. It will be shown that it is quite possible in this case that if the countries operate jointly, the quantity of fish diverges to infinity, while if they operate in the Cournot-Nash mode, the quantity of fish converges to zero.

The growth function for fish is now assumed to be of the form

$$x_{t+1} = r(x_t - c_1(x_t) - c_2(x_t)), \qquad \text{with} \qquad r > 1. \qquad (35)$$

---

[12] If both countries have $\beta_1 = \beta_2 = \beta$, then, as in the case of cooperative exploitation, $\hat{x} > \bar{x}$, hence *a fortiori* $\hat{x} > \hat{\hat{x}}$, i.e., $((\alpha\beta)^{\alpha/(1-\alpha)} > (\alpha\beta)^{2\alpha/(1-\alpha)})$. Whether the "leader" country consumes more under cooperation depends on whether $(\alpha\beta)^{\alpha/(1-\alpha)} < \frac{1}{2}$.

332 / THE BELL JOURNAL OF ECONOMICS

The infinite horizon Cournot-Nash policies are

$$\bar{c}_1 = \frac{(1 - \beta_2)\beta_1}{1 - (1 - \beta_1)(1 - \beta_2)} x, \qquad \bar{c}_2 = \frac{(1 - \beta_1)\beta_2}{1 - (1 - \beta_1)(1 - \beta_2)} x, \text{[13]} \quad (36)$$

and

$$x - \bar{c}_1 - \bar{c}_2 = \frac{\beta_1\beta_2 x}{1 - (1 - \beta_1)(1 - \beta_2)} = \frac{\beta_1\beta_2}{\beta_1 + \beta_2 - \beta_1\beta_2} x = \frac{x}{\dfrac{1}{\beta_1} + \dfrac{1}{\beta_2} - 1}. \quad (37)$$

Hence, from equations (37) and (38), the dynamic equation is easily found to be

$$x_{t+1} = r\left(\frac{x}{\dfrac{1}{\beta_1} + \dfrac{1}{\beta_2} - 1}\right) = \frac{r}{\dfrac{1}{\beta_1} + \dfrac{1}{\beta_2} - 1} x_t.$$

Assuming $0 < \beta_1 < 1, 0 < \beta_2 < 1$, it is clear that

$$\frac{1}{\dfrac{1}{\beta_1} + \dfrac{1}{\beta_2} - 1} < 1.$$

If

$$\frac{r}{\dfrac{1}{\beta_1} + \dfrac{1}{\beta_2} - 1} > 1, \qquad \text{then} \qquad x_t \to \infty \qquad (38)$$

for any $x_0 > 0$, while if

$$\frac{r}{\dfrac{1}{\beta_1} + \dfrac{1}{\beta_2} - 1} < 1, \qquad \text{then} \qquad x_t \to 0$$

for any $x_0 > 0$. In the borderline case, i.e.,

$$\frac{r}{\dfrac{1}{\beta_1} + \dfrac{1}{\beta_2} - 1} = 1,$$

x_t remains at a neutral steady state at the level of the initial stock of fish x_0.

If both countries have the same rate of time discount $\beta_1 = \beta_2 = \beta$, we have:

$$x_t \to \infty \qquad \text{if} \qquad \frac{r\beta}{2 - \beta} > 1, \qquad \text{while} \qquad x_t \to 0 \qquad \text{if} \qquad \frac{r\beta}{2 - \beta} < 1.$$

If both countries combine their resources to exploit the sea, the limiting policy is given by

$$c = \frac{(1 - \beta)}{2} x, \quad (39)$$

and the dynamics is given by

---

[13] Note that the consumption policy is independent of the rate of return, r. This results from the logarithmic form of the utility function.

$$x_{t+1} = r[x_t - 2c] = r\beta x_t.$$

If $r\beta > 1$, then $x_t \to \beta$; if $r\beta < 1$, then $x_t \to 0$.[14]

Note that $(1 - \beta)/(2 - \beta) > (1 - \beta)/2$. Hence the proportion consumed is smaller under joint maximization. It is, of course, conceivable that $r\beta > 1$, while $r\beta/(2 - \beta) < 1$. Thus it is quite possible that when the countries cooperate, the "quantity" of fish will diverge to infinity, while under the Cournot-Nash solution the number of fish tends to zero.

4. Summary and conclusion

■ In this paper we have studied a model incorporating both dynamic and strategic aspects. The dynamics enters through the natural biological dynamics of the fish population, and the strategic aspect enters since there are two countries competing for fish. The equilibrium concept used to study the strategic aspect is a Cournot-Nash equilibrium in which each country maximizes the sum of discounted utilities subject to the actions of the other country. We studied only responses in the form of the size of the catch which depend upon the size of the fish population. Therefore threat strategies are excluded. We derived closed-form expressions for the Cournot-Nash policies for an example and studied the dynamics of the fish population under the influence of these Cournot-Nash policies. We also obtained closed-form expressions for the long-run or steady-state properties of these dynamics for this example. Then we compared these expressions with the values which result when the countries combine their resources and maximize a convex combination of their discounted utilities. We showed that the Cournot-Nash equilibrium leads to greater consumption as a function of the size of the fish population and to a smaller steady-state consumption. In fact, for a linear model, it is possible for the stock of fish to tend to extinction in the Cournot-Nash equilibrium, while with a cooperative regime the stock of fish tends to infinity.

Although we have derived our results only for an example, we may use them to study many important economic problems which require strategic behavior of the participants in a dynamic framework, e.g., limit-pricing models. Moreover, in the fishing or natural resource area other aspects of the problem can be studied: for example, the effects of profit maximization, when costs of extraction depend upon the size of the fish stock, or the effects of fixed costs (or nonconvexities) on the optimal policies. However, it should be noted that one is not likely to be able to derive results as simple as those in our example when more general models are considered. In fact, even under the rather simplified assumptions made in this paper, the Cournot-Nash equilibrium need not be unique, as is shown in Mirman (1978). But local properties for these nonunique equilibria should yield interesting results.

References

BROCK, W. "Differential Games with Active and Passive Variables." University of Chicago, 1975.

CLARK, C.W. "Economically Optimal Policies for the Utilization of Biologically Renewable Resources." *Mathematical Biosciences* (1971).

---

[14] If $r = 1$, then the stock of "fish" can be interpreted as an exhaustible resource in which case depletion occurs at a slower rate under joint maximization.

334 / THE BELL JOURNAL OF ECONOMICS

————. *Mathematical Bioeconomics*. New York: Wiley-Interscience, 1976.

———— AND MUNRO, G.R. "The Economics of Fishing and Modern Capital Theory: A Simplified Approach." *Journal of Environmental Economics and Management*, Vol. 2 (1975), pp. 92–106.

CLEMHOUT, S. AND H.Y. WAN, JR. "Interactive Economic Dynamics and Differential Games: A Survey," Cornell University, 1978.

FISHER, A. AND PETERSON, F. "Natural Resources and the Environment in Economics." University of Maryland, 1975; published in part as "The Environment in Economics: A Survey," *Journal of Economic Literature*, Vol. 14 (1976), pp. 1–33.

GORDON, H.S. "The Economic Theory of a Common Property Resource: The Fishery." *Journal of Political Economy*, Vol. 62 (1954), pp. 124–142.

KYDLAND, F. "Equilibrium Solutions in Dynamic Dominant Player Models." *Journal of Economic Theory*, Vol. 15, No. 2 (August 1977a).

————. "A Dynamic Dominant-Firm Model of Industry Structure." University of Minnesota, 1977b.

———— AND PRESCOTT, E. "Rules Rather than Discretion: The Inconsistency of Optimal Plans." *Journal of Political Economy*, Vol. 85 (1977), pp. 473–491.

LEVHARI, D., MICHENER, R., AND MIRMAN, L.J. "Dynamic Programming Models of Fishing." University of Illinois, 1978.

LONG, N. "Optimal Exploitation and Replenishment of a Natural Resource" in J. Pitchford and S. Turnovsky, eds., *Applications of Control Theory to Economic Analysis*, Amsterdam: North-Holland Publishing Co., 1977, pp. 81–106.

LOTKA, A.J. *Elements of Mathematical Biology*. New York, 1956.

MIRMAN, L.J. "Dynamic Models of Fishing: A Heuristic Approach" in P.T. Liu and J.G. Sutinen, eds., *Control Theory in Mathematical Economics*, New York: Dekker, 1979, pp. 39–73.

SCOTT, A. "The Fishery: The Objectives of Sole Ownership." *Journal of Political Economy*, Vol. 63 (1955), pp. 16–124.

SMITH, V.L. "Economics of Production from Natural Resources." *American Economic Review*, Vol. 58 (1968), pp. 409–431.

————. "Control Theory Applied to Natural and Environmental Resources: An Exposition." *Journal of Environmental Economics and Management*, Vol. 4 (1977), pp. 1–24.

TAKAYAMA, T. "Dynamic Theory of Fisheries Economics—II; Differential Game Theoretic Approach." University of Illinois, 1977.

[25]

The optimal management of transboundary renewable resources

GORDON R. MUNRO / University of British Columbia

Abstract. This paper investigates the question of the optimal management of renewable resources jointly owned by two states. A dynamic model of fisheries is combined with Nash's theory of two-person co-operative games. Conflicts in the management strategies of the two states arising from differences in perceptions of the social rate of discount, fishing effort costs, and consumer preferences are examined. Cases are considered in which the two partners can and cannot make side or transfer payments to one another. It is concluded that side payments greatly ease the resolution of resource management conflicts.

La gestion optimale de ressources renouvelables trans-nationales. Ce mémoire examine le problème de la gestion optimale de ressources renouvelables possédées conjointement par deux états. Un modèle de pêcheries fondé sur la théorie du capital est combiné par l'auteur avec certains éléments de la théorie des jeux coopératifs à deux personnes de Nash. Le mémoire examine les conflits entre les stratégies de gestion des deux états qui émanent de différences dans les perceptions du taux social d'escompte, des coûts de l'effort de pêche et des préférences des consommateurs. L'auteur étudie les cas où les deux partenaires peuvent et ne peuvent pas effectuer des paiements de compensation ou de transfert entre eux. Il appert que la possibilité de faire par à côté des paiements de compensation facilite beaucoup la résolution de conflits dans la gestion des ressources.

An earlier and shorter version of this work was presented to the Fifth Pacific Regional Science Conference in Vancouver in August 1977 under the title 'Canada and Extended Fisheries Jurisdiction in the Northeast Pacific: some issues in optimal resource management.' Research for this paper has been funded both by the Donner Canadian Foundation as part of the project 'Canada and the international management of the oceans' sponsored by the Institute of International Relations, University of British Columbia and by the Social Sciences and Humanities Research Council of Canada through the Programme in Natural Resource Economics, Department of Economics, University of British Columbia.

I have received helpful comments on earlier drafts of this paper from fellow economists Peter Pearse and Anthony Scott. Above all I must acknowledge a debt of gratitude to my friend and colleague Colin Clark of the Department of Mathematics, UBC, who has been unstinting in his assistance. I must also acknowledge the helpful comments of two anonymous referees.

356 / Gordon R. Munro

INTRODUCTION

The widespread implementation of Extended Fisheries Jurisdiction (EFJ) has meant that many of the world's fishery resources hitherto the object of unregulated or weakly regulated international exploitation will become subject to the management of individual coastal states. Saetersdal (1977) and other official observers have noted, however, that several of these resources present difficult management problems because they are 'transboundary,' that is, they straddle the boundaries of the Exclusive Economic Zones of two or more coastal states. Obviously joint coastal state management must be applied to these resources. However, there is no reason to suppose that the goals and interests of the relevant coastal states will coincide.

While the problem of transboundary fishery resources is hardly a new one, it is clear that the scope and the urgency of the problem have been greatly magnified by the advent of EFJ. This paper investigates some of the aspects of the question of optimal management of such resources with the aid of a dynamic model of fisheries combined with a theory of bargaining, specifically Nash's theory of co-operative games.[1] Earlier attempts to investigate this question have relied upon static fisheries models and have not addressed the problem of resolving conflicts of interest between the joint owners of the resources (Anderson, 1975a and b).

It is assumed that within each coastal state the fisheries management policy is the responsibility of a single social manager and for any transboundary fishery the manager's goal will be to maximize his country's benefits from the fishery. Conflicts in management strategies between the coastal state managers arise through differences in perceptions of the social rate of discount, fishing effort costs, and consumer tastes between countries. When considering the resolution of management conflicts, cases in which the prospective partners can make transfer or side payments to one another are constrasted with cases in which political or other constraints make such payments impossible.

The discussion will be confined to two-country, or two-entity, joint ownership of resources. While this may appear restrictive, it can be argued that with the advent of EFJ there will be many cases in which the number of prospective managers will be no more than two. Several examples are provided by transboundary fishery resources on the Pacific and Atlantic coasts of North America and, one might add, in the North Sea. In time, of course, the analysis

1 For a discussion of Nash's theory of co-operative games see Bacharach (1976), Intriligator (1971), and Luce and Raiffa (1967). For an application of Nash's theory of co-operative games in the area of exhaustible resources see Hnyilicza and Pindyck (1976). Hnyilicza and Pindyck discuss an optimal pricing policy for OPEC given that the organization is split into two groups which differ in terms of their perception of the appropriate social rate of discount. The problem they discuss thus bears some similarity to the one we consider below in the third section.

The optimal management of transboundary renewable resources / 357

should be extended to cover transboundary resources involving several countries. That however, would be the subject of another paper.

THE BASIC MODEL

For ease of exposition a single fishery is considered. The relevant underlying biological model is assumed to be the well-known Schaefer (1957) model. The population dynamics are modelled by the equation

$$dx/dt = F(x) - h(t), \tag{1}$$

where $h(t)$ and $x(t)$ are the harvest rate and the population biomass respectively at time t, while $F(x)$ is the natural growth function. Specifically in the Schaefer model,

$$F(x) = rx(1 - x/K), \tag{2}$$

where K is the maximum biomass size and r, a constant, is the intrinsic growth rate, and

$$h(t) = qE(t)x(t), \tag{3}$$

the harvest production function, where $E(t)$ is the rate of fishing effort at time t and q, a constant, is the 'catchability' coefficient. The harvest production functions relevant to the two countries are assumed to be identical. We shall also assume for the time being that both face a world demand for the harvested fish which is infinitely elastic and that the effort input supply functions are also infinitely elastic.

If the number of owners of the resource were reduced to one, the optimization model could then be formulated as follows (see Clark and Munro, 1975; Clark, 1976). The objective functional is the discounted net cash flow from the fishery:

$$PV = \int_0^\infty e^{-\delta t}[p - c(x)]h(t)dt, \tag{4}$$

subject to

$$dx/dt = F(x) - h(t), \tag{5}$$

$$0 \le h(t) \le h_{max}, \tag{6}$$

$$0 \le x(t), \tag{7}$$

where δ is the instantaneous social rate of discount, p is the price of fish, h_{max} is the maximum feasible harvest rate, and $c(x)$ is the unit cost of harvesting. Unit costs of harvesting are derived as follows. Let the total cost of fishing

358 / Gordon R. Munro

effort be $C(E)$. By assumption, $C(E) = aE$, where a, a constant, is the unit cost or price of fishing effort. From (3) we have

$$E = h/qx. \tag{8}$$

Total harvesting costs can thus be expressed as

$$C(x, h) = ah/qx \tag{9}$$

and unit harvesting costs as

$$c(x) = a/qx. \tag{10}$$

The problem can now be seen to be a linear optimal control problem with $x(t)$ as the state variable and $h(t)$ as the control varible. The Hamiltonian of the problem is

$$H = e^{-\delta t}\{[p - c(x)]h(t)\} + \lambda(F(x) - h(t)), \tag{11}$$

where λ, the adjoint or costate variable, can be viewed as the shadow price of the resource discounted back to $t = 0$.

A straightforward application of the maximum principle permits the determination of the optimal equilibrium biomass x^* and hence the optimal equilibrium harvest rate h^*. The optimal equilibrium biomass is given by the following modified golden rule equation:

$$F'(x^*) - \frac{c'(x^*)F(x^*)}{p - c(x^*)} = \delta. \tag{12}$$

The left hand side of (12) is the 'own rate of interest' of the biomass. It is equal to $d/dx^*\{[p - c(x^*)]F(x^*)\}/[p - c(x^*)]$, i.e. the marginal sustainable net return or rent from the fishery divided by the supply price of the resource. The supply price in turn is simply the forgone marginal current rent from harvesting. As expressed in (12), the 'own rate of interest' consists of two components: the instantaneous marginal product of the resource ($F'(x)$) and what Clark and Munro (1975) refer to as the marginal stock effect.

Thus (12) states simply that the optimal biomass level is that level at which the own rate of interest of the resource is equal to the social rate of discount. The equilibrium solution is unique and is not a function of time.[2] Hence the equilibrium harvest policy is given by

$$h^*(t) = F(x^*). \tag{13}$$

With regard to the optimal approach path to x^*, the steady-state solution, it

---

2 When the underlying biological model is the Schaefer model and the optimal control model is linear, the solution will indeed be unique. If other biological models were used or if the control model were non-linear, we could not be assured of a unique solution. These points are discussed in Clark and Munro (1975).

The optimal management of transboundary renewable resources | 359

can be observed that the model is linear in the control variable h. Hence the optimal approach is the so-called bang-bang approach:

$$h^*(t) = h_{max} \quad \text{whenever } x(t) > x^*$$

$$= h_{min} \quad \text{whenever } x(t) < x^*. \tag{14}$$

For a more detailed discussion of the optimal approach path see Clark and Munro (1975).

If two countries rather than one had claims to the resource but had identical effort costs and were identical in their perception of the appropriate social discount rate, the above model would be applicable without modification. Bargaining would have to take place with respect to the harvest shares or, what amounts to the same thing, the proceeds shares. The relative size of the shares would, however, have no impact on the optimal management policy (see Munro, 1977).

INCONSISTENT VIEWS ON THE SOCIAL RATE OF DISCOUNT

Let us now consider the consequences if the social managers take different views of the appropriate social rate of discount.[3] Designate the perceived discount rates of Country 1 and Country 2 as δ_1 and δ_2. Assume that $\delta_1 < \delta_2 < \infty$ and further that no other differences exist between the two countries.

Although the relevant costs and prices are identical, the desired social management policies of the two countries will be quite different. It can be demonstrated that Country 1, the low-discount-rate country, will be the more 'conservationist' of the two. If we let $x_{\delta_1}^*$ denote the optimal biomass as perceived by Country 1 and $x_{\delta_2}^*$ the optimal biomass perceived by Country 2, then $x_{\delta_1}^* > x_{\delta_2}^*$. Country 1, having the lower discount rate, has a greater incentive to invest in the resource than Country 2.

The Informal Composite Negotiating Text (United Nations, 1977) arising from the third Law of the Sea Conference admonishes coastal states with transboundary stocks to co-operate with neighbouring coastal states in the management of such resources. We shall suppose, therefore, that the two countries contemplate a binding agreement.

The first question to be considered is that of harvest or proceeds shares. Since the optimal management problem is being examined in a dynamic framework, we allow for the harvest or distributional shares to vary over time and attempt to determine the optimal time path of the shares.

Yet some of the optimal policies which emerge when the harvest shares are time-variant (particularly when side payments cannot be made) are extreme in that a partner is called upon to endure extended periods with no return

3 There is a good deal of controversy over what constitutes the appropriate rate of discount for management of fisheries. See Copes (1977) and Munro (1977). Hence there is no reason why the two social managers must agree on the rate of discount.

360 / Gordon R. Munro

whatsoever from the resource. Clearly it may be difficult to reach permanent agreement when such policies are called for. We therefore allow for the possibility that because of political or other considerations the constraint of time-invariant distributional shares may be imposed. Incidentally, virtually all agreements pertaining to transboundary stocks that do not ignore harvest shares entirely call for time-invariant shares (Saetersdal, 1977).

Suppose at the outset, then, that the harvest shares are in fact time-invariant and the two countries first settle on such shares for any forthcoming management program. The shares might, for example, be based upon the historical catch records of the two countries. The consequences of relaxing the constraint of time-invariant shares will be investigated later.

Denote Country 1's harvest share as α, where $0 \leq \alpha \leq 1$ and α is independent of time. The objective functionals of the two countries can thus be expressed as follows:

$$PV_1 = \int_0^\infty e^{-\delta_1 t} \alpha[p - c(x)]h(t)dt, \tag{15}$$

$$PV_2 = \int_0^\infty e^{-\delta_2 t}(1 - \alpha)[p - c(x)]h(t)dt. \tag{16}$$

The management preferences of Country 1 and Country 2 will be independent of α.[4]

We now turn to the problem of determining a management agreement. To characterize a potential agreement between the two countries we adopt an intuitively appealing method used by Hnyilicza and Pindyck (1976; see n. 1). Let such an agreement be one in which the harvesting policy through time is set to maximize a weighted sum of the objective functionals of the two countries. Thus we have

$$\text{maximize } PV = \beta PV_1 + (1 - \beta)PV_2, \qquad \text{where } 0 \leq \beta \leq 1. \tag{17}$$

It is easiest to think of β as a bargaining parameter which permits the establishment of a tradeoff between the management preferences of the two countries. Thus if β were equal to 1 the preferences of Country 1 would be totally dominant.

To determine which β is most likely to arise out of the negotiations between the two countries a theory of bargaining is required. For this we turn to Nash's theory of two-person co-operative games (Nash, 1953). It is assumed initially

---

4 Consider the determination of the optimal management policy as perceived by Country 1. By a straightforward application of the maximum principle we can determine $x_{\delta_1}^*$. This is given by the following equation:

$$F'(x_{\delta_1}^*) - [\alpha c'(x_{\delta_1}^*)F(x_{\delta_1}^*)]/\{\alpha[p - c(x_{\delta_1}^*)]\} = \delta_1.$$

Obviously the αs cancel out. Intuitively this means that, while Country 1 will be happier with a large α than with a small one, its perception of the optimal biomass will not change. The optimal approach path will be the most rapid or 'bang-bang' approach regardless of the size of α. The trivial exception to this rule arises if $\alpha = 0$. Then Country 1 would have no management policy. What applies to Country 1 applies with equal force to Country 2.

The optimal management of transboundary renewable resources | 361

that side payments cannot be made between the two countries, i.e. payoffs are non-transferable.

Denote the payoffs for Country 1 and Country 2 as π and θ respectively, where these variables are seen to represent the present values of the streams of returns enjoyed by the two countries given the implementation of a specific harvest policy. By varying β between 0 and 1 and determining for each β the harvest policy that will maximize $\beta \text{PV}_1 + (1 - \beta)\text{PV}_2$, the Pareto-efficient frontier in the space of realized outcomes or payoffs can be obtained. By solving the game and thus deciding at what point on the frontier the two players are most likely to settle, the question of which β is most likely to arise out of the negotiations is answered.

To provide a measure of the relative bargaining power of the players, Nash (1953) introduces the concept of a 'threat point' consisting of a set of payoffs π° and θ° that the prospective partners would enjoy if no co-operation were forthcoming. These could be payoffs arising from the solution of a two-person *competitive* game (e.g. Nash, 1951).[5] Clearly these are the minimum payoffs the two countries would accept under an agreement.

Nash (1953) sets forth six readily acceptable assumptions.[6] He then proves that if these assumptions are satisfied a unique solution to the game can be obtained by maximizing the following expression:

$$\text{maximize } (\pi^* - \pi^\circ)(\theta^* - \theta^\circ), \tag{18}$$

where π^* and θ^* are the solution payoffs. Thus if π^* and θ^* corresponded, say, to $\beta = \frac{1}{2}$, the co-operative agreement would be such that equal weight would be given to the two countries' management preferences.[7]

5 Clark (1978) in a somewhat different context proves that, given a linear model such as this, two players in a Nash competitive game would drive the fishery to bionomic equilibrium.

6 The assumptions are that the solution is contained within the feasible set of solutions, is rational in outcome in the sense that each player receives a payoff not less than its threat point payoff, is independent of irrelevant alternatives and of linear transformations of the set of payoffs, is unaffected by the numbering of the players, and finally is Pareto-optimal. For a detailed discussion of these assumptions see Luce and Raiffa (1967).

Obviously there may be bargaining solutions other than the Nash one. For the analysis which follows it is in fact of less than critical importance how the tradeoff between management preferences, i.e. the size of β, is determined. We have chosen to employ the Nash model as a bargaining framework because it combines the generality and robustness referred to by Hnyilicza and Pindyck with the virtues of simplicity and strong intuitive appeal.

7 Two points need to be raised at this juncture. First we have assumed implicitly that the frontier is such that the feasible or attainable set (of payoffs) is convex. If in fact this were not the case we could convexify the set by introducing correlated mixed strategies and go on to obtain a solution to the game (see Bacharach, 1976; Luce and Raiffa, 1967).

How likely is it that such problems will arise? While I cannot prove they will never arise, upon consideration of the rather bizarre implications of a frontier being of the sort to give rise to non-convexity problems, I believe the likelihood is small. Secondly we must allow for the possibility of significant bargaining and enforcement costs. If we can assume that these costs are independent of the harvest rate and hence of the biomass level, they can be easily incorporated in the model. One way of dealing with them would be to add them in capitalized form to the threat point payoffs as forgone costs. High bargaining and/or enforcement costs could well mean that the solution would violate the rationality assumption and no agreement would be forthcoming.

362 / Gordon R. Munro

We now examine the implications for management policy on the assumption that bargaining leads to a β, such that $0 < \beta < 1$,[8] that is, the preferences of both partners carry weight in the management program. The objective functional in (17) can be expressed more fully as

$$\text{PV} = \int_0^\infty \{\beta\alpha e^{-\delta_1 t} + (1 - \beta)(1 - \alpha)e^{-\delta_2 t}\}[p - c(x)]h(t)dt. \qquad (19)$$

The Hamiltonian of our problem is

$$H = \{\beta\alpha e^{-\delta_1 t} + (1 - \beta)(1 - \alpha)e^{-\delta_2 t}\}[p - c(x)]h(t) + \lambda(F(x) - h(t)). \qquad (20)$$

The optimal biomass time path can be obtained by a routine application of the maximum principle. This is given by the following modified golden rule equation:

$$F'(x^*) - \frac{c'(x^*)F(x^*)}{p - c(x^*)} = \frac{\delta_1\beta\alpha e^{-\delta_1 t} + \delta_2(1 - \beta)(1 - \alpha)e^{-\delta_2 t}}{\beta\alpha e^{-\delta_1 t} + (1 - \beta)(1 - \alpha)e^{-\delta_2 t}}. \qquad (21)$$

The right hand side of (21) can be viewed as a complex weighted average of the two discount rates δ_1 and δ_2. Denote the weighted average by $\delta_3(t)$. Observe that $\delta_3(t)$, and hence $x^*(t)$, are not independent of time. Note further that

$$\lim_{t \to \infty} \delta_3(t) = \delta_1. \qquad (22)$$

One such possible outcome is illuatrated in Figure 1. Suppose we commence at bionomic equilibrium so that $x(0)$ is below the perceived optimal biomass levels of both countries, levels which are denoted by $x_{\delta_1}^*$ and $x_{\delta_2}^*$. The biomass time path $x_{\delta_3}^*(t)$ then represents the optimal biomass time path resulting from bargaining between the two partners. Because the control model is linear, the optimal approach path is the most rapid or 'bang-bang' one. Thus the nature of the tradeoff is to give the high-discount-rate country's management preferences relatively strong weight in the present and near future but to allow the low-discount-rate partner's preferences to dominate the more distant future.

Now let us alter the assumption and suppose that α is not constrained with respect to time. Both countries are prepared to see α vary over the life of the management or planning program. We continue to assume that side payments are not possible.

Since α is now a control variable rather than a parameter, the Hamiltonian H must be maximized with respect to α (as well as with respect to h) at each moment. Upon differentiating (20) with respect to α we have

$$\partial H/\partial\alpha = \{\beta e^{-\delta_1 t} - (1 - \beta)e^{-\delta_2 t}\}[p - c(x)]h(t). \qquad (23)$$

---

8 If $\beta = 1$ or $\beta = 0$, the problem reduces to that of a single owner of the resource and hence is uninteresting at this stage.

The optimal management of transboundary renewable resources / 363

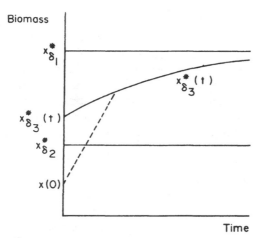

FIGURE 1   Optimal biomass time path, bargaining without side payments

Since $e^{-\delta_2 t} < e^{-\delta_1 t}$, a $\beta \leq \frac{1}{2}$ would certainly mean that $\partial H/\partial\alpha$ was positive for all $t > 0$.[9] This in turn would call for setting $\alpha = 1$ for all $t > 0$, which would imply that $\theta^* = 0$, i.e. Country 2 would get no return from the fishery. Conversely, if $\beta = 0$, this would mean that $\partial H/\partial\alpha$ was negative for all $t > 0$, implying that $\pi^* = 0$.

On the other hand if $0 < \beta < \frac{1}{2}$ and if β is such that at the beginning of the management program the expression in braces in (23) is negative, the resultant optimal harvest policy will be such that both π^* and θ^* will be positive, that is, both countries will enjoy a return from the fishery. At the beginning of the management program $\partial H/\partial\alpha < 0$, which means setting $\alpha = 0$. There must, however, be some moment $t = T$, where $0 < T < \infty$, at which a switch will occur, i.e. $\partial H/\partial\alpha$ will become positive. Hence the optimal policy will be to allow the high-discount-rate country to enjoy the *entire* proceeds of the fishery from the beginning of the planning period until time T. From that point on the entire proceeds would be enjoyed by the low-discount-rate country.[10] When the switch will occur (given $e^{-\delta_1 t}$ and $e^{-\delta_2 t}$) will depend upon the size of β.

The time path of $x_{\delta_3}^*(t)$ associated with this optimal harvest policy is easy to describe. It would coincide with $x_{\delta_2}^*$ up to $t = T$ and with $x_{\delta_1}^*$ thereafter.[11]

It would seem reasonable to suppose that, if a solution to the co-operative game exists (i.e. the threat point does not lie above the Pareto-efficient

---

9 It will be recalled that β is the weight given to Country 1's management preferences.
10 Interesting complications could arise if we had to bring correlated mixed strategies into the game. However it seems unlikely the need would arise (see n. 7).
11 One complication likely to arise is the fact that x almost certainly cannot be instantly increased from $x_{\delta_2}^*$ to $x_{\delta_1}^*$ at $t = T$. Recall that h has a lower bound, i.e. $h \geq 0$. This will give rise to interesting so-called blocked interval problems. For a discussion of such problems see Clark and Munro (1975).

364 / Gordon R. Munro

frontier), the solution to the game would be such that both countries would enjoy a return from the resource, i.e. $\pi^* > 0$, $\theta^* > 0$. An outcome where either $\theta^* = 0$ or $\pi^* = 0$ is not abolutely impossible but is exceedingly unlikely. Consider, for example, an outcome in which $\alpha = 1$ for $t > 0$, and hence in which $\theta^* = 0$. This would imply first that $\theta^° = 0$ (we ignore the possibility of $\theta^°$ < 0). Secondly such an outcome would imply that in maximizing (18) we would have $(\pi^* - \pi^°)(\theta^* - \theta^°) = 0$. This is possible if $\pi^* = \pi^°$; in other words if Country 1 were to receive anything less than exclusive control over the fishery it would be worse off than if it had refused to co-operate. These two conditions being met would suggest that with no co-operation Country 1 could effectively deny Country 2 access to the resource and could manage the resource as it saw fit. Hence co-operation would be pointless. Similar arguments apply to the opposite extreme, in which we have $\pi^* = 0$, $\theta^* > 0$ as the solution payoffs.

If side payments between the two prospective partners are permitted, the establishment of an optimal management program is much simplified. The objective will now be simply to seek out that policy which will maximize PV_1 + PV_2. In other words PV_1 and PV_2 would be given equal weight, and bargaining would be confined to the division of the total return from the fishery.

Note the obvious fact that giving PV_1 and PV_2 equal weight is the equivalent of setting $\beta = \frac{1}{2}$. If we continue to assume that α is time-variant, our previous discussion implies that the optimal policy will be that of setting $\alpha = 1$ for $t > 0$. The optimal policy is that of making Country 1, the partner placing the highest value upon the resource, its sole owner. Side payments will be made to Country 2, these payments being concentrated at the beginning of the program. Any planned side payment to be made at time $t > 0$ would have a smaller present value for Country 2 than it would for Country 1. Thus the optimal policy calls for the conservationist partner to buy out its less conservationist partner at the beginning of the program.

The problem of bargaining over the division of the total or global return from the fishery is easily solved. Let the threat point payoffs, as before, be $\pi^°$ and $\theta^°$. Let the solution payoffs now be $\pi^\dagger$ and $\theta^\dagger$, where $\pi^\dagger + \theta^\dagger = \omega$ and where ω, a constant, denotes the global return from the resource. It can be shown (Luce and Raiffa, 1967; Munro, 1977) that

$$\pi^\dagger = (\omega - \theta^° + \pi^°)/2$$

and

$$\theta^\dagger = (\omega - \pi^° + \theta^°)/2. \tag{24}$$

Luce and Raiffa (1967, 138) interpret this result by saying each player receives an average of the return it could expect without an agreement and the marginal contribution it makes by accepting the agreement with side payments.

UNEQUAL HARVESTING COSTS

Consider now the case in which $\delta_1 = \delta_2$ but the fishing effort costs, and hence the harvesting costs, of the two partners differ. Let a_1 and a_2 be the unit costs

The optimal management of transboundary renewable resources / 365

of fishing effort in Country 1 and Country 2. Assume that Country 1 is the high-cost country, i.e. $a_1 > a_2$, and that a_1 and a_2 are independent of the levels of fishing effort. We shall suppose that barriers to factor mobility are sufficient to maintain this inequality. From equation 10 it can be seen that unit harvesting costs in the two countries can now be expressed as

$$c_1(x) = a_1/qx,$$ (25)

$$c_2(x) = a_2/qx.$$ (26)

Thus for any finite, positive biomass level $c_1(x) > c_2(x)$.

We commence as before by assuming no side payments and by assuming that α is time-invariant. If harvesting costs are sensitive to the size of the biomass, the high-cost country will be more conservationist than its low-cost partner. This is simply because Country 1 will be more sensitive to the impact of the size of the biomass upon harvesting costs.[12] We again turn to Nash's two-person, co-operative game theory to achieve a resolution of the conflict, i.e. to decide the weight to be given to each country's management preferences. The threat point, as before, can be seen in terms of the payoffs arising from the solution to a two-person competitive game. I shall demonstrate that, if harvesting costs are insensitive to the size of the biomass, the differences in harvesting costs will be irrelevant, and no bargaining will be required to determine an optimal management policy.

If the β arising from the solution to the co-operative game is such that $0 < \beta < 1$, our objective functional becomes

$$\text{PV} = \int_0^\infty e^{-\delta t}[(\alpha\beta + (1 - \alpha)(1 - \beta))p - (\alpha\beta c_1(x)$$
$$+ (1 - \alpha)(1 - \beta)c_2(x))]h(t)dt, \quad (27)$$

while the Hamiltonian for the problem is

$$H = e^{-\delta t}\{[(\alpha\beta + (1 - \alpha)(1 - \beta))p - (\alpha\beta c_1(x)$$
$$+ (1 - \alpha)(1 - \beta)c_2(x))]h(t)\} + \lambda(F(x) - h(t)). \quad (28)$$

Once again a straightforward application of the maximum principle gives the following modified golden rule for determining the optimal biomass:

$$F'(x^*)$$
$$- \frac{\{\alpha\beta c_1'(x^*) + (1 - \alpha)(1 - \beta)c_2'(x^*)\}F(x^*)}{(\alpha\beta + (1 - \alpha)(1 - \beta))p - (\alpha\beta c_1(x^*) + (1 - \alpha)(1 - \beta)c_2(x^*))} = \delta. \quad (29)$$

The second term on the left hand side of (29) is a complex version of what

12 If we return to equations (8) to (10) and observe the relationships between the costs of fishing effort and of harvesting it can be easily shown for any given biomass level x^t that

$$\left\{-\frac{c_1'(x^t)F(x^t)}{p - c_1(x^t)}\right\} > \left\{-\frac{c_2'(x^t)F(x^t)}{p - c_2(x)}\right\},$$

i.e. the 'marginal stock effect' perceived by Country 1 will be higher than that perceived by Country 2.

366 / Gordon R. Munro

Clark and Munro (1975) refer to as the marginal stock effect. The larger β, the larger will be the marginal stock effect, and the closer we will be pushed to the biomass level perceived by Country 1 as being optimal. Regardless of the size of x^*, however, the optimal approach path to x^* continues to be the 'bang-bang' one.

There is no necessary reason why harvesting costs must be sensitive to the biomass level. If in fact they are not, then (29) reduces to

$$F'(x^*) = \delta, \tag{30}$$

and thus no bargaining will required to determine the optimal management policy. Moreover, the questions whether α is or is not to be time-variant and whether side payments are or are not allowed become irrelevant. Bargaining will be required only to determine the size of α.

Next let us investigate the consequences of permitting α to be time-variant, given that the marginal stock effect is significant and no side payments can be made. Differentiating (28) with respect to α gives

$$\partial H/\partial \alpha = e^{-\alpha t}\{\beta[p - c_1(x)] - (1 - \beta)[p - c_2(x)]\}h(t). \tag{31}$$

Whether $\partial H/\partial \alpha$ is positive (negative)will depend upon whether the following inequality holds (is reversed):

$$\beta + [p - c_1(x)]/[p - c_1(x)) + (p - c_2(x))] > 1. \tag{32}$$

The second term on the left hand side of the above inequality can be referred to as Country 1's marginal rent ratio MRR_1. This is a measure of the relative marginal contribution to global current rent from the fishery generated by Country 1's fishing activities. Since $c_1(x) > c_2(x)$, $MRR_1 < \frac{1}{2}$. Thus $\partial H/\partial \alpha$ will be positive only if the weight given to Country 1's preferences is great enough (i.e. β must at minimum exceed $\frac{1}{2}$) to offset the country's cost disadvantages.

While $MRR_1 < \frac{1}{2}$, it is also true that MRR_1 is an increasing function of x. Indeed it can be seen that

$$\lim_{x \to \infty} MRR_1 = 1/2. \tag{33}$$

This gives rise to the possibility of an optimal harvest policy in which both countries will enjoy a return from the fishery. This can be seen most easily if we commence with a virgin fishery $x(0) = K > x_1^*$, where x_1^* is the optimal biomass as perceived by Country 1. If $\beta \le \frac{1}{2}$, then

$$\beta + [p - c_1(K)]/[(p - c_1(K)) + (p - c_2(K))] < 1. \tag{34}$$

This inequality will hold for all biomass levels, and the implied optimal harvest policy will be to have Country 2 harvest the resource exclusively for all time.

If on the other hand $\beta > \frac{1}{2}$ and the above inequality is reversed, it is possible that the optimal harvest policy will call for both countries to harvest the resource in turn. Since $\beta + MRR_1 > 1$, $\partial H/\partial \alpha > 0$, and the appropriate policy

The optimal management of transboundary renewable resources | 367

will be to have Country 1 harvest the resource exclusively. If we define $x_1\infty$ as the biomass level at which $p - c_1(x_1\infty) = 0$, there will be a biomass level x_0, $x_1\infty \leq x_0 < K$, at which $\beta + \text{MRR}_1 = 1$. When x is reduced below x_0, $\partial H/\partial \alpha$ will become negative, which in turn implies that $\alpha = 0$. Country 2 will now be the exclusive harvester until $t = \infty$. The biomass will be reduced to x_2^*, the optimal biomass level as perceived by Country 2. From that point on the optimal harvest rate will be $h^* = F(x_2^*)$.

If $x_0 < x_1^*$, there is no basis for assuming that a switch will occur. With $\alpha = 1$, the optimal harvest policy at any stock level below x_1^* will be $h = h_{\min}$. However, x_0 is among other things a function of β. Thus for both countries to enjoy a return from the resource, β must be large enough that at the commencement of the program $\partial H/\partial \alpha > 0$ but small enough that $x_0 > x_1^*$.

We assumed that $x(0) = K$. Obviously it is possible to have an optimal harvest policy in which both countries enjoy a return from the resource with $x(0) < K$, given that $x(0) > x_1^*$. The sense of this policy is simply that the high-cost country's harvesting activities should be concentrated in those periods in which the activities are least disadvantageous in terms of the global return from the fishery.

If $x(0) < x_2^*$, it will also be possible to have an optimal harvest policy in which $\pi^* > 0$, $\theta^* > 0$. A problem does exist however. Since in this instance the stock is being rebuilt, $\partial H/\partial h < 0$ for all stock levels below the optimum. The optimal harvest policy, while $\partial H/\partial h < 0$, is that of setting $h = h_{\min}$. Only if $h_{\min} > 0$ would Country 2 enjoy a return from the resource.[13]

If side payments are possible, the nature of the outcome will be straightforward and obvious. All the harvesting will be done by Country 2, which in turn will pension off its high-cost partner.[14] As for the division of the

---

13 Suppose that we had $x_2^* < x(0) < x_1^*$. Then if β is such that at the commencement of the program $\beta + \text{MRR}_1 > 1$, Country 2 will get no return from the resource. Conversely, if β is such that at the commencement of the program $\beta + \text{MRR}_1 < 1$, Country 1 will get no return. Only if β is such that at the program's commencement $\beta + \text{MRR}_1 = 1$ (implying that $\partial H/\partial \alpha = 0$) will both countries enjoy a return. This curious result is possible only because by assumption no side payments are permitted. Once side payments are permitted the problem is greatly simplified.

14 If we let $\beta = \frac{1}{2}$ and permit α to be time-variant, it must be true, as we have seen, that $\partial H/\partial \alpha < 0$ for $t \geq 0$. Consequently Country 1's vessels must be barred from harvesting from the beginning of the management program. It seems implausible that Country 1 would find it politically possible to accept such an agreement. Yet one can point to an actual co-operative arrangement with side payments in which certain partners agreed to the complete elimination of their harvesting activity. In 1911, the United States, Russia, Canada, and Japan signed the North Pacific Fur Seal Convention. The first two countries harvested the seals on land, while the latter harvested the seals at sea – a higher-cost operation. Under the terms of the convention Canada and Japan agreed to reduce their sealing activities to zero in return for a yearly share of the harvest. The convention has been renewed several times and remains in force today.

　　If Country 1 insists on maintaining a minimum fleet, in other words if α is made time-invariant and greater than zero, the equilibrium equation becomes:

$$F'(x^*) - [\{\alpha c_1'(x^*) + (1 - \alpha)c_2'(x^*)\}F(x^*)]/[p - \alpha c_1(x^*) - (1 - \alpha)c_2(x^*)] = \delta.$$

It is not clear who would 'bribe' whom in this case. It is quite possible that Country 1 would 'bribe' Country 2 into accepting a more conservationist policy than Country 2 would otherwise accept.

368 / Gordon R. Munro

return from the fishery, we apply the bargaining theory used in the divergent discount rate case. The payoff to each partner or player will be an average of the payoff it would enjoy without co-operation and the marginal contribution it makes by accepting an agreement with side payments.

CONSUMER TASTES, NON-CONVEXITIES, AND MULTIPLE EQUILIBRIA

So far, dissimilarities in consumer tastes between the two countries have been ignored as a source of conflict. This was done by assuming that the two countries faced an infinitely elastic world demand for harvested fish. This assumption is now replaced by its polar opposite, namely that it is not feasible to market the fish outside the region comprised by the two adjoining states. It is also assumed that because of cultural factors consumer preferences in fish differ markedly between the two states. Let us select a relatively simple case in which differences in consumer preferences are unambiguous and lead clearly to a divergence in perceived optimal management strategies. This simple case leads to some interesting complexities, one of which is that an optimal control may fail to exist.

We assume the following. No barriers to trade in fish exist between the two countries. The two countries are identical with respect to harvesting costs, and the unit cost of fishing effort in each is independent of the level of fishing effort. Furthermore, $\delta_1 = \delta_2$. In Country 1 consumers view the fish as a highly desirable food fish. Country 1's (national) demand function for the fish is assumed to be linear and to exhibit finite price elasticity. In Country 2 on the other hand consumers of fish and fish products regard the fish as fit only for reduction and see it as a perfect substitute for other 'trash' fish. Furthermore, the price at which the fish would sell in Country 2 at the reduction plants falls below unit harvesting costs at $x = K$. All fish caught by Country 2 fishermen, therefore, are exported to Country 1.

The social manager of Country 1 tries to maximize the discounted flow through time of *net* social benefits enjoyed by residents of Country 1 from the fishery.[15] The social manager of Country 2 on the other hand tries to maximize the discounted stream of profits from his country's export of fish to Country 1.

We commence, as before, by assuming that α is time-invariant and no side payments are possible. The objective functionals of the two countries can be written as:

$$PV_1 = \int_0^\infty e^{-\delta t}\{U(h) - (\alpha c(x) + (1 - \alpha)p(h))h(t)\}dt, \tag{35}$$

where (see Clark and Munro, 1975, 103)

$$U(h) = \int_0^h p(h)dh$$

15 If Country 1 were the sole owner of the resource, the appropriate model for optimal exploitation of the resource would be that described in Clark and Munro (1975, 102–4).

The optimal management of transboundary renewable resources / 369

and

$$PV_2 = \int_0^\infty e^{-\delta t}(1 - \alpha)(p(h) - c(x))h(t)dt. \tag{36}$$

Country 1 faces two sets of costs: harvesting costs for the segment of the harvest taken by its vessels αh, and what might be termed import procurement costs for the remainder of the harvest $(1 - \alpha)h$.

As before, the combined objective functional is

$$PV = \int_0^\infty e^{-\delta t}\{\beta U(h) + [(1 - \alpha)(1 - 2\beta)p(h)$$
$$- (1 - \alpha + 2\alpha\beta - \beta)c(x)]h(t)\}dt. \tag{37}$$

The Hamiltonian of the problem is

$$H = e^{-\delta t}\{\beta U(h) + [(1 - \alpha)(1 - 2\beta)p(h)$$
$$- (1 - \alpha + 2\alpha\beta - \beta)c(x)]h(t)\}dt + \lambda(F(x) - h(t)). \tag{38}$$

By a straightforward application of the maximum principle a modified golden rule equation is obtained giving the optimal biomass time path:

$$F'(x^*) - [\Psi c'(x^*)F(x^*)]/\{\beta p(F(x^*))$$
$$+ \Gamma\partial[p(F(x^*)) \cdot F(x^*)]/\partial F(x^*) - \Psi c(x^*)\} = \delta, \tag{39}$$

where $\Psi = (1 - \alpha + 2\alpha\beta - \beta)$ and $\Gamma = (1 - \alpha)(1 - 2\beta)$.

Country 2 will be more conservationist than Country 1, because gross marginal benefit to Country 2 of harvested fish is measured by marginal revenue, whereas price serves as the measure for Country 1. Since marginal revenue is less than price, the marginal stock effect at any level of x will appear to be larger to Country 2 than it will to Country 1. From all this it follows that the tradeoff of the management preferences of the two countries shows up through a complex weighting of the components of the marginal stock effect.[16]

Two areas of difficulty require investigation. First, in certain instances, which shall be specified, optimal controls will not exist because of non-convexities. Clark (1976) and Lewis and Schmalensee (1977) have pointed out that one condition for convexity is that the integrand of the objective functional, the net return or benefit function, be non-convex downwards with respect to the control variable h (for a discussion of this point see Clark, 1976, 171). If we denote the net return function by ϕ, it must be found that $\partial^2\phi/\partial h^2 \not> 0$. There is no guarantee that this condition will be met for every pair of α and β. On the contrary, it will be shown that the condition will certainly be violated with combinations of relatively low αs and relatively high βs. Con-

16 One unusual aspect of the model is the approach path to x^*. The optimal approach will be asymptotic, rather than 'bang-bang' as was the case when the model was linear. The decision rule to be applied along the approach path can be expressed (see Clark and Munro, 1975) as

$$F'(x) - [\Psi c'(x)F(x)]/\{\beta p(F(x)) + \Gamma\partial[p(F(x)) \cdot F(x)]\partial F(x) - \Psi c(x)\} + (d\lambda/dt)/\lambda = \delta.$$

370 / Gordon R. Munro

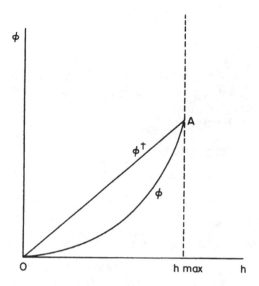

FIGURE 2   Net return function ϕ as a convex downwards function of the harvest rate h for a given t and x

sider the extreme case in which $\alpha = 0$ and $\beta = 1$ (Country 1 has complete control over the management of the fishery, but all fish consumed in Country 1 are imported from Country 2). The relevant objective functional now reduces to

$$\text{PV} = \int_0^\infty e^{-\delta t}\{U(h) - p(h)h(t)\}dt. \tag{40}$$

Since Country 1's demand function for fish is linear,

$$\partial^2\phi/\partial h^2 = - e^{-\delta t}p'(h). \tag{41}$$

Since $p'(h) < 0$, $\partial^2\phi/\partial h^2 > 0$. The root of the problem lies in the fact that marginal import procurement costs $\partial[p(h) \cdot h]/\partial h$ constitute a monotonically decreasing function of h (see Figure 2).

We can convexify by introducing so-called chattering controls (Clark, 1976, 172 ff). Instead of harvesting at a steady rate in equilibrium, we switch back and forth, i.e. chatter, between $h = 0$ and $h = h_{max}$, yielding an *average* harvest rate per period of chattering. Depending upon how the period of chattering is divided between $h = 0$ and $h = h_{max}$, equilibrium can be anywhere along the straight line OA in Figure 2. If one denotes the new return function by $\phi^\dagger$, it is clear that $\phi^\dagger$ is superior to ϕ, except at O and at A. Moreover, it is also clear that:

$$\partial^2\phi^\dagger/\partial h^2 = 0. \tag{42}$$

The optimal management of transboundary renewable resources / 371

Adapting Clark's (ibid.) analysis to our problem, we have

$$h(t) = \begin{cases} h_{max} & \text{for } 0 \leq t \leq \gamma\epsilon \\ 0 & \text{for } \gamma\epsilon < t < \epsilon, \end{cases} \tag{43}$$

where ϵ, a small positive number, is the period of chattering and γ is a constant such that $0 \leq \gamma \leq 1$. Obviously γ, which is to be determined, is the fraction of each chatter spent harvesting. Let $h_\epsilon(t)$ be continued periodically for all $t \geq 0$. The change in the biomass during a single chatter is approximately:

$$\Delta x^* \simeq F(x^*)\epsilon - h_{max}\gamma\epsilon.$$

The equilibrium condition $\Delta x^* = 0$ implies that

$$\gamma = F(x^*)/h_{max} = h^*/h_{max}, \tag{44}$$

where h^* is the average harvest rate in equilibrium.

The total net return for one chatter, or ϕ^t, is therefore

$$\phi(h_{max})\gamma\epsilon + 0(1 - \gamma) = e^{-\delta t} \left[\frac{U(h_{max})}{h_{max}} - p(h_{max}) \cdot h \right], \tag{45}$$

where h should be interpreted as the average harvest rate over the period of the chatter.

While an optimal control in the normal sense does not exist, the maximum principle does remain valid for the *average* of a chattering control (Warga, 1962). Drawing upon this fact, Clark (1976, 174) shows that in a problem such as this one can express the equation for the equilibrium biomass x^* as the familiar modified golden rule, given that the equilibrium harvest rates are interpreted as average harvest rates.

In the example, $\beta = 1$. Now observe that in (45) the unit net benefit derived from the harvest at any point in time,

$$\left[\frac{U(h_{max})}{h_{max}} - p(h_{max}) \right],$$

is independent both of the harvest rate (average) and of the biomass. Returning to the discussion of the optimal management of a fishery subject to a single owner, it can be seen from equations (4) and (12) that if the unit net benefit enjoyed from the harvest is independent of h and x the modified golden rule equation reduces to $F'(x^*) = \delta$. Hence in our example of $\beta = 1$, $\alpha = 0$, the equation for the optimal stock level x^* is

$$F'(x^*) = \delta, \tag{46}$$

where $h^* = F(x^*)$ is the optimal average harvest rate.

Earlier Country 1 was described as being less conservationist than its partner. Now observe that if $\alpha = 0$ Country 1 could become the ultimate

372 / Gordon R. Munro

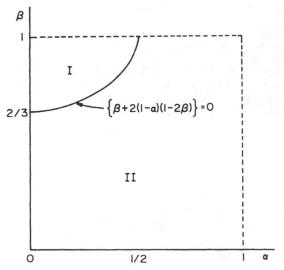

FIGURE 3   Combinations of α and β and non-convexities.
I: $\{\beta + 2(1 + \alpha)(1 - 2\beta)\} < 0$.   II: $\beta + 2(1 - \alpha)(1 - 2\beta)\} > 0$.

non-conservationist. If $F'(0) < \delta$, its object would be to drive the resource to extinction![17]

If we relax the extreme assumptions and allow for $\beta < 1$ and $\alpha > 0$, we have

$$\partial^2\phi/\partial h^2 = e^{-\delta t}\{\beta + 2(1 - \alpha)(1 - 2\beta)\}p'(h). \tag{47}$$

Thus all combinations of α and β for which $\{\beta + 2(1 - \alpha)(1 - 2\beta)\} < 0$ will confront us with the problem of non-convexity. It can easily be seen that the above inequality cannot hold unless $\alpha < \frac{1}{2}$. If the predetermined α is less than $\frac{1}{2}$, the above expression can be used to determine the range of β's for which non-convexity will arise for the given α. It is easy to demonstrate that non-convexity problems cannot arise for $\beta \leq \frac{2}{3}$.

If non-convexity does arise, we convexify as before by using chattering controls. The equation for the optimal biomass level becomes (Clark, 1976, 174)

$$F'(x^*) - [\Psi c'(x^*)F(x^*)] \Big/ \left\{\beta \frac{U(h_{\max})}{h_{\max}} + \Gamma p(h_{\max}) - \Psi c(x^*)\right\} = \delta. \tag{48}$$

If $\alpha = 0$ and $\beta = 1$, (48) reduces to (46).[18]

It may be very difficult to implement a policy of chattering controls,

17 We must heavily qualify this statement by noting that the model is autonomous. Clark and Munro (1978) have shown that once the assumption of autonomy is relaxed the sufficiency conditions for the optimality of extinction become much more stringent.
18 One interesting aspect of chattering control is that if its implementation is appropriate the optimal approach path is the 'bang-bang' one familiar from the linear model. For a discussion of this point see Clark (1976, 174).

The optimal management of transboundary renewable resources / 373

especially since in theory the rate of chattering should be infinitely fast. Nevertheless, it may be possible to approximate the policy through pulse fishing.[19]

One problem that will not be encountered if $\{\beta + 2(1 - \alpha)(1 - 2\beta)\} < 0$ is that of multiple equilibria. Since the underlying biological model is the Schaefer model (see n. 5 and equation 48), it becomes clear that, if chattering controls are feasible, a unique x^* will exist. Observe that

$$\left\{\beta \frac{U(h_{max})}{h_{max}} + p(h_{max}) + \Gamma p(h_{max}) - \Psi c(x^*)\right\}$$

in equation (48) is independent of h. If on the other hand the above inequality does not hold, so that non-convexity problems do not arise, we shall indeed be confronted with the possibility of multiple equilibria because the model will be non-linear in the control variable h (see Clark and Munro, 1975).

This can be most easily seen if we let $\alpha > \frac{1}{2}$ and assume that harvesting costs are independent of the size of the biomass x. The equation for the optimal biomass level then becomes

$$\{\beta p(F(x^*)) + \Gamma \partial[p(F(x^*)) \cdot F(x^*)]/\partial F(X^*) - \Psi c\}(F'(x^*) - \delta) = 0. \quad (49)$$

The possibility of multiple equilibria for some or all β's becomes transparent.

This may or may not create difficulties. With our problem, if multiple equilibria at no point exceed three it should be possible through bargaining to arrive at a compromise management program (ibid., 102). However, if the multiple equilibria exceed three it may prove impossible for a single owner to determine an optimum optimorum. If joint owners of a resource are unable to specify an optimum optimorum from their own perspectives, it will obviously be difficult, if not impossible, to arrive at a satisfactory program for joint management of the resource.

The pre-bargaining over α can now be seen to have an important bearing on the difficulties to be encountered in establishing a co-operative management arrangement. There is no α which will guarantee an absence of serious difficulties. In a sense the two partners may be caught between the Scylla of non-convexities and the Charybdis of multiple equilibria. If α is small but most of the bargaining power rests with Country 1, non-convexities can arise. If on the other hand α is large and/or Country 1's bargaining power is moderate, the problem of multiple equilibria might have to be faced.

Consider next the consequences of allowing α to be time-variant. We observe that, while the objective functional is not linear in h, it is linear in α.

19 See Clark's discussion of pulse fishing (ibid.). For another example of a potential difficulty suppose that fishermen in Country 2 could store their fish. Rather than throw their fish on the Country 2 market when caught, receiving a unit return of $p(h_{max})$, they would smooth out their sales and receive a unit return of $p(F(x^*))$. The non-convexity problem would remain. But the attempted convexification by Country 1 would be thwarted.

374 / Gordon R. Munro

Upon differentiating equation (38) with respect to α we have

$$\partial H/\partial \alpha = e^{-\alpha}(2\beta - 1)(p(h) - c(x))h(t). \tag{50}$$

If we cannot have a management agreement in which one or both countries is harvesting at a loss, i.e. in which $(p(h) - c(x)) < 0$, it becomes obvious that $\partial H/\partial \alpha$ will be positive or negative depending upon whether β is greater or less than $\frac{1}{2}$.[20]

No switch can occur. Nevertheless, it is still possible to have both countries enjoy benefits from the fishery. It is true that if $\partial H/\partial \alpha > 0$ the resultant optimal management policy would be one in which Country 2 got no return from the fishery. However, if $\partial H/\partial \alpha < 0$, so that $\alpha = 0$ for all time, both countries will enjoy a return from the fishery. While Country 1 will do no harvesting, it will benefit because its citizens consume the harvested fish. In light of the previous discussion, this is the outcome we should expect from the co-operative game.

Permitting α to be time-variant does not allow us to escape the problems we have discussed. It is true that if both countries enjoy benefits from the fishery ($\beta < \frac{1}{2}$) we do not encounter difficulties of non-convexity. But potential multiple equilibria problems remain.

If side payments are allowed, some of our difficulties vanish. We proceed as before. Equal weights are assigned to the individual objective functionals, and α is treated as a control variable. Upon differentiating with respect to α we get the following disconcerting result

$$\partial H/\partial \alpha = 0, \tag{51}$$

which appears to indicate that it does not matter what value is assigned to α at any time. The problem is resolved by considering the equilibrium equation for the optimal biomass, given $\beta = \frac{1}{2}$. The equation reduces to

$$F'(x^*) - [c'(x^*)F(x^*)]/[p(F(x^*)) - c(x^*)] = \delta. \tag{52}$$

This, of course, is the modified golden rule equation that would be forthcoming if $\beta = \alpha = 1$. In other words, with side payments the optimal policy focuses upon the gross social benefits accruing to Country 1 and the actual harvesting costs, i.e. $c(x)$. How the harvesting is in fact divided between the two countries is indeed a matter of indifference since the harvesting costs of the two countries are identical.

Since Country 1's policies are dominant, it will make transfer payments to Country 2. The division of the global returns between the two countries will be determined in the same manner as in the other cases discussed involving side payments.

It should be self-evident that non-convexity problems cannot arise if side

20 If β is identically equal to $\frac{1}{2}$, then $\partial H/\partial \alpha = 0$, which suggests that in terms of the optimality of the harvesting program it does not matter whether α is set equal to 0 or 1 or any value in between.

The optimal management of transboundary renewable resources / 375

payments can be made. But multiple equilibria can still arise and can certainly be a source of difficulty. At least we have the small consolation that the difficulties would be reduced to those associated with a single owner. The optimal management policy is, after all, to act as if Country 1 were the sole owner of the resource.

CONCLUSIONS

The dynamic analysis of the optimal management of renewable resources owned by single states is now well developed. This paper attempts to extend such analysis to the rather more difficult problem of optimal management of renewable resources subject to joint ownership.

The conflict in management strategies between the co-owners of the resource is viewed in terms of the propensity of each of the owners to invest (disinvest) in the resource. Conflicting views on the optimal stock or biomass, of course, imply conflicting views on optimal harvest rates.

Three major sources of management conflict were considered, differences in perceived social rates of discount, harvesting costs, and consumer preferences. No serious difficulties were encountered in the first two cases, but interesting analytical difficulties presented themselves in the consumer preferences case. There the possibility exists that an optimal control will fail to exist. Furthermore, the road to management conflict resolution may be blocked by multiple equilibria. If there is one general conclusion arising from the discussion, it is the not surprising one that it becomes far easier to resolve management conflicts if it is feasible for one co-owner of the resource to make transfer (side) payments to the other. In all the cases considered in this paper the use of side payments causes the resource to be managed as if it were under the control of a sole owner.

Much scope remains for additional research on the optimal management of transboundary renewable resources. The most obvious avenue to be explored is that of transboundary renewable resources having a large number of co-owners (e.g. fishery resources off West Africa or tuna in the Pacific). Secondly, the consequences of allowing parameters to fluctuate through time should be investigated; in other words the assumption of autonomy should be relaxed.[21] Furthermore, we should consider the implications for joint management programs of limited alternative uses for the capital employed by the owners in harvesting the resource.[22] Finally, the greatest scope for research

21 Two comments are in order here. The first is that it should be noted that the model employed in analysing the case of differing social discount rates with no side payments became non-autonomous willy-nilly. Recall that the 'compromise' discount rate $\delta_3(t)$ was a function of time. Secondly, making the model employed in the last section non-autonomous would present serious problems, because non-autonomous, non-linear models are notoriously difficult. For a discussion of non-autonomous models in the case of resources subject to single ownership see Clark and Munro (1975).

22 This issue is analysed for renewable resources owned by a single state in Clark, Clarke, and Munro (1979).

376 / Gordon R. Munro

lies in applying our analytical tools to the many actual cases of jointly owned renewable resources presented by the emerging regime of 200-mile Exclusive Economic Zones.

REFERENCES

Anderson, Lee G. (1975a) 'Criteria for maximizing the economic yield of an internationally exploited fishery.' In H.G. Knight, ed., *The Future of International Fisheries Management* (St Paul: West)
Anderson, Lee G. (1975b) 'Optimum economic yield on an internationally utilized common property resource.' *Fishery Bulletin* 73, 51–6
Bacharach, Michael (1976) *Economics and the Theory of Games* (London: Macmillan)
Clark, Colin W. (1976) *Mathematical bioeconomics: the optimal management of renewable resources* (New York: Wiley)
Clark, Colin W. (1978) 'Restricted entry to common-property fishery resources: a game-theoretic analysis.' Institute of Applied Mathematics and Statistics, Technical Report No. 78–9, University of British Columbia
Clark, Colin W., Frank H. Clarke, and Gordon R. Munro (1979) 'The optimal exploitation of renewable resource stocks: problems of irreversible investment.' *Econometrica* 47, 25–47
Clark, Colin W. and Gordon R. Munro (1975) 'The economics of fishing and modern capital theory: a simplified approach.' *Journal of Environmental Economics and Management* 2, 92–106
Clark, Colin W. and Gordon R. Munro (1978) 'Renewable resource management and extinction.' *Journal of Environmental Economics and Management* 5, 198–205
Copes, Parzival (1977) 'On growth, conservation and order.' *Annals of Regional Science* 11, 1–10
Hnyilicza, Esteban and Robert S. Pindyck (1976) 'Pricing policies for a two-part exhaustible resource cartel: the case of OPEC.' *European Economic Review* 8, 139–54
Intriligator, Michael D. (1971) *Mathematical Optimization and Economic Theory* (Englewood Cliffs: Prentice Hall)
Lewis, Tracey R. and Richard Schmalensee (1977) 'Non-convexity and optimal exhaustion of renewable resources.' *International Economic Review* 18, 535–52
Luce, R.D. and H. Raiffa (1967) *Games and Decisions* (New York: Wiley)
Munro, Gordon R. (1977) 'Canada and extended fisheries jurisdiction in the Northeast Pacific: some issues in optimal resource management.' Mimeo. Paper presented to the Fifth Pacific Regional Science Conference, Vancouver
Nash, J.F. (1951) 'Non-cooperative games.' *Annals of Mathematics* 54, 286–95
Nash, J.F. (1953) 'Two-person cooperative games.' *Econometrica* 21, 128–40
Saetersdal, Gunnar (1977) 'Problems of managing and sharing the fishery resources under the new ocean regime.' United Nations, Food and Agriculture Organization, Committee on Fisheries, Information Bulletin 11
Schaefer, M.B. (1957) 'Some considerations of population dynamics and economics in relation to the management of the commercial marine fisheries.' *Journal of the Fisheries Research Board of Canada* 14, 669–81
United Nations (1977) 'Informal composite negotiating text.' Third Conference on the Law of the Sea, Working Paper 10
Warga, J. (1962) 'Relaxed variational problems.' Parts 1 and 2. *Journal of Mathematical Analysis* 4, 111–45

[26]

JOURNAL OF ENVIRONMENTAL ECONOMICS AND MANAGEMENT **32**, 309–322 (1997)
ARTICLE NO. EE970971

Fishing as a Supergame

Rögnvaldur Hannesson

*The Norwegian School of Economics and Business Administration, Helleveien 30, N-5035
Bergen-Sandviken, Norway*

Received February 6, 1996; revised May 15, 1996

This paper considers how cooperative solutions to games of sharing fish resources can be
supported by threat strategies. With highly mobile fish stocks, the number of agents
compatible with a cooperative self-enforcing solution is not very high for reasonable values of
the discount rate, but sensitive to changes in the discount rate and costs and to cost
heterogeneity. With migrating stocks, where growth and reproduction depend on how much
all agents leave behind after harvesting, the likelihood of a cooperative, self-enforcing
equilibrium is increased. With a dominant player and a competitive fringe the rents and
optimum stock level of the dominant player fall quickly as the share of the competitive fringe
increases. © 1997 Academic Press

1. INTRODUCTION

Exploitation of renewable resources such as fish stocks shared by a limited
number of agents involves strategic choices. Should the agents cooperate and
maximize their aggregate returns, or is cooperation a futile exercise undertaken by
the naive, who will be outsmarted by the realistic? There exists by now a volumi-
nous literature on this subject. Much of this literature was inspired by the
international disputes over fishing limits in the 1970s and the deliberations of the
third United Nations Conference on the Law of the Sea 1973–1982 which endorsed
the 200-mile exclusive economic zone. Two early papers in this genre are Munro
[11] and Levhari and Mirman [9]. Recently Fischer and Mirman [2, 3] have
extended the analysis to interacting species. Three of these papers compare Nash
equilibria and global optima while Munro is concerned with bargaining solutions in
cooperative games. All these papers deal with the case of two agents, which is true
of most of the literature on game theory and fisheries; few authors have considered
explicitly the importance of the number of agents for obtaining a cooperative
solution. An exception is Clark [1] who does not, however, consider the implications of threat strategies in repeated games. Hämäläinen *et al.* [6] do so, but again
limit themselves to the case of two agents.

The motivation to explore the importance of the number of agents is provided by
the fact that many important fish stocks enclosed by the 200-mile limit are shared
by two or more coastal states. What is the likelihood that they will cooperate in
setting the rate of exploitation of the stocks and how does it depend on the number
of states sharing a stock? A further motivation is provided by the straddling of
some fish stocks outside the 200-mile limit where they are accessible by fleets of
any nationality. What are the chances of cooperation under those circumstances, or
more to the point perhaps, how will the strategy of the coastal state(s) controlling
the main part of a stock be affected by this competition?

309

This latter case is a topical one. In 1993 the UN convened a conference on the exploitation of straddling and highly migratory fish stocks. The conference was concluded in August 1995 with an agreement that authorizes regional organizations to manage fisheries outside the 200-mile limit. All who fish in a given area will have to abide by the rules agreed by the relevant organization. It is unclear, however, what if anything limits the membership of such organizations; the agreement says only that member nations must have a "real interest" in fishing in the area. Neither is it clear how decisions will be taken, whether this will be by majority vote, a qualified majority, or a consensus. Last but not least, it is unclear what sanctions can and will be applied to those who do not cooperate.

In this paper we consider how critical the number of agents sharing a fish stock is for realizing the cooperative (globally optimal) solution. The problem is formulated as a repeated game of an infinite duration (a supergame). We begin by considering a game of N identical agents. Then we look into the importance of cost differences among agents. In both these formulations agents are assumed to exploit a truly common stock; i.e., the density of the stock will be the same for all. We then consider a somewhat different problem which may be more relevant for countries sharing a fish stock. Each agent is assumed to exploit a certain portion of the stock being accessible for exploitation only in that agent's territory but nevertheless linked to the remainder of the stock through a common growth function. The cooperative solution is defined as the global optimum, while the noncooperative solution is the one where each agent realizes his individual optimum without taking into account the stock growth externality. Finally we look at a case meant to reflect the straddling stock situation where a dominant agent maximizes his profits, taking into account a fringe of exploiters who have access to a portion of the stock outside his territory. The focus here is on the efficiency losses vis-à-vis the global optimum.

2. THE BASIC MODEL

Assume that the growth of the fish stock is determined by how much is left behind after harvesting; i.e., the stock at the beginning of period t is a function $G(S_{t-1})$, where S_{t-1} is the stock left behind after harvesting in period $t - 1$ (the size of the stock left behind will be referred to as the abandonment level). Ignoring natural mortality of the stock while it is being fished, the amount caught in period t will be $G(S_{t-1}) - S_t$. At a given price (p) the revenue (R) obtained in period t will be

$$R_t = p[G(S_{t-1}) - S_t].\tag{1}$$

Assume that the marginal cost of fish is inversely proportional to the size of the stock at any point in time.[1] The total cost (C) in period t will then be

$$C_t = \int_{S_t}^{G(S_{t-1})} \frac{c}{x}\, dx = c[\ln G(S_{t-1}) - \ln S_t],\tag{2}$$

---

[1] This cost function obtains if the cost per unit of fishing effort is constant and the catch per unit of effort is proportionate to the size of the exploited stock. The latter obtains if the stock is always evenly distributed over a given area. While popular and not unreasonable, this is obviously a special case. The emphasis here is on obtaining numerical results, which makes it necessary to use a simple but not unreasonable cost function. For a further discussion, see [7].

where c is a cost parameter. Since the quantity caught (Q) is $Q = G - S$, with G being given at the beginning of each period, this function has the usual properties $C_Q = -C_S > 0$ and $C_{QQ} = -C_{SS} > 0$ (subscripts denote derivatives).

The present value (V) of fishing rent $(R - C)$, for an infinite time horizon, is

$$V = \sum_{t=0}^{\infty} \delta^t \{ p[G(S_{t-1}) - S_t] - c[\ln G(S_{t-1}) - \ln S_t] \}. \tag{3}$$

Maximizing V with respect to S_t gives the first order condition[2]

$$-(p - c/S_t^0) + \delta[p - c/G(S_t^0)]G'(S_t^0) = 0, \tag{4}$$

where $\delta = 1/(1 + r)$ is the discount factor, r being the discount rate, with G' denoting the first derivative of G and S^0 the optimum value of S.

As an illustration in numerical calculations below we shall use the discrete variant of the logistic growth function

$$G(S) = S[1 + a(1 - S/K)], \tag{5}$$

where a and K are parameters (intrinsic growth rate and carrying capacity, respectively). This gives

$$G'(S) = 1 + a(1 - 2S/K). \tag{6}$$

3. COOPERATIVE EQUILIBRIUM: IDENTICAL AGENTS

Suppose there are N identical agents who share a fish stock. Suppose further that they plan to harvest the stock for an indefinite period of time. If they cooperate in realizing the optimal solution (which is identical from everybody's perspective) each will get $1/N$th of the total profits in each period. If one of them deviates from the optimal solution he will get more, as long as the deviation has not been discovered and punished. Assume that deviation would be detected after one period and that the other agents then would retaliate by fishing down the stock in each period until further depletion becomes unprofitable, i.e., until the marginal cost of fish caught has risen to equal the price (cf. Eq. (2)). This is in fact the best strategy they could follow, as long as the deviating agent depletes the stock to the level where fishing becomes unprofitable (an alternative trigger strategy will be commented upon below). In view of the above cost and revenue functions (Eqs. (1) and (2)), the abandonment level of the stock (S^*) would then be

$$S^* = c/p. \tag{7}$$

---

[2] The same first order condition also obtains for a finite time horizon except for the last period where the optimal abandonment level is $S^* = c/p$ when the stock beyond the horizon has no value. If the initial stock is less than the optimal stock it will be necessary to leave it unfished for one or more periods, until $G(S_{t-1}) > S^0$.

For an infinite time horizon, the present value of the cooperative strategy (V^0) for a typical agent is

$$V^0 = \frac{\pi^0}{N} \frac{1}{1 - \delta}, \tag{8}$$

where

$$\pi^0 = p[G(S^0) - S^0] - c[\ln G(S^0) - \ln S^0] \tag{9}$$

and S^0 is the abandonment level of the stock along the optimal (cooperative) stationary path.

The present value of the payoff for an agent that deviates from the cooperative solution and is then punished by all other agents playing noncooperatively forever is

$$V^d = \frac{\pi^0}{N} + \pi^d + \frac{\pi^*}{N} \frac{\delta}{1 - \delta}, \tag{10}$$

where

$$\pi^d = p(S^0 - S^*) - c(\ln S^0 - \ln S^*) \tag{11}$$

and

$$\pi^* = p[G(S^*) - S^*] - c[\ln G(S^*) - \ln S^*]. \tag{12}$$

In the first period, the defector gets the same profit as in the cooperative solution, as all other participants play cooperatively, and in addition he gets the profit of driving the stock down to the noncooperative abandonment level. In the second and all later periods he will be punished by all other agents playing noncooperatively and gets only the profit obtained in the noncooperative solution (cf. Eq. (7)).

If defection is not profitable, $V^0 > V^d$, which implies

$$N < \frac{\delta}{1 - \delta} \frac{\pi^0 - \pi^*}{\pi^d}. \tag{13}$$

As $\delta \to 1$ the right-hand side of (13) approaches infinity and defection will never be profitable; the losses from being punished will always outweigh the temporary gains of defecting. For a positive discount rate $(\delta < 1)$ the temporary gains of defecting may outweigh the long term loss of playing noncooperatively rather than cooperatively. How likely this is depends on N, the number of players. The gains from defecting accrue to the defector while the losses from playing noncooperatively rather than cooperatively are shared by all participants. The temptation of defecting therefore becomes greater the more participants there are. Specifying the parameters of the growth equation $G(S)$ makes it possible to find the critical value of N for alternative values of the discount factor. A selection of results is presented in Fig. 1.

The values of the parameters a, c, and δ in Fig. 1 are meant to reflect a realistic range. The parameter a is the maximum relative rate of growth (cf. Eq. (6)). Figure 1 shows solutions for two alternative values of a, 20 and 50%. The ratio c/S shows the marginal cost of fish (cf. Eq. (2)). The maximum size of the stock is K (cf. Eq.

FISHING AS A SUPERGAME   313

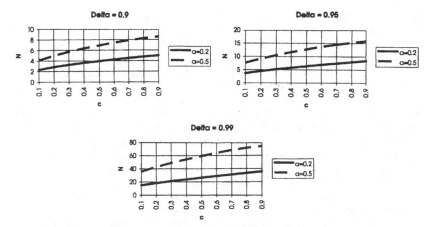

FIG. 1. Relationship between the maximum number of players in a cooperative solution and the minimum marginal cost of fish.

(5)), so c/K can be interpreted as the minimum marginal cost of fish, prevailing at the beginning of exploitation of a pristine stock. Since K has been set equal to 1, c shows the lowest marginal cost of fish as a fraction of the pristine stock. The values of c in Fig. 1 range from 0.1 to 0.9. A value of 1 would imply that fishing would never be profitable while a value of 0 would imply no cost of fishing. Finally, the three values of δ in Fig. 1 imply discount rates of approximately 1, 5, and 11%.

Figure 1 indicates that N, the maximum number of participants in the fishery compatible with a self-enforcing cooperative solution, will be highly sensitive to the discount factor and the cost of fishing. As already mentioned, a higher discount rate (lower value of δ) makes the cooperative solution less likely. For $c = 0.1$ and $a = 0.2$ N is only 2 when $\delta = 0.9$, but rises to 3 and 15, respectively, as δ increases to 0.95 and 0.99. A higher marginal cost of fish makes it more likely that the cooperative solution will prevail; the abandonment level of the stock in the noncooperative solution will be higher and the gains from defection smaller. This is readily seen from the figure; as c increases from 0.1 to 0.9 N approximately doubles. Finally, raising a from 0.2 to 0.5 approximately doubles N for any given c.

If we interpret N as the number of countries which share a resource and have the necessary control over their fishing fleets, the results in Fig. 1 are not too discouraging. Whenever stocks are fully contained within the 200-mile zone but migrate between different national zones, the number of countries with access rights is usually highly restricted. Quite often there are just two countries sharing a stock, like Canada and the United States on George's Bank, while several countries share the stocks of the North Sea. A discount rate of the order of 5–10% appears realistic and would accommodate a cooperative solution among a few countries.

For the case of stocks located outside the 200-mile zone the conclusion is much more pessimistic. The number of potential exploiters of such stocks is high, as witnessed by the fact that some boat owners use flags of convenience when fishing on the high seas. Up to now this number has in fact been indefinite, as interna-

tional law has not recognized any mechanism to limit access to fishing outside the
200-mile zone. It remains to be seen whether the agreement reached by the UN
conference on highly migratory and straddling fish stocks will change this.

The results in Fig. 1 in part corroborate a tendency which Mesterton-Gibbons
[10] has termed "comedy of the commons," namely that cooperation can arise
spontaneously and be self-enforcing if the commons are sufficiently unproductive.
In Fig. 1 we see that the number of participants that can be accommodated in a
self-enforcing cooperative solution increases with the cost of exploitation. However, the more productive stock (the one with an intrinsic growth rate of 0.5)
accommodates more participants in a self-enforcing cooperative solution than the
less productive stock, which would seem to contradict the comedy of the commons.
The reason why the more productive stock accommodates more participants in a
self-enforcing cooperative equilibrium is that the losses from playing noncooperatively are greater, being associated with foregone potential growth, while the gains
from defecting are due to fishing down the stock in one period.

It is possible to give an alternative interpretation of the results in Fig. 1. These
results were derived on the assumption that the punishment of a defector would go
on forever. This is unnecessarily heavy handed. The defector might promise to
mend his ways, and all participants might be well advised to revert to the
cooperative solution. However, in order to be an effective deterrent the punishment must go on for a sufficiently long time to make the cooperative strategy more
attractive for a potential defector. To accomplish this the $N - 1$ participants might
threaten to play the noncooperative strategy for a finite number of periods. We
may ask what is the minimum number of periods (T) in which the noncooperative
strategy must be played in order to deter a potential defector. Instead of (10),
define $V^d(T)$ as[3]

$$V^d(T) = \frac{\pi^0}{N} + \pi^d + \sum_{t=1}^{T} \delta^t \frac{\pi^*}{N} + \frac{\delta^{T+1}}{1 - \delta} \frac{\pi^0}{N}. \tag{10'}$$

An effective punishment requires $V^0 > V^d(T)$, which implies

$$T > \frac{\ln F}{\ln \delta}, \tag{14}$$

where

$$F = 1 - \frac{1 - \delta}{\delta} \frac{N\pi^d}{\pi^0 - \pi^*}. \tag{15}$$

As $F \to 0$, the right-hand side of (14) approaches infinity. The value of N which
gives $F = 0$ is precisely the same as would make (13) an equality. Hence it is
possible to interpret the figures in Fig. 1 as the number of participants that will
make it possible to deter a defector by threatening to play the noncooperative
strategy for a finite number of periods.

---

[3] I am grateful to an anonymous referee for pointing this out. Punishment strategies that last for a
finite period are discussed in [12], Chapter 8.6.

4. COOPERATIVE EQUILIBRIUM WHEN COSTS DIFFER AMONG AGENTS

The analysis so far has been based on the assumption that all agents have identical cost functions. However, what if some agents are more efficient than others? Such agents might be tempted to "undercut" high cost agents and fish down the stock to a level where the high cost agents are barred from entry.[4] The incentive to do so will be strongest when there are many high cost agents and few low cost agents. Here we shall look at an example with one low cost agent and $N - 1$ high cost agents, with cost parameters c_l and c_h, respectively, but otherwise identical cost functions.

The low cost agent will be able to fish down the stock to a lower abandonment level than the high cost agents without incurring losses. In the noncooperative solution two cases may arise. (i) The low cost agent depletes the stock to the level c_l/p, but since the stock at the beginning of each period is $G(c_l/p)$, the high cost agents can still do some profitable fishing if $G(c_l/p) > c_h/p$. (ii) The cost difference is so great that $G(c_l/p) < c_h/p$. In that case the low cost agent can undercut the high cost agents and exclude them from the fishery altogether. It would not, however, be profitable for the low cost agent to leave behind a smaller stock than that which gives $G(S^*) = c_h/p$ (i.e., exactly undercuts the high cost agents), unless the optimum stock for the low cost agent alone is less than this.

Hence the abandonment level of the stock in a noncooperative equilibrium will be

$$S^* = \max\left[\frac{c_l}{p}, G^{-1}\left(\frac{c_h}{p}\right)\right].$$ (16)

The profit of the low cost agent in a noncooperative play may be split into two parts. The first is the profit of fishing down the stock from $G(S^*)$ to c_h/p. In this phase (which may be nonexistent) the low cost agent gets only $1/N$th of the catch. In the second phase the low cost agent drives the stock further down to S^* and has all this catch to himself. The two parts of the profit of the low cost agent thus are

$$\pi_1^* = p[G(S^*) - c_h/p] - c_l[\ln G(S^*) - \ln(c_h/p)]$$ (17a)

$$\pi_2^* = p[c_h/p - S^*] - c_l[\ln(c_h/p) - \ln S^*].$$ (17b)

The gain from defection will be greater than before, because some and possibly all of the catch in the noncooperative equilibrium does not have to be shared with all agents but will be taken by the low cost agent alone. Equation (10) becomes

$$V^d = \frac{\pi^0}{N} + \pi^d + \left[\frac{\pi_1^*}{N} + \pi_2^*\right]\frac{\delta}{1 - \delta}$$ (10″)

and (13) becomes

$$N < \frac{\delta(\pi^0 - \pi_1^*)}{\delta\pi_2^* + (1 - \delta)\pi^d}.$$ (13″)

---

[4] This is what happens in the closed loop solution derived by Clark [1].

The question arises as to what would be the optimum stock level in the cooperative solution, since the high cost agents and the low cost agent will not agree on this. A global optimum would entail fishing by the low cost agent only, with side payments to the high cost agents. If this is not possible, the agents might agree on some mutually acceptable abandonment level of the stock that allows both low cost and high cost agents to make a profit. Here it will be assumed that the cooperative solution is determined with reference to the cost of the low cost agent, as this will minimize the likelihood that the low cost agent will want to defect. Note that this solution does not impose losses on the high cost agents as long as $pS^0 > c_h$.

As before, we can calculate the number of participants accommodated by a self-enforcing cooperative equilibrium (N). A sample of such results is shown in Fig. 2 and compared to the case with cost homogeneity. The dashed curves in Fig. 2 show N when all agents have the same cost (same as in Fig. 1). The solid curves show what happens when one agent has a lower cost parameter (c) than the rest. As the cost of one agent starts to fall below that of the rest N drops quickly, particularly when the cost is high. Note that when the cost is high N is also high, and the losses from ending up in a noncooperative solution are low compared to the gains from defecting, since the losses are spread among all players. This is why it takes only a slight difference in costs to produce a steep fall in N when N is high initially.

The effect of cost heterogeneity is seen to be substantial; it does not take a great difference in costs to reduce N to a number not much higher than two, irrespective of the discount rate, cost level, or productivity of the fish stock. There may be reason, therefore, for less optimism than expressed above for the likelihood of a cooperative solution emerging among a limited number of countries sharing a fish stock.

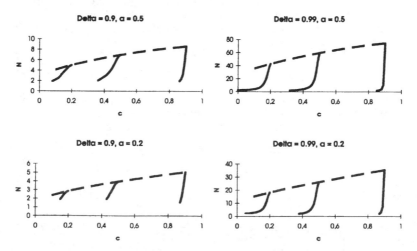

FIG. 2. How the minimum number of players in a cooperative solution depends on the cost difference between a low cost player and high cost players. Dashed line, equal cost for all players.

5. COOPERATIVE EQUILIBRIUM WITH LIMITED MIGRATION OF STOCKS ACROSS BOUNDARIES

In the above examples it was implicitly assumed that the agents fish simultaneously on the same stock or, what amounts to the same thing, that each agent fishes in his own area but that the fish redistribute themselves instantaneously between the areas so that the distribution is always the same in all areas. A more realistic example could be where each agent fishes in his own area and the fish migrate only slowly between areas or visit these areas sequentially with a seasonal pattern.[5] The growth of the stock could still be dependent on the aggregate size of the stock, because of a seasonal pattern of breeding migration or because the eggs and larvae are distributed over the entire habitat of the stock irrespective of where they are spawned.

Here we shall look at a stylized version of this. Let the stock be measured as density, i.e., units of fish per unit area (tonnes per square km, for example). The reason for taking this approach is that the marginal cost of fish depends on the density of the stock and thus indirectly on the size of the stock, provided that the area it occupies does not shrink in proportion to the depletion of the stock. To make the comparison with previous cases easier we define the unit area as the entire area that the stock occupies, assumed always to be of the same size. Assume that the area is divided among N agents. In each period we start with a stock (G) that is uniformly distributed over the area. The amount available for agent i will be $\gamma_i G$, where γ_i is agent i's share of the area where the stock is located.[6] The fish are assumed not to migrate between the agents' subareas during each fishing period, so each agent has full control over the abandonment level of the stock in his area ($\gamma_i S_i$). After each fishing period the stock grows and redistributes itself randomly over the entire area. This leads to the growth function

$$G(\Sigma \gamma_i S_{i,t-1}),$$

where

$$\Sigma \gamma_i = 1. \tag{18}$$

The growth of the stock depends on what all agents leave behind ($\Sigma \gamma_i S_i$), and all agents start fishing a stock of the same density at the beginning of each period $[G(\Sigma \gamma_i S_{i,t-1})]$.

The globally optimal solution will be the same as derived above (Eq. (4)). For this solution to be realized, each agent would have to leave behind a stock compatible with the global optimum, and he would get a share in the total profits equal to his share of the area where the stock is located. If, on the other hand, each agent optimizes for himself, the optimum stock to be left behind in period t

---

[5] The case of sequential fishing is dealt with in [8] and will not be considered here.

[6] The assumption of uniform distribution is unnecessarily strong. All that is required is that the stock always redistribute itself in the same way after each fishing period. The share parameters γ_i will then be constant.

will be given by maximizing a modified version of (3):

$$V_i = \sum_{t=0}^{\infty} \delta^t \gamma_i \{ p[G(\Sigma \gamma_j S_{j,t-1}) - S_{i,t}] - c[\ln G(\Sigma \gamma_j S_{j,t-1}) - \ln S_{i,t}] \}. \quad (19)$$

Note that the profit function is multiplied by γ_i because the stock is measured as density, with the unit area being equal to the entire habitat of the stock.

The first order condition for maximum is

$$-[p - c/G(\Sigma \gamma_j S_j^0)] + \delta \gamma_i G'(\Sigma \gamma_j S_j^0)[p - c/S_i^0] = 0. \quad (20)$$

We shall now define "defection" as optimization of each agent for himself, i.e., choosing the abandonment level which is optimal according to Eq. (20). We denote this by S^* and reserve the notation S^0 for the globally optimal abandonment level. We continue to assume that defection will be discovered after the period in which it occurs. In the period after defection occurs, other agents adjust their harvesting to attain the abandonment level (S^*), so in this case the punishment strategy does not begin to bite until two periods after defection begins. The stock returning at the beginning of the period after defection will be $G(S^0 \gamma_{-i} + S^* \gamma_i)$, where the notation "$-i$" means all agents except agent i, the defecting agent. The condition for defection not being profitable is

$$\frac{\pi^0}{1-\delta} > \pi_1^d + \delta \pi_2^d + \frac{\delta^2}{1-\delta} \pi^*, \quad (21)$$

where

$$\pi_i^0 = \gamma_i \{ p[G(\Sigma \gamma_j S_j^0) - S_i^0] - c[\ln G(\Sigma \gamma_j S_j^0) - \ln S_i^0] \} \quad (22a)$$

$$\pi_{1,i}^d = \gamma_i \{ p[G(\Sigma \gamma_j S_j^0) - S_i^0] - c[\ln G(\Sigma \gamma_j S_j^0) - \ln S_i^0] \}$$
$$+ \gamma_i \{ p(S_i^0 - S_i^*) - c(\ln S_i^0 - \ln S_i^*) \} \quad (22b)$$

$$\pi_{2,i}^d = \gamma_i \{ p[G(\Sigma S_j^0 \gamma_{-i} + S_i^* \gamma_i) - S_i^*] - c[\ln G(\Sigma S_j^0 \gamma_{-i} + S_i^* \gamma_i) \text{-} \ln S_i^*] \} \quad (22c)$$

$$\pi_i^* = \gamma_i \{ p[G(\Sigma \gamma_j S_j^*) - S_i^*] - c[\ln G(\Sigma \gamma_j S_j^*) - \ln S_i^*] \}. \quad (22d)$$

The expression on the left-hand side of (21) is the present value of playing cooperatively while the expression on the right-hand side is the present value of defecting. In the period when defection occurs the defector gets the extra benefit of reducing the density of the stock in his area to the level S^* (Eq. (22b)). This benefit is, of course, less than in previous cases where the defector could reduce the overall density of the stock, due to its instantaneous migration. At the end of this period the defector is found out and everyone depletes his stock to density S^* from then onward, but the stock emerging at the beginning of the period after defection has not yet reached the long term equilibrium (Eq. (22c)). The noncooperative equilibrium is reached in the second period after defection (Eq. (22d)).

The maximum number of players that can be accommodated in a cooperative equilibrium is defined implicitly by (21). Table I shows some solutions for the case with identical agents ($\gamma_i = 1/N$). In general it is much more likely that a coopera-

TABLE I

Maximum Number of Agents (N) Compatible with Sustaining the Global Optimum as a
Cooperative Self-Enforcing Equilibrium ($a = 0.2$)

| | | $c = 0.1$ | | | | | $c = 0.05$ | | | |
		Global	optimum	Individual	optimum		Global	optimum	Individual	optimum
δ	N	S^0	G^0-S^0	S^*	G^*-S^*	N	S^0	G^0-S^0	S^*	G^*-S^*
0.8	4	0.212	0.0335	0.105	0.0187	2	0.133	0.0231	0.057	0.0108
0.85	7	0.267	0.0391	0.102	0.0184	2	0.195	0.0314	0.058	0.0109
0.9	∞	0.347	0.0453	0.100	0.0180	11	0.291	0.0413	0.051	0.0097

tive solution will emerge in this case than in the case with instantaneous redistribution of fish analyzed in Section 4. It was noted in Section 4 that N increases with c,
a, and δ. Here we get $N = \infty$ for $c = 0.1$, $\delta = 0.9$, and $a = 0.2$. N falls quickly,
however, when c or δ are reduced below these values; reducing δ to 0.85 (18%
rate of discount) or c to 0.05 (lowest marginal cost of fish equal to 5% of the
pristine stock) reduces N to 7 and 11, respectively. Even with this highly restricted
migration, a cooperative solution is not guaranteed.[7]

Table I also shows figures that indicate the waste associated with agents
choosing the individually optimal rather than the globally optimal solution (note
that the pristine stock and its density are equal to 1). Superscript "o" refers to the
global optimum while "*" refers to the individual optimum; if the number of
participants were to increase by one the equilibrium solution would go from the
global optimum to one very close to the individual optimum for the number of
agents shown in the table. The individually optimal stock level is in all cases very
much lower than the globally optimal level and in fact not much greater than the
abandonment level with free access (c). The amount harvested is equal to the
difference $G - S$. In the two cases shown in Table I the amount harvested in the
noncooperative solution (the individual optimum) is only a quarter to one-half of
the cooperative (globally optimal) harvest.

6. STRADDLING STOCKS

The final case we shall consider is one where a single agent has a high degree of
control over a stock which is partly accessible to an indefinite number of agents.
The case we have in mind is the one where a fish stock is largely contained within
the 200-mile zone but straddles into the high seas. There are many real world cases
of this kind, e.g., the groundfish stocks of the Grand Banks of Newfoundland, the
Alaska pollack which is shared between the United States and Russia and straddles
an area popularly known as the Donut Hole, and the Arcto-Norwegian cod which is
shared by Norway and Russia and straddles an area similarly known as the
Loophole.

---

[7] If the other players discover the defection immediately and start their punishment strategy in the
same period as the defection occurs, their stock abandonment level in the period of defection will be
the same as the defector's. The first term on the right-hand side of (21) will be divided by N, the second
drops out and the last term will be multiplied by δ instead of δ^2. Expression (21) becomes a variation
around an optimal path and is always negative. Hence defection will never be profitable and the
cooperative equilibrium will be compatible with any number of agents.

320   RÖGNVALDUR HANNESSON

We shall model this in a manner similar to the previous section. A certain fraction of the stock is supposed to be located within the "dominant agent's" area and the remainder outside and accessible to an indefinite number of agents. We ignore transboundary migration during the phase of exploitation; the dominant agent can choose the abandonment level of the stock within his area, while the remaining agents are assumed not to be interested in a cooperative solution, their number being indefinite, and hence choose an abandonment level for the remainder of the stock equal to the break-even marginal cost of fish ($p = c/S^*$). The transboundary externality is captured by letting the growth of the stock depend on total abandonment in the two areas, like in Eq. (18) above.

The dominant agent is assumed to choose an optimal abandonment level for his part of the stock, taking into account that the "fringe" agents choose the abandonment level $S^* = c/p$ for "their" part of the stock. The optimal steady-state abandonment level of the stock for the dominant agent will be given by a modified version of Eq. (20). With α being the share of the stock controlled by the dominant agent, and letting S and S^* denote the abandonment level of the dominant agent and the fringe agents, respectively, the objective function of the dominant agent becomes

$$V = \sum_{t=0}^{\infty} \delta^t \alpha \{ p [G(\alpha S_{t-1} + (1 - \alpha)S^*) - S_t]$$

$$- c [\ln G(\alpha S_{t-1} + (1 - \alpha)S^*) - \ln S_t] \}, \qquad (19')$$

with the optimum abandonment level being given by

$$-(p - c/S^0) + \delta \alpha G'(\alpha S^0 + (1 - \alpha)S^*) [p - c/G(\alpha S^0 + (1 - \alpha)S^*)] = 0. \qquad (20')$$

A set of optimum solutions for the dominant agent are shown in Table II. The solution for $\alpha = 1$ is, of course, identical to the globally optimal solution. A striking feature of the solutions shown in Table II is how quickly the total present value and the optimum abandonment level of the stock fall with the dominant agent's share of the total stock. A dominant agent who controls 95% of the stock chooses an optimum stock level which is only 75% of the globally optimal level, and the total present value of profits (fringe plus dominant agent) is only 67% of the global optimum. This indicates that only a minor straddling of fish stocks into

TABLE II

Optimum Stock Level, Growth, and Present Value (PV) of Profits for Different Stock Shares of the Dominant Agent ($a = 0.2$, $c = 0.1$, and $\delta = 0.9$)

α	S^0 (dominant agent)	G	PV dominant agent	PV fringe	Sum PV
1	0.347	0.392	0.330	0	0.330
0.95	0.260	0.291	0.179	0.042	0.221
0.9	0.198	0.219	0.097	0.040	0.138
0.8	0.144	0.158	0.039	0.025	0.064
0.5	0.111	0.125	0.010	0.013	0.023
0.1	0.102	0.118	0.001	0.01	0.015

the high seas may result in substantial losses of efficiency. As the dominant agent's share of the stock is reduced further, the optimum abandonment level of the stock falls rapidly and so does the present value of aggregate profits; when the dominant agent controls one-half of the stock, the optimum abandonment level is only slightly above the free access abandonment level (0.1), and the aggregate profits are only about 5% of the global maximum.

Another interesting feature of the results in Table II is that beyond a certain point the fringe is not better off by having access to a larger part of the stock; the fringe's profit actually decreases slightly as its share of the stock increases from 5 to 10%. The reason is that the dominant agent maintains a large and productive stock as long as he has a high degree of control and some of the benefits of a large stock spills over to the fringe due to the growth externality.

7. CONCLUSION

The above analysis indicates that the number of agents that will cooperate in setting the exploitation rate for a shared fish stock is quite limited, but probably not much lower than the number of countries typically sharing a fish stock contained within 200 miles, and quite possibly higher. Cost heterogeneity will greatly limit this number, but in repeated games with an indefinite horizon low cost agents will not necessarily outfish high cost agents as in Clark's closed loop differential game [1]. The number of agents compatible with a self-enforcing cooperative solution is highly sensitive to the discount rate.

For stocks that are contained within separate national fishing zones but the growth of which is interrelated, depending on the aggregate stock size, the number of agents compatible with a self-enforcing cooperative equilibrium is much higher, but becomes very sensitive to fishing costs and the discount rate as these fall below a certain value. It is not at all unlikely that this critical number of agents will be comparable with or higher than the number of countries sharing stocks that are fully contained within 200 miles. Letting one of these zones represent the high seas with an indefinite number of agents, with one agent acting as a "leader" in the remaining zone, it appears, however, that even an ostensibly insignificant "fringe" may lead to significant losses of efficiency.

REFERENCES

1. C. W. Clark, Restricted access to common-property fishery resources: A game–theoretic analysis, *in* "Dynamic Optimization and Mathematical Economics" (P.-T. Liu, Ed.), pp. 117–132, Plenum, New York (1980).
2. R. D. Fischer and L. J. Mirman, Strategic dynamic interaction. Fish wars. *J. Econom. Dynamics Control* **16**, 267–287 (1992).
3. R. D. Fischer and L. J. Mirman, The compleat fish wars: Biological and dynamic interactions, *J. Environ. Econom. Management* **30**, 34–42 (1996).
4. D. Fudenberg and J. Tirole, "Game Theory," MIT Press, Cambridge, MA (1991).
5. R. Gibbons, "A Primer in Game Theory," Harvester Wheatsheaf, London (1992).
6. R. P. Hämäläinen, V. Kaitala, and A. Haurie, Bargaining on whales: A differential game model with Pareto optimal equilibria, *Oper. Res. Lett.* **3**, 5–11 (1984).
7. R. Hannesson, "Bioeconomic Analysis of Fisheries," Fishing News Books, Oxford (1993).

322   RÖGNVALDUR HANNESSON

8. R. Hannesson, Sequential fishing: Cooperative and non-cooperative equilibria, *Natural Resource Modeling* **9**, 51–59 (1995).
9. D. Levhari and L. J. Mirman, The great fish war: An example using a dynamic Cournot–Nash solution, *Bell J. Econom.* **11**, 322–334 (1980).
10. M. Mesterton-Gibbons, Game–theoretic resource modeling, *Natural Resource Modeling* **7**, 93–147 (1993).
11. G. R. Munro, The optimal management of transboundary renewable resources, *Canad. J. Econom.* **12**, 355–376 (1979).
12. M. J. Osborne and A. Rubinstein, "A Course in Game Theory," MIT Press, Cambridge, MA (1994).
13. J. Vislie, On the optimal management of transboundary renewable resources: A comment on Munro's paper, *Canad. J. Econom.* **2**, 870–875 (1987).

Uncertainty

[27]

Stochastic Bioeconomics: A Review of Basic Methods and Results

Peder Andersen

Aarhus University
Aarhus, Denmark

Jon G. Sutinen

University of Rhode Island
Kingston, Rhode Island

Abstract   Basic bioeconomic models which incorporate uncertainty are reviewed to show and compare the principal methods used and results reported in the literature. Beginning with a simple linear control model of stock uncertainty, we proceed to discuss more complex models which explicitly recognize risk preferences, firm and industry behavior, and market price effects. The effects of uncertainty on the results of bioeconomic analysis are rarely unambiguous, and in some instances differ little from corresponding deterministic results. This review is presented to enhance readers' appreciation of the papers to follow in this and the next issue of the journal.

1. Introduction

As with most scientific endeavors, the study of fisheries is shaped by the intellectual currents of the times. One such current in recent years has been the concern with behavior under uncertainty. After Bernoulli's (1738) paper formulating the expected utility hypothesis, the study of uncertainty languished for

Marine Resource Economics, Volume 1, Number 2
0738-1360/84/020117-00$02.00/0

two centuries before being rekindled by von Neuman and Morgenstern (1944). Their work provided the foundation for an intellectual revolution that has brought operational analyses of
uncertainty to several practical endeavors in commerce, government, and law (Hirshleifer and Riley 1979).

The study of fisheries economics, or, more specifically, of
bioeconomics, has not escaped this intellectual revolution. Biologists, mathematicians, and economists began developing in
the 1970s the foundations for analyzing the behavior of bioeconomic systems under uncertainty. The literature in this area
has grown dramatically during the last few years. Of the nearly
three-score papers we examined for this review, only seven were
published before 1975. Uncertainty now is clearly established in
the bioeconomics literature as a principal field of inquiry.

A comprehensive review of all important aspects of bioeconomics and uncertainty is impossible in this limited space. Our
modest aim here is to review the basics of bioeconomic analysis
under uncertainty, especially to report and compare principal
methods and results found in the literature. Where possible, we
identify gaps and inconsistencies where further investigation appears fruitful.

Our review begins in section 2 with one of the simplest models
of bioeconomics under uncertainty, distinguished by being linear
in the control variable. In section 3 we review models which are
nonlinear in the control variable, including two which take risk
preferences into account. Formal analyses of firm and industry
behavior under uncertainty are reviewed in section 4. Section 5
contains brief reviews of assorted analyses. We sum up our review in section 6.

2. Linear Control Models

The first model we examine is a direct extension of Clark's (1976)
familiar bioeconomic framework to include stock uncertainty.
First developed by Ludwig (1979), the model is in continuous
time and is linear in the control variable. Stock uncertainty is
incorporated by the stochastic differential equation

$$dx = [g(x_t) - h_t] \, dt + \sigma x_t \, dz \tag{1}$$

Stochastic Bioeconomics: Methods and Results   *119*

where x_t is the size of the resource stock in period t, $g(x_t)$ is the natural growth rate of x_t, h_t is the harvest rate, σx_t reflects the level of random fluctuation in x_t, and dz is an increment of a stochastic process.[1] This specification assumes that the size of the resource stock in the current period is known without error and that the change in stock size is composed of a deterministic part, $[g(x_t) - h_t]dt$, and a random part, $\sigma x_t\, dz$. The expected rate of change in the stock size is $[g(x_t) - h_t]$.

The net revenue from harvesting the resource in each period is given by $[p - c(x_t)]h_t$, where p is the unit price of harvest, assumed constant, and $c(x_t)$ is the unit cost of harvest, $c'(x_t) < 0$. The harvest rate is constrained by upper and lower bounds, that is, $h_l(x_t) \leq h_t \leq h_u(x_t)$.[2] The expected present value of net revenue from harvest is given by

$$\epsilon\left\{ \int_0^\infty e^{-\delta t}[p - c(x_t)]h_t\, dt \right\} \tag{2}$$

where ϵ is the expectation operator, and δ is the discount rate. Optimal harvest policy is found by maximizing equation 2 subject to equation 1 and $h_l(x_t) \leq h \leq h_u(x_t)$.

The basic necessary condition for this stochastic optimal control problem (see Kamien and Schwartz 1981) is

$$\max_h \left\{ e^{-\delta t}[p - c(x)]h + V_x[g(x) - h] + \frac{\sigma^2 x^2}{2}V_{xx} \right\} \tag{3}$$

where $V(x_0)$ is the maximum of equation 2 at $t = 0$ and, therefore, $V_x = \partial V/\partial x$ is the marginal expected present value of (or shadow price on) the resource stock.

From equation 3 it follows that the optimal feedback policy at $t = 0$ is given by

$$h^* = \begin{cases} h_u(x_0), & \text{if } [p - c(x_0)] > V_x(x_0) \\ \bar{h}, & \text{if } [p - c(x_0)] = V_x(x_0) \\ h_l(x_0), & \text{if } [p - c(x_0)] < V_x(x_0) \end{cases}$$

The feedback policy specifies the optimal harvest rate as dependent on the current state, that is, $h_t^* = h(x_t)$.[3] The optimal policy for the stochastic linear control model has the same bang-bang feature as its deterministic counterpart. In both stochastic and deterministic models, optimal policy attempts to attain a steady state at the equilibrium stock size x^* where $p - c(x^*) = V_x(x^*)$. For the stochastic case, we denote this stock size by x_s^*, and for the deterministic case by x_d^*. In each case, harvest policy switches at the respective x^*. When $x_t > x^*$, h_t is set at its upper bound, and when $x_t < x^*$, h_t is set at its lower bound. When $x_t = x^*$, h_t is set equal to the natural growth rate, that is, $\overline{h} = g(x^*)$.[4]

Ludwig (1979) shows that $x_s^* > x_d^*$ under certain conditions.[5] Ludwig and Varah (1979) use numerical methods to show that increased noise (here, a larger σ) causes x_s^*/x_d^* to increase for large values of h_u. At low values of h_u and the discount rate, increased noise causes x_s^*/x_d^* to decrease with $x_s^* < x_d^*$ in some situations.[6] Thus the effect of stock uncertainty in this simple model is ambiguous.

Reed (1979) removes some of this ambiguity using a discrete-time, stock-recruitment model. He shows how the form of the harvest cost function $c(x)$ affects the level of x_s^* relative to x_d^*. The stochastic difference equation used is

$$x_{t+1} = z_t f(x_t - h_t) \tag{4}$$

where $f(\cdot)$ is the expected stock-recruitment function, and $\{z_t\}$ is a sequence of independent, identically distributed (iid) random variables with unit mean. Optimal policy is obtained by maximizing expected present value of net revenue,

$$\epsilon\left\{\sum_{t=1}^{\infty} \alpha^t \left[ph_t - \int_{x_t - h_t}^{x_t} c(r)\, dr \right]\right\}$$

subject to equation 4 and $0 \leq h_t \leq x_t$, where α is the discount factor. Note that the constraint on the harvest rate is not as

Stochastic Bioeconomics: Methods and Results   *121*

general as Ludwig's, but it is this constraint that results in a constant escapement policy. That is,

$$h_t^* = \begin{cases} 0, & \text{if } x_t \leq x_s^* \\ x_t - x_s^*, & \text{if } x_t > x_s^* \end{cases}$$

where x_s^* is the optimal escapement level and is approximately equivalent to Ludwig's continuous time x^*.

Reed shows that $x_s^* \geq x_d^*$ when $xc(x)$ is strictly concave or linear and that $x_s^* \leq x_d^*$ when $xc(x)$ is strictly convex. He also asserts it can be shown that x_s^* increases with an increase in uncertainty (i.e., the spread of the z_t) when $c(x) = k/x^\theta$, $k > 0$ and $0 \leq \theta \leq 1$.[7]

The models discussed to this point allow no price effects and implicitly assume risk-neutral preferences and costless changes in the level of harvesting effort. Models which allow for risk-averse preferences and price effects are discussed in the next two sections. As reported by Ludwig (1980), costly changes in effort eliminate the bang-bang policy feature of the linear model. Using a numerical example, he demonstrates that the model with costly changes in effort yields a present value of net revenue that lies between the present values for the feedback and open-loop (constant effort) policies of the linear model.[8]

The structure of harvesting costs significantly affects the character of optimal policy. In an earlier paper, Reed (1974) develops a model which is similar to the one above but in which total costs also involve a fixed setup cost K incurred only if harvest is undertaken in the period. That is, total costs are given by

$$TC(h_t, x_t) = h_t c(x_t) + \gamma(h_t) \cdot K$$

where

$$\gamma(h_t) = \begin{cases} 0, & \text{if } h_t = 0 \\ 1, & \text{if } h_t > 0 \end{cases}$$

Under certain curvature properties of the growth function and the cost function, an (S, s) policy is shown to maximize the

present value of net revenue. That is, optimal harvest policy is
given by

$$
h_t^* = \begin{cases} 0, & \text{if} \quad x_t \leq s \\ x_t - S, & \text{if} \quad x_t > s \end{cases}
$$

where $S < s$. With no setup cost ($K = 0$), $S = s = x_s^*$, as above.

The intuition behind this result is straightforward. When $K = 0$, it is optimal to harvest x_t down to S whenever $x_t > S$. However, with positive setup costs, small harvests would not cover the setup costs and should not be undertaken. A harvest should be undertaken only if it is large enough to cover all costs. The level s is the smallest stock size at which positive net revenue is realized.

Spulber (1982) extends Reed's setup cost model in two ways. The first involves allowing the random disturbances in resource growth $\{z_t\}$ to follow a general Markov process. When $K > 0$, optimal harvest policy is now given by

$$
h_t^* = \begin{cases} 0, & \text{if} \quad x_t \leq s(z) \\ x_t - S(z), & \text{if} \quad x_t > s(z) \end{cases}
$$

where z is the last observed disturbance. Therefore, the two critical stock sizes (S, s) are not constant. Instead, expected future disturbances are taken into account which, in the Markovian framework, depend on previously observed disturbances. Spulber also shows that when $K = 0$, $S(z) = s(z)$ and when the $\{z_t\}$ are iid, S and s are constants as in Reed (1974, 1979).

Spulber's second extension is to show that the resource stock and harvest rate converge to unique, time-invariant probability distributions. He observes that the time-invariant probability distributions on harvest and stock size constitute the stochastic analogue to the steady-state equilibrium found in deterministic models. Spulber explicitly derives these probability distributions for the case $K = 0$ using the logistic model. Since his harvest cost function, $c \cdot h_t$ (c = constant), is independent of x_t, the expected stock size and harvest rate he computes for the stochastic steady state are equal to their deterministic equilibrium values.

Stochastic Bioeconomics: Methods and Results 123

Taken together, the analyses of Reed and Spulber show that the structure of harvest costs is a principal determinant of the expected difference between stochastic and deterministic outcomes. The lack of an expected difference does not mean that optimal stochastic policy is the same as optimal deterministic policy. Indeed, they can be quite different in any given period. Deterministic harvest policy is constant in equilibrium, while stochastic harvest policy (of the feedback type) varies with the state of the stock.

3. Nonlinear Control Models

We now consider a more complex set of models, distinguished principally by being nonlinear in the control variable. Curvature properties of the criterion function are especially important for problems involving uncertainty, so one might argue that the following models are somewhat more appropriate.

The first nonlinear model we consider was developed by Gleit (1978), who maximizes expected utility in order to account for risk preferences. The Ludwig model above is therefore modified to be

$$\max_h \epsilon \left\{ \int_0^\infty e^{-\delta t} U[ph - c(x)h] \, dt \right\}$$

subject to equation 1, where $U(\cdot)$ is the utility function, $U' > 0$, $U'' < 0$.[9] The basic necessary condition analogous to equation 3 is

$$\max_h \left\{ e^{-\delta t} U[ph - c(x)h] + J_x[g(x) - h] + \frac{\delta^2 x^2}{2} J_{xx} \right\}$$

where J_x and J_{xx} are analogous to V_x and V_{xx} in equation 3. Therefore, the optimal harvest rate h_i^* is where

$$e^{-\delta t} U'(\cdot)[p - c(x)] = J_x$$

About all we can say about optimal policy in this model is that it is not bang-bang.[10] It can also be shown that for certain conditions, h_t^* and x_t are directly related.[11] A direct relationship also exists in the linear model, but it is discontinuous. Here there is a smooth, direct relationship. To this point, it is not possible to establish any qualitative difference between stochastic and deterministic policies in the nonlinear control case.

The work of Lewis (1981, 1982) is an elaboration of the above nonlinear model. The stochastic growth relationship used by Lewis is given by

$$x_{t+1} = x_t + \eta_{1t}g(x_t) - \eta_{2t}h(E_t, x_t)$$

where

$g(\cdot)$ = expected rate of change in the stock x resulting from natural growth,

$h(E,x)$ = production function,

η_1 = random variation in growth caused by changes in recruitment, growth, and natural mortality, and

η_2 = random variation in harvest caused by changes in environmental conditions, stock distribution, catchability, etc.

Both η_1 and η_2 are nonnegative random variables, independently distributed through time.

For his calculations, Lewis uses the logistic growth law for $g(x)$ and a Cobb-Douglass production function $h(E,x) = qEx$, linear in E and x. Net returns are given by $R(E_t, x_t) = p_t h_t \eta_{2t} - C(E_t)$, where p_t = the exvessel price, a nonnegative random variable, and $C(E_t)$ = the total cost of effort. Lewis considers three specifications for $C(E)$:

$$C(E) = 0 \tag{5a}$$

$$C(E) = c_1 E + c_2 E^2, \qquad c_1, c_2 > 0 \tag{5b}$$

$$C(E) = c_3 E^{1/2}, \qquad c_3 > 0 \tag{5c}$$

He imagines a social manager who has a utility function

Stochastic Bioeconomics: Methods and Results   *125*

$U[R(x_t, E_t)]$ where $U' > 0$, $U'' \leq 0$. For a risk-neutral social manager, Lewis lets $U(R) = R$, and for the risk-averse social manager, $U(R) = \ln(R + G)$, where G is a large enough constant to insure $R + G > 0$ always. The social manager is assumed to

$$\max \sum_{t=0}^{\infty} \alpha^t \; \epsilon\{U[R(x_t, E_t)]\}$$

subject to $x_{t+1} = x_t + \eta_{1t} g(x_t) - \eta_{2t} h_t$, where α is the riskless discount factor.

Using Howard's stochastic dynamic programming algorithm, numerical solutions are generated for this model with parameter estimates from the Eastern Pacific yellowfin tuna fishery. Lewis's results are especially interesting because they describe behavior along the optimal trajectory to the steady state as well as at the steady state. We summarize Lewis's results as follows:

Effects of Risk-Bearing Attitudes. Optimal effort and harvest levels for the risk-averse manager are greater (less) than the optimal levels for the risk-neutral manager at small (large) stock sizes. Both manager's programs converge to the same steady-state stock size, but the risk-averse program converges at a slower rate.

Effects of Increased Uncertainty. For the risk-neutral manager, increased variation in price has no effect as long as expected price remains the same, but increased variation in the harvest rate parameter η_2 decreases optimal effort corresponding to each stock size. For the risk-averse manager, increased variation in price or in η_2 leads to increases or no change in optimal effort at small stock sizes and to decreases in effort at large stock sizes. Increased variation in the growth rate parameter η_1 alone has only a negligible effect on optimal effort levels, a surprising result in light of Ludwig's and Réed's analyses.

Stochastic Versus Deterministic Analysis. The difference (in present value terms) between the stochastic and deterministic results was negligible for the more conventional cost function (equation 5b) and small for the other two cost functions (5a and 5c). Therefore, in some cases solutions to deterministic problems yield good approximations for solutions to stochastic problems.

For his study of lobster, Smith (1980) also finds that stochastic growth is not significant. He uses a production function of the form $h_t = qE_t^\beta x_t$, where E_t is fishing effort and $0 < \beta < 1$. Like Spulber, Smith derives the time-invariant probability distribution for the resource stock for a growth process similar to equation 1. For a constant (i.e., open-loop) effort policy, Smith shows that the average, stochastic steady-state stock size is given by

$$\epsilon\{x\} = x_d - \frac{\sigma^2}{2(\rho/k)}$$

where x_d is the deterministic steady-state stock size, ρ is the intrinsic growth rate, and k is the carrying capacity for the logistic growth law. For the lobster fishery he studied, there was no significant difference between $\epsilon\{x\}$ and x_d.

Mendelssohn (1982) also examines the effects of risk preferences on optimal harvest policy. In a stochastic stock recruitment model which maximizes expected discounted utility of harvest, Mendelssohn's numerical analysis shows that risk aversion results in more harvested at small stock sizes than with a risk-neutral utility function. This same result is reported by Lewis. Mendelssohn and Lewis also conclude that adjusting the discount rate is not a satisfactory means of accounting for different attitudes toward risk.

4. Firm and Industry Analysis

In contrast to the models reviewed above, Andersen (1982) explicitly models individual firms and examines both open-access and optimal fisheries exploitation under price uncertainty. Individual firms have profits given by

$$\pi^i = p_t h_t^i - c(E_t^i)$$

where p_t is random with mean $\bar{p}$, $h_t^i = qE_t^i x_t$, $c(E_t^i)$ represents

Stochastic Bioeconomics: Methods and Results   127

total costs, and $c'(E^i) > 0$. Each firm maximizes the expected utility of profits $\epsilon\{U(\pi^i)\}$, where $U'(\pi^i) > 0$ and $U''(\pi^i) < 0$. Since firms are risk-averse, they operate where the expected marginal value product of effort exceeds the marginal factor cost of effort. That is, their effort level is determined by the condition

$$\bar{p}qx = c'(E^i) + \gamma$$

where $\gamma = \gamma(\sigma_p^2) > 0$ is the marginal cost of risk bearing. If the variance of price σ_p^2 is zero, $\gamma(0) = 0$, firms are facing a deterministic price and apply more effort. And since $\gamma'(\sigma_p^2) > 0$, increases in price variation induce less effort to be applied. He also assumes that firms enter and exit the fishery as

$$\bar{p}qx \gtreqless m + \gamma(\sigma_p^2)$$

where m is the minimum average cost of effort.
    The growth of the fish resource stock is governed by

$$\dot{x} = g(x) - \sum_{i=1}^{N} h^i$$

where N is the number of firms fishing. The logistic growth model is assumed for $g(x)$.
    Andersen proceeds to show that in an open-access equilibrium, total effort and the number of fishing firms are less and the stock size greater with price uncertainty than without (where the deterministic price equals $\bar{p}$). Also, increases in the variance of price reduce total effort and the number of firms and increase the stock size.
    For his first-best optimum, Andersen assumes that society (in the form of a managing authority) is willing to bear risk at zero cost. That is, society is risk-neutral, attaching importance to risk only through the costs of risk borne by individuals. It is also assumed that the managing authority is able to vary the price variance (at the exvessel level) without cost and faces the same expected price, $\bar{p}$. Under these conditions, it is shown that the

first-best optimum *may* have more total effort applied than would result under open access if the price variance is high. Furthermore, the only regulation method that produces the first-best optimum is proved to be a fixed price system. The second-best regulation method is a tax on revenue, which is shown to be superior to both transferable quotas and a tax on catch.

In Andersen (1981b), fishing firms' behavior and characteristics of the open-access and optimal fishery under stock uncertainty are examined. He makes the same economic assumptions as above, except price is now constant. The growth of the resource stock (with exploitation) is given by the stochastic differential equation

$$dx_t = \left[g(x_t) - \sum_{i=1}^{N} h_t^i \right] dt + \sigma x_t \, dz_t$$

where z_t is assumed to describe a white-noise stochastic process (i.e., $dz \sim N[0, \sigma]$). As before, the logistic form of $g(x)$ is used. This stochastic growth relationship is almost identical to equation 1 discussed above. Andersen shows that if the stock size is known by fishing firms at fishing time, the optimal levels of catch and effort are less than in the deterministic case. If the stock size is not known by firms at fishing time, the optimal levels of effort and catch are less than if the stock size is known at fishing time.

Comparing open access and optimum fisheries, he shows that if the stock level is known at fishing time, the optimal effort level is less than the expected effort level under open access—although, in some periods, optimal effort may be larger than under open access. If firms know only the mean and variance of the stock, effort is always larger under open access than in the optimally managed fishery.

Pindyck (1984) represents the only stochastic bioeconomic analysis which explicitly incorporates a downward-sloping demand function. Pindyck treats the case where resource markets are competitive, property rights to the resource stock are well defined and enforced, and firms are risk-neutral. The source of

Stochastic Bioeconomics: Methods and Results 129

uncertainty is the stochastic growth of the resource stock, given by equation 1 where $\sigma(x)$, $\sigma'(x) \geq 0$, replaces σx. Equilibrium harvest (and, under these conditions, optimal harvest) is obtained by

$$\max_h \int_0^h p(r) \, dr - c(x)h + V_x[g(x) - h] + \frac{\sigma^2(x)}{2} V_{xx}$$

where $p(h)$ is the inverse market demand curve. Among several interesting results is that for a convex $c(x)$, stock uncertainty increases the expected rate of growth in price. This does not imply, however, that the harvest rate is greater than in the deterministic case, as Pindyck demonstrates with some examples. Harvest rates, for any x, can be increased, decreased, or unchanged by changes in $\sigma(x)$.

5. Additional Issues

Several issues beyond those covered in previous sections are treated in the literature. A selected few follow.

Reed (1974) and Andersen (1981a) address the issue of extinction under uncertainty. For the case $K > 0$, Reed derives two interest rate values i_1 and i_2 ($i_1 > i_2$). If the actual interest rate $i > i_1$, extinction is optimal; if $i < i_2$, survival is optimal. It is not clear what happens when $i_1 > i > i_2$. For the case $K = 0$ and a constant marginal harvest cost (i.e., $c[x] = \bar{c}$), extinction is optimal if the expected population growth rate is always less than the rate of interest. Andersen examines the implications price uncertainty has for extinction. The same assumptions are made as in his papers cited above. His principal result is that the deterministic results regarding extinction (e.g., see Clark 1973; Clark and Munro 1975) cannot be carried over to the case of price uncertainty. With price uncertainty, the conditions for extinction can be less restrictive in an optimal fishery than in an open-access fishery. That is, if extinction is optimal (in a first-best sense), it will not necessarily occur under open access. This is impossible in the deterministic case. Also, the conditions under which extinction is optimal depend on the method of regulation. For a tax on revenue, the conditions are less restrictive than with a tax on catch or individual quotas.

High variation in equilibrium (or steady-state) harvests has been established by a number of studies (e.g., Beddington and May 1977; Sissenwine 1977; and May et al. 1978), especially when harvested at high effort levels. Since the manager's utility function usually is not known, it is not clear how much variation in harvest should be permitted. Mendelssohn (1980a) approaches this problem by devising a numerical method which computes the trade-off between the mean and variance of the return. He applied this technique to the Bristol Bay, Alaska, salmon fishery to generate a trade-off schedule for the mean and standard deviation of long-run harvest.

So far in this review we have discussed results based on lumped parameter models (e.g., the logistic growth law). Two studies have examined optimal policy based on the cohort model.

Dudley and Waugh (1980) develop a numerical model for determining optimal harvest policies for a single-cohort fishery under uncertainty. Their application is to an Australian prawn fishery. A Beverton-Holt growth law is assumed where recruitment, natural mortality, and catchability are random variables. Expected net revenue for each policy, state, and period, plus the transition probabilities, were generated by simulation and used as the data in a stochastic dynamic programming model. The stochastic DP generated policies which maximize expected net revenue over the season (12 months). They find that with a high level of harvesting capacity, the optimal procedure is not to harvest at all until the biomass reaches its peak, and then to harvest at the maximum rate until the minimum profitable biomass is reached. This policy has an (S, s) character, as in Reed and Spulber. The stochastic and deterministic results differ by little and in no apparent systematic way given the tabular results presented. With a low level of harvesting capacity, optimal harvest begins a month earlier but is not applied at its maximum until the peak biomass time arrives. The presence of uncertainty results in more fishing effort in the first month and spreads it out over more months (in most, but not all, cases). A demand relationship instead of a constant price tends, as one would expect, to spread effort and catch over more of the year. As with most numerical results, one cannot confidently generalize Dudley and Waugh's results. Their results also consider the three random variables in only three combinations, and they do not present results with different levels of variation in the random parameters.

Mendelssohn (1978) develops single-species, multicohort optimal harvesting models in which recruitment and age-dependent survivorship rates are random. His two models assume perfect selection in harvesting the cohorts, or age classes. In the first model where he assumes that recruitment is independent of total stock size, Mendelssohn derives a "Fisher rule" for harvest. That is, do not harvest a cohort until it reaches an optimum age, then harvest the entire cohort. While there is no comparison with the deterministic policy, it appears qualitatively similar to the deterministic analysis of Clark, Edwards, and Friedlander (1973). In his second model, recruitment is assumed to depend on the total stock size and yields a policy that the oldest are always harvested first. Using similar methods, Mendelssohn (1980b) derives the qualitative properties of optimal policy for a stochastic multispecies fishery.

6. Concluding Remarks

In this brief review we have described the basic methods and results found in the stochastic bioeconomics literature. Where the dynamics of the resource stock are given by a stochastic differential or difference equation, stochastic dynamic programming methods are used to derive optimal policy. In several cases, optimal policy under stochastic conditions is qualitatively different from optimal policy under deterministic conditions. Such differences, however, are not unambiguous. Two empirically based studies, Lewis (1982) and Smith (1980), conclude that deterministic policies are reasonably good substitutes for stochastic policies on average. One cannot easily generalize from these results, but this does raise the question of whether uncertainty is significant. Future studies hopefully will show more clearly

132   *Peder Andersen and Jon G. Sutinen*

the significance (or insignificance) of uncertainty for policy analysis and other studies of the fishery.

Acknowledgments

An earlier version of this paper was presented to the Workshop on Uncertainty and Fisheries Economics, University of Rhode Island, November 1981. Support for this research was provided by Sea Grant, University of Rhode Island, and the Carlsberg Foundation, Denmark. Contribution No. 2191 of the Rhode Island Agriculture Experiment Station.

Notes

1. In addition, $g(x)$ is strictly concave, $\sigma > 0$, and $\epsilon\{dz\} = 0$.

2. The control variable in Ludwig's model actually is effort E in the production function $h = qEx$, and his constraint is $E_l \leq E \leq E_u$. Therefore, in the present specification, the constraint on harvest is where $h_l(x_t) = qE_l x_t$ and $h_u(x_t) = qE_u x_t$.

3. A feedback, or closed-loop, policy prescribes a rule for specifying future harvest rates when future information is given. Feedback policies are generally superior to open-loop and revised open-loop policies (cf. Ludwig 1980). An open-loop policy is one for which all present and future harvest rates are determined once and for all in the initial period. This policy is often fully appropriate in a deterministic setting where all future states are known with perfect certainty. Under a revised open-loop policy, estimates of the state and parameters of the system are periodically updated and lead to revisions in the open-loop policy.

4. Ludwig also examines the consequences of noise for depensation models with two or more switching points.

5. The conditions include a small σ^2, a large h_u, and $h_l = 0$.

6. Their model allows x_d^* to increase with σ^2 also, since they use a deterministic growth function which has σ^2 as an argument. Their justification is that this specification gives better results when formulating deterministic policy. Unfortunately, this choice seems to confound their results somewhat.

7. A number of not unreasonable conditions must be satisfied for these results to hold. Reed notes that a constant escapement policy is not optimal if the demand curve is not perfectly elastic or if the total cost function is not linear in h.

Stochastic Bioeconomics: Methods and Results   *133*

8. Clark (1979) suggests extending Ludwig's model to allow h_u to be determined optimally, thereby also determining the optimal level of excess capacity above $h = g(x_s^*)$ that should be constructed to take advantage of stock sizes above x_s^*.

9. Gleit actually solves a finite time horizon problem; hence, we do not present his precise results.

10. Gleit derives an explicit solution for h^* where $g(x)$ is assumed linear. This case, of course, is not appropriate for fisheries problems.

11. The conditions are that $J_{xx} < 0$ and the Arrow-Pratt measure of relative risk aversion is less than unity. This result is obtained by totally differentiating the above condition.

References Cited

Andersen, P. 1981a. *Extinction and price uncertainty* (Staff Paper No. 81–09). Kingston, R.I.: University of Rhode Island, Department of Resource Economics.

———. 1981b. *The exploitation of fish resources under stock uncertainty* (Staff Paper No. 81–14). Kingston, R.I.: University of Rhode Island, Department of Resource Economics.

———. 1982. Commercial fisheries under price uncertainty. *J. Environ. Econ. Manage.* 9(1): 11–28.

Beddington, J. R., and R. M. May. 1977. Harvesting natural populations in a randomly fluctuating environment. *Science* 197: 463–465.

Bernoulli, D. 1738. Specimen theoriae novae de mensura sortis. *Commentarii Academiae Scientiarum Imperialis Petropolitanae* 5: 175–192. Translated into English by L. Sommer, 1954, Exposition of a new theory on the measurement of risk. *Econometrica* 22: 23–36.

Clark, C. W. 1973. Profit maximization and the extinction of animal species. *J. Polit. Econ.* 81: 950–961.

———. 1976. *Mathematical bioeconomics*. New York: Wiley.

———. 1979. Mathematical models in the economics of renewable resources. *SIAM Rev.* 21(1): 81–99.

Clark, C. W., G. Edwards, and M. Friedlaender. 1973. Beverton-Holt model of a commercial fishery: optimal dynamics. *J. Fish. Res. Board Can.* 30: 1629–1640.

Clark, C. W., and G. R. Munro. 1975. Economics of fishing and modern capital theory: A simplified approach. *J. Envir. Econ. Manage.* 2: 92–106.

Dudley, N., and G. Waugh. 1980. Exploitation of a single-cohort fishery under risk: A simulation-optimization approach. *J. Envir. Econ. Manage.* 7: 234–255.

Gleit, A. 1978. Optimal harvesting in continuous time with stochastic growth. *Math. Biosci.* 41: 111–123.

Hirshleifer, J., and J. G. Riley. 1979. The analytics of uncertainty and information—An expository survey. *J. Econ. Lit.* 17: 1375–1421.

Kamien, M. I., and N. L. Schwartz. 1981. *Dynamic optimization: The calculus of variations and optimal control in economics and management.* New York: Elsevier North-Holland.

Lewis, Tracy R. 1981. Exploitation of a renewable resource under uncertainty. *Can. J. Econ.* 14(3): 422–439.

———. 1982. *Stochastic modeling of ocean fisheries resource management.* Seattle: University of Washington Press.

Ludwig, D. 1979. Optimal harvesting of a randomly fluctuating resource. I. Application of perturbation methods. *SIAM J. Appl. Math.* 37: 166–184.

———. 1980. Harvesting strategies for a randomly fluctuating population. *J. Cons. Perm. Int. Explor. Mer* 39: 168–174.

Ludwig, D., and J. M. Varah. 1979. Optimal harvesting of a randomly fluctuating resource. II. Numerical methods and results. *SIAM J. Appl. Math.* 37: 185–205.

May, R. M., J. R. Beddington, J. W. Horwood, and T. G. Sherpherd. 1978. Exploiting natural populations in an uncertain world. *Math. Biosci.* 42: 219–252.

Mendelssohn, R. 1978. Optimal harvesting strategies for stochastic single-species, multiage class models. *Math. Biosci.* 41: 159–174.

———. 1979. Determining the best trade-off between expected economic returns and the risk of undesirable events. *J. Fish. Res. Board Can.* 36: 939–949.

———. 1980a. A systematic approach to determining mean-variance trade-offs when managing randomly varying populations. *Math. Biosci.* 50: 75–84.

———. 1980b. Managing stochastic multispecies models. *Math. Biosci.* 49: 249–261.

———. 1982. Discount factors and risk aversion in managing random fish populations. *Can. J. Fish. Aquat. Sci.* 39: 1252–1257.

Pindyck, Robert S. 1984. Uncertainty in the theory of renewable resource markets. *Rev. Econ. Stud.* 51:289–303.

Reed, W. J. 1974. A stochastic model for the economic management of a renewable animal resource. *Math. Biosci.* 22: 313–337.

Stochastic Bioeconomics: Methods and Results 135

Reed, W. J. 1979. Optimal escapement levels in stochastic and deterministic models. *J. Envir. Econ. Manage.* 6: 350–363.

Sissenwine, M. 1977. The effect of random fluctuations on a hypothetical fishery. *ICNAF Sel. Pap.* 2: 137–144.

Smith, J. B. 1980. Replenishable resource management under uncertainty: A re-examination of the U.S. northern fishery. *J. Envir. Econ. Manage.* 7: 209–219.

Spulber, D. F. 1982. Adaptive harvesting of a renewable resource and stable equilibrium. In *Essays in the economics of renewable resources*, ed. L. J. Mirman and D. F. Spulber, 117–139. New York: North-Holland.

von Neuman, J., and O. Morgenstern. 1944. *Theory of games and economic behavior*. New York: Wiley.

Related References

Allan, K. R. 1973. The influence of random fluctuations in the stock-recruitment relation on the economic return from salmon fisheries. *Cons. Perm. Int. Explor. Mer Rapp.* 164: 351–359.

Anderson, David R. 1975. Optimal exploitation strategies for an animal population in a Markovian environment: A theory and an example. *Ecology* 56: 1281–1297.

Clark, C. W. 1981. Structural uncertainty in multispecies fisheries. Background paper, University of British Columbia, Vancouver.

Doubleday, W. G. 1976. Environmental fluctuation and fisheries management. *ICNAF Sel. Pap.* 1: 141–150.

Everitt, R. R., N. C. Sonntag, M. L. Puterman, and P. Whalen. 1978. A mathematical programming model for the management of a renewable resource system: The Kemano II development project. *J. Fish. Res. Board Can.* 35: 235–246.

Gaither, N. 1980. A stochastic constrained optimization model for determining commercial fishing seasons. *Manage. Sci.* 26: 143–154.

Gatto, M., and S. Rinalch. 1976. Mean value and variability of the catch in fluctuating environments. *J. Fish. Res. Board Can.* 33: 189–193.

Goel, N. S., and N. Richter-Dyn. 1974. *Stochastic models in biology*. New York: Academic Press.

Hilborn, R. 1976. Optimal exploitation of multiple stocks by a common fishery: A new methodology. *J. Fish. Res. Board Can.* 33: 1–5.

Huang, C. C., I. B. Vertinsky, and N. J. Wilimovsky. 1976. Optimal controls for a single species fishery and the economic value of research. *J. Fish. Res. Board Can.* 33: 793–809.

Hutchinson, C. E., and T. R. Fischer. 1979. Stochastic control applied to fishery management. *IEEE Trans. Syste., Man, Cybern.* 9: 253–259.

Jaquette, David L. 1972. A discrete time population control model. *Math. Biosci.* 15: 231–252.

———. 1974. A discrete-time population-control model with setup cost. *Oper. Res.* 22: 298–303.

Ludwig, D., and C. J. Walters. 1981. Optimal harvesting with imprecise parameter estimates. *Can. J. Fish. Aquat. Sci.* 38: 711–720.

May, R. M. 1974. *Stability and complexity in model ecosystems.* Princeton, N.J.: Princeton University Press.

Mendelssohn, R. 1980. Using Markov decision models and related techniques for purposes other than simple optimization: Analyzing the consequences of policy alternatives on the management of salmon runs. *Fish. Bull.* 78: 35–50.

Mendelssohn, R., and M. J. Sobel. 1980. Capital accumulation and the optimization of renewable resource models. *J. Econ. Theory* 23: 243–260.

Reed, W. J. 1978. The steady-state of a stochastic harvesting model. *Math. Biosci.* 41: 273–307.

Sherpherd, T. G., and J. W. Horwood. 1979. The sensitivity of exploited population to environmental "noise" and the implications for management. *J. Cons. Perm. Int. Explor. Mer* 38: 318–323.

Smith, A. D. M., and Carl J. Walters. 1981. Adaptive management of stock-recruitment systems. Can. J. Fish. Aquat. Sci. 38: 690–703.

Tautz, A., P. A. Larkin, and W. E. Ricker. 1969. Some effects of simulated long-term environmental fluctuations on maximum sustained yield. *J. Fish. Res. Board Can.* 26: 2715–2786.

Turelli, M. 1977. Random environments and stochastic calculus. *Theor. Popul. Biol.* 12: 140–178.

Walters, C. J. 1975. Optimal harvest strategies for salmon in relation to environmental variability and uncertain production parameters. *J. Fish. Res. Board Can.* 32: 1777–1784.

Walters, C. J., and R. Hilborn. 1976. Adaptive control of fishing systems. *J. Fish. Res. Board Can.* 33: 145–159.

[28]

JOURNAL OF ENVIRONMENTAL ECONOMICS AND MANAGEMENT **20**, 71–91 (1991)

How to Set Catch Quotas: Constant Effort or Constant Catch?

RÖGNVALDUR HANNESSON AND STEIN IVAR STEINSHAMN

Norwegian School of Econonomics and Business Administration, Helleveien 30,
N-5035 Bergen-Sandviken, Norway

Received June 16, 1989; revised January 15, 1990

This paper considers whether the total allowable catch from a fish stock should be a fixed annual quantity or based on constant fishing effort. It consists of two parts, a theoretical part and an empirical part based on data from the Arcto-Norwegian cod stock. In the theoretical part it is shown that realistic cost and revenue functions have opposite effects on whether a constant quota or a constant effort yields the highest expected profit. A concave revenue function implies that a constant quota will be preferable, while a stock-dependent unit cost of landed fish has the opposite implication. The empirical part investigates how large the difference between the average profit yielded by the two strategies is likely to be, on the basis of some stylized facts about the Arcto-Norwegian cod stock. The size of this stock fluctuates considerably over time, due mainly to fluctuations in the size of year classes. Spectral analysis indicates cyclical movements, and so a sine curve was used to generate recruitment cycles. The difference in average profit yielded by the two harvest strategies is very small in most cases, or of the order of 1–2%. This result is relatively robust with respect to alternative specifications of the cost and the revenue functions, but a maximum difference of 20% was produced by a non-stock-dependent unit cost of fish and a kinked revenue function, where catches exceeding a certain quantity are worthless. © 1991 Academic Press, Inc.

1. INTRODUCTION

Since the introduction of the 200-mile economic zone, it has become increasingly common to control fish stocks by setting an upper limit on the catch to be taken each year. This "total allowable catch" (TAC) usually is determined on the basis of advice provided by fisheries biologists.

The basic philosophy behind this advice usually is that a fixed fishing mortality ought to be maintained, either the one that provides maximum sustainable yield, or a lower value implying a 10% less sustainable yield (known as $F_{0.1}$). The application of this principle means that catches will fluctuate with the size of the fish stocks, which is known to fluctuate for reasons apparently beyond man's control. As large year to year variations in catches have usually proven to be unacceptable for the decisionmakers, fisheries biologists have begun to calculate the consequences of alternative quota options rather than recommending a single quota based on a given fishing mortality. Nevertheless, the notion of the best fishing mortality to be maintained over time is still influential.

The fishing industry has not been happy about the fluctuations in catches resulting from the advice provided by fisheries biologists, even after the modifications made by the decisionmakers. As a result, the idea that the annual catch should be stabilized has been gaining ground in the industry. Would this be a better strategy, provided it is biologically feasible? In this paper we attempt to throw some light on this issue. We compare the profitability of two fishing strategies, one in which the annual catch is held constant and another in which

71

fishing effort is held constant. We have chosen to focus on constant effort rather than constant fishing mortality, because the activity of the fishing fleet seems to be as likely a target for stabilization as the annual catch. However, in the câse when fishing effort is proportional to fishing mortality, as is widely believed to be approximately correct for many demersal fish stocks, the constant effort strategy is identical to a constant fishing mortality strategy. This case is considered extensively in the applied section of the paper.

The optimal fishing strategy is likely to be neither of the two considered here. In Hannesson [3] it is shown that the optimal fishing pattern in one particular fishery model involves periodic fishing with no fishing at all in between. This implies wide fluctuations in both the annual catch and the annual effort. That notwithstanding, we see some merit in discussing the two cases at hand, as such less than fully optimal rules of thumb have been suggested as principles of management and have in fact played a role in the setting of catch quotas. To this it may be added that the 200-mile economic zone has severely restricted the use of highly mobile fishing fleets and made periodic fishing a less feasible strategy.

The contents of the paper are as follows. In the next section we start with some general theory on the profitability of the two fishing strategies, a constant annual catch and annual catch quotas based on constant effort. In Section 3 the theoretical results are illustrated by simulations of a multi-year class model of an imaginary fish stock resembling the Arcto-Norwegian cod. In particular we focus on fluctuations in the annual recruitment of young fish to the stock, which are quite substantial for the Arcto-Norwegian cod stock. As many other fish stocks in temperate and Arctic waters are characterized by similar fluctuations which may, however, differ in magnitude and regularity, this exercise should be of interest for the management of such stocks. The paper concludes with some remarks on the relevance of our results for practical fisheries management.

2. CONSTANT QUOTAS OR CONSTANT EFFORT: SOME GENERAL THEORY

Let x denote the catch rate (the quantity of fish caught per unit of time) and y the size of the stock from which it is taken. In this section it is assumed that y varies stochastically over time and that the probability distribution of y is constant over time; that is, the probability of a given y occurring at any given point in time is independent of all previous values of x and y.

Let z denote fishing effort, defined as an index of the amount of factors of production used to catch the fish. Assuming technical efficiency in production, the catch rate at any given time will be given by the production function

$$x = F(z, y).$$

As a special case, the catch rate may depend on the effort only.

Given the assumption of technical efficiency, we may also define an effort requirement function, determining the minimum effort needed to obtain a given catch rate from a stock of a given size:

$$G(x, y) = \min z; \qquad x = F(z, y).$$

Provided $F(\cdot)$ is monotonically increasing in z for any given y,

$$G(x; y) = F^{-1}(x; y).$$

In this section we compare the profitability of two quota strategies, a constant effort strategy in which the catch rate varies with the size of the fish stock, according to the production function, and a constant quota strategy in which the catch rate is fixed and equal to the expected catch rate under the constant effort strategy while the annual effort varies according to the effort requirement function. More precisely, the two strategies are defined as follows:

(i) The constant effort strategy:

$$x = F(y; \bar{z}), \qquad \bar{z} \text{ constant};$$

(ii) the constant quota strategy:

$$x = \bar{x}, \qquad \bar{x} = EF(y; \bar{z}), \qquad z = G(y; \bar{x}),$$

with E henceforth denoting the expectation operator.

To compare the profitability of the two strategies, we define the profit function

$$\pi(x, y, z) = R(x) - C(z);$$

that is, the revenue depends on the quantity caught, while the cost depends on the effort applied. The argument y appears in the profit function because the value of x realized for any given value of z depends on y through the production function, and the amount of z needed to realize a given x depends on y through the effort requirement function. If x depends on z only and not on y, the argument y drops out of the production function. In that case there will be no difference between the two quota strategies defined above; a constant effort will then always result in a constant catch rate and vice versa. Ignoring price uncertainty as we do here, the profit function will in that case be deterministic. Otherwise π will be stochastic because y is stochastic.

Provided π is stochastic, the expected profit of the two strategies under consideration will be as follows:

(i) Constant effort:

$$E\pi(x, y; \bar{z}) = ER(F(y; \bar{z})) - C(\bar{z}).$$

(ii) Constant quota:

$$E\pi(z, y; \bar{x}) = R(\bar{x}) - EC(G(y; \bar{x})).$$

Ignoring costs for a moment, a concave revenue function implies $\pi(z, y; \bar{x}) > E\pi(x, y; \bar{z})$; that is, the (certain) profit of the constant quota strategy will be greater than the expected profit of the constant effort strategy.

Taking costs into account may reverse this conclusion. Suppose, first, that $G(y; \bar{x})$ is linear in y and that $EG(y; \bar{x}) = \bar{z}$. A convex cost function will then imply $EC(G(y; \bar{x})) > C(\bar{z})$; that is, the expected cost of the constant quota strategy will be greater than the constant cost of the constant effort strategy.

But the effort requirement function is most likely not linear but rather is convex in y. This means that $EG(y; \bar{x}) > G(Ey; \bar{x})$. Then, if $\bar{z} = G(Ey; \bar{x})$, the average effort needed to maintain a given catch rate $\bar{x}$ will be greater than the constant effort yielding an average catch rate equal to $\bar{x}$. This increases the cost advantage of the constant effort strategy even further. Consider, therefore, how the shape of the effort requirement function is related to the shape of the production function and under what conditions $\bar{z} = G(Ey; \bar{x})$. On this we have the following two theorems:

THEOREM 1. *If $F_{zz} \leq 0$, $F_{yy} \leq 0$, and $F_{zy} = F_{yz} \geq 0$ (subscripts denote partial derivatives), with at least one strict inequality, the function $G(y, x)$ is convex with respect to y.*

Proof. By the implicit function theorem,

$$G_y = \frac{dz}{dy}\bigg|_{F(z,\,y)=\bar{x}} = -F_y/F_z < 0$$

and

$$G_{yy} = \frac{d^2z}{dy^2}\bigg|_{F(z,\,y)=\bar{x}} = F_z^{-2}\left(-F_z F_{yy} + 2F_y F_{yz} - F_y^2 F_{zz} F_z^{-1}\right) > 0$$

Q.E.D.

THEOREM 2. *If $F(y, z)$ is linear and increasing in y and $F_{zy} > 0$, then $EG(y; \bar{x}) > \bar{z}$.*

Proof. $\bar{x} = EF(y; \bar{z})$ by definition. $EF(y; \bar{z}) = F(Ey; \bar{z})$ by the linearity of $F(\cdot)$. By monotonicity of $F(\cdot)$, $\bar{z} = F^{-1}(Ey; \bar{x})$ and $G(Ey; \bar{x}) = F^{-1}(Ey; \bar{x})$. Since $G(y; \bar{x})$ is convex with respect to y, it follows from Jensen's inequality that $EG(y; \bar{x}) > G(Ey; \bar{x}) = \bar{z}$. Q.E.D.

Theorem 2 is illustrated in Fig. 1.

On the other hand, if $F(y; \bar{z})$ is concave with respect to y, it is possible that $EG(y; \bar{x}) < \bar{z}$. In this case $EG(y; \bar{x}) > G(Ey; \bar{x})$ because of the convexity of $G(\cdot)$.

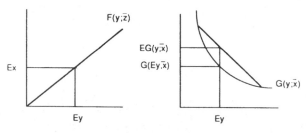

FIG. 1. The production function $F(\cdot)$ and the effort requirement function $G(\cdot)$ when the catch rate (x) is a linear function of the stock (y) for a given level of effort (z). Suppose y can assume two alternative values. The expected effort, $EG(y; \bar{x})$, needed to maintain the catch rate $\bar{x}$ will lie on the straight line connecting the two possible y-values on the curve $G(\cdot)$ above the point Ey, the expected size of the stock.

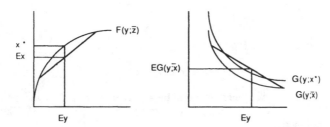

FIG. 2. The production function $F(\cdot)$ and the effort requirement function $G(\cdot)$ when $F(\cdot)$ is concave with respect to y. The effort required to take the given catch $\bar{x}$ is given by $G(y; \bar{x}) = F^{-1}(y; \bar{x})$. Suppose the stock can assume two alternative values with given probabilities. The expected catch, Ex, will lie on the straight line connecting the two alternative values of y on the curve $F(\cdot)$. Hence, $\bar{x} = Ex < x^* = F(Ey; \bar{z})$. The expected effort needed to maintain a constant catch $\bar{x}$ lies on the straight line connecting the two alternative values of y on the curve $G(y; \bar{x})$ above the point Ey. Note, however, that $\bar{z} = F^{-1}(Ey; x^*) = G(Ey; x^*)$, and, since $x^* > \bar{x}$, $G(Ey; x^*) > G(Ey; \bar{x})$. Hence $Ez > G(Ey; \bar{x})$ does not necessarily imply $Ez > \bar{z} = G(Ey; x^*)$.

But $G(Ey; \bar{x}) = F^{-1}(Ey; \bar{x}) = F^{-1}(Ey, Ex)$, while $\bar{z} = F^{-1}(Ey, x^*) > F^{-1}(Ey, Ex)$ because of the concavity of $F(\cdot)$. Figure 2 illustrates this.

It is, perhaps, most likely that $EG(y; \bar{x}) > \bar{z}$, however, We provide a popular example in which this is true and then produce an apparently much less realistic counterexample.

An Example with $Ez > \bar{z}$

Let $F(y, z)$ be specified as

$$x = y^n z^m,$$

which implies the effort requirement function

$$z = x^{1/m} y^{-n/m}.$$

Let y be evenly distributed over the interval $1 - \epsilon, 1 + \epsilon$. This implies $Ey = 1$ and a constant probability density of $1/2\epsilon$. The expected catch, for a given level of effort, is

$$\bar{x} = Ex = (\bar{z}^m/2\epsilon) \int_{1-\epsilon}^{1+\epsilon} y^n \, dy = (\bar{z}^m/2\epsilon)\left[(1 + \epsilon)^{n+1} - (1 - \epsilon)^{n+1}\right]\big/(n + 1).$$

The expected effort needed to maintain the constant catch rate $\bar{x}$ is

$$Ez = \bar{x}^{1/m} Ey^{-n/m} = (\bar{x}^{1/m}/2\epsilon) \int_{1-\epsilon}^{1+\epsilon} y^{-n/m} \, dy$$

$$= \bar{z}(2\epsilon)^{-(1+1/m)}(1 + n)^{-1/m}$$

$$\times \left[(1 + \epsilon)^{n+1} - (1 - \epsilon)^{n+1}\right]^{1/m}$$

$$\times \left[(1 + \epsilon)^{1-n/m} - (1 - \epsilon)^{1-n/m}\right][m/(m - n)].$$

By computation it can be verified that $Ez > \bar{z}$ for $0 < n \le 1, 0 < m \le 1, n \ne m$.

An Example with $Ez < \bar{z}$

Let $F(y, z)$ be specified as

$$x = \min(zy, 1).$$

For $z > (\max y)^{-1}$ the function F will be a piecewise linear function with a kink at $y = \bar{z}^{-1}$, and it will be concave with respect to y (see Fig. 3). The effort requirement function will be

$$z = x/y.$$

Assuming that y is distributed as in the previous example, the expected catch for a given level of effort, for $(1 - \epsilon)^{-1} > \bar{z} > (1 + \epsilon)^{-1}$, will be

$$Ex = (1/2\epsilon)\left[\int_{1-\epsilon}^{1/\bar{z}}\bar{z}y\,dy + \int_{1/\bar{z}}^{1+\epsilon}dy\right] = \left[-1/\bar{z} - \bar{z}(1 - \epsilon)^2 + 2(1 + \epsilon)\right]\Big/4\epsilon,$$

while the average effort needed to maintain a given catch rate $\bar{x} = Ex$ will be

$$Ez = (1/2\epsilon)\bar{x}\int_{1-\epsilon}^{1+\epsilon}y^{-1}\,dy$$

$$= \left[-1/\bar{z} - \bar{z}(1 - \epsilon)^2 + 2(1 + \epsilon)\right]\left[\ln(1 + \epsilon) - \ln(1 - \epsilon)\right]\Big/8\epsilon^2.$$

By computation it can be verified that $Ez < \bar{z}$ for high values of $\bar{z}$. For example, with $\epsilon = 0.5$, $Ez < \bar{z}$ for $\bar{z} > 0.9291$.

A Summary of this Section

To summarize the results of this section, we have found that realistic assumptions about revenue and cost relationships have opposite effects on the comparative profitability of constant quotas and constant effort strategies. A concave revenue function implies that a constant quota equal to the expected catch of a constant effort will yield a greater revenue than the constant effort does on the average. On the other hand, provided that the catch depends on the stock as well

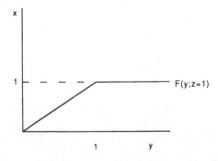

FIG. 3.   The production function $x = \min(zy, 1)$ for $z = 1$.

HOW TO SET CATCH QUOTAS 77

as on the fishing effort, an effort cost function that is convex with respect to effort implies that a constant effort will be less costly than a variable effort that maintains a steady catch equal to the expected catch of the constant effort. If the catch increases linearly with the stock for a constant effort, the average effort needed to take a constant catch equal to the expected catch of a given level of effort will be greater than that effort. This also makes the constant effort strategy less costly than the constant catch strategy. This will most likely be true as well if the catch increases at a diminishing rate with the stock, but counterexamples can be found.

3. A QUASI-EMPIRICAL MODEL OF THE ARCTO-NORWEGIAN COD

In this section we investigate the comparative profitability of a constant catch quota versus a constant effort in a quasi-empirical setting. We use a discrete fishery model based on some vital characteristics of the Arcto-Norwegian cod. The results produced in the preceding section are useful to identify circumstances under which each strategy is likely to be preferable, but they do not carry over to this framework unmodified, because we are dealing here with a stock the size of which at the beginning of any given period depends in part on how much is left at the end of the previous period. Another departure from the framework set out in the previous section is that we do not model the recruitment of young fish to the stock as a stochastic process but as a deterministic process that follows a regular cycle. The results derived in the previous section are useful nevertheless for this case; the probability density function implicit in that analysis corresponds to a frequency function derived from the cyclical behavior of the fish population in this deterministic environment.

The Arcto-Norwegian cod has a moderately long life span; approximately 10-year classes are of importance in the fishable part of the stock. Spawning and recruitment to the stock take place at discrete intervals, and the recruitment of young fish to the stock available for fishing varies quite substantially from year to year. The causes of these fluctuations are believed to be environmental, such as changes in sea temperature, but are not yet fully understood. Figure 4 shows the estimated recruitment to the stock since the mid-1940s. The recruitment is characterized by a few extraordinarily strong year classes appearing at somewhat irregular intervals. In Appendix 2 the periodicity of these fluctuations is analyzed. There is a strong indication of an 8-year cycle, with a weaker 11-year cycle.[1]

The said fluctuations in recruitment generate fluctuations in the size of the stock. The Arcto-Norwegian cod is not unique in this respect; many and probably most fish stocks exhibit such fluctuations, but possibly with a lesser amplitude and regularity. Using the Arcto-Norwegian cod stock as an illustrative example therefore seems to be a worthwhile undertaking.

Even if the recruitment fluctuations in the Arcto-Norwegian cod stock seem fairly regular, they appear to be influenced by stochastic events to some extent. We stop short of analyzing the implications of stochastic recruitment, but focus instead on the general implications of fluctuations in recruitment on the choice of fishing

---

[1]Attempts have been made to relate the fluctuations in recruitment to cyclical oscillations in the Earth's rotation axis. These oscillations would imply a 7-year cycle, see Larraneta and Vazquez [7].

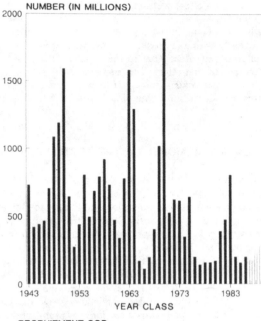

NUMBER (IN MILLIONS)

YEAR CLASS

RECRUITMENT COD

FIG. 4. Estimated strength of year classes (millions of individuals) of 3-year old fish in the Arcto-Norwegian cod stock. Courtesy of the Institute of Marine Research, Bergen.

strategy. We look at rather simple strategies where the periodicity and the nature of these fluctuations are unimportant.

We assume that the recruitment process is governed by the sinusoidal function

$$R_t = R_{min} + u(1 + \sin v\pi t),$$

where R is recruitment at time t, R_{min} is a minimum level of recruitment, u and v are parameters, and π (here) has the usual geometric meaning. The parameters u and v govern the amplitude and the length of the recruitment cycle. Maximum recruitment is

$$R_{max} = R_{min} + 2u,$$

while the length of the recruitment cycle is

$$T_r = 2/v.$$

We investigate the profitability of a constant quota versus that of a constant effort strategy by calculating the average profit per year over a recruitment cycle, on the basis of a stable pattern of fishing in the sense that either the catch or the fishing effort follows cycles of the same length as the recruitment cycle, but somewhat out of phase because of the growth and the mortality of individual fish. The model used is a standard Beverton–Holt model (Beverton and Holt [2], Hannesson [4]). The parameter values assumed are listed in Appendix 1.

The results in the previous section showed that the constant effort strategy will be better than the constant catch strategy only if the catch rate depends on the size of the exploited stock, provided the cost and the revenue functions have the shape usually assumed in economic theory. In the simulations to be discussed it is assumed that fishing mortality is proportional to fishing effort. This makes the catch per unit of effort proportional to the size of the available stock, while the unit cost of fish landed is inversely proportional to the stock. This is arguably the best possible case for the constant effort strategy, and is somewhat unrealistic. In the real world the catch per unit of effort is likely to increase with the stock at a decreasing rate, and it could possibly be independent of the stock altogether. This last case is by contrast the most favorable one for a constant quota strategy and can be dealt with by assuming a zero cost of fishing mortality. This is also considered below.

Constant Price and Cost of Effort

We first compare the two strategies with linear revenue and cost of effort functions. As a base case, we set $u = 600$, $v = 0.25$, and $R_{min} = 100$. This implies a recruitment cycle of 8 years and a maximum recruitment of 1300 (million fish), or 13 times the minimum recruitment. The actual variability of the recruitment to the Arcto-Norwegian stock is even greater than this (see Fig. 4).[2] We normalize the price at $p = 1$ and look at a "high" cost case ($c = 5000$; c, cost per unit of fishing mortality) and a "low" cost case ($c = 2000$). In these two cases the resource rent, or the net profit, is about one-half and two-thirds of the catch value, respectively.

Tables I and II show the result of optimizing constant fishing mortality (the constant effort strategy) and constant annual catch (the constant quota strategy), respectively, over the recruitment cycle. The difference in profitability of the two strategies is small to negligible. The greatest difference between the two strategies occurs when the fishery is "selective," that is, when the age at first capture is 9 years. In that case the average annual profit yielded by the constant quota strategy is only about 2% less than that of the constant effort strategy.[3]

The fact that the comparative profitability of the constant effort strategy increases with the degree of selectivity of the fishery (age at first capture) may be

---

[2] Figure 4 shows the strength of year classes of 3-year old fish. Here we implicitly define the age of recruitment as 4 years. The size of the year classes shown in Fig. 4 should therefore be reduced by a percentage reflecting the mortality of 3-year old fish. This is typically assumed to be close to 20%.

[3] Comparing the two policies by looking at the average profit over one cycle rather than the discounted present value may seem out of place, but was chosen for its simplicity. This allows us to ignore the initial state of the stock and instead to look at the average profit once the stock has settled down to a dynamic equilibrium of repeated cycles. Looking at present values makes it necessary to specify an initial stock, as the discounting procedure affects the optimal approach to equilibrium, and this approach also depends on the initial stock. The important question, in this context, is, however, whether discounting distorts the comparison between the two quota policies. We have done some trial runs for this, starting both policies from the same initial stock, and found that the relative difference between the two policies remains similar to the undiscounted case. Nevertheless it cannot be ruled out that discounting would reverse the order of profitability of the two policies, but to verify this would involve long and cumbersome calculations which we happily leave to others.

HANNESSON AND STEINSHAMN

TABLE I

Constant Price of Fish ($p = 1$) and Cost per Unit of Fishing Mortality ($c = 5000$)

s	4	5	6	7	8	9
	\multicolumn{6}{c}{Variable quotas and constant effort}					
F	0.075	0.082	0.088	0.093	0.096	0.096
NR	413.1	426.9	430.8	419.7	389.0	341.3
AC	788.1	836.9	870.8	884.7	869.0	821.3
C_{max}	838.1	912.6	971.0	1004.7	1005.1	968.8
C_{min}	738.1	761.2	770.7	764.8	732.9	673.8
S_{min}	7290	7420	7710	8161	8818	9583
COV	−0.757	−1.709	−3.126	−4.925	−6.942	−8.928
	\multicolumn{6}{c}{Constant quotas and variable effort}					
Q	790	840	870	880	860	810
NR	413.3	426.6	430.0	417.0	384.2	334.5
AF	0.075	0.083	0.088	0.093	0.095	0.095
F_{max}	0.080	0.090	0.099	0.107	0.112	0.114
F_{min}	0.070	0.075	0.078	0.081	0.081	0.079
S_{min}	7262	7330	7635	8142	8946	9841

Note. s = selectivity (age at first capture, knife edge selectivity); F = fishing mortality; NR = net revenue (profit); AC = average catch per year; C_{max} = maximum catch per year; C_{min} = minimum catch per year; S_{min} = minimum size of spawning stock; Q = annual catch quota; AF = average fishing mortality per year; F_{max} = maximum fishing mortality per year; F_{min} = minimum fishing mortality per year, COV = covariance of the amount caught and cost per unit caught.

TABLE II

Constant Price of Fish ($p = 1$) and Cost per Unit of Fishing Mortality ($c = 2000$)

s	4	5	6	7	8	9
	\multicolumn{6}{c}{Variable quotas and constant effort}					
F	0.123	0.138	0.156	0.174	0.190	0.202
NR	700.4	745.4	781.9	801.3	793.0	758.4
AC	946.4	1021.4	1093.9	1149.3	1173.0	1162.4
C_{max}	1029.0	1146.8	1266.1	1375.3	1452.8	1482.0
C_{min}	863.8	896.0	921.7	923.2	893.3	842.7
S_{min}	4709	4738	4828	5113	5677	6475
COV	−0.946	−2.190	−4.436	−7.774	−12.070	−16.848
	\multicolumn{6}{c}{Constant quotas and variable effort}					
Q	950	1030	1100	1150	1160	1150
NR	701.8	746.5	781.8	798.7	786.4	746.7
AF	0.124	0.142	0.159	0.176	0.187	0.202
F_{max}	0.135	0.161	0.188	0.216	0.237	0.264
F_{min}	0.114	0.125	0.134	0.143	0.146	0.152
S_{min}	4626	4535	4690	5088	6046	6977

Note. See note to Table I.

HOW TO SET CATCH QUOTAS 81

explained by increased variability in the unit cost of fishing caused by increasing the degree of selectivity. For a constant price, the difference in profitability between the constant effort and a constant catch equal to the expected catch of the constant effort is due to difference in the expected costs of the two regimes. That is,

$$E\pi(x; \bar{z}) - E\pi(z; \bar{x}) = -Ec^*(\bar{z})Ex - \text{Cov}(c^*, x) + Ec^*(\bar{x})Ex; \qquad \bar{x} = Ex,$$

where c^* denotes the cost per unit of fish caught.

Provided that $Ec^*(\bar{z}) = Ec^*(\bar{x})$, that is, the expected unit cost of fishing is equal for both strategies, the constant effort strategy is seen to be more profitable than the constant quota strategy if $\text{Cov}(c^*, x) < 0$. This is indeed the case; for constant effort, the catch will vary with the size of the fish stock, while the unit cost will vary inversely with the stock, total costs being constant. In Tables I and II the covariance of unit cost and catch is seen to increase in absolute value as the fishery becomes more selective. The fact that the constant quota strategy turns out to be slightly more profitable than the constant effort strategy in some cases, even if the said covariance is negative, is due to two phenomena. First, the optimal constant quota is sometimes greater than the expected catch under a constant effort strategy and hence yields a greater revenue.[4] Second, the average stock also is sometimes greater under the constant quota strategy than under the constant effort strategy, which implies $Ec^*(\bar{x}) < Ec^*(\bar{z})$. We encounter the first phenomenon again when we consider maximum sustainable yield under the two quota regimes.

Furthermore, it may be noted that the constant effort strategy becomes relatively more profitable as the cost of effort increases. This is clearly shown by Tables I and II, even if the differences are very small. This reflects the fact that the attraction of the constant effort strategy is due to its average fishing costs being lower than those under the constant quota strategy. This effect is stronger the higher the cost per unit of effort.

As mentioned in the Introduction, neither a fixed effort nor a constant annual catch need be the optimal fishing strategy. As shown in [3], the optimal fishing strategy in the particular type of model considered in this section involves "pulse fishing;" that is, no catch and no effort for a number of periods followed by an intense fishing pulse in one period. Optimizing fishing mortality over two recruitment cycles (16 years) confirmed this to be true here as well, under the weak constraints of a high upper value of F (2.0) and a final stock that is at least as large as the initial stock. As an example, for the selectivity $s = 4$ and $c = 2000$ (see Table II), the optimal pattern of F is $F = 0$ for all years, except $F = 0.449$ in year 1, $F = 1.538$ in year 8, and $F = 0.872$ in year 14. The net profit per year was increased from 700 to 886, which is considerably more than the difference between the fixed effort and the fixed catch strategies.[5] However, constraints as weak as those used in formulating this problem are hardly likely to be relevant in any real

---

[4]The optimal constant quota was found by varying the quota in intervals of 10 units (a unit is 1000 metric tonnes), except in Table III, with a precision of 0.005 units. The sensitivity of the objective function to a variation in the quota over an interval of 10 is slight, never exceeding one unit.

[5]Why the pulse fishing strategy is advantageous is explained in [3]. Briefly, this is due to two effects: (i) there are economies of scale with respect to biomass and effort, and (ii) pulse fishing compensates for unselective fishing. The figures quoted in the text refer to undiscounted profits. Discounting makes the pulses less intense and more frequent.

82   HANNESSON AND STEINSHAMN

world fishery, particularly not since the 200-mile limit became established. In practice decisionmakers are likely to require that some minimum catch·be taken from their national stocks in any particular year. This reduces the difference between the two strategies considered in this paper and the optimal strategy. The precise nature of the optimal strategy will vary according to how constraints on minimum annual catch, maximum effort, etc., are perceived in each particular case, and we do not pursue that matter further here.

Finally, it may be noted that the constant quota strategy is just about as biologically feasible as the constant effort strategy. Since a constant quota implies a pattern of fishing mortality that varies inversely with the stock, there is a potential danger that the mature stock might occasionally be severely depleted before it gets a chance to reproduce. In the cases shown in Tables I and II this does not occur. The minimum spawning stock is seen to be not very different under the two strategies. This is, however, the fortuitous outcome of the optimal fishing mortality being quite low in all the cases presented.[6] If, on the other hand, the cost per unit of fishing mortality is low, which could, for example, be caused by a negligible "stock effect" (that is, the size of the stock has a minimal effect on the catch for a given level of effort), things might turn out to be different. It could, for example, happen that the fishing mortality necessary to take the optimum constant catch in years when fish are scarce would be so high that most of the mature stock would be wiped out before it gets a chance to reproduce. This occurs in one case with a zero fishing cost, presented below.

Maximum Sustainable Yield under Constant Effort and Constant Quota

Consider now the optimal constant effort and the constant quota when the cost of fishing effort is zero. In the absence of discounting of the future the optimal constant effort will be the one which provides the maximum average catch of fish over the recruitment cycle. The optimal constant quota will be the maximum constant catch that can be maintained indefinitely. Note that this zero cost of effort case can be used to represent the more realistic case in which the catch is independent of the stock. Fishing costs will then be related only to the amount caught and can be taken into account directly in the price of fish.

Table III shows the optimal constant effort and the optimal constant quota. In five of the six cases considered the constant quota is in fact greater than the average catch resulting from a constant effort strategy. The variable fishing mortality implied by the constant quota strategy thus results in a more efficient utilization of the growth potential of the fish. The phenomenon that a variable rate of exploitation may result in a greater yield of fish over time was first noted by Pope [8].

---

[6]These fishing mortalities are very low, one-fifth to one-tenth of the estimated fishing mortalities in recent years. There are two reasons to caution against taking this uncritically as an advice on the management of the stock. First, the implied change in the average stock level would probably produce density-dependent effects, such as decreased individual growth and increased predation on juveniles, which would diminish the profitability of decreasing the rate of exploitation. Second, the relationship between fishing mortality and fishing effort is likely to be complex and non-proportional, such that the impact of the size of the stock on the catch per unit of effort will be less than proportional. Both of these effects imply a higher optimal fishing mortality than that shown in the tables accompanying the text. For a thorough discussion of fishing mortality and effort, see Arnason [1].

TABLE III

Constant Price of Fish ($p = 1$) but Zero Cost per Unit of Fishing Mortality ($c = 0$)

s	4	5	6	7	8	9
	Variable quotas and constant effort					
F	0.210	0.262	0.342	0.491	0.820	Infinite
NR	1013.2	1111.7	1215.6	1317.9	1401.6	1588.9
C_{max}	1152.8	1332.6	1588.9	1910.0	2237.7	2950.8
C_{min}	873.6	890.7	842.2	725.7	565.4	227.0
S_{min}	2254	1938	1617	1266	1100	836
	Constant quotas and variable effort					
Q	1018	1118	1224	1325	1422	1464
AF	0.209	0.255	0.351	3.973	4.530	4.259
F_{max}	0.240	0.315	0.487	22.855	26.728	26.772
F_{min}	0.181	0.206	0.252	0.422	0.408	0.391
S_{min}	2227	1936	1433	0	1433	2953

Note. See note to Table I. Note that AC = NR or Q = NR, and $F = F_{msy}$.

One of the cases shown in Table III clearly illustrates the dangers inherent in a constant quota strategy when the quota is close to the maximum yield potential of the stock. In one case the year classes are fished out before they reach maturity. In the other cases there is not any major difference between the two strategies on this account; the minimum spawning stock is similar under both strategies. It may be noted that for a high degree of selectivity the fluctuations in annual catch become extremely violent in the constant effort case; when the age at first capture is 9 the optimum fishing mortality becomes infinite and the largest annual catch is more than 10 times greater than the smallest.

Impact of Amplitude and Length of Recruitment Cycles

Next, let us consider briefly the impact of changing the amplitude and the length of the recruitment cycle. Changing first the value of v to 0.2 lengthens the recruitment cycle to 10 years. The two strategies are compared in Table IV. The profitability of the constant effort strategy is the same as before while the profitability of the constant quota strategy is decreased slightly. The reason for this is that the variability of the stock is increased. This is reflected in a greater covariance of unit cost and catch. As discussed above, this makes the constant effort strategy relatively more attractive. Again it may be noted that the minimum size of the spawning stock is almost the same under both strategies.

Reducing the amplitude of the recruitment cycle is accomplished by changing the parameter u. Lowering u from 600 to 300 and increasing R_{min} to 400 reduces the amplitude of the recruitment cycle by one-half while the average recruitment is the same as before. The result of this is reported in Table V. The profitability of the constant effort strategy is the same as before, while the profitability of the constant quota strategy is increased and is virtually the same as that of the constant effort strategy. This is as expected; the reduced amplitude of the fluctuations in recruitment moderates the fluctuations in the unit cost, which is reflected in smaller covariance terms in Table V. It may be noted that this result is probably more typical of demersal, multi-year class stocks than the ones produced so far, as

TABLE IV

Constant Price of Fish ($p = 1$) and Cost per Unit of Fishing Mortality ($c = 5000$):
10-Year Recruitment Cycle

s	4	5	6	7	8	9
	Variable quotas and constant effort					
F	0.075	0.082	0.088	0.093	0.096	0.096
NR	413.1	426.9	430.8	419.7	389.0	341.3
AC	788.1	836.9	870.8	884.7	869.0	821.3
C_{max}	852.7	935.0	1002.7	1046.3	1052.3	1022.9
C_{min}	723.6	738.8	738.9	723.2	685.6	619.8
S_{min}	7415	6941	7210	7631	8239	8927
COV	−1.310	−2.848	−5.204	−8.406	−12.248	−16.120
	Constant quotas and variable effort					
Q	790	840	870	880	850	790
NR	413.1	426.0	428.2	414.1	379.5	327.6
AF	0.075	0.083	0.088	0.093	0.094	0.092
F_{max}	0.081	0.093	0.099	0.113	0.118	0.119
F_{min}	0.069	0.073	0.075	0.077	0.075	0.071
S_{min}	7354	6804	7086	7568	8428	9358

Note. See note to Table I.

the recruitment fluctuations of the Arcto-Norwegian cod stock seem to be exceptionally great.

A Concave Revenue Function

So far the difference in the profitability of a constant effort and that of a constant quota has turned out to be very small. In the previous section it was noted

TABLE V

Constant Price of Fish ($p = 1$) and Cost per Unit of Fishing Mortality ($c = 5000$): Reduced Amplitude of
Recruitment Cycle by 50%, with Average Recruitment Unchanged

s	4	5	6	7	8	9
	Variable quotas and constant effort					
F	0.075	0.082	0.088	0.093	0.096	0.096
NR	413.1	426.9	430.8	419.7	389.0	341.3
AC	788.1	836.9	870.8	884.7	869.0	821.3
C_{max}	813.1	874.7	920.9	944.7	937.0	895.1
C_{min}	763.1	799.0	820.8	824.8	800.9	747.6
S_{min}	8226	8047	8387	8901	9634	10,468
COV	−0.189	−0.425	−0.775	−1.217	−1.708	−2.186
	Constant quotas and variable effort					
Q	790	840	870	890	870	820
NR	413.2	426.8	430.5	419.0	387.8	339.6
AF	0.075	0.083	0.088	0.094	0.096	0.096
F_{max}	0.077	0.086	0.093	0.101	0.105	0.105
F_{min}	0.073	0.078	0.082	0.088	0.088	0.087
S_{min}	7841	7971	8341	8803	9638	10,554

Note. See note to Table I.

HOW TO SET CATCH QUOTAS 85

TABLE VI

Constant Elasticity of Demand (2.0), with $p = 1$ for Catch = 1000 Units:
Constant Cost per Unit of Fishing Mortality ($c = 2000$)

s	4	5	6	7	8	9
		Variable quotas and constant effort				
F	0.091	0.099	0.106	0.113	0.118	0.123
NR	743.7	755.9	762.1	759.5	745.4	720.8
AC	857.5	911.2	950.9	974.3	967.3	940.2
C_{max}	900.6	1022.2	1070.7	1118.8	1136.7	1131.7
C_{min}	796.7	820.2	831.1	829.8	797.9	748.7
S_{min}	7867	6452	6779	7226	7957	8599
		Constant quotas and variable effort				
Q	860	910	950	970	960	930
NR	744.3	756.6	762.6	759.6	744.8	719.3
AF	0.092	0.099	0.106	0.113	0.118	0.123
F_{max}	0.098	0.106	0.121	0.132	0.141	0.150
F_{min}	0.085	0.089	0.093	0.096	0.097	0.099
S_{min}	6227	6419	6716	7220	8053	8895

Note. See note to Table I.

that a concave revenue function makes a constant quota relatively more attractive. How does this affect the relative profitability of the two strategies?

Consider, first, the case with a demand function with a constant elasticity, that is,

$$R = x^{1-1/\epsilon}.$$

The maximum relative difference between the expected revenue of a variable quota and the constant revenue of a fixed quota $\bar{x} = Ex$ will occur when the relative curvature of the revenue function at $\bar{x}$ is as great as possible. This will occur for ϵ that maximizes $R''(\bar{x})/R(\bar{x}) = -\epsilon^{-1}(1 - \epsilon^{-1})\bar{x}^{-2}$, which implies $\epsilon = 2$. Using this elasticity of demand with the low cost of effort produced the results shown in Table VI. This is clearly not enough to make a constant annual quota any more attractive than a constant effort; the profit yielded by the two strategies is virtually the same in all cases, even if the concavity of the revenue function has made the constant quota strategy slightly better (see Table II). The impact on optimal fishing effort and landings is on the other hand quite substantial. The low elasticity of demand makes it profitable to reduce landings by 10–20% in order to get a higher price, and the optimal fishing effort (mortality) is reduced further still, or by 25–40% (see Table II). The results for the high cost of effort case are similar and will not be reproduced here; in this case the constant quota strategy typically results in a slightly lesser profit than the constant effort strategy.

From the above it is clear that a sharper concavity of the revenue function than that produced by a constant elasticity demand function will be necessary to make the constant quota strategy more profitable than the constant effort strategy. Such sharp concavity would, for example, result from a raw fish price that falls sharply if landings exceed a certain level. That seems quite possible; catches in excess of what the processing industry could accommodate with normal capacity utilization

TABLE VII

Constant Cost per Unit of Fishing Mortality ($c = 5000$)

s	4	5	6	7	8	9
	Variable quotas and constant effort					
F	0.067	0.072	0.074	0.075	0.076	0.074
NR	410.0	421.8	422.5	408.7	376.4	329.0
AC	745.0	782.9	793.5	783.9	757.5	699.1
C_{max}	789.5	849.6	878.2	881.2	864.4	811.3
C_{min}	700.5	716.3	708.8	686.5	712.7	586.9
S_{min}	7855	8056	8532	9130	9790	10,539
	Constant quotas and variable effort					
Q	790	840	870	880	860	810
NR	413.3	426.6	430.0	417.0	384.2	334.5
AF	0.075	0.083	0.088	0.093	0.095	0.095
F_{max}	0.080	0.090	0.099	0.107	0.112	0.114
F_{min}	0.070	0.075	0.078	0.081	0.081	0.079
S_{min}	7262	7330	7635	8142	8946	9841

Note. Kinked revenue curve ($R = x$ for $x < Q$, $R = Q$ for $x > Q$). Otherwise variables are defined as in the note to Table I.

are likely to lead to progressively increasing unit costs, through overtime, work in shifts, etc., or to make it necessary to resort to less profitable products and markets. This would reduce the ability of the processing industry to pay for additional supplies of raw fish. Lesser landings would similarly introduce spare capacity, but the ability of the industry to compete for the raw fish and to pay a higher price need not increase much as the landings fall; it would not if the product were sold at a constant price and the cost savings per unit reduction in output were constant.

In Table VII an extreme case of this kind is shown. Here the price of fish in excess of the level corresponding to the optimum constant quota is zero, but is constant below that level. This is, of course, not very realistic and only serves the purpose of giving an idea of how great the difference in profitability between the two quota strategies might possibly be. For the high cost case shown in Table VII the difference is in fact only 1–2%. In Table VIII we show the results produced in the zero cost case. This case is of interest because it is the most favorable one for the constant quota strategy, for any given assumption about the relation between the net price of fish and quantity landed, and corresponds to the case in which the catch depends on the effort only and not on the stock. In this case the annual net revenue of the constant quota strategy is up to 20% greater than the average annual net revenue of the constant effort strategy, but the difference varies with the degree of selectivity of the fishery and is greatest for the most selective fishing profile ($s = 9$).

Asymmetry in the Cost of Effort

When the catch per unit of effort depends on the size of the exploited stock, stability in catches is achieved through variations in fishing effort. Variability of effort is no less likely to be economically disadvantageous than variability of catch.

TABLE VIII
Zero Cost per Unit of Fishing Mortality ($c = 0$)

s	4	5	6	7	8	9
		Variable quotas and constant effort				
F	0.176	0.200	0.228	0.262	0.323	0.380
NR	973.9	1047.0	1179.9	1179.9	1221.9	1227.5
AC	1005.9	1095.9	1184.3	1261.9	1328.5	1360.9
C_{max}	1123.8	1272.4	1432.3	1606.9	1797.8	1931.3
C_{min}	887.9	919.4	936.3	917.0	859.2	790.7
S_{min}	2983	2987	3076	3273	3416	3878
		Constant quotas and variable effort				
Q	1018	1118	1224	1325	1422	1464
AF	0.209	0.255	0.351	3.973	4.530	4.259
F_{max}	0.240	0.315	0.487	22.855	26.728	26.772
F_{min}	0.181	0.206	0.252	0.422	0.408	0.391
S_{min}	2227	1936	1433	0	1433	2953

Note. Kinked revenue curve ($R = x$ for $x < Q$, $R = Q$ for $x > Q$). Otherwise, variables are defined as in the note to Table I.

If stability at one level is to be achieved by destabilizing the other, it is not clear a priori which is to be preferred.

In order to attain some insights into the importance of this, we assumed that the inconvenience of varying effort takes the form of asymmetry in costs, such that the cost of expanding effort from some reference level is higher than the cost saved through decreasing effort from the same level. The rationale behind this is obvious enough; some part of the cost of effort is fixed cost which cannot be saved by decreasing effort temporarily, while expanding effort temporarily beyond normal utilization of capacity is likely to generate some cost over and above that normally incurred. Here we model this by imposing a cost penalty per unit of effort in excess of some reference level, while letting cost savings for reducing effort below that level amount to less than the full cost per unit of effort. In symbols,

$$C(z) = cz^* + kc(z - z^*), \qquad z \geq z^*,$$
$$C(z) = cz^* - kc(z^* - z), \qquad z \leq z^*,$$

where k is some fraction of the cost per unit of effort (c). Other ways of modeling this are possible, such as assuming that a higher cost is incurred by expanding effort from its previous level than that saved by contracting it from that level. While this is perhaps more reasonable, we chose the former for its simplicity. It will, at any rate, give some idea of what such cost asymmetry means in this model.

To make this a meaningful inquiry we must study a case where the constant quota strategy has some advantage over the constant effort strategy. At the same time, the unit cost of fishing must depend on the stock, so that the constant effort strategy is potentially better. Such a case is provided above, where the value of landings in excess of some reference level is zero (Table VII). In Table IX we show the results of setting z^* equal to the level of effort needed on the average to take the constant quota, with $k = 0.2$. This is enough to erode the profitability advantage of the constant quota strategy. It may be noted that the cost penalty of varying

HANNESSON AND STEINSHAMN

TABLE IX

Cost Penalties for Varying Fishing Effort (Mortality)

s	4	5	6	7	8	9
	Variable quotas and constant effort					
F	0.067	0.072	0.074	0.075	0.076	0.074
NR	410.0	421.8	422.5	408.7	376.4	329.0
Q^*	790	840	870	880	860	810
AC	745.0	782.9	793.5	783.9	757.5	699.1
C_{max}	789.5	849.6	878.2	881.2	864.4	811.3
C_{min}	700.5	716.3	708.8	686.5	712.7	586.9
S_{min}	7855	8056	8532	9130	9790	10,539
	Constant quotas and variable effort					
Q	790	840	870	870	850	790
NR	410.3	422.0	422.9	408.4	373.9	323.0
F^*	0.075	0.083	0.088	0.093	0.095	0.095
AF	0.075	0.083	0.088	0.091	0.093	0.091
F_{max}	0.080	0.090	0.099	0.104	0.109	0.109
F_{min}	0.070	0.075	0.078	0.078	0.085	0.076
S_{min}	7262	7330	7635	8240	9033	9994

Note. $c = 5000$, with an additional cost of 1000 per unit for $F - F^*$, and fixed cost of 4000 per unit for $F^* - F$. Kinked revenue curve ($R = x$ for $x < Q^*$, $R = Q^*$ for $x > Q$). Otherwise, variables are defined as in the note to Table I.

effort makes it worthwhile, for some selectivity profiles, to lower the constant annual quota from what it would otherwise be.

4. CONCLUSIONS

In accordance with the theoretical results in Section 2, the results of simulating the fishing of the Arcto-Norwegian cod show that the relative profitability of constant quotas versus constant effort depends on the shape of the revenue and the cost functions and whether the abundance of fish affects the catch for any given level of effort. If the catch depends on the effort only and not on the stock, a constant catch will be preferable to a constant effort, provided either the cost or the revenue function (or both) is non-linear and shaped as is usually assumed in economic theory.

It is, however, quite unlikely that the catch will not be at all affected by the abundance of fish. In the simulations above we have assumed what appears to be maximum stock-dependence in this respect; that is, the catch per unit of effort increases linearly with the size of the fish stock. This enhances the relative profitability of a constant effort. The difference between the constant effort and the constant catch quota strategies is nevertheless minimal in this case, for linear revenue and cost of effort functions. The assumptions about the revenue and the cost of effort functions needed to make the constant catch quota strategy more profitable in this case were rather extreme and the resulting difference in profitability small.

The general conclusion to emerge from this is, first, that the most important determinant of which strategy is better is the strength of the stock effect in the production function; that is, how sensitive the catch per unit of effort is to changes in the stock level. If there is only a weak link between the abundance of fish and the catch per unit of effort, the constant quota strategy seems likely to be preferable, particularly because of non-linearities in cost and revenue. The difference in profitability between these strategies seems likely to be small, however.

Second, the wisdom of setting catch quotas on the basis of some reference level of fishing mortality is certainly not beyond doubt. From the economic point of view this strategy is not necessarily better than a constant quota strategy, and from the biological point of view it may not be much less risky. For lightly exploited stocks consisting of many year classes, the minimum spawning stock will not necessarily be much smaller under a constant quota strategy than under a constant effort (fishing mortality) regime.

This leads us to conclude that the best fishing strategy will probably be neither of the two, but an adjustment of catch quotas that does not blindly follow from the maintaining of a constant fishing mortality and varies over time as economic and other circumstances dictate, provided the future recruitment to the stock is not endangered. This is particularly likely to be the case if investment and disinvestment in non-malleable capital and how they could be adapted to changes in the abundance of fish are taken into account.

APPENDIX 1

The data used in the model were taken from the reports of the Arctic fisheries working group of ICES (The International Council for the Exploration of the Sea). The weight at each age is as follows, starting with age group 4 (weight in kilograms):

    1.00, 1.55, 2.35, 3.45, 4.70, 6.17, 7.70, 9.25, 10.85, 12.50, 13.90, 15.00.

Fish older than 15 years are assumed to weigh 15 kg.

A natural mortality of 0.2 is assumed. Selectivity is assumed to be knife edge; that is, an age group is assumed to be either fully vulnerable to the fishing gear or not at all vulnerable. The mature part of the stock is assumed to consist of fish 8 years and older.

A standard Beverton–Holt model with constant weight at age and constant fishing mortality throughout the year was used in the simulations. This model is documented elsewhere (see, for example, [2] or [4]). The annual catch for a constant effort (fishing mortality) is readily calculated over one recruitment cycle. Finding the annual necessary effort (fishing mortality) to take a constant annual catch is more complicated. It is not enough to consider only one recruitment cycle, as it may take a long time for the stock to attain a dynamic equilibrium of identical, repetitive cycles. The results reported in the text are the ones emerging after the stock has settled down to a dynamic equilibrium. Because of the large number of year classes needed to conduct simulations with an acceptable degree of precision (we have used 30 year classes), it takes a long time for the fluctuations in these variables to settle down to identical, repetitive cycles (hundreds of years). Whether the model converges to the dynamic equilibrium with the maximum constant quota depends on the starting position (or the adjustment path); if the initial stock is too small or the initial fishing mortality too high, the maximum sustainable quota cannot be attained.

APPENDIX 2

In this appendix we analyze the power spectrum of the recruitment time series data in Fig. 4. This analysis takes place in the frequency domain instead of in the time domain; that is, interest is here centered on the contributions made by the various periodic components in the series which we call λ. $\lambda = v\tau$ is the frequency in radians and $T_r = 2\pi/\lambda$ is the recruitment period as before.

The power spectrum of λ is often denoted by $f(\lambda)$ and is symmetric about $\lambda = 0$. $\lambda \in [-\pi, \pi]$, but because of the symmetry all relevant information from the power spectrum is contained in the range $[0, \pi]$.

The power spectrum of any indeterministic process of the form

$$y_t = \sum_{j=0}^{\infty} \psi_j \epsilon_{t-j} \tag{A1}$$

can be defined by the continuous function

$$f(\lambda) = (2\pi)^{-1} \left[\gamma(0) + 2 \sum_{\tau=1}^{\infty} \gamma(\tau) \cos \lambda\tau \right]. \tag{A2}$$

ψ and ϵ_t in (A1) are fixed parameters and white noise, respectively; i.e., ϵ_t is a sequence of independent, identically distributed random variables with constant mean and variance. In (A2) γ denotes the autocovariance. For the mathematical rationale behind formula (A2), see, e.g., Harvey [6].

The area under the power spectrum in the range $[-\pi, \pi]$ is equal to the variance σ^2. If the time series consists of a white noise process, all λ's will have the same power and theoretically the power spectrum will be rectangular at $f(\lambda) = \sigma^2/2\pi$.

An estimate of the power spectrum, i.e., the sample spectral density, is given by the equation

$$I(\lambda) = (2\pi)^{-1} \left[k(0)c(0) + 2 \sum_{\tau=1}^{T-1} k(\tau)c(\tau) \cos \lambda\tau \right], \tag{A3}$$

where T is the number of observations and $c(\ )$ denotes the sample autocovariances. In our case $T = 42$. Here we implicitly assume that the recruitment series is real and not complex.

We have estimated the spectral density (A3) at the frequencies

$$\lambda_j = 2\pi j/T, \quad j = 0, 1, \ldots, T/2.$$

The $k(\tau)$ is a weighting function or a lag window in the time domain. Some weighting is necessary since choosing a unit weighting function, $k(\tau) = 1 \; \forall \; \tau$, will have undesirable properties, as it causes the resulting estimator to lack statistical consistency. A number of weighting functions have been suggested, and we have chosen one of the most widely employed, the Tukey–Hanning lag window,

$$k_M(\tau) = [1 + \cos(\pi\tau/M)]/2, \quad \tau < M,$$
$$k_M(\tau) = 0, \quad \tau \geq M,$$

with lag value $M = 6$.

If the spectrum changes slowly over the bandwidth we consider our spectrum averages to be representative of the spectrum and the averages are said to be well resolved. The bandwidth is the width of the window in radians. There is a trade-off

HOW TO SET CATCH QUOTAS   91

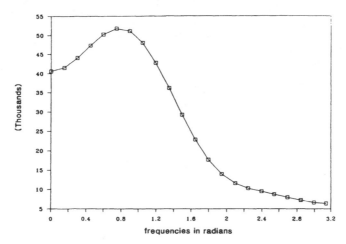

FIG. 5.   Sample spectral density for recruitment to the Arcto-Norwegian stock (cf. Fig. 4).

between resolution and variance; more lags improve the resolution but increase the variance.

The graph of the spectral density distribution is shown in Fig. 5. The density has a maximum at $\lambda_s = 0.748$, suggesting a recruitment period of 8.4 years. From this we can conclude that there is a relatively strong degree of periodicity in the recruitment time series, and that an irregular cycle of about 8 years seems to be likely.

ACKNOWLEDGMENTS

We are grateful to Trond Bjørndal, Norwegian School of Economics and Business Administration; John Conrad, Cornell University; and Ola Flaaten, University of Tromsø, for comments on an earlier draft. The research was supported by the Norwegian Fisheries Research Council.

REFERENCES

1. R. Arnason, "Efficient Harvesting of Fish Stocks: The Case of the Icelandic Demersal Fisheries," PhD thesis, University of British Columbia (1984).
2. R. J. H. Beverton and S. J. Holt, On the dynamics of exploited fish populations, *in* "Fishery Investigations," Series II, Vol. 19, Her Majesty's Stationery Office, London (1957).
3. R. Hannesson, Fishery dynamics. A North Atlantic cod fishery, *Canad. J. Econom.* **8**, 151–173 (1975).
4. R. Hannesson, "Economics of Fisheries," Universitetsforlaget, Bergen (1978).
5. R. Hannesson, Fixed or variable catch quotas? The importance of population dynamics and stock dependent costs, *in* "Rights Based Fishing" (P. Neher, R. Arnason, and N. Mollett, Eds.), Kluwer Academic Press, Dordrecht (1989).
6. A. C. Harvey, "Time Series Models," Oxford Univ. Press, Oxford (1981).
7. M. G. Larraneta and A. Vazquez, On a possible meaning of the polar motion in the ecology of the North Atlantic cod (*Gadus morhua*), International Council for the Exploration of the Sea, C.M. 1982/G:14 (1982).
8. J. G. Pope, An investigation into the effects of variable rates of the exploitation of fishery resources, *in* "The Mathematical Theory of the Dynamics of Biological Populations" (M. S. Bartlett and R. W. Hiorns, Eds.), pp. 23–34, Academic Press, London (1973).

[29]

Ecological Applications, 8(1) Supplement, 1998, pp. S72–S78
© 1998 by the Ecological Society of America

IMPLEMENTING THE PRECAUTIONARY PRINCIPLE IN FISHERIES MANAGEMENT THROUGH MARINE RESERVES

Tim Lauck,[1] Colin W. Clark,[1,4] Marc Mangel,[2] and Gordon R. Munro[3]

[1]Department of Mathematics, University of British Columbia, Vancouver, British Columbia, Canada V6T 1Z2
[2]Environmental Studies Board, University of California, Santa Cruz, California 95064 USA
[3]Department of Economics, University of British Columbia, Vancouver, British Columbia, Canada V6T 1Z1

Abstract. Overexploitation of marine fisheries remains a serious problem worldwide, even for many fisheries that have been intensively managed by coastal nations. Many factors have contributed to these system failures. Here we discuss the implications of persistent, irreducible scientific uncertainty pertaining to marine ecosystems. When combined with typical levels of uncontrollability of catches and incidental mortality, this uncertainty probably implies that traditional approaches to fisheries management will be persistently unsuccessful. We propose the use of large-scale protected areas (marine reserves) as major components of future management programs. Protected areas can serve as a hedge against inevitable management limitations, thus greatly enhancing the long-term sustainable exploitation of fishery resources. Marine reserves would also provide an escape from the need of ever more detailed and expensive stock assessments and would be invaluable in the rehabilitation of depleted stocks.

Key words: bet hedging; controlling overexploitation; diversification; fisheries; irreducible scientific uncertainty; marine protected areas; marine reserves; precautionary principle in fisheries; risk aversion.

Introduction

In 1982, the United Nations (U.N.) Third Conference on the Law of the Sea closed and formally ushered in a new era in world fisheries management through the Law of the Sea Convention (United Nations 1982). Under the Convention, coastal states throughout the world were enabled to extend their management jurisdiction over fishery resources from 12 to 200 (22.2 to 370.4 km) nautical miles. It was estimated that 90% of the harvests of marine fishery resources would be accounted for by resources encompassed by the coastal state 200-mile (370.4-km) zones—Exclusive Economic Zones (EEZs; Kaitala and Munro 1995). The objective in establishing the EEZ regimes was to enhance the conservation and economic management of world marine fishery resources. In 1980, a Food and Agricultural Organization of the United Nations (FAO) publication, in anticipation of the EEZ regime, stated that: "the opportunity exists, as never before, for the rational exploitation of marine fisheries.... The 1980s provide the threshold for a new era in the enjoyment of the ocean's wealth in fisheries" (cited in United Nations Food and Agriculture Organization 1992).

The hopes and expectations of the early 1980s have

Manuscript received 20 February 1996; accepted 15 December 1996; final version received 13 February 1997. For reprints of this Special Issue, see footnote 1, p. S1.
[4] Address correspondence to this author.

not been realized. The same FAO recently reported that "69% of the world's marine [fish] stocks . . . are either fully to heavily exploited, overexploited, depleted . . . and therefore are in need of urgent conservation and management measures" (United Nations Food and Agriculture Organization 1995). Coastal state fishery management programs have proven, in far too many instances, to be seriously deficient.

One of the most dramatic and depressing examples of fishery management failure under the EEZ regime is provided by the large and extremely productive groundfish resources on the famous Grand Banks of Newfoundland, which constituted Canada's main bonanza under the EEZ regime. These resources had been overexploited while international common property. Under conservative Canadian management, it was hoped that fish stocks would be rebuilt, to the benefit of the Canadian fishing industry. The single most important of these resources, a cod stock complex extending from southern Labrador to southeastern Newfoundland, popularly known as Northern cod (*Gadus morhua*), was expected to yield sustainable annual harvests of 4×10^8 kg by the late 1980s (Canada 1983).

These sustainable harvests were not achieved. In the late 1980s, the Canadian government introduced drastic cuts in the Northern cod total allowable catches (TACs). The drastic TAC cuts were not enough. In 1992, the Canadian authorities felt compelled to impose a temporary 2-yr harvest moratorium on Northern cod.

The authorities were horrified to find that the resource continued to decline after the moratorium had been imposed. The harvest moratorium still remains in place (in 1996) and has ceased to be temporary. It is now indefinite. To compound the misery, the Canadian authorities have had to impose harvest moratoria on several neighboring groundfish stocks.

The causes of the fishery resource collapse off Atlantic Canada are now the subject of an intense debate (Myers et al. 1997). What is clear is that the collapse came as a stunning shock to the authorities. One commentator remarked that the resource collapse would have had no credibility as a worst-case scenario, even a few years prior to the imposition of the moratorium (Roy 1996). What is equally clear is that the management of even seemingly stable fishery resources, such as groundfish, is subject to a far greater degree of uncertainty than heretofore had been realized and appreciated (Gordon and Munro 1996).

STANDARD PRESCRIPTIONS FOR CORRECTING MANAGEMENT FAILURES

Suggestions for improving the management of marine fisheries have not been in short supply. We will not review here the long history of discussion of the "problem of overfishing," but will concentrate instead on the implications of uncertainty in fisheries management.

We take as an underlying assumption that fishery declines and collapses result in large part from overfishing, that is to say, from a level of fishing intensity that is excessive in terms of maintaining a sustainable population and fishery. We nevertheless recognize that changes in the marine environment are also often involved in the decline or collapse of any particular fishery. Levels of catch that may be sustainable under normal or favorable environmental conditions may prove not to be sustainable under abnormal conditions. Many fish populations that have suddenly collapsed under intensive exploitation had presumably persisted for thousands of years in spite of fluctuations in the marine environment. The parsimonious assumption is, therefore, that fishing decreased the resilience of these populations, rendering them more vulnerable to environmental change. From our perspective, this still constitutes overfishing.

Environmental fluctuations are but one of many sources of major uncertainty in fisheries. It is now widely accepted that management must somehow allow for uncertainty and potential inaccuracy in projected sustainable catch levels. It is our contention in this paper, however, that the full implications of uncertainty have not been recognized in the design and implementation of fisheries management strategies. This shortcoming, we believe, has been a major factor in the decline and collapse of many fisheries.

It is often suggested that uncertainty could be reduced if more research were to be undertaken. For example, increased stock assessment activity should keep management informed as to the current population size and its changes over time. While no one can dispute the need for stock assessment, it must be recognized that it is often very costly and that the estimates of stock abundance are almost always subject to considerable uncertainty. Many fish populations have become severely depleted before clear signals have appeared in the stock estimates. Fish stock assessment is now a highly developed and sophisticated science, but it is doubtful whether the levels of uncertainty can be greatly reduced by further refinements of technique. Fisheries science can and doubtlessly will continue to improve, but management decisions must depend on currently available methodology.

Accepting the inevitability of errors in sustainable catch estimates, fishery biologists have recently adopted management criteria that presumably err on the side of caution, a common example being the so-called $F_{0.1}$ criterion widely used in Canada and elsewhere. ($F_{0.1}$ is defined as the level of fishing mortality F at which the slope of the yield-per-recruit curve equals 0.1 times the slope at $F = 0$.) This criterion is more conservative than the maximum sustained yield criterion formerly favored, and as such allows for some degree of error.

But is the $F_{0.1}$ criterion sufficiently conservative? Walters and Pearse (1995) calculated that the $F_{0.1}$ value used in the Northern cod fishery, namely $F_{0.1} = 0.2$, would have to be reduced by 50%, to $F = 0.1$, in order to incorporate even a moderate degree of risk aversion in that fishery. ($F = 0.1$ means that 10% of the stock would be taken each year.)

Recommendations of this kind are based on the principle of erring on the side of caution, whether by maintaining catch levels or fishing mortality below estimated Maximum Standard Yield (MSY) levels, or maintaining stocks above the estimated MSY level (Roughgarden and Smith 1996). In essence these prescriptions are equivalent. Provided that such objectives can be reliably achieved in practice over the long term, sustainable fisheries will result. The question then becomes one of method and degree: how great a safety margin should be allowed, and which methods of management are most likely to achieve the objective of sustainable fisheries? For example, suppose that in a certain fishery stock estimates are considered valid to within ±30%, that annual productivity varies unpredictably over a range of ±50% from the mean, and that fishing plus incidental mortality varies within ±25% of the TAC. In the worst-case scenario, stocks are 30% below the mean estimate, productivity is 50% below the mean, and fishery-induced mortality is 25% greater than the TAC. To ensure sustainability, the TAC should then be set at 28% of the mean estimated value (this does not allow for any stock rehabilitation in the event that the stock is, in fact, below the mean estimate).

S74 TIM LAUCK ET AL. Ecological Applications
Special Issue

Such a safety margin may seem extreme. Indeed, the industry might argue in favor of the best-case scenario, with a TAC equal to 244% of the mean estimate. From this point of view, the original mean estimate TAC doesn't seem so bad, yet it is undeniably fraught with risk. This fanciful, but perhaps not quantitatively unrealistic illustration, raises several interesting issues. What levels of uncertainty exist in particular fisheries? How much can these uncertainties be reduced by additional research, or by tighter control of fishing operations? What is an appropriate safety margin? Will episodes of overfishing and underfishing balance out over the long run? Do estimation errors tend to be unbiased, in retrospect, or are worst cases more common than best cases? We know of no literature addressing such questions, which seem fundamental for the transition to sustainable fisheries when managed by traditional methods based on catch or effort quotas.

UNCERTAINTY AND UNKNOWABILITY IN COMPLEX SYSTEMS

In this era of scientific wonders it is hard to avoid the "world view" of science as being ultimately capable of fully revealing and understanding the complexities of nature. This view is encountered frequently in fisheries in terms of recommendations for more scientific research into the functioning of marine ecosystems. Thus, we are repeatedly admonished to graduate from single-species fisheries models to multispecies or full-ecosystem models, presumably represented as computer code. That the data requirements needed to validate any such model are vastly beyond our current capacity is seen only as the result of insufficient research funding.

An alternative, and we believe much more realistic view, is that there are limitations, both practical and theoretical, to what science can accomplish (Mangel et al. 1996). Full understanding and predictability of anything as complex (and, we should add, as unobservable) as a marine ecosystem will forever remain a chimera. The implications seem obvious. Progress in fisheries management will now proceed most rapidly, not from vastly increased research effort in marine biology, but from research into ways to deal with this irreducible uncertainty, or as it might be called, unknowability. This is a topic that has hardly ever been studied in the fisheries literature, to our knowledge, so that progress might be quite rapid.

Fisheries managers (whether individuals or committees) are regularly faced with the problem of setting catch quotas on the basis of current information. They may be quite aware of the fact that this information is incomplete, so that sustainable catch levels cannot be determined with a high degree of certainty. But how are the managers to take this uncertainty into account? Should they simply ignore it and base quotas on the "best scientific estimates" currently available? Our perception is that most management decisions are made in this way—and with good reason. Any admission of uncertainty only encourages the fishing industry to demand quotas at the upper limit of the confidence interval, on the grounds that science has not "proved" that lower quotas are necessary.

This approach would perhaps be workable if the system were self correcting, in the sense that excessive quotas in one year would have immediately detectable effects on the fish population, leading to reduced quotas in the next year. The truth is that overfishing, unless it is extreme, often takes years to detect. Moderate overfishing may lower the resilience of the population, but the impending collapse cannot be predicted from the available data. Also, reductions in quotas are always politically difficult to achieve, especially given the all but universal tendency towards overcapitalization in commercial fisheries.

In addition to these biases, actual fishing mortality often greatly exceeds the targeted level, from a variety of causes including unreported catches, by-catches, discards of small fish, and incidental mortality. Moreover, the productivity of marine ecosystems may be disrupted to an unknown extent as the result of habitat damage by fishing gear, or from pollution, as well as from the capture of species that serve as food for other commercial species. Little if any of this incidental impact is quantifiable in any scientific sense.

Given all these sources of uncertainty, error, and bias, is it any wonder that valuable fish populations continue to disappear at an alarming rate? What, if anything, can be done to reverse the trend?

BIOLOGICAL AND ECONOMIC RESPONSES TO UNCERTAINTY: BET HEDGING

Both the world of biology and that of economics possess a variety of techniques for dealing with uncertainty. Of particular interest in the fisheries context is the use of "bet hedging" strategies in biology and economics.

Bet hedging is a form of diversification of activities, having the purpose of reducing risk through pooling or averaging of (at least partially) independent random events. In biology, various types of reproductive strategies are thought to constitute bet hedging in uncertain environments (Seger and Brockmann 1987, Yoshimura and Clark 1993). Examples include multiple episodes of reproduction (iteroparity), dispersal of progeny, and delayed germination of seeds. At the population level, metapopulation structures may increase the chances that a species will survive in spite of local extinctions (Pulliam 1988).

In the financial world, bet hedging can be observed in the common practice of portfolio diversification, and also in the purchase of accident and liability insurance. Both of these practices serve to reduce the risk of a severe loss of financial assets. Bet hedging is usually thought to involve a cost, or "premium," in terms of a decrease in expected benefits, which is accepted in order to achieve a reduction in risk.

BET HEDGING IN FISHERIES: PROTECTED MARINE RESERVES

How can fisheries management strategies be redesigned so as to include a bet hedging component? The risk that is to be avoided, of course, is a collapse or severe decline in the fish population as a result of overfishing.

The current "world view" of fisheries management is that every commercially valuable stock should be exploited at the optimal level. Given the large uncertainties and biases of management, overfishing of every stock seems almost predetermined. This practice, clearly the opposite of bet hedging, suggests what a bet-hedging management strategy would consist of: different stocks, or substocks, would be managed in different ways.

The simplest way to diversify the management of a given fishery resource would be to exploit part of the resource while protecting the remainder. We therefore propose that Protected Marine Reserves (sometimes called Marine Protected Areas, or "no-take" areas; Shackell and Willison 1995) should become an integral component in the management of all marine fisheries. The actual design and implementation of marine reserves would depend on what is known about the biological characteristics of each particular species or species complex. For the purposes of discussion, we will here consider the case of a demersal species inhabiting a large area of the ocean floor. The design of marine reserves for highly migratory species will obviously involve additional complications.

Desirable features of a program of Protected Marine Reserves are:

1) The area included in the reserve should be large enough to protect the resource in the event of overfishing in the unprotected area. Several mathematical models (see Appendix) suggest that reserves need to include up to 50% of the original population in order to hedge successfully against overfishing.

2) The reserve area should serve as a "source" (in the sense of metapopulation theory: Pulliam 1988) capable of replenishing the exploited stock in the event of its depletion (Brown and Roughgarden 1995). In particular, reserves should protect spawning grounds and other areas critical to the viability of the population.

3) The reserve areas should be rigorously and completely protected. Typically, reserve areas will contain greater concentrations of fish than exploited areas, making them prime targets for poaching. As in terrestrial reserves, poaching must be treated as a criminal activity.

Protected marine reserves would provide benefits over and above protection of the resource. In general terms, reserves would preserve marine biodiversity by protecting intact marine ecosystems. They would also facilitate scientific research, in that the unexploited area would play the role of a control in the "experiment" of fishing (Lindeboom 1995).

PROBLEMS OF RESERVE DESIGN

Many practical issues will arise in the design of marine reserves. How large should the reserve be and where should it be located? Should there be one large, or several small protected areas? Should the reserve be tailored for individual species, or for the protection of an entire marine ecosystem? Economic as well as biological aspects may influence reserve design. For example, a large reserve encompassing traditional fishing grounds may unfairly affect local fishing communities. Fragmented reserves may have fewer economic impacts, but may be less effective and more difficult to manage than one or two larger reserves. Reserves should be permanent, but this requirement will have to be balanced with the need for flexibility in reserve design. Because of the very uncertainties that underlie the need for reserves, the concept of an "optimal" reserve may be meaningless. As in other instances of bet hedging, adopting a diversified strategy is the important step; the exact allocation of total assets to different types of investment is then largely a matter of judgment.

COMPARISON OF PROTECTED RESERVES WITH OTHER STRATEGIES

Protected marine reserves are not at present a common component of fisheries management programs. Indeed, many fisheries biologists behave as if they consider reserves as unnecessary or unworkable. Others have asserted that reducing catch or effort levels would have the same effect as a reserve. This claim is erroneous and can only arise from a misunderstanding of the role of uncertainty and uncontrollability in fisheries management.

Opening the entire population to exploitation exposes it to the risk of depletion, even if inadvertent. While it is obviously true that this risk would be reduced with reduction of the allowable catch, the uncertainties and biases associated with setting quotas and determining actual fishing mortality imply that the fishery would probably remain vulnerable unless the quotas were set far below the "best point estimates." As noted, target fishing mortality in the Northern cod fishery should have been reduced from 0.2 to 0.1 if even a moderate degree of risk aversion were to be included. Given that actual fishing mortality often exceeded the targeted value by up to 200% (Myers et al. 1997), even this unheard-of reduction might not have been sufficient to save the cod fishery. In any event, achieving a given target fishing mortality, whether 0.2

or 0.1, has two critical prerequisites: first, stock assessments must be accurate and up-to-date, and second, all sources of fishing mortality must be accurately accounted for. As we have already noted, often neither of these prerequisites holds true.

Other management practices, such as mesh or other gear restrictions, may also reduce the risk of overfishing, but, like reductions in catch quotas, they do not amount to a diversification of management strategy, but only to a switch to an apparently more conservative strategy. The possibility of biases, errors, and excessive catch rates remains in effect under such restrictions.

An important aspect of bet hedging is that risk reduction can be achieved at minimal cost, and our models suggest that this may be true for reserves. For example, placing 50% of a population's natural marine habitat into a reserve does not necessarily imply a 50% reduction in long-term catches, particularly if the reserve is highly productive and operates as a source. Also, because of the safety aspect of the reserve, the exploited area probably can be fished somewhat more intensively than would be desirable in the absence of the reserve.

Maintenance and protection of marine reserves will incur certain costs. If successful, the reserve areas would contain higher concentrations than the exploited areas. In addition, fish inside the reserve would tend to be larger. Poaching would therefore be especially attractive in the short term. It might be argued that reserves would interfere with economic efficiency (Walters and Pearse 1995), but reducing the risk of collapse by maintaining an adequate reserve has to be weighed against the short-term gains of "creaming the top" off the reserved stocks. It is certainly clear that reserve areas would need to be rigorously policed to prevent poaching; present satellite technology would make it easy to accomplish the necessary monitoring.

A MODEL OF UNCERTAIN HARVESTS

It is probably not useful to attempt developing a general model of marine protected areas, given the great variety of marine ecosystems and conceivable management regimes. To illustrate our ideas, we model a single harvested stock that grows according to a discrete logistic (Ricker) equation. Thus, in the absence of harvest, the stock in year t, $N(t)$, and the stock in year $(t + 1)$, $N(t + 1)$ are related by:

$$N(t + 1) = N(t)\exp\left[r\left(1 - \frac{N(t)}{K}\right)\right] \quad (1)$$

where r and K have the usual interpretations. In particular, K is carrying capacity, in the sense of a stable steady state, and e^r is maximum per capita growth rate of the population. The role of a reserve is to prevent part of the stock from being harvested. In particular, we assume that a fraction A of the area in which the stock exists is available for harvesting and that the

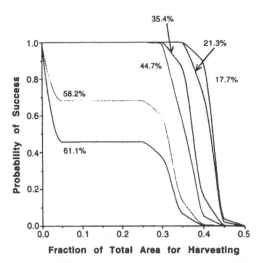

FIG. 1. The probability that the stock $[N(t)]$ remains $>0.6K$ for a 20-yr time horizon, as a function of the fraction A of area available for harvesting, for six different coefficients of variation in the harvest distribution. The model assumes that $N(1) = K$ and uses beta distributions for the harvest, all with the same mean, 0.5.

harvest fraction in this area is targeted at u. However, we assume that the target harvest is uncontrollable. This lack of controllability is captured by assuming that the harvest fraction has a probability distribution. We further assume that the mean of the distribution is fixed at the target harvest fraction, but that the actual harvest varies about this mean.

As a criterion for successful management, we assume that the stock starts at carrying capacity and consider the probability that it remains >60% of carrying capacity over a time horizon of T yr. The notion of maintaining the stock $>0.6K$—which puts it in the so-called "Optimal Sustainable Population" region—appears in legislation such as the U.S. Marine Mammal Protection Act (MMPA) of 1972 (16 U.S. Code 1361 et seq., Publ. L. 92-522, as amended) and Magnuson Fishery Management and Conservation Act (16 U.S. Code 1801 et seq., amended 104 Congress, 'Sustainable Fisheries Act').

In the Appendix, we show how the probability of successfully achieving this goal can be computed. For computations, we used $r = 0.5$ (so that this is in the non-chaotic regime of the stock dynamics), $K = 80$, and $T = 20$. We used six different frequency (beta) distributions of catch, each with a mean of 0.5 (so that half of the animals in the harvested region are captured on average) and with coefficient of variation (CV = standard deviation of the harvest fraction/mean of the harvest fraction) ranging from ~18% to ~61%.

The results (Fig. 1) are striking. Even when the CV is moderate (say, <50%), the chance of success drops

TABLE 1. Fraction (*A*) of the fishing ground available for harvest to ensure a given level of protection for 40 yr, and the associated expected total catch (*C*) (see Appendix).

Mean harvest rate†	Level of stock protection			
	95%		99%	
	A	C	A	C
0.5	0.304	13.12	0.271	11.82
0.4	0.367	12.71	0.347	12.11
0.3	0.472	12.30	0.443	11.62
0.2	0.686	11.93	0.634	11.14

† $cv = 50\%$.

rapidly from 1 once the fraction of the total area available for harvesting becomes greater than ~30%. When the cv is larger, the result is even more impressive: even at very low areas available for harvest (5%), the chance of success is <1. That is, a strategy that is very conservative on average is still likely to fail if it is too difficult to control.

We also experimented with unknowable carrying capacity. Interestingly, the results are not nearly as striking, as long as the carrying capacity is not too far off. Other models (J. Roughgarden, *personal communication*) have also shown this effect.

One conceivable alternative to a reserve is simply to lower the catch level. For example, if the mean catch is 10% of the stock, with a 50% cv, then there is >99% probability of keeping the stock $>0.6K$ for a 40-yr time horizon. The problem, of course, is that catch suffers. The methods described in the Appendix allow us to determine the size of reserve required to ensure a given level of stock protection and the catch associated with that reserve size. Typical results are shown in Table 1.

Two important points emerge from this table. First, a reserve can simultaneously lead to stock protection and a higher level of catch. For example, at a 95% level of stock protection over the 40-yr time horizon, a reserve of 70% of the potential fishing ground and a catch rate of 0.5 both protects the stock and gives an expected catch that is nearly 50% larger than the expected catch if the mean catch rate were reduced to 0.1 and the entire fishing ground fished. Second, it is possible to maximize catch while protecting the stock. For example, at a 99% level of protection, a mean catch rate of 0.4 provides slightly better expected catch than any of the alternatives.

We thus conclude that a system based on reserves may simultaneously provide protection of the stock and a higher long-term catch by allowing greater intensity of fishing in the fraction of the potential fishing ground in which fishing is allowed.

Finally, and not obvious from the figures or tables but consistent with our notions of fundamental uncertainties, the reserve provides insurance against errors in the model. That is, any real stock is managed with estimates of growth rates and carrying capacities. Furthermore, actual mean catches may exceed targeted val-

ues. A protected reserve provides a buffer against many of these uncertainties, without necessarily leading to great reductions in catch.

DISCUSSION

Widespread concern has been expressed over the failure to manage the world's ocean fisheries in a sustainable way, in spite of the opportunities provided by the 1982 Law of the Sea, and by EEZs. Recent conferences with titles such as "Re-inventing Fisheries Management" (held in Vancouver, British Columbia, Canada in February 1996) and "Ecosystem Management for Sustainable Marine Fisheries" (held in Monterey, California, USA, in 1996) attest to the desire for new approaches that would improve the dismal record. The most important component now needed is an operational admission of the limitations of science in comprehending and controlling as complex and unobservable a system as the marine environment.

Novaczek (1995) lists eight important advantages of Marine Protected Areas (MPAs). She says that MPAs can be used:

1) to protect biomass and population structure of commercial species,
2) to limit by-catch of juveniles,
3) to protect ocean biodiversity,
4) to protect essential life stages of commercial species,
5) to protect and enhance productivity,
6) to provide a location for marine research,
7) to protect artisanal and community fisheries, and
8) to enhance public education and encourage nondestructive enjoyment of the sea.

We would only add that MPAs can serve to hedge against inevitable uncertainties, errors, and biases in fisheries management. Marine Protected Areas (or as we have called them, simply, protected reserves) may well be the simplest and best approach to implementing the precautionary principle and achieving sustainability in marine fisheries.

LITERATURE CITED

Brown, G., and J. Roughgarden. 1995. An ecological economy: notes on harvest and growth. Pages 150–189 *in* C. Perrings, K.-G. Maler, C. Folke, C. S. Holling, and B.-O. Jansson, editors. Biodiversity loss: economic and ecological issues. Cambridge University Press, Cambridge, UK.

Canada. 1983. The Task Force on Atlantic Fisheries, Report. Supply and Services Canada, Catalogue Number CP32-43/1983. Ottawa, Ontario, Canada.

Gordon, D. V., and G. R. Munro, editors. 1996. Fisheries and uncertainty: a precautionary approach to resource management. University of Calgary Press, Calgary, Canada.

Kaitala, V., and G. Munro. 1995. The management of transboundary resources and property rights systems: the case of fisheries. Pages 69–84 *in* S. Hanna and M. Munasinghe, editors. Property rights and the environment. Beijer International Institute of Ecological Economics and the World Bank, Washington, D. C., USA.

Lindeboom, H. J. 1995. Protected areas in the North Sea: an

absolute need for future marine research. Helgoländer Meeresunters **49**:591–602.

Mangel, M., et al. 1996. Principles for the conservation of wild living resources. Ecological Applications **6**:338–362.

Martz, H. F., and R. A. Waller. 1982. Bayesian Reliability Analysis. John Wiley and Sons, New York, New York, USA.

Myers, R. A., J. A. Hutchings, and N. J. Barrowman. 1997. Why do fish stocks collapse? The example of cod in Atlantic Canada. Ecological Applications **7**:91–106.

Novaczek, I. 1995. Possible roles for marine protected areas in establishing sustainable fisheries in Canada. Pages 31–36 *in* N. L. Shackell and J. H. M. Willison, editors. Marine protected areas and sustainable fisheries. Centre for Wildlife and Conservation Biology, Acadia University, Wolfville, Nova Scotia, Canada.

Pulliam, H. R. 1988. Sources, sinks and population regulation. American Naturalist **132**:652–661.

Roughgarden, J., and F. Smith. 1996. Why fisheries collapse and what to do about it. Proceedings National Academy of Sciences USA **93**:5078–5083.

Roy, N. 1996. What went wrong and what can we learn from it? Pages 15–26 *in* D. V. Gordon and G. R. Munro, editors. Fisheries and uncertainty: a precautionary approach to re-

source management. University of Calgary Press, Calgary, Canada.

Seger, J., and J. Brockmann. 1987. What is bet-hedging? Oxford Surveys in Evolutionary Biology **4**:182–411

Shackell, N. L., and J. H. M. Willison, editors. Marine protected areas and sustainable fisheries. Centre for Wildlife and Conservation Biology, Acadia University, Wolfville, Nova Scotia, Canada.

United Nations. 1982. Convention on the Law of the Sea. United Nations Document A/Conf.61/122.

United Nations Food and Agriculture Organization. 1992. Marine Fisheries and the Law of the Sea: a decade of change. U. N. Food and Agriculture Organization Fisheries Circular Number 853, Rome, Italy.

———. 1995. The state of world fisheries and aquaculture. Rome, Italy.

Walters, C. J., and P. H. Pearse. 1996. Stock information requirements for quota management systems in commercial fisheries. Reviews in Fish Biology and Fisheries **6**:21–42.

Yoshimura, J., and C. W. Clark, editors. 1993. Adaptation in stochastic environments. Lecture notes in biomathematics. Volume 98. Springer-Verlag, Berlin, Germany.

APPENDIX

In this appendix, we provide details for the computation of the success probability for the model of a stock growing according to the discrete logistic equation with unknowable harvest. We also clarify the assumptions used in the calculation.

Assume the size of the stock in the current year is $N(t)$. If the reserve fraction is $1 - A$, then the stock size on the reserve:

$$N_r(t) = (1 - A)N(t) \qquad (A.1)$$

is untouched. The stock available for fishing is

$$N_f(t) = AN(t) \qquad (A.2)$$

and a fraction u of this stock is harvested. Thus, the total stock remaining after fishing is

$$N_r(t) + (1 - u)N_f(t) = (1 - A)N(t) + (1 - u)AN(t)$$

$$= (1 - uA)N(t). \qquad (A.3)$$

We assume that this stock is well mixed over the combined reserve and fishing areas in order to determine the stock size in the next year. That is, the reserve boundaries are set for harvesting but the stock moves smoothly across the boundary and fills the entire fishing ground.

Define the probability of success by

$$p(n, t) = \Pr\{N(s) > n_c \text{ for } t < s \le T \mid N(t) = n\}. \qquad (A.4)$$

Here, $n_c = 0.6K$ is the critical level and T is the length of time over which we focus protection.

This function can be evaluated by a dynamic iteration equation. At $t = T$,

$$p(n, T) = \begin{cases} 1 & \text{if } n > n_c \\ 0 & \text{otherwise.} \end{cases} \qquad (A.5)$$

If the stock at the start of the fishing season in year t is n, then the size of the stock at the end of the fishing season is $n_u = (1 - uA)n$ and the stock at the start of the next season will be $n_u f(n_u)$, where $f(n_u) = \exp(r(1 - n_u/K))$. Consequently,

$$p(n, t) = E_u\{p(n_u f(n_u), t + 1)\} \qquad (A.6)$$

where E_u denotes the expectation over the distribution on the harvest rate.

One can also compute the total catch this way by defining

$$c(n, t) = E_u\left\{\sum_{s=t}^{T-1} u(s)AN(s)\right\} \qquad (A.7)$$

so that $c(n, T) = 0$ and

$$c(n, t) = E_u\{Aun + c(n_u f(n_u), t + 1)\}$$

$$= A\bar{u}n + E_u\{c(n_u f(n_u), t + 1)\}. \qquad (A.8)$$

We assume that the harvest fraction u follows the beta density (Martz and Waller 1982)

$$g(u) = c_u u^{\alpha-1}(1 - u)^{\beta-1} \qquad (A.9)$$

where c_u is a normalization constant (which can be written in terms of gamma functions) and α and β are parameters. For the density of Eq. A.9, the mean and square of the coefficient of variation of u are:

$$\bar{u} = \frac{\alpha}{\alpha + \beta}$$

$$cv^2 = \frac{\beta}{\alpha(\alpha + \beta + 1)}. \qquad (A.10)$$

Thus, one can specify the mean and coefficient of variation of u and determine the values of α and β by solving Eq. 11.

[30]

Marine Resource Economics, Volume 13, 159–170
Printed in the U.S.A. All rights reserved

0738-1360/98 $3.00 + .00
Copyright © 1998 Marine Resources Foundation

Marine Reserves:
What Would They Accomplish?

RÖGNVALDUR HANNESSON
The Norwegian School of Economics and Business Administration

Abstract *A marine reserve is defined as a subset of the area over which a fish stock is dispersed and closed to fishing. This paper investigates what will happen to fishing outside the marine reserve and to the stock size in the entire area as a result of establishing a marine reserve. Three regimes are compared: (i) open access to the entire area, (ii) open access to the area outside the marine reserve, and (iii) optimum fishing in the entire area. Two models are used: (i) a continuous-time model, and (ii) a discrete-time model, both using the logistic growth equation. Both models are deterministic equilibrium models. The conservation effect of a marine reserve is shown to be critically dependent on the size of the marine reserve and the migration rate of fish. A marine reserve will increase fishing costs and overcapitalization in the fishing industry, to the extent that it has any conservation effect on the stock, and in a seasonal fishery it will shorten the fishing season. For stocks with moderate to high migration rates, a marine reserve of a moderate size will have only a small conservation effect, compared with open access to the entire area inhabited by a stock. The higher the migration rate of fish, the larger the marine reserve must be in order to achieve a given level of stock conservation. A marine reserve of an appropriate size would achieve the same conservation effect as optimum fishing, but with a smaller catch.*

Key words Bioeconomic analysis, fisheries economics, fisheries management, marine reserve.

Introduction

Recently, the idea that certain areas be closed to fishing has gained popularity. This idea has developed in the wake of persistent overfishing and occasional stock collapses, the most recent of which is the northern cod disaster on the Grand Banks. The northern cod disaster is particularly disturbing since it took place despite a high degree of control over the harvest by the Canadian government, which was committed to a moderate rate of exploitation and whose marine science and scientists must be ranked as world class. Seen against this background, it would clearly be desirable to use fisheries management strategies that would work independently of incomplete information on stocks and catches and less-than-fully-effective enforcement policies.

Rögnvaldur Hannesson is a professor at the Centre for Fisheries Economics, The Norwegian School of Economics and Business Administration, Helleveien 30, N-5035 Bergen-Sandviken, e-mail: rognvaldur.hannesson@nhh.no.

I am grateful to Daniel V. Gordon, University of Calgary, for discussions on the issue of marine reserves; Frank Asche, The Norwegian School of Economics and Business Administration; and Jon Conrad, Cornell University, for comments.

How effective would the protected areas, or marine reserves as they are often called, be in protecting fish populations? Would this not depend on the size of the area protected and the rate at which fish disperse? It would appear that a marine reserve need not be very effective if the mobility of fish in and out of the protected area is high. How would the industry respond? If the conservation policy is successful, would not excessive fishing capacity be built up in response to improved conditions in the area still open to fishing, defeating, at least, the economic gains of this policy and possibly even the conservationist advances as well?

In this paper, we investigate the economic and conservationist effects of marine reserves and how they depend on the migration rate of fish, the cost of fishing, and the size of the marine reserve. We use the logistic growth model and start with the continuous-time formulation, as this is the most simple and elegant approach. Some features of seasonal fisheries are, however, necessarily ignored in this formulation, and, so, we also investigate a discrete-time model. To isolate the effects of marine reserves, it is assumed that a marine reserve is the only form of management imposed and that there is open access to the area outside the reserve.

The approach taken in this paper is deterministic. This is quite sufficient to capture the response of fishing effort and capacity to the establishment of marine reserves, and how the effectiveness of marine reserves depends on the migration rate of the fish. Previous investigations of this, such as Polacheck (1990), DeMartini (1992), and Holland and Brazee (1996) have concentrated on biological aspects, such as yield per recruit and changes in spawning stock biomass, using age-structured models of real-world fish populations, but taking fishing effort as given. The main focus in this study is on the reactions of the industry to the effects of marine reserves under an open-access regime. For this purpose, general biomass models seem adequate, despite their limited empirical applicability.

A Continuous Time Model

Consider a fish stock located in an area of unit size. Suppose the stock obeys the logistic law of growth, so that, in the absence of exploitation,

$$\frac{dS}{dt} = rS(1 - S) \tag{1}$$

where S is the size of the stock, and r is the intrinsic rate of growth. The stock is measured as a fraction of the carrying capacity of the area, which, thus, implicitly is set equal to unity.

Now let a fraction, m, of the area be set aside as a marine reserve. Let the stock in the two sub-areas be measured as densities, so that the carrying capacity of each subarea is also equal to unity. With the fish moving between the two sub-areas, equation (1) must be modified to take account of this. The size of the stock in the marine reserve is mS_m, where S_m is the density of the stock, and m is the size of the marine reserve. With fish moving at the rate z, the migration rate of the stock in the marine reserve will be zmS_m. The probability that a fish will migrate out of the reserve is $1 - m$, so the migration rate out of the reserve will be $(1 - m)zmS_m$. To translate this into change in stock density in the area outside the reserve, we divide by the size of that area, $1 - m$, so that the increase in the density of fish in the area outside the marine reserve due to migration from the marine reserve will be zmS_m. Similarly, $mz(1 - m)S_o$ is the rate of migration into the marine reserve, and the change in the density of fish outside the marine reserve due to this out-migration will be $-zmS_o$, where o denotes the "other" sub-area, *i.e.*, the area outside the marine reserve.[1] Thus, the rate of change in the density of fish outside the marine reserve is

$$\frac{dS_o}{dt} = rS_o(1 - S_o) + zm(S_m - S_o) - Y \tag{2a}$$

where Y is the catch rate of fish outside the marine reserve, expressed as density. By a similar reasoning, the rate of change in the density of fish in the marine reserve is

$$\frac{dS_m}{dt} = rS_m(1 - S_m) + z(1 - m)(S_o - S_m). \tag{2b}$$

In equilibrium, the density of the stock in the marine reserve will be

$$S_m = -\frac{1}{2}\left[\frac{(1 - m)z}{r} - 1\right] + \sqrt{\frac{1}{4}\left[\frac{(1 - m)z}{r} - 1\right]^2 + \frac{(1 - m)zS_o}{r}} \tag{3}$$

with open access prevailing outside the marine reserve, $S_o = c$, where c is the stock density at which the fishery breaks even.[2] Substituting c for S_o in equation (3) we can find S_m, and from equation (2a), the equilibrium catch rate, the catch in weight units being $(1 - m)Y$.

In the following, we shall compare a policy of marine reserve with open access outside the reserve to two alternative regimes, open access to the entire area, and optimum fishing in the entire area. Under open access, the equilibrium stock density will be c, as previously stated. We define optimum fishing as that which maximizes sustainable rent per year, thus ignoring discounting of the future

$$\max rS(1 - S)\left(p - \frac{c}{S}\right) \tag{4}$$

where p is the price of fish, and c is a cost parameter. This formulation implies that the unit cost of fish is inversely proportional to the exploited stock, or that the catch per unit of effort is proportional to the stock, with a constant unit cost of effort. With $p = 1$, the optimum stock[3] will be

---

[1] This migration function is similar to the one used by Conrad (1997). He considers two areas with different carrying capacities, K_1 and K_2, and uses the migration function $s(S_1/K_1 - S_2/K_2)$. Here the carrying capacity is unity everywhere, and the diffusion rate, s, is the product of z and $m(1 - m)$.

[2] With the catch equation $Y = EqS$, E denoting effort and q being normalized at unity, profits in equilibrium will be $pES - cE$, so with zero profits, and p also normalized at unity the zero-profit stock will be $S = c$.

[3] Discounting the future at the rate δ would affect the optimum stock size. The formula for the optimum stock size with discounting is, in this case,

$$S = \frac{1}{4}\left[c + 1 - \frac{\delta}{r} + \sqrt{\left(c + 1 - \frac{\delta}{r}\right)^2 + \frac{8c\delta}{r}}\right].$$

Hence, for a given cost of effort, a positive discount rate would lower the optimum stock level. On the other hand, a positive discount rate raises the cost of fishing effort, c; the more it is raised, the more capital intensive the fishing technology is (Hannesson 1987). This increases the optimum stock level, so the impact of the discount rate on the optimum stock level is, in fact, ambiguous.

$$S = \frac{1 + c}{2}. \tag{5}$$

Figures 1 through 3 compare the three regimes, (*i*) open access to the entire area, (*ii*) open access to the area outside the marine reserve, and (*iii*) optimum exploitation in the entire area. Figure 1 shows the impact of fishing costs for a given size of the marine reserve (40% of the whole area) and a given migration rate ($z = 0.5$). The size of the stock in the entire area [$mS_m + (1 - m)S_o$] is remarkably similar under both open-access regimes, with and without the marine reserve. As fishing costs approach zero, the population becomes extinct under both open-access regimes. This is a more serious case than the no-cost assumption might suggest, as this also covers the case where the unit cost of fish is insensitive to the size of the exploited stock (Hannesson 1993).

The catch level is also remarkably similar under both open-access regimes; if fishing costs are low, the stock is overexploited to the point that catch diminishes as the cost falls, as would, for example, happen as a result of technological progress. The maximum catch with a marine reserve is only slightly lower than when the entire area can be fished, and occurs at a slightly lower cost level. The equilibrium exploitation rate under both open-access regimes is also similar. The exploitation rate is an expression of the total effort applied, or the total cost of fishing. The total cost of fishing is about twice as high under both open-access regimes as with an optimal exploitation, without necessarily resulting in greater catch.

Figure 2 shows the effect of varying the size of the marine reserve for a given and relatively low cost of fishing (under open access, the stock would be reduced to 15% of its pristine level). With a relatively large reserve (80% of the entire area), the equilibrium stock would be the same as the optimum one, even if there was open access outside the reserve.[4] The maximum catch with a marine reserve and open access outside is not quite as great as the optimum one, but the stock is smaller and the exploitation rate (total costs) higher (the maximum catch is obtained when the marine reserve is about 75% of the total area).

Finally, figure 3 shows the effect of varying the migration rate. Not surprisingly, as the migration rate increases, the marine reserve solution approaches the solution with open access to the entire area. With the fish redistributing themselves rapidly, it makes no difference whether a part of the area is closed, or open, to exploitation. If the fish do not move around at all ($z = 0$), the total stock is almost as large as with optimal exploitation, but the catch is much lower. The exploitation rate is also lower, but since the regime is open access, there are no rents, and the revenue is fully absorbed by fishing costs.

A Discrete Time Model

Considering the problem in discrete time brings out distinctions which one may expect to encounter in fisheries with a seasonal character. The discrete-time analogue of the above model is the following

---

[4] From the top panel of figure 2, we see that the effect of the size of the marine reserve on the equilibrium stock is not very great until the size of the reserve is quite substantial (about one-half of the area). Hence, even if a positive discount rate would reduce the optimum standing stock, the effect on the size of the reserve needed to achieve this would not be very great. For example, with $c = 0.15$, a discount rate of 5%, and ignoring the effect of the discount rate on c, the optimum stock would be 0.48, compared to 0.57 in the absence of discounting, using the formula in footnote 3. The marine reserve would still need to be almost 80% of the entire area to achieve this stock level.

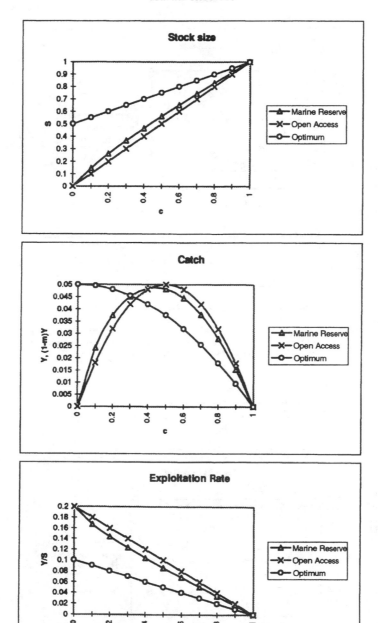

Figure 1. Effect of Varying Cost ($m = 0.4$; $z = 0.5$; $r = 0.2$)

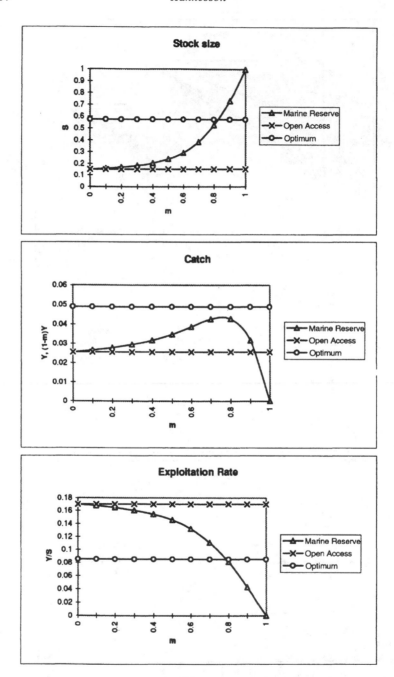

Figure 2. Effect of Varying Size of Reserve ($c = 0.15$; $z = 0.5$; $r = 0.2$)

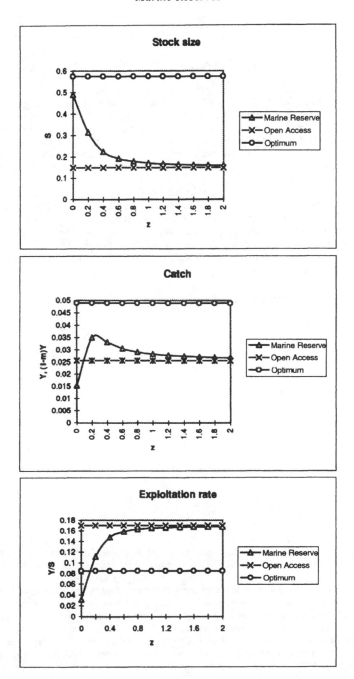

Figure 3. Effect of Varying Rate of Migration ($c = 0.15$; $m = 0.4$; $r = 0.2$)

$$R_{o,t} = S_{o,t-1} + rS_{o,t-1}(1 - S_{o,t-1}) + zm(S_{m,t-1} - S_{o,t-1}) \qquad (6a)$$

$$R_{m,t} = S_{m,t-1} + rS_{m,t-1}(1 - S_{m,t-1}) + z(1 - m)(S_{o,t-1} - S_{m,t-1}). \qquad (6b)$$

This formulation implies that fishing is concentrated at the beginning of each period, during which the stock is fished down from R_t, the level at the beginning of period t, to S_t. During the harvesting phase, the fish do not move into or out of the marine reserve, but the stock that remains outside the marine reserve after fishing intermingles with the stock inside the marine reserve. This aggregate stock determines the net growth of the stock, and the resulting stock disperses itself randomly over the entire area. These two processes determine the stock available at the beginning of the next period.

In equilibrium, $R_m = S_m$, and the solution for S_m is the same as in equation (3). With the unit operating cost of fish inversely proportional to S and equal to c/S, the stock density outside the marine reserve will be depleted to $S_o = c$. Again, we compare this regime to open access to the entire area (in which case the stock density at the end of the fishing season will be c) and optimum fishing in the entire area. Maximizing sustainable rent in this model, including the contribution to capital costs (fixed costs), entails

$$\max R(S) - S - c[\ln R(S) - \ln S] \qquad (7)$$

as the total operating cost is

$$\int_S^R \frac{c}{x}\, dx = c(\ln R - \ln S).$$

The first order condition is

$$R'(S) - 1 - c\left[\frac{R'(S)}{R(S)} - \frac{1}{S}\right] = 0. \qquad (8)$$

From this we can find the optimum density as

$$S = \frac{1}{2}\left(\frac{2 + 3r}{2r}\right) - \sqrt{\frac{1}{4}\left(\frac{2 + 3r}{2r}\right)^2 - \frac{1 + r + c}{2r}}. \qquad (9)$$

The results, with respect to equilibrium stock and yield, are similar to the continuous model. What is new here is that intra-seasonal rents emerge; instead of keeping the stock always at the equilibrium level, continuously creaming off the surplus growth, we start with a bigger stock at the beginning of each season and deplete it until the operating cost per unit of fish has risen to the level where it is equal to the price. The top panel of figure 4 shows the intra-seasonal rents with open access to the area outside the marine reserve, open access to the entire area, and optimal fishing. Under open access, the intra-seasonal rents are higher with the marine reserve. Due to migration from a more plentiful stock in the reserved area, we always start with a higher density outside the marine reserve than we would with open access to the entire area.

The intra-seasonal rents are quasi-rents; *i.e.*, revenues exceeding operating

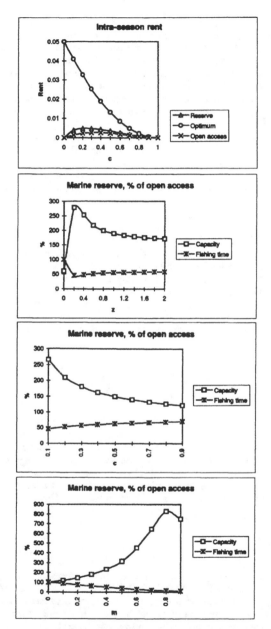

Figure 4. Effect of Marine Reserve on Intraseason Rent,
Fishing Capacity, and Season Length

Note: The second, third, and forth panel diagrams show the results of varying c, z, and m.
The values of these when held fixed are $c = 0.15$, $m = 0.4$; and $z = 0.5$, while $r = 0.2$.

costs. It is to be expected that these rents will attract investment in fixed capital, the cost of which is independent of the rate of utilization in the fishing season. Assuming that investments in fishing capacity are proportional to these quasi-rents, we get the results shown in figure 4, where we compare the fishing capacity with open access outside the marine reserve to open access to the entire area. The excess capacity generated by the marine reserve is sensitive to the cost of fishing, migration rate of the fish, and the size of the marine reserve. For a marine reserve of 40% of the entire area, the fishing capacity is twice that under open access to the entire area when the cost of fishing is low. For a relatively low cost ($c = 0.15$), the fishing capacity with a marine reserve reaches a maximum of about eight times what it would be with open access to the entire area when the marine reserve is about 80%. For a marine reserve of 40% and a cost of $c = 0.15$, the fishing capacity is two to three times the open-access level, except for very low migration rates. In these scenarios, one of the main results of establishing the marine reserve is to encourage overinvestment in fishing capacity.

An associated effect is a shortening of the fishing season, as it will take less time to deplete the population to the break-even level the larger the fishing capacity is. The time it will take to deplete the stock to the level c is given by $Re^{-FT} = c$, where F is the fishing mortality produced by the fishing fleet, and the length of the fishing season (T) is $T = \ln(R/c)/F$. By assumption, the capacity of the fishing fleet is proportional to the intra-season rent. Now the fishing mortality produced by a unit of fishing capacity is inversely related to the size of the area in which the stock is confined, so if one unit of capacity produces fishing mortality, F^*, over the entire area, the mortality produced in the area outside the marine reserve will be $F^*/(1 - m)$. Hence, the relative length of the fishing season under open access with and without the marine reserve will be

$$\frac{T_{MR}}{T_{OA}} = \left[\frac{\ln R_{MR}}{\ln R_{OA}}\right]\left(\frac{V_{OA}}{V_{MR}}\right)(1 - m) \tag{10}$$

where V is rent, the subscript MR refers to the regime with open access to the area outside the marine reserve, and OA to open access to the entire area. Figure 4 shows the ratio T_{MR}/T_{OA} as a percentage. The higher capacity buildup generated by the higher intra-season rents with the marine reserve, results in a very substantial shortening of the fishing season. For example, when varying m in figure 4, the fishing capacity under open access with a marine reserve rises to a peak of more than eight times the level when the entire area is open, while the fishing season shrinks by more than 80%.[5]

Conclusion

The foregoing analysis suggests that little would be gained by establishing marine reserves without applying some measures that constrain fishing capacity and effort. The migration of fish ensures that the fish stock to be protected would be depleted despite the existence of a marine reserve. The catch might be larger than under open access to the entire area, but this gain would be nullified by increasing cost. The re-

---

[5] The seasonal model in this section is in many ways similar to the approach taken by Homans and Wilen (1997). In their model, intraseason rents attract fishing capacity which regulators counteract by shortening the fishing season, as needed, to take a given total allowable catch.

sults also suggest that marine reserves would have to be very large, maybe 70% to 80% of the entire fishing area, in order to achieve yield and conservation effects on par with an optimally controlled fishery. However, the difference would be the erosion of economic benefits in the absence of any controls that reign in fishing capacity and effort. In a seasonal fishery, the increase in intra-season rents would lead to a buildup of greater fishing capacity, which, in turn, would lead to a shorter fishing season. These effects could be very substantial. As the existing literature indicates, marine reserves might provide a hedge against stock collapses (Lauck 1996; Lauck *et al.*, 1998), but only if they supplement other management measures that keep effort and capacity in check. Marine reserves by themselves, without any measures to restrain fishing effort and capacity, might achieve little other than increasing the costs of fishing.

References

Clark, C.W. 1976. *Mathematical Bioeconomics*. New York: John Wiley & Sons.
Conrad, J.M. 1997. *The Bioeconomics of Marine Sanctuaries*. Working Paper, Cornell University.
Conrad, J.M., and C.W. Clark. 1987. *Natural Resource Economics: Notes and Problems*. Cambridge, UK: Cambridge University Press
DeMartini, E.E. 1992. Modeling the Potential of Fishery Reserves for Managing Coral Reef Fishes. *Fishery Bulletin* 91:414–27.
Hannesson, R. 1987. The Effect of the Discount Rate on the Optimal Exploitation of Renewable Resources. *Marine Resource Economics* 3:319–29.
____. 1993. *Bioeconomic Analysis of Fisheries*. Oxford: Fishing News Books.
Holland, D.S., and R.J. Brazee. 1996. Marine Reserves for Fisheries Management. *Marine Resource Economics* 3:157–71.
Homans, F.R., and J.E. Wilen. 1997. A Model of Regulated Open Access Resource Use. *Journal of Environmental Economics and Management* 32:1–21.
Lauck, T. 1996. Uncertainty in Fisheries Management. In *Fisheries and Uncertainty. A Precautionary Approach to Resource Managament*, D.V. Gordon and G.M. Munro, eds., pp. 91–105. University of Calgary Press.
Lauck T., C.W. Clark, M. Mangel, and G.R. Munro. 1998. Implementing the Precautionary Principle in Fisheries Management Through Marine Reserves. *Ecological Applications* 8:S72–S78.
Polacheck, T. 1990. Year Around Closed Areas as a Management Tool. *Natural Resource Modeling* 4:327–54.
Sumaila, U.R. 1997. *Protected Marine Reserves as Fisheries Management Tools: A Bioeconomic Analysis*. Paper given at the Symposium on Fisheries Management and Uncertainty, Bergen, June 3–5, 1997.

Appendix

Stability of Equilibrium With Open Access Outside Marine Reserve

Continuous Model

Consider a perturbation of the linearized system [equations (2a) and (2b)] around the equilibrium point

$$\frac{dS_o}{dt} = \left[r(1 - 2S_o) - zm - \frac{\partial Y}{\partial S_o}\right]\Delta S_o + zm\Delta S_m = a_{11}\Delta S_o + a_{22}\Delta S_m$$

$$\frac{dS_m}{dt} = z(1 - m)\Delta S_o + \left[r(1 - 2S_m) - z(1 - m)\right]\Delta S_m = a_{21}\Delta S_o + a_{22}\Delta S_m.$$

We consider only the biological part of the system and hold the fishing effort constant. It may be noted, however, that the human part of the system could destabilize an otherwise stable biological system if the reaction to positive or negative profits is strong enough. With the catch function implicit in the previous analysis (cf. footnote 2), we have $Y = ES_o$, and so $\partial Y/\partial S_o = E$. Now, in equilibrium, $S_o = c$, and so $E = Y/c$, and $\partial Y/\partial S_o = Y/c$.

The characteristic equation is $\lambda^2 - (a_{11} + a_{22})\lambda + (a_{11}a_{22} - a_{21}a_{12}) = 0$ with roots

$$\lambda = \frac{1}{2}\left(\alpha \pm \sqrt{\alpha^2 - 4\beta}\right)$$

where $\alpha = a_{11} + a_{22}$ and $\beta = a_{11}a_{22} - a_{21}a_{12}$. For stability, we need at least one negative root, or a negative real part. In all cases reported, there is at least one negative root, but for some low values of m or c, there is one positive root, implying that the equilibrium is a saddle point. See Clark (1976, ch. 6), or Conrad and Clark (1987, pp. 45–48).

Discrete Model

A perturbation of the linearized system around equilibrium gives

$$\Delta S_{o,t+1} = \left[1 + r(1 - 2S_o) - zm - \frac{\partial Y}{\partial S_o}\right]\Delta S_{o,t} + \left(zm - \frac{\partial Y}{\partial S_m}\right)\Delta S_{m,t} = a_{11}\Delta S_{o,t} + a_{12}\Delta S_{m,t}$$

$$\Delta S_{m,t+1} = z(1 - m)\Delta S_{o,t} + \left[1 + r(1 - 2S_m) - z(1 - m)\right]\Delta S_{m,t} = a_{21}\Delta S_{o,t} + a_{22}\Delta S_{m,t}$$

As for the continuous model, we consider stability of the biological system, holding effort constant. The catch is given by $Y = R_o(1 - e^{-E})$. In equilibrium, $S_o = c = R_o e^{-E}$. Hence, $E = -\ln(c/R_o)$, and $\partial Y/\partial S_o = (\partial R_o/\partial S_o)(1 - c/R_o)$, and analogously for $\partial Y/\partial S_m$.

Stability of the equilibrium, S_o, S_m, requires, in this case, that the roots of the characteristic equation be less than 1 in absolute value, or have real parts that are less than 1 in absolute value. In the cases discussed in this paper, all roots are less than 1 in absolute value.

Beyond the Schaefer Model

[31]

Beverton-Holt Model of a Commercial Fishery: Optimal Dynamics

Colin Clark, Gordon Edwards, and Michael Friedlaender

Department of Mathematics
University of British Columbia, Vancouver, B.C.

Clark, C., G. Edwards, and M. Friedlaender. 1973. Beverton-Holt model of a commercial fishery: optimal dynamics. J. Fish. Res. Board Can. 30: 1629–1640.

The problem of optimal regulation of a fishery is discussed. Of special interest is the problem of regulating an overexploited fishery by reducing effort to allow the fish population to build up to a suitable level.

We first argue that the problem requires an economic analysis based on the concept of maximization of present value. From this concept we then deduce a simple, general rule, the "Fisher Rule," which we subsequently use to determine optimal exploitation. Among the principal results are the following: (a) an optimal mesh-size is determined, which, because of the discounting of future revenues, is smaller than the size corresponding to maximum sustainable yield; (b) the optimal recovery policy for an overexploited fishery is deduced; it consists of a fishing closure permitting the fish population to reach an optimal age; (c) the optimal development of an unexploited fishery is deduced; an initial development stage characterized by large landings and profits is rapidly transformed into a situation of optimal sustained yield, in which both landings and profits may be significantly reduced; (d) the optimal policy is deduced for a fishery in which gear limitations are impractical; the result may be a strongly unstable fishing industry; (e) the effect of high discount rates, which might be employed by private fishing interests, is discussed; such rates may result in overfishing similar to the case of a common-property fishery.

Clark, C., G. Edwards, and M. Friedlaender. 1973. Beverton-Holt model of a commercial fishery: optimal dynamics. J. Fish. Res. Board Can. 30: 1629–1640.

Nous traitons de la question du contrôle optimal d'une pêcherie. Un cas d'intérêt particulier est le réglage d'une pêcherie surexploitée par réduction de l'effort afin de permettre à la population d'atteindre un palier convenable.

Nous soutenons tout d'abord que le problème exige une analyse économique fondée sur l'idée de la maximisation de la valeur présente de la pêcherie. Nous déduisons de ce concept une règle simple et générale, appelée "règle de Fisher," qui nous permet de préciser par la suite l'exploitation optimale. Les principaux résultats sont: (a) détermination de la grandeur optimale de la maille; à cause de l'escompte de revenus futurs, cette maille est plus petite que celle qui correspond au maintien constant des prises maximales; (b) définition d'une politique optimale de redressement d'une pêcherie surexploitée; ceci comporte une suspension de la pêche pour permettre aux poissons d'atteindre un âge optimal; (c) détermination du développement optimal d'une pêcherie vierge; une phase initiale de développement, caractérisée par des débarquements et des profits élevés, se transforme rapidement en une situation de rendement optimal soutenu, avec débarquements et profits appréciablement diminués; (d) mise au point d'une stratégie optimale pour une pêcherie où il n'est pas pratique d'imposer des restrictions sur l'équipement; il peut en résulter une industrie très instable; (e) discussion des conséquences du taux d'escompte élevé que peuvent adopter des intérêts privés et qui pourraient conduire à une surexploitation semblable à celle d'une pêcherie de propriété publique.

Received March 19, 1973

The subject of this article is the optimal management of a commercial fishery, whether privately or publicly owned, and the first question that must be faced is the meaning of the word "optimal." The most obvious and most frequently encountered interpretation of optimality is embodied in the concept of maximum sustained yield. This interpretation has much to recommend it, particularly in view of the increasing prevalence of severe overexploitation in fisheries and other renewable resources.

Nonetheless maximum sustained yield has serious limitations as a criterion for resource management.

1630   JOURNAL FISHERIES RESEARCH BOARD OF CANADA, VOL. 30, NO. 11, 1973

Consider, for example, the case of a heavily over-exploited fishery. It is clear that any management program designed to increase yield will involve at least a temporary reduction in landings before the desired increase is achieved. Conservation inevitably requires some degree of current sacrifice to obtain future benefits, so that we are immediately faced with the basic economic question of "trade-offs" between income and investment.

The concept of maximum sustained yield, being a noneconomic concept, has nothing to contribute to this basic economic problem. Nor does it have anything to say about the development of a virgin resource stock, since it is not a dynamic concept — it specifies only the final state, not the path by which such a state is reached.

A possible suggestion for the optimal development — or recovery — of a fishery would be that the level of maximum sustained yield should be reached as rapidly as possible. An overexploited fishery, for example, would be closed until the stock had recovered to the desired level, as is currently the situation in the blue whale fishery.

Can such a policy be considered optimal in every case? If, for example, there are fishermen whose very survival depends upon the fishery in question, or fishing companies holding large investments in boats and gear, would the increased revenues to be obtained from building up the fishery justify a scheme of temporary compensation?

Another question closely related to the above concerns a single company (or public agency) that owns complete property rights in a certain fishery. It can sell its rights to another company, or it can operate the fishery itself. If it sells, what is a fair price? If it operates the fishery itself, what management scheme will it adopt?

All these questions can be treated by considering not the maximization of physical yield, but rather the maximization of economic benefits. Specifically, a rational management policy, whether private or public, will attempt to maximize the total "present value" of all profits, present and future, from the resource in question.

Such a definition, while still open to question on various technical grounds (such as the problem of uncertainty, or the distinction between private and social discount rates), is at least a complete definition in the sense that it determines in principle the entire management program. It provides the basis for cost–benefit analysis by means of which trade-offs of all kinds can be quantified; it also determines a fair selling price, or economic value, of the resource, namely the maximum present value itself.

The definition, however, raises a critical question: what is the correct discount rate to utilize in the computation of present values? This question is beyond the scope of this paper (see Baumol 1968), and the discount rate λ will therefore be taken as "given." The cases of small discount rates ($\lambda \simeq 0$) as well as large ones ($\lambda \rightarrow \infty$) are particularly interesting, since (as we shall explain) they correspond, respectively, to the maximization of sustained economic yield, and the "dissipation of economic rent" as described by economists for the case of common-property fisheries (Gordon 1954).

In the remainder of this article we shall first introduce and analyze our basic model, following Beverton and Holt (1957), for the case of a single year-class of fish. The analysis will then be extended to the case of conglomerate year-class structures. The entire analysis is based on a single rule that we call the "Fisher Rule" after the economist I. Fisher (1930).

Mathematically speaking, the problems to be analyzed are problems in optimal control theory, which can be dealt with by means of the maximum principle of L. S. Pontrjagin (1964). Rigorous mathematical details, however, are relegated to the Appendix, and the main body of the paper is devoted to the discussion and intuitive interpretation of the results of the theory.

Single Year-Class Model

In fisheries economics, and in the literature of fisheries management generally, there are two basic and quite different types of models, the "logistic" models in which the population is treated as a homogeneous quantity, and the more refined models based on "year-class" structures.

Logistic models are normally based on a differential equation of the form $dN/dt = f(N)$, for the total population $N = N(t)$. The function $f(N)$ thus combines the effects of recruitment, growth, and natural mortality into a single relationship between the population N and its net growth rate dN/dt. Fishing in such a model is assumed to counteract natural growth, and the equation

$$\frac{dN}{dt} = f(N) - h(t)$$

is taken to describe the combined process. Here $h(t)$ the "fishing rate," is the control variable for optimization problems. The economic equilibrium analysis of this model is due to Gordon (1954) and Scott (1955). A complete dynamic analysis, based upon present-value maximization, has recently been given by Clark (1973a,b).

In the Beverton-Holt model, on the other hand, recruitment is taken to be constant, and the population numbers are reduced instead by a constant mortality rate. Thus, if $x = x(t)$ denotes the number of fish in a given year-class (recruited at time $t = T$), then we suppose that under natural conditions we have

$$\frac{dx}{dt} = -Mx, \quad x(T) = R \qquad (1)$$

where the recruitment R and the natural mortality rate M are both positive constants. A variable "fishing mortality" $F(t)$ is assumed to affect the population of any particular year-class $x(t)$ by simply increasing the natural mortality:

$$\frac{dx}{dt} = -(M + F(t))x \qquad (2)$$

This function $F(t)$ is now the control variable; it specifies the age at which fish should be caught and the intensity of fishing at each age.

The model also accounts for the growth in weight of the fish, by means of a function $w(t)$ representing the (average) weight of a fish of age t. We shall assume that $w(t)$ is an increasing, bounded function of t, with $w(t) \rightarrow w_\infty$ as $t \rightarrow \infty$. For technical reasons, we shall also assume that $w'(t)/w(t)$ is decreasing. The latter is a sort of convexity assumption; it is valid for any concave growth curve $w(t)$ and also for the standard von Bertalanffy curve $w(t) = a(1-e^{-kt})^3$.

Restricting our attention henceforth to the Beverton-Holt model, we shall begin our analysis by considering the problem of maximizing the present value of the harvest from a single year-class, recruited at time $t = 0$.

Let $B(t)$ denote the biomass of the given year-class:

$$B(t) = x(t)w(t). \qquad (3)$$

From (2) we have $B' = x'w + xw' = -(M + F)xw + xw'$, so that

$$B' = B(\frac{w'}{w} - M - F). \qquad (4)$$

In the time interval from t to $t + dt$ the yield in biomass is given by

$$Y(t)dt = F(t)x(t)w(t)dt = F(t)B(t)dt.$$

This completes the description of the biological aspects of our model, and we now turn to its economic aspects.

First we shall assume a fixed price p per unit weight of fish. Next we assume that fishing costs are proportional to fishing effort, which in turn is proportional to the fishing mortality rate $F(t)$. This assumption, sometimes referred to as "mass contact fishing technology," has a simple probabilistic interpretation in terms of the expected time to catch one fish from a population of x fish, uniformly distributed, see Clark (1973a). (For the sake of clarity we shall henceforth refer to $F(t)$ itself as fishing effort rather than fishing mortality.) Consequently the net revenue, or profit, derived during the time interval from t to $t + dt$ is given by

$$\begin{aligned} R(t)dt &= [pY(t)-cF(t)]dt \\ &= F(t)[pB(t)-c]dt \end{aligned} \qquad (5)$$

where c is a constant denoting the cost of a unit fishing effort.

If $\lambda > 0$ represents the given (instantaneous) discount rate, then the total present value of all profits is given by the expression

$$PV = \int_0^\infty e^{-\lambda t} F(t)[pB(t)-c]dt \qquad (6)$$

The problem (for a single year-class) is therefore to determine the fishing-effort function $F(t) \geq 0$ that maximizes the expression (6), where $B(t) = x(t)w(t)$ and $x(t)$ satisfies the basic equation (2). A special case, as we shall see, is the problem considered by Beverton and Holt of simply maximizing the total revenue.

In the absence of fishing, the biomass satisfies the equation

$$B'(t) = B(t)(\frac{w'}{w} - M); \qquad (7)$$

we shall refer to this as the "natural biomass equation." Since w'/w is assumed to be decreasing, we see that the natural biomass increases to a maximum value, which occurs at time t_0 determined by

$$\frac{w'(t_0)}{w(t_0)} = M \qquad (8)$$

and subsequently $B(t)$ decreases — see Fig. 1. If equation (8) has no solution, then the biomass $B(t)$ has its maximum value when $t_0 = 0$. Notice that even if the population has been reduced by fishing, once fishing ceases the biomass will continue to satisfy equation (7), and will therefore follow the corresponding curve of Fig. 1. This fact plays an important role in our description of optimal fishing.

According to equation (5) the quantity

$$V(t) = pB(t)-c \qquad (9)$$

represents the net revenue generated by a unit fishing effort per unit time. The quantity $V(t)$ will be called the "net biovalue" of the fish population at time t.

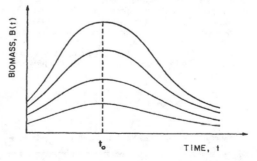

FIG. 1. Natural biomass curves.

1632   JOURNAL FISHERIES RESEARCH BOARD OF CANADA, VOL. 30, NO. 11, 1973

In the absence of fishing, the natural rate of change in the net biovalue is

$$V'(t) = pB'(t) = pB(t)(\frac{w'}{w} - M) \qquad (10)$$

where we have used the natural biomass equation (7). We can now describe the optimal fishing policy.

At any given time t, the decision must be made as to the optimal fishing effort $F(t)$. This decision does not depend upon the past history of the fishery, but only upon the existing net biovalue $V(t)$ and its potential future growth, given by equation (10). First of all, it is clear from the expression (6) and the differential equation (2) that

$$F(t) = 0 \text{ unless } V(t) > 0 \qquad (11)$$

for otherwise there would result an unnecessary negative contribution to present value. On the other hand, if $V(t) > 0$, the decision to fish or not to fish will depend on a comparison of the rate $V'(t)$ at which the net biovalue is changing as a result of natural growth, with the rate of growth of an equivalent monetary investment at the given rate of interest λ. Specifically, if $V'(t) > \lambda V(t)$ (with $V(t) > 0$) then it is better to postpone fishing, since the net biovalue of the fish population is increasing faster than its monetary equivalent. Conversely, a population whose net biovalue is growing more slowly than its monetary equivalent should be harvested and converted into cash value. Thus for $V(t) > 0$, we have the following rule:

Fisher Rule — The optimal fishing effort satisfies

$$F(t) = \begin{cases} 0 & \text{if } V'(t) > \lambda V(t) \\ F_{max} & \text{if } V'(t) < \lambda V(t) \end{cases} \qquad (12)$$

where F_{max} denotes the maximum possible fishing effort, and $V'(t)$ is given by equation (10). Marginal rules of this kind are common throughout investment theory (see Hirshleifer 1970).

The Fisher Rule, as we shall see later, does not necessarily apply in all possible cases. It does apply, however, provided that the function $V'(t) - \lambda V(t)$ has at most one change of sign, as is the case in this section (but not in the next section). An elementary rigorous proof of the rule is given in the appendix.

To see how the Fisher Rule operates in our case, we consider the inequality $V'(t) > \lambda V(t)$, which by (10) is equivalent to

$$pB(t)(\lambda + M - \frac{w'}{w}) < c\lambda$$

This inequality is automatically satisfied for $\frac{w'}{w} > \lambda + M$, while for $\frac{w'}{w} < \lambda + M$ it is equivalent to the inequality

$$B(t) < \Phi_\lambda(t) = \frac{c\lambda}{p(\lambda + M - \frac{w'}{w})} . \qquad (13)$$

Note that the curve $\Phi_\lambda(t)$ is monotone decreasing (since $\frac{w'}{w}$ is) with a vertical asymptote at time t_λ determined by the condition[1]

$$\frac{w'(t_\lambda)}{w(t_\lambda)} = \lambda + M. \qquad (14)$$

If equation (14) has no solution t_λ, then there is no vertical asymptote. Note also that $\Phi_\lambda(t_0) = \frac{c}{p}$ where t_0 is the time of "maximum natural biomass" determined by equation (8). These relationships are illustrated in Fig. 2, with time along the horizontal axis and biomass along the vertical (as in Fig. 1). The shaded area in Fig. 2 represents the "fishing zone," where both $V(t) > 0$ and $V'(t) < \lambda V(t)$, or (in terms of biomass) where $B(t) > c/p$ and $B(t) > \Phi_\lambda(t)$. The Fisher Rule requires maximum-intensity fishing in this zone. Similarly, the Fisher Rule requires zero fishing in each of the zones defined by the inequalities $t < t_\lambda$, $B(t) < \Phi_\lambda(t)$, $B(t) < c/p$. The boundary of the fishing zone is defined by the condition $V'(t) = \lambda V(t)$ (or $B(t) = \Phi_\lambda(t)$) subject to the restraint $V(t) \geq 0$ (or $B(t) \geq c/p$).

Let us now follow the biomass curve $B(t)$ corresponding to optimal fishing. Figure 2 shows the case in which the initial biomass lies in the no-fishing zone. In this case $B(t)$ will follow the natural biomass curve until that curve hits the boundary $B = \Phi_\lambda(t)$ of the fishing zone. If we assume that F_{max} is large, then the biomass curve under optimal fishing will never actually enter the fishing zone, since if it did so, the maximum fishing effort would immediately reduce biomass to the level $\Phi_\lambda(t)$. Optimal fishing could not reduce the biomass below this curve, for then fishing would cease, and the

---

[1]Our notation for t_0, t_λ, and T differs from that of Beverton and Holt (1957).

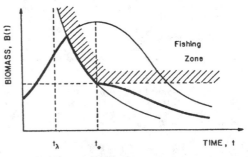

FIG. 2. Optimal fishing — the Fisher Rule.

natural growth would force B(t) back towards $\Phi_\lambda(t)$. Consequently the optimal fishing effort maintains the biomass B(t) precisely on the boundary curve $\Phi_\lambda(t)$, wherever possible.

Eventually, however, fishing reduces the biomass to the level of zero revenue, V(t) = 0. Since $\Phi_\lambda(t_0) = c/p$, it follows that optimal fishing just succeeds in reducing biovalue to the zero level at the exact instant when natural biomass reaches its maximum value. In other words, since the curve $\Phi_\lambda(t)$ is characterized by the "Fisher equation"

$$V'(t) \;=\; \lambda V(t),$$

at $t = t_0$ we have $V'(t_0) = V(t_0) = 0$.

At $t = t_0$ fishing therefore ceases, and the biomass subsequetly follows the natural curve. In Fig. 2, the heavy curve shows the overall biomass curve B(t) as affected by optimal fishing.

Although the optimal fishing policy just described is obviously too complex to be of much practical value in fisheries management, it does possess certain pertinent features. It shows for example, that regardless of the costs of fishing (which are difficult if not impossible to determine in practice), optimal fishing ta es place only when

$$t_\lambda \;<\; t \;<\; t_o$$

where t_λ is determined by equation (14). Indeed, optimal fishi g begins at some time $t^* > t_\lambda$ depending on the economic and biological parameters; for example, t^* increases if the cost parameter c increases.

Three particular cases, illustrated in Fig. 3, are especially easy to describe.

Suppose first that the costs of fishing are negligible, c = 0. Then $V'/V = B'/B = (xw'-Mxw)/xw = \lambda$ if and only if $t = t_\lambda$, so that the curve Φ_λ degenerates to the vertical line $t = t_\lambda$. Hence optimal fishing applies a maximum possible effort at the time $t = t_\lambda$ (Fig. 3a). (Figure 3a and b are drawn under the assumption that the maximum fishing effort F_{max} is large, so that the fish population can be rapidly reduced by fishing. The case of a limited maximum rate F_{max} has been studied by Goh (1972), who shows that (assuming c = λ = 0) optimal fishing will begin at the age t^* determined by the equation B(t^*) = $\int_{t^*}^{\infty} F_{max}$ B(t)dt. It can be shown that $t^* \rightarrow t_o$ as $F_{max} \rightarrow +\infty$.)

Consider next the case of a zero discount rate, $\lambda = 0$. Then $V'/V = 0$ if and only if $t = t_o$, and the curve Φ_λ becomes the vertical line $t = t_o$. Optimal fishing applies a maximum effort at the instant t_o of maximum natural biomass. Optimal fishing with no discounting of future revenues is therefore equivalent simply to the maximization of total revenue. This was the case considered in the book of Beverton and Holt (1957).

Finally we consider the limiting case of an infinite discount rate. From equation (13) it follows that $\Phi_\lambda(t) \rightarrow c/p$ as $\lambda \rightarrow \infty$. Fishing therefore takes place whenever

pB(t) > c. Since the net biovalue never attains a positive value, the total present value is equal to zero in this limiting case. This is similar to the classical economic model of a common-property fishery (Gordon 1954), which predicts that zero "economic rent," i.e. zero net profits, will be derived by the competing fishermen. It has been observed previously by Clark (1973a,b) that the common-property fishery is equivalent mathematically to the limiting case of maximization of present value with an infinite rate of discount.

To explain this observation more fully, we remark that, in the case of a common-property resource, the individual exploiter can hold no confidence that the investment stock (i.e. the fish population) will remain intact. Whenever the resource has a positive net value, the danger exists that a competitor will harvest it. The

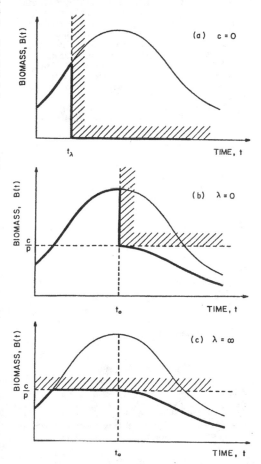

FIG. 3. Optimal fishing — special cases: (a) zero cost; (b) zero discount rate; (c) infinite discount rate

individual fisherman cannot afford to leave the investment to increase in value since he may thereby lose it entirely. Potential future revenues must therefore be discounted to zero since they cannot be realized in a competitive situation. The effective discount rate for common-property exploitation is consequently infinite.

Conglomerate Year-Classes

Most commercial fisheries do not exploit a single year-class. The Pacific halibut (*Hippoglossus* spp.) fishery, for example, normally covers an age-span from 7 to 18 years (Crutchfield and Zellner 1962).

Theoretical investigation of the optimal dynamics of conglomerate year-class structures presents many challenging difficulties. If the effort applied to each year-class could be selected independently, there would of course be no new problem — the optimal overall policy would simply consist of separate optimization for each year-class. Any practical fishing method, however, will generally affect many year-classes simultaneously.

It is fortunate, therefore, that the analysis of the preceding section, and in particular the Fisher Rule, can be extended to give considerable information regarding conglomerate age structures.

We begin by considering the case of two populations $x_1(t)$ and $x_2(t)$. The populations have initial values

$$x_i(T_i) = R_i, \quad i = 1,2$$

where T_i is the time of recruitment. The fishing effort $F(t)$ is assumed to affect both populations equally; this assumption will be modified later. We therefore obtain the following mathematical model.

$$\frac{dx_i}{dt} = -(M + F(t))x_i, \quad x_i(T_i) = R_i \ (i = 1,2) \quad (15)$$

For simplicity we take $T_1 = 0$. The present-value expression is

$$PV = \int_0^\infty e^{-\lambda t} \, F(t)[pB_1(t) + pB_2(t)-c] \, dt \quad (16)$$

where $B_i = x_i(t)w(t-T_i)$.

As in the previous section, the expression

$$V(t) = pB_1(t) + pB_2(t)-c \quad (17)$$

will be called the "net biovalue" of the combined population, and $V'(t)$ will denote the natural rate of change of $V(t)$ in the absence of fishing. It can be shown (see appendix) that the optimal fishing effort $F(t)$ satisfies the following weak form of the Fisher Rule:

$$F(t) = \begin{cases} 0 \text{ if } V'(t) > \lambda V(t) \\ 0 \text{ or } F_{max} \text{ if } V'(t) < \lambda V(t). \end{cases} \quad (18)$$

The optimal effort $F(t)$, however, cannot be unambiguously deduced from this rule. The mathematical reason for the difficulty lies in the fact that $V'(t)-\lambda V(t)$ may have several changes of sign, so that refraining from fishing even when $V'(t) < \lambda V(t)$ may result in increased profits at a later time.

There is one case, however, that can be analyzed in detail, namely the case in which costs of fishing are negligible: $c = 0$. Notice that in this case the present-value expression (16) depends linearly upon the biomasses $B_1(t)$ and $B_2(t)$. It follows that an optimal fishing policy must maximize the contribution from each population separately, at least inasmuch as separation can be achieved. Let us consider two concrete examples.

First we consider the case in which the two biomass curves are "nonoverlapping" (Fig. 4a). We know from the previous section that the maximum present value is obtained from the population B_1 by harvesting the entire population at an optimal age $t_{\lambda 1}$. Once the first population has been harvested, the second population is considered independently, and harvested at its corresponding optimal age $t_{\lambda 2}$, i.e. at time $t = T_2 + t_{\lambda 2}$. This argument is valid provided

$$t_{\lambda 1} < T_2 \quad (19)$$

and we may take this inequality as the (economic) definition of "nonoverlapping."

Suppose next that the two population curves do overlap (Fig. 4b). In this case there is no incentive to harvest the first population before the time $t = T_2$ when the second population enters the fishery. For $t > T_2$, the complete Fisher Rule of the previous section can be used, thereby determining an optimal time t_λ ($t_{\lambda 1} < t_\lambda < t_{\lambda 2} + T_2$) at which both populations are harvested completely.

It is important to note that, in the case of overlapping year-classes, the optimal harvest policy for the combined population does not generally result in the optimal harvest for each year-class separately. The purpose of mesh-size limitations is precisely to increase the total harvest by allowing the escape of immature (x_2) fish while the mature (x_1) fish are being harvested.

Let us now consider a mesh-size limitation, which we shall assume simply prevents the capture of all fish less than some critical age T^*. Since fishing decisions are made only in terms of the fish available to the fishery, it is sufficient to consider the modified biomass curves (Fig. 5):

$$B_i^*(t) = \begin{cases} 0 & \text{if } t \le T_i + T^* \\ B_i(t) & \text{if } t > T_i + T^* \end{cases} \quad (20)$$

It is easy to see now that the optimal age T^* is the same as the optimal age t_λ for a single year-class. To explain this conclusion, we note that for this choice, the modified biomass curves $B_i^*(t)$ are nonoverlapping according to the definition (19). Optimal fishing therefore harvests each population at the optimal age t_λ; by the mesh-size limitation, harvesting the first population has no effect on the immature fish of the second population.

It follows that optimal fishing for the combined population automatically maximizes the present value from each separate population. Obviously no better policy than this can exist!

To summarize the results of this section, we have shown that, if fishing costs are neglected, then optimal fishing requires the use of a mesh-size S corresponding to the optimal age t_λ determined by the condition

$$\frac{w'(t_\lambda)}{w(t_\lambda)} = M + \lambda.$$

Moreover, the optimal fishing policy harvests each population as it reaches the age t_λ.

If fishing costs are significant, then it appears that the optimal age T^* will be larger than t_λ, and optimal fishing will spread the effort more uniformly over the fishing season. As stated earlier, however, we have not been able to solve this problem in general.

Sequential Year-Classes

Most of the analysis of the previous section can be extended easily to the case of several year-classes. In this section we shall briefly consider the case of a continuing sequence (x_n) of year-classes, all assumed to belong to the same species. If no mesh-size limitation

is in effect, then the appropriate mathematical model is as follows.

$$\frac{dx_n}{dt} = -(M + F(t))x_n \quad (t \geq T_n), \qquad (21)$$

$$x_n(T_n) = R = \text{constant} \qquad (22)$$

$$PV = \int_0^\infty e^{-\lambda t} F(t)[p \sum_{n=0}^\infty w(t-T_n)x_n(t)-c] \, dt \qquad (23)$$

where R is a constant recruitment rate, T_n is the time of recruitment of the nth year-class, and $w(t-T_n)$ represents the weight of a member of the nth year-class at time $t > T_n$. Because $x_n(t)$ decreases at least as fast as e^{-Mt} the series appearing in (23) converges uniformly, and there are no problems with convergence of the integral.

As in the previous section, we shall consider only the case of negligible fishing costs, $c = 0$. The expression

$$V(t) = p \sum_{n=0}^\infty B_n(t)$$

again denotes the net biovalue of the fishery at time t.

We begin with the case of an unexploited fishery (Fig. 6). In general, such a fishery will consist of a mixture of several year-classes. The total unharvested biovalue $V(t)$ will be large and relatively constant over time.

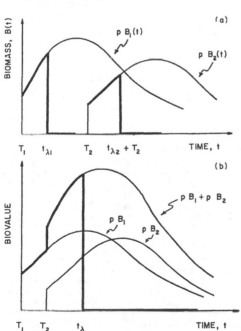

FIG. 4. Optimal fishing; two year-classes: (a) "non-overlapping" case; (b) "overlapping" case.

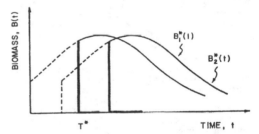

FIG. 5. Mesh-size limitations.

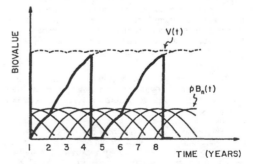

FIG. 6. Optimal fishing — no mesh limitation.

1636 JOURNAL FISHERIES RESEARCH BOARD OF CANADA, VOL. 30, NO. 11, 1973

Unless λ is small, the condition $V'/V > \lambda$ may never hold, and will surely not hold throughout the year. Consequently optimal fishing will initially utilize the maximum effort F_{max}. But since fishing reduces all existing year-classes in equal proportion the value

$$\frac{V'(t)}{V(t)} = \frac{\sum B_i(t)\,(w_i'/w_i - M)}{\sum B_i(t)}$$

will not be changed as a result of fishing. Consequently fishing will continue until the entire population has been harvested (or, in practice, until fishing costs become significant and fishing becomes unprofitable).

Because of the assumed constant recruitment rate,[2] however, new year-classes will enter the fishery and the fish population will begin to build up again (Fig. 6). Since the fishery now contains primarily immature fish with large growth rates, the inequality $V'/V > \lambda$ may hold, and may persist for several years. Eventually, however, the average growth rate will decline and the equality $V'(t) = \lambda V(t)$ will occur. The Fisher Rule then requires maximum intensity fishing at this time, again reducing the population to zero and starting the process once more.

Thus with no mesh limitations, the optimal fishing policy may result in a highly unstable fishery, in which, for the case of a long-lived species, several seasons with no fishing are followed by a season with intense fishing. In practice this ideal optimum might be realized by a fishing fleet that alternated its effort between several separate fishing grounds. The technique of periodic, or "pulse," harvesting in order to maximize the average long-term yield from a fishery (the case $\lambda = 0$), has been suggested in simulation studies by Allen (1972) and Walters (1969).

If mesh limitations are taken into consideration, the situation changes remarkably. As in the case of two year-classes, a mesh-size limitation determined by the optimal age t_λ effectively prevents the overlapping of year-classes available to the fishery, and permits separate optimization for each year-class (Fig. 7), thereby attaining the optimum optimorum.

Notice that the optimal mesh-size limitation has the additional effect of stabilizing fishing effort to a constant total seasonal effort. At the beginning of the development stage, however, there occurs (in our idealized situation, at least) an initial season in which a very large fishing effort results in a large harvest, as all the mature fish are removed from the previously unexploited fishery. It

---

[2]Our conclusion that in a low-cost fishery, optimal fishing will reduce the standing population to zero depends strongly upon the assumption that recruitment and stock size are independent. It is probably fair to assume that for many fisheries the stock–recruitment relation can be virtually ignored, but certainly not for all cases. Fisheries in which recruitment is of dominant importance may be analyzed by means of the logistic model. We hope in a future publication to be able to discuss the case in which growth and recruitment are both of importance.

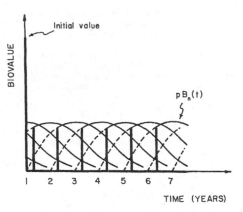

Fig. 7. Optimal mesh-size limitation.

appears to be a characteristic property of virgin renewable resource stocks that an initial "mining" stage is later transformed into a sustained yield situation, with a significantly reduced standing population. This development does not necessarily imply overexploitation. There is, however, a danger that the initial developers, failing to understand the dynamics of the situation, may adopt an unwarranted optimism regarding the long-range productivity of the resource. Other considerations such as stock-recruitment relationships and ecological stability mechanisms may also be involved in the complex problem of overfishing.

Finally let us mention a problem left unanswered by Beverton and Holt (1957, Sec. 19.2.2.), the regulation of an overexploited fishery. From the single year-class model we see that a common-property fishery will always operate at or near the level of vanishing profits. If fishing costs are low, few fish will survive beyond the minimum marketable size. The overexploited fishery will thus consist primarily of immature fish.

Suppose now that a government agency considers the possibility of imposing a mesh-size limitation in order to reduce overfishing and to increase yield. As observed by Beverton and Holt there will necessarily be an initial reduction in fish landings owing to the shortage of mature fish. The greater the mesh-size regulation that is chosen, the more prolonged will be the recovery period. What is the proper trade-off between increased yields and the length of the recovery stage?

This question can be answered immediately on the basis of our theory. The overexploited fishery has been operating as if the discount rate were infinite, whereas the regulated fishery will adopt a finite "social" discount rate λ. The Fisher Rule implies that fish should not be harvested below the optimal age t_λ. Consequently the optimal regulation of the fishery will consist of the immediate adoption of a mesh-size limitation determined by this optimal age. Landings will be significantly reduced (to zero in our idealized model) until the first year-class attains the optimal age, at which time the fishery will enter the stage of optimal sustained yield[3] shown in Fig. 7. This yield, and the profits that it generates, will be greater than before, and the increased profits will be sufficiently great to offset the losses incurred during the recovery stage.

We remark again, however, that the optimal sustained yield (for $\lambda > 0$) will be less than the maximum sustainable yield, and the fishery will reach the optimal level sooner than it would reach the level of maximum yield. This is a result of the economic trade-off between yield and length of the recovery stage. This conclusion has been overlooked in most of the conservation literature, and has also been generally misunderstood in the literature on the economics of fisheries (see Clark 1973b, for further details). The greater the discount rate λ, the smaller will the sustained yield be, and in fact (as we have seen in the first section), especially large discount rates may result in sustained yield levels approximately equal to the "rent-dissipating" yields of a common-property fishery.

Since optimal management of a fishery with any given positive discount rate will result in a sustained yield smaller than the maximum sustained yield — and, indeed, smaller than the "eumetric" yield of Beverton and Holt — the question arises of defining the concept of "overfishing." We believe the answer should be given along the following lines. First, there is assumed to exist a socially optimal discount rate λ. Correspondingly there is an optimal sustained economic yield $Y\lambda$, where "economic yield" refers to (average) annual net revenue rather than physical yield by weight.

Overfishing can then be defined as a state in which the existing sustained yield is less than $Y\lambda$, and can be increased only by means of a temporary reduction of yield. Similarly underfishing can be defined as a state in which yields can be permanently increased without a period of temporary reduction. It is clear that, barring "catastrophes of nature," a state of overfishing can be reached only through a process of excessive harvests. A common-property fishery is always overfished according to this definition, but so is any fishery operated for profit, whenever the owners adopt a discount rate greater than the social rate of discount.

A Numerical Example

To exhibit the effect of time-discounting, we consider the North Sea plaice (*Pleuronectes platessa*) fishery considered by Beverton and Holt. The following parameters apply to this species.

Natural mortality rate, $M = 0.10$
Weight curve $w(t) = w_\infty(1-e^{-K(t-\tau)})^3$
  where $w_\infty = 2867$ g
    $\tau = -0.815$ years
    $K = 0.095$
By equation (14) the optimal age $t\lambda$ satisfies

---

[3]The term "optimal sustained yield" as used here does not involve any stock–recruitment relationship, but refers only to optimization of the size of fish.

$$\frac{w'(t\lambda)}{w(t\lambda)} = \frac{3Ke^{-K(t\lambda-\tau)}}{1-e^{-K(t\lambda-\tau)}} = M + \lambda,$$

and this has the solution

$$t\lambda = \tau + \frac{1}{K} \ln\left(1 + \frac{3K}{M+\lambda}\right). \qquad (24)$$

When $\lambda = 0$ we obtain the age of maximum biomass, $t_o = 13.37$ years. If $\lambda = 10\%$, for example, then $t\lambda = 8.51$ years; the general dependence of $t\lambda$ upon λ is indicated in Fig. 8.

The optimal length $l\lambda$ can be calculated from $t\lambda$ by means of the formula (Beverton and Holt 1957, p. 34 and 285)

$$l\lambda = L_\infty (1-e^{-K(t\lambda-\tau)}); \qquad L_\infty = 68.5 \text{ cm}$$

We see that the optimal age $t\lambda$ is quite sensitive to the values of the instantaneous discount rate λ. This would be expected from equation (24), since the natural mortality rate $M = 10\%$ is of the same order of magnitude as normal discount rates.

How does the sustained yield under optimal fishing compare to the maximum sustainable yield? Let

$$\rho\lambda = \frac{B(t\lambda)}{B(t_o)}.$$

If fishing costs are negligible, then most of the biomass will be harvested at age $t\lambda$, so that $\rho\lambda$ represents the sustained yield as a proportion of the maximum yield. For $\lambda = 10\%$ we have $\rho\lambda = 81.4\%$; in general the dependence of $\rho\lambda$ upon λ is shown in Fig. 8.

This simple example emphasizes the fact that optimal sustained yield is not the same as maximum sustained yield.

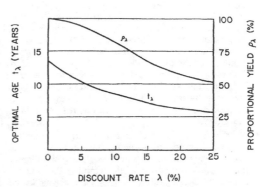

FIG. 8. Optimal age and yield for plaice.

1638   JOURNAL FISHERIES RESEARCH BOARD OF CANADA, VOL. 30, NO. 11, 1973

Limitation of the Analysis

As an introductory theoretical treatment, the present analysis has been forced to adopt certain unrealistic assumptions, the most serious of which are (a) independence of stock and recruitment, and (b) negligibility of costs. These assumptions have been dictated largely by our desire to present the dynamic aspects of fishery economics in the simplest possible form. Once the basic principle of the Fisher Rule is understood, there is nothing to prevent the inclusion of more realistic assumptions in computer analysis of specific fisheries problems.

If fishing costs are taken into consideration, our theory can still be used to obtain certain results. First, the optimal mesh size will be determined by some age t^* such that

$$t_\lambda < t^* < t_o.$$

Since t^* depends in a complex way upon all the parameters of the problem, and since it appears that t^* will usually be closer to t_λ than to t_o, except for marginally profitable fisheries, it would seem reasonable to choose t_λ itself as an optimal age of harvest. It is possible, however, that the age t_λ would result in a negative biovalue $V(t_\lambda)$, in which case a greater age of entry would clearly be preferable.

The question remains of determining the optimal fishing effort. In this regard, it should be remarked that our analysis has proceded along lines quite different from the analysis of Beverton and Holt, who based their recommendations for controlling fisheries on the concept of "eumetric" fishing (Beverton and Holt 1957; Section 19), which appears to be a purely equilibrium concept.

Beverton and Holt observed, however, that if a mesh-size limitation corresponding to the age t_o of maximum biomass is used, then sustained yield is an increasing function of fishing mortality F (which they assume to be constant over time). The same conclusion may be expected to hold, at least approximately, for the case of the optimal age t_λ, provided that reasonably small discount rates λ are utilized. If so, it follows that physical yields could not be increased by limiting fishing effort. There may still be valid economic reasons, however, for controlling the effort.

Certain fisheries, such as trawl fisheries, may not be amenable to any form of limitation upon the age of first capture. If fishing costs are significant, the theoretical problem of determining the optimal choice of fishing effort in this case seems to be very difficult. To optimize the recovery stage of an overexploited fishery, the best procedure would probably consist of estimating numerically the total present values for a number of alternate choices of regulatory schemes. Examples of the latter are given in Section 19 of the Beverton and Holt work.

Another practical shortcoming of our analysis is its disregard of random variations. Variations in the annual recruitment R should not seriously affect our conclusions; variations in the growth rate of the individual fish would have the effect of smoothing out the theoretically optimal fishing-effort function shown in Fig. 7. Random variations in the economic parameters could also have important implications, but we have not studied this problem.

Finally it must again be emphasized that our study has neglected any possible stock–recruitment relationship, and this could have a serious effect upon the description of an optimal harvest policy, particularly for heavily exploited fish populations. Further research appears to be required to unravel the combined effects of growth and recruitment on optimal fishing policies.

Conclusions

In this article we have discussed a dynamic version of the Beverton-Holt model of a commercial fishery in which stock–recruitment effects can be neglected. We have suggested that optimization in fisheries management would require maximization of the total present value of net revenues. The implications of such a management criterion, in the case of a fishery with negligible costs, include the following:

1. There is an optimal age of harvest t_λ that depends upon the given discount rate λ. If λ is positive then t_λ is less than the age t_o of maximum biomass.

2. If an optimal mesh limitation, determined by the optimal age t_λ, is adopted, then the optimal fishing effort will capture most of the mature fish during each fishing season.

3. The optimal recovery of an overexploited fishery requires an immediate mesh-size limitation as determined by the optimal age of harvest t_λ. The additional revenues from the regulated fishery will be sufficient to cover losses incurred during the recovery stage.

4. If mesh limitations, or other gear limitations affecting the age of harvest, are impractical, then optimization may be achieved through seasonal closures of the fishery. These closures may result in an unstable "pulse" fishery, with large efforts occuring once every several years.

5. Optimal development of an unexploited fishery, particularly of a long-lived species pos-

sessing a large natural population, will result in very large landings during the initial stage of development. These landings cannot be maintained in perpetuity by any regulatory mechanism, however.

6. The physical results of optimal regulation may depend quite strongly upon the value of the given discount rate λ. The use of high discount rates would lead to overfishing, at least in the usual "biological" sense in which sustained yields are significantly lower that maximum sustainable yields.

Mathematical Appendix

In this appendix we prove the complete Fisher Rule under the hypothesis that the expression $V'(t)-\lambda V(t)$ has at most one change of sign. When this hypothesis does not hold, the proof given still suffices to obtain a weak form of the rule. Our proof is elementary and self-contained, and does not rely on variational techniques.

We begin with the single year-class model. The problem then is to maximize the integral

$$PV(F) = \int_0^\infty e^{-\lambda t} F(t)[p\, x_F(t)w(t)-c]dt \qquad (25)$$

where

$$\frac{dx_F}{dt} = -(M + F(t))\, x_F(t), \qquad x_F(0) = R \quad (26)$$

$$0 \le F(t) \le F_{max} \qquad (t \ge 0). \qquad (27)$$

We assume that F_{max} is sufficiently large that the path $B(t)$ determined by the Fisher Rule can be followed.

Henceforth let $F(t)$ denote the effort function specified by the Fisher Rule, and let $F^*(t)$ denote an arbitrary admissible effort function — that is, $F^*(t)$ is a piecewise continuous function $(t>0)$ satisfying (27). We will argue by contradiction that $F^*(t)$ is not optimal unless $F^*(t) = F(t)$.

The curve $B_F(t)$ is as indicated in Fig. 2. Suppose for example that the curve $B_{F^*}(t)$ lies below the curve $B_F(t)$ — Fig. 9. If $F^*(t)$ is optimal, we know that $B_{F^*}(t)$ cannot fall below the level $B(t) = c/p$ for $t < t_o$. Hence there exist $T_1 < T_2$ with the property that

$$B_{F^*}(t) < B_F(t), \qquad T_1 < t < T_2;$$

$$B_{F^*}(T_i) = B_F(T_i), \quad i = 1,2.$$

We will establish the following inequality

$$\int_{T_1}^{T_2} e^{-\lambda t} F^*(t)[pB_{F^*}(t)-c]dt$$

$$< \int_{T_1}^{T_2} e^{-\lambda t} F(t)[pB_F(t)-c]dt, \qquad (28)$$

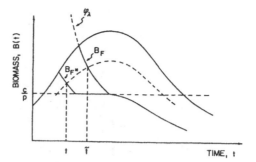

FIG. 9. Proof of the Fisher Rule.

which shows that $F^*(t)$ cannot be an optimal solution, since the principle of optimality (Pontrjagin et al. 1964) states that any subpath of an optimal path must itself be optimal. Inequality (28) shows that $F^*(t)$ is not optimal for $T_1 \le t \le T_2$, and hence not optimal overall.

To prove (28) we introduce a transformation $t \rightarrow \tilde{t}$ (for $T_1 \le t \le T_2$) by following the natural biomass curve through the point $(t, B_{F^*}(t))$ until it meets the curve $\phi_\lambda(t)$ at the point $(\tilde{t}, B_F(\tilde{t}))$ — see Fig. 9.

Now the basic equation (26), $dx/dt = -(M + F(t))x$, can be solved explicitly:

$$x_F(t) = x_F(T_1)exp[-M(t-T_1)-\int_{T_1}^t F(\tau)d\tau].$$

If $x(t) = x_o(t)$ denotes the natural population curve, then the transformation $t \rightarrow \tilde{t}$ is determined by the conditions

$$x(t) = e^{-M(\tilde{t}-t)}\, x(t),$$

$$x(t) = x_{F^*}(t); \qquad x(\tilde{t}) = x_F(\tilde{t}).$$

Combining these equations and simplifying, we obtain

$$\int_{T_1}^t F^*(\tau)a\tau = \int_{T_1}^{\tilde{t}} F(\tau)d\tau. \qquad (29)$$

By our hypothesis, the "no-fishing" zone is characterized by the inequality $V'(\tau) > \lambda V(\tau)$ where $V(\tau)$ is the natural biovalue; integrating between t and $\tilde{t}$ we conclude that

$$V_F(\tilde{t}) = V(\tilde{t}) > e^{\lambda(\tilde{t}-t)}\, V(t) = e^{\lambda(\tilde{t}-t)}\, V_{F^*}(t). \qquad (30)$$

Inequality (30) has the intuitive interpretation that a unit of fishing effort applied at the point $(\tilde{t}, x_F(\tilde{t}))$ contributes more to present values than a unit of effort applied at $(t, x_{F^*}(t))$. For a "unit of effort" corresponds to $F(t)\Delta t = 1$, and this contribues $e^{-\lambda t} F(t)$ $V(t)\Delta t = e^{-\lambda t} V(t)$ to present values. Inequality (30) asserts simply that $e^{-\lambda t}\tilde{V(t)} > e^{-\lambda t} V(t)$. Since (29) implies that $F^*(t)at = F(\tilde{t})at$, we obtain

$$\int_{T_1}^{T_2} e^{-\lambda t}\, F^*(t) V_{F^*}(t) dt < \int_{T_1}^{T_2} e^{-\lambda t} F^*(t) V_F(\tilde{t}) at$$

$$= \int_{\tilde{T}_1}^{T_2} e^{-\lambda \tilde{t}} F(\tilde{t}) V_F(\tilde{t}) d\tilde{t}$$

$$= \int_{T_1}^{T_2} e^{-\lambda \tilde{t}} F(\tilde{t}) V_F(\tilde{t}) d\tilde{t}$$

(since $F(\tilde{t}) = 0$ for $T_1 < \tilde{t} < \tilde{T}_1$) and this proves (28).

A similar argument can be used in case $B_{F^*}(t) > B_F(t)$, and we therefore conclude that $F(t)$ must be optimal.

For the case of a population con isting of several year-classes $x_1(t)$, $x_2(t)$, ..., let

$$x(t) = \sum_i x_i(t).$$

Then $\dfrac{dx}{dt} = -(M + F(t))x$, and the foregoing argument can be invoked verbatim to show that $F(t) = 0$ whenever $V'(t) > \lambda V(t)$. Ambiguous cases arise, however, when $V'(t) < \lambda V(t)$ if this inequality can subsequently reverse direction. It follows from the Pontrjagin maximum principle that either $F(t) = 0$ or $F(t) = F_{max}$ except when $V'(t) = \lambda V(t)$, but we shall not pursue the analysis.

Recently completed computer studies (unpublished) of the North Atlantic cod fisheries by Mr R. Hannesson of the University of B. C. Economics Department have verified and extended the analytical results presented in this paper. Hannesson's model allows for stock-sensitive recruitment relationships, imperfect gear selectivity, and non-negligible costs.

Acknowledgments

We wish to express our gratitude to Professor P. Larkin and to Mr S. Holt for their valuable suggestions. The research has been supported in part by the National Research Council of Canada, grant number A-3990.

ALLEN, R. L. 1972. Properties of age structure models for harvested populations. Ph.D. Thesis. Univ. British Columbia, Vancouver, B.C. 111 p.

BAUMOL, W. J. 1968. On the social rate of discount. Amer. Econ. Rev. 58: 788–802.

BEVERTON, R. J. H., AND S. J. HOLT. 1957. On the dynamics of exploited fish populations. Minist Agr. Fish. Food London Fish. Invest. Ser. 2 (19): 533 p.

CLARK, C. W. 1973a. Profit maximization and the extinction of animal species. J. Polit. Econ. (In press)
1973b. The economics of overexploitation. Science 181, No. 4100: 630–634.

CRUTCHFIELD, J. A., AND A. ZELLNER. 1962. Economic aspects of the Pacific halibut fishery. U.S. Dep. Inter. Fish. Ind. Res. Vol. 1: 173 p.

FISHER, I. 1930. The theory of investment. MacMillan, New York, N.Y. 566 p.

GOH, B. S. 1972. Optimal control of renewable resources and pest populations. 6th Hawaii Int. Conf. Syst. Sci. Jan. 8–12, 1972. (In press)

GORDON, H. S. 1954. Economic theory of a common-property resource: the fishery. J. Polit. Econ. 62: 124–142.

HIRSHLEIFER, J. 1970. Investment, interest and capital. Prentice-Hall, New York, N.Y. 320 p.

PONTRJAGIN, L. S. et al. 1964. The mathematical theory of optimal processes. Pergamon Press, Oxford. 438 p.

SCOTT, A. D. 1955. The fishery: the objectives of sole ownership. J. Polit. Econ. 63: 116–124.

WALTERS, C. J. 1969. A generalized computer simulation model for fish population studies. Trans. Amer. Fish. Soc. 98(3): 505–512.

[32]

Journal of Environmental Economics and Management 37, 129–150 (1999)

Article ID jeem.1998.1060, available online at http://www.idealibrary.com on IDEAL®

Bioeconomics of Spatial Exploitation in a Patchy Environment*

James N. Sanchirico[†]

Resources for the Future, 1616 P Street NW, Washington, DC 20036

and

James E. Wilen

Department of Agricultural and Resource Economics, University of California, Davis, California 95616

Received October 9, 1997; revised October 1998

This paper presents a model of renewable resource exploitation that incorporates both intertemporal dynamics and spatial movement. The model combines the H. S. Gordon–Vernon Smith hypothesis of a rent dissipation process with Ricardian notions that resources are exploited across space in a pattern dependent upon relative profitabilities. The population structure is characterized in a manner consistent with modern biological ideas that stress patchiness, heterogeneity, and interconnections among and between patches. Generally, we find the equilibrium patterns of biomass and effort across the system to be dependent upon bioeconomic conditions within each patch and the nature of the biological dispersal mechanism between patches. We use simple examples to illustrate how the distribution of effort throughout the system reflects the heterogeneity and the spatial biological linkages. © 1999 Academic Press

I. INTRODUCTION

Economists' models of renewable resource exploitation have utilized various abstractions for both analytical tractability and to focus attention on variables of particular interest in the policy process. The most prominent and widely cited examples of bioeconomic analysis begin, for example, with the assumption that the exploitable biomass is enumerable in terms of either total weight or numbers, and that the principal birth–death–growth processes can be adequately described by a small number of parameters. These "lumped parameter" models include, for example, the widely used Shaefer [32] model in which biomass dynamics are characterized by the intrinsic growth rate and the carrying capacity of the environ-

*The authors are Fellow, Quality of the Environment Division, Resources for the Future, and Professor, Department of Agricultural and Resource Economics, University of California, Davis, respectively. We thank participants at the European Association of Environmental and Resource Economists meeting (Lisbon, June 1996), the Association of Environmental and Resource Economists Workshop (Annapolis, MD, June 1997), and at seminars at U.C. Davis, Oregon State, Resources for the Future, and North Carolina State for helpful comments. This research was partially funded by a grant from the National Sea Grant College Program, NOAA, U.S. Dept. of Commerce, under Grant No. NA36RG0537, Project No. 89-F-N; and by the Division of Agricultural and Natural Resources and the Giannini Foundation, University of California.

[†] E-mail address: sanchirico@rff.org.

129

130   SANCHIRICO AND WILEN

ment. The logistic–Shaefer model is the foundation for the bulk of fisheries economics literature.

While economists have been content to stick to these simple models as descriptions of biological processes, there has been a subtle paradigm shift in biology away from these simplifications. There is, in fact, a well-developed literature that incorporates many more realistic features of populations than those embedded in the simple lumped parameter abstractions. Biologists are well versed, for example, in the intricacies introduced by considering age- and size-structured populations and no serious modeler would ignore these aspects in slow-growing, broad age distribution populations where exploitation targets particular sizes, ages, or sexes either for market reasons or because of regulations. In addition, biologists have begun to focus much more seriously than economists on the implications of *space* in their models of populations.[1,2] The newest paradigms in biology, from the genetic to the population level, recognize that real populations are heterogeneously distributed across space and that the spatial dimension adds some new insights to our understanding of biological processes.[3]

Aside from keeping up with these new developments in biology, there are several other reasons why resource economists might want to devote some attention to the spatial dimensions of bioeconomic activities. First, in systems in which the resource is distributed heterogeneously in space, we are most likely missing a considerable amount of interesting behavior by aggregating out the spatial aspects of economic activity. To take a simple example, suppose vessels are moving back and forth between several patches in response to changes in relative profit opportunities. It is easy to conceive of circumstances where much of the between patch movement cancels out when aggregated, suggesting very little behavioral responsiveness when in fact there is a considerable amount of response. If we are interested in estimating behavioral elasticities, we would uncover little with aggregated data but we would uncover much more with data disaggregated over space in this example. In order to structure empirical analysis, however, we need a theory which deduces testable hypotheses about how decision makers behave over space. A second reason for incorporating space into models of exploitation is that there is a considerable amount of new interest in spatial management options, particularly among biologists and fisheries and wildlife managers. For example, biologists have promoted natural refuges as management tools, under the expectation that permanent closures of certain areas might enhance overall biological productivity of an exploited system.[4] Similarly, others have proposed rotating harvest zones, under which areas are closed for a period and then reopened for exploitation after they have been allowed to recover back to a natural state [9]. In order to analyze these

---

[1]See, Skellam [33]; Levin [23, 24]; Allen [2]; Hastings [17, 18]; Holt [19]; Possingham and Roughgarden [26]; Huffaker, Bhat, and Lenhart [21].

[2]Among the few examples of work by economists incorporating space are the paper by Brown and Roughgarden [10] examining larval pools in a metapopulation model, papers by Huffaker, Bhat, and Lenhart [5, 6, 21] examining spatial–intertemporal control of a pest population, and papers by Schulz and Skonhoft [31], Skonhoft and Solstad [34] analyzing exploitation of transboundary terrestrial species. In addition, there has been some work linking landscape ecology features to economic behavior. See, for example, a paper by Albers [1] examining tropical forest management and Bockstael [7] and Geoghegan, Wainger, and Bockstael [15] reporting interesting new empirical relationships in a rural–suburban setting.

[3]See, Holt [19]; Levin [23]; Man, Law, and Polunin [25].

[4]See, Bohnsack [8]; Botsford *et al.* [9]; Davis [13]; Dugan and Davis [14]; Roberts and Polunin [28].

kinds of new options, we need a conceptual model that explicitly considers spatial characteristics of the resource base as well as spatial dimensions of the exploiting industry.

In this paper, we present a model of renewable resource exploitation that incorporates both intertemporal dynamics and spatial movement. The model combines both the H. S. Gordon [16]–Vernon Smith [35] hypothesis of a rent dissipation process with Ricardian notions that resources are exploited across space in a pattern depending upon relative profitabilities. We characterize the population structure in a manner consistent with modern biological ideas that stress patchiness, heterogeneity, and interconnections among and between patches. For the purpose of this paper we assume a "patch" is a location in space that contains or has the potential to contain an aggregation of biomass. We also assume that patches are located a fixed and discrete distance from one another. The biological interconnections between patches may be of many varieties and we develop a general model which nests several cases and then we explore the implications of several of the most common from the biological literature.

In the next section we begin by describing a biological model of a spatially interconnected population system. For simplicity, we first ignore the possibility of harvesting in order to focus attention on the natural biological forces inducing population movement across space and time. We also elaborate on the biological side of the system by showing how various descriptions of biological systems in the ecological literature can be reduced to special cases of a general model through specifications of restrictions on the dispersal parameters. In Section III, we sketch out a simple open access harvesting model which captures essential features of the Gordon rent dissipation model in a patchy environment.[5] Then in Section IV we pull together the biological and economic components and we solve an example of a spatially explicit bioeconomic model. We discuss equilibrium properties and we show how they reflect the specific nature of the biological and economic system. Section V discusses some of the important insights that emerge when we add the spatial dimension to bioeconomic models of exploitation.

II. A SPATIAL MODEL OF POPULATION BIOLOGY

Biologists have explored a range of population models that incorporate spatial features, starting with the paper by Skellam [33]. These models are sometimes referred to as metapopulation models, where a metapopulation is defined as a group of linked subpopulations distributed across a set of spatially discrete habitats or patches. The number of organisms in each patch is assumed to depend upon

---

[5]A referee raised a question about whether institutional configurations other than open access ought to be considered, because most fisheries are essentially limited (or restricted) access. This is a good point, and one that had been made persuasively by Homans and Wilen [20]. Our response is twofold. First, as a matter of modeling philosophy, developing more realistic models often benefits by starting with simple models and adding structure, once the simpler model is understood. In this case, adding spatial structure complicates the model considerably and hence starting from the simple polar extreme is a sensible first step along the way to developing more encompassing models. Second, and more to the point of the referee's concerns, unless real world regulatory structures establish property rights (i.e., ITQs), there are still open-access incentives operating which affect decisions across many margins, including location choice in a spatial system. Hence our approach would apply, even in more general circumstances with some forms of regulations and limited entry.

"own" density-dependent growth processes as well as dispersal from and to other patches in the system. The dispersal process is important because it allows for the possibility of temporary local extinction without driving the whole population to extinction. This could occur, for example, if the population in one patch is temporarily driven to zero and subsequently recolonized via dispersal from other patches.

Following Levin [23, 24], Hastings [17, 18], and others [19, 38], most analytical spatial population models are structured as follows,

$$\dot{x}_i = f_i(x_i)x_i + d_{ii}x_i + \sum_{\substack{j=1 \\ i \neq j}}^{n} d_{ij}x_j, \qquad i = 1, \ldots, n, \tag{1}$$

where $\dot{x}_i$ is the instantaneous rate of change of biomass in patch i, x_i is the biomass in patch i in time t, $f_i(x_i)$ is the per capita growth rate in patch i, d_{ii} is the rate of emigration from patch i ($d_{ii} \leq 0$), and d_{ij} is the dispersal rate between patches i and j ($d_{ij} \geq 0$). In this system there are n patches, each of which has own population dynamics described by the per capita growth function $f_i(x_i)$.[6] In addition to density-dependent own growth processes, each patch also contributes and receives organisms via dispersal to and from other patches. In the previous formulation, dispersal is written as separable from own growth processes and linear in population levels. At any point in time a given patch may be a net contributor to, or a net receptor from, the rest of the system. The net dispersal associated with patch i is ($ND_i = d_{ii}x_i + \sum_{j, i \neq j}^{n} d_{ij}x_j$) in the preceding model.

The dispersal representation in (1) is capable of depicting a range of qualitatively different dispersal processes via appropriate choice of the coefficients d_{ij}. There are various assumptions that have been made in the biological literature regarding the nature of the dispersal process, depending upon the assumed behavior of the organism and the nature of the habitat the population inhabits. A common assumption is that population biomass flows in directions over space that depend upon *relative densities*. Perhaps the simplest representation of this type of density-dependent dispersal is with dispersal out of patch 1 as $d_{11}x_1 + d_{12}x_2 \equiv b(x_2/k_2 - x_1/k_1)$ and dispersal into patch 2 as $d_{22}x_2 + d_{21}x_1 \equiv -b(x_2/k_2 - x_1/k_1)$ in a two patch model.[7] In this simple case, there is a common marginal dispersal rate b, and population disperses in a manner dependent upon the population densities relative to their natural carrying capacities. Hence if the population in patch 2 is 90% of its carrying capacity and the population in patch 1 is at 85% of carrying capacity, dispersal flows from patch 2 to patch 1 regardless of absolute population levels. Dispersal is thus an equilibrating force here that augments natural growth processes; when populations are low relative to their carrying capacities, both own growth and dispersal from other patches operate in a complementary fashion to bring populations in each patch to their carrying capacities. It is also the case that the directional gradient of the dispersal process is endogenous at any point in time, and the system responds in a gravity flow manner over space to bring the whole

---

[6] If one assumes that the biological own production is characterized by a logistic growth curve then $f_i(x_i) = r_i(1 - x_i/k_i)$ for each patch i.

[7] In the case of three patches (i, j, k), we can represent dispersal out of i as $d_{ii}x_i + d_{ij}x_j + d_{ik}x_k \equiv b(x_j/k_j - x_i/k_i) + b(x_k/k_k - x_i/k_i)$ so that $d_{ii} = -2b/k_i$.

BIOECONOMICS OF SPATIAL EXPLOITATION   133

system into equilibrium. This type of system approaches what we call a *homogeneous equilibrium* in which each population reaches its carrying capacity, at which point system dispersal goes to zero. It is homogeneous in the sense that the system is in a space–time steady-state; population levels are unchanging over time, and there is no dispersal or movement *over space* in equilibrium.[8]

While most spatial ecology literature focuses on relative density-dependent dispersal mechanisms, there have been some formulations based on *unidirectional* processes.[9] This special subclass of models, often referred to as sink-source models, depicts populations flowing from a source to a sink regardless of the population or density level in the sink. For example, a two patch sink-source model might have growth in the source area equal to $\dot{x}_i = r_i x_i (1 - x_i/k_i) + d_{ii} x_i$ and might have growth in the sink area equal to $\dot{x}_j = r_j x_j (1 - x_j/k_j) + d_{ji} x_i$ (recall $d_{ii} < 0$). This type of model exhibits qualitatively different behavior compared with the general density dependent formulation in its unexploited equilibrium. In a sink-source model, the equilibrium may be a *nonhomogeneous* (or flux) *equilibrium*, in the sense that while population levels in each patch are unchanging over time, there is actually movement over space between patches in equilibrium. With this type of dispersal process, biomass continues to flow into the sink even after both populations reach their equilibria, at which point, the source patch has births exactly offset by deaths and migration and the sink patch has deaths exactly offset by births and immigration.[10]

The general model in Eq. (1) is thus capable of depicting a broad spectrum of biological conditions in a system distributed over space. If we are content to characterize the dispersal mechanism as linear and separable from the own growth process, we can stack Eq. (1) for all n patches and we can express the whole system in matrix fashion as

$$\dot{\mathbf{x}} = \mathbf{F}(\mathbf{x})\mathbf{x} + \mathbf{D}\mathbf{x}. \tag{2}$$

The biological system in (2) is comprised of n dynamic equations where $\dot{\mathbf{x}}$ is a $n \times 1$ vector whose components $(\dot{x}_i)$ are the instantaneous rate of change of biomass in patch i in time t, $\mathbf{x}$ is a $n \times 1$ vector whose components (x_i) are the level of biomass in patch i in time t, $\mathbf{F}(\mathbf{x})$ is a $n \times n$ diagonal growth matrix where the diagonal components $f_i(x_i)$ are the average growth rates in patch i, and $\mathbf{D}$ is a $n \times n$ matrix of dispersal rates $(D_{ij} = d_{ij}$ for all $i, j)$. The biological model in (2) depicts the dispersal of elements of the population differentiated by location through $\mathbf{D}$, and heterogeneous own-patch-specific biological conditions such as intrinsic growth rates and carrying capacities through $\mathbf{F}(\mathbf{x})$. This system behaves over time and space in a manner reflecting the dual effects of own-patch-specific growth processes and dispersal processes, the former captured in the first component of (2) whereas the latter captured by the matrix $\mathbf{D}$. While biologists have characterized various natural systems in many ways, most can be thought of as

[8]Note that homogeneity in this case stems from two assumptions. First, dispersal is assumed to depend upon differences in relative densities $(x_i/k_i - x_j/k_j)$. Second, we are considering an unexploited equilibrium where population in each patch approaches carrying capacity. For both reasons, in this system spatial movement goes to zero in equilibrium.

[9]See, for example, Pulliam [27]; Tuck and Possingham [37]; Sanchirico and Wilen [30]; Sanchirico [29].

[10]This may lead to an unexploited equilibrium population in the source patch that is less than k_i and similarly, an equilibrium population in the sink greater than k_j.

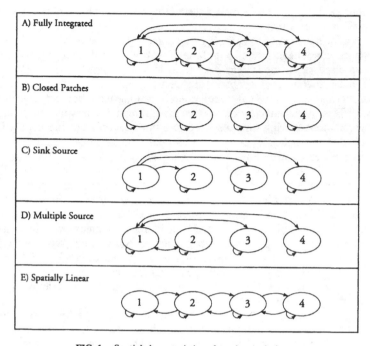

FIG. 1. Spatial characteristics of patch populations.

special cases of the general system in (2), depicted with restrictions on the **D** matrix.[11]

In Figure 1, we illustrate five general types of biological systems nested in (2) (discussed separately in Carr and Reed [11]). See Table I. The five types of biological systems are; fully integrated, closed, sink-source, multiple source, and spatially linear. A *fully integrated* system allows for the biomass to disperse directly from any patch to any other patch throughout the system. The dispersal matrix for this system would have nonzero elements in each column–row position. A *closed* system is the opposite case where maintenance of biomass density within each patch is only determined by own production and there is no dispersal out of or into the system (**D** is the null matrix). A *sink-source* system occurs when one (or multiple) patch(es) provides unreciprocated biomass replenishment to other patches. A sink-source system with one patch feeding all others would have

---

[11]Ecologists generally impose some structure on the dispersal process, either to ensure sensible interpretation or for analytical convenience. In this paper we will impose the following restrictions on the **D** matrix: (i) $d_{ii} \leq 0$, (ii) $d_{ij} \geq 0$, and (iii) $\sum_{k=1}^{n} d_{ki} = 0$ $i = 1, 2, \ldots, n$ (column sums to zero). Assumptions (i) and (ii) are essentially accounting restrictions that allow us to separate population fractions that are leaving one area from those that are arriving. Assumption (iii) is an "adding up" restriction which ensures that whatever leaves a patch during dispersal from a group of patches also shows up in the receptor patches. A stronger version of this adding up restriction that is sometimes imposed is the symmetry condition $d_{ij} = d_{ji}$, which ensures that whatever leaves patch i specifically for patch j also arrives in patch j specifically from patch i. However, this symmetry condition excludes sink-source dispersal processes and hence we impose the weaker condition (iii).

TABLE I

Restrictions on the Dispersal Parameters

Spatial configuration	Restrictions
Fully integrated	$d_{ii} < 0, d_{ij} > 0$ for all $i, j = 1, \ldots, n$
Closed	$d_{ij} = 0$ for all $i, j = 1, \ldots, n$
Sink-source	source (patch k): $d_{kk} < 0, d_{ki} = 0$
	sink: $d_{ik} > 0, d_{ij} = 0$ for $(i, j) \neq k$ where $i, j = 1, \ldots, n$
Multiple-source	source (patch k): $d_{kk} < 0, d_{ki} > 0$
	sink: $d_{ik} > 0, d_{ij} = 0$ for $(i, j) \neq k, i, j = 1, \ldots, n$
Spatially linear	$d_{ii} < 0, d_{ij} > 0, d_{ik} = 0$ where (i, j) are neighbors for $i = 1$ and n
	$d_{ii} < 0, d_{ij} > 0, d_{ik} > 0, d_{ik+1} = 0$ where (i, j, k) are neighbors for $1 < i < n$

nonzero elements only in the column corresponding to the source. The *multiple source* case occurs when many patches contribute biomass to one common pool which then is redistributed among the patches, as, for example, with a breeding ground or a juvenile nursery area. The system would have nonzero elements in the common pool patch column, and the system would have zeros in every row element that was not linked (either as contributor or receptor) to the source patch. Lastly, the *spatially linear* case nests various spatial configurations (e.g., linear, circular, or square) and assumptions regarding the edge effects of neighboring patches. For example, if patches are arrayed in a line, patch dispersal may only occur in a pairwise manner between adjacent patches and this can be denoted by a dispersal matrix which is band diagonal.

Note that each of these cases involves dispersal structure between patches which reflects inherent behavioral characteristics of the population and/or oceanographic features of the spatial system. For example, a pelagic species with high mobility inhabiting a relatively homogeneous marine environment might be depicted with a fully integrated metapopulation system with relatively high dispersal rates. This system would equilibrate faster than a corresponding system with slow dispersal rates and/or blocked dispersal paths caused by currents, prevailing winds, or geographic and oceanographic barriers. A relatively sedentary species inhabiting a substrate with a high degree of variability might be characterized by blocked dispersal paths between certain groups of patches and low dispersal rates. Similarly, we might have a species characterized by a lifecycle involving larval dispersal into a juvenile area, which subsequently disperses to various patches in a manner dependent upon currents and prevailing winds. This could be depicted with a sink-source structure involving unidirectional and unreciprocated biomass replenishment from one or several source patches. The important point is that even in its linear and separable formulation, (2) is relatively general and capable of capturing a broad range of ecological circumstances.

III. OPEN ACCESS EXPLOITATION IN A SPATIAL SETTING

In this section we develop a model of an industry exploiting a patchy resource, under conditions of open access. While the model is basically an economic model reflecting economic behavior, it is also qualitatively symmetric with the population model discussed in the foregoing text. In the spatial population model, the key state variables are patch-specific population levels, which may adjust across space and time in response to relative densities, both within the patch and across patches. In the economic model, the key state variables (effort) adjust in response to rent differentials across space and time, both within each patch and between patches.

The model we develop here is a generalization of two important papers characterizing open access resource exploitation. The first is the paper by H. S. Gordon, published in 1954 [16]. Gordon hypothesized that under open access, any rents that would normally accrue to a resource owner would instead accrue to entrants. This additional incentive, over and above entrants' normal opportunity costs, would attract excess effort until average (and not marginal) returns were driven down to opportunity costs. Hence open access conditions would attract excess effort and rents would be dissipated.[12] Gordon's model was extended in 1968 by Vernon Smith [35], who hypothesized an explicit dynamic entry–exit process in addition to a biological growth mechanism.

Both the Gordon and Smith models begin by assuming that a composite effort variable E (which we refer to as vessels, for simplicity) exists so that total industry harvest can be written $H(E, X)$ where X is the population biomass. Assume output price p is given, industry level operating costs are $C(E, X)$, and industry level opportunity costs associated with other uses of the composite input outside this particular fishery are πE. We can thus write average net rents generated per vessel as $NR(E, X) = [pH(E, X) - C(E, X) - \pi E]/E$. In Gordon's story, under open access effort continues to enter until average gross operating profits equal opportunity costs, thereby dissipating all potential rents. Smith used the same structure, but in addition, hypothesized a dynamic process describing entry–exit behavior.[13] Smith's model depicted the interaction between the harvesting industry and the population as

$$\dot{E} = \Phi(NR),$$
$$\dot{X} = F(X) - H(E, X). \tag{3}$$

In this generalized model, the rate of change of effort is assumed increasing in average net rents per vessel such that when average net rents are positive entry

---

[12] Interestingly, Gordon [16] also briefly addressed the issue of how effort might distribute itself across space, with an analysis of different fishing grounds. In his (static) model, effort would distribute itself over different grounds until average rent in each patch was equalized at some common opportunity cost. As he suggested, this allocation would be inefficient, because it would be desirable to allocate effort over space to equalize marginal and not average rents.

[13] Smith's model is more complete than Gordon's in several ways. First, it replaces Gordon's awkward representation of yield–population relationships with an explicit density-dependent biological growth function. Second, it introduces variable factors and distinguishes them from vessel capital, which is assumed to enter sluggishly in response to rents. Finally, it brings the entry–exit and biological dynamics to the forefront of attention, showing how open-access exploitation dynamics can generate oscillatory overshoot. This possibility provoked several subsequent papers that examined whether an open-access industry might drive a fishery to extinction during the initial phase of oscillatory overshoot (Berck [3], and Leung and Wang [22]).

occurs and when negative exit occurs. Smith's dynamic model nests Gordon's static model because, in equilibrium, average net rents are driven to zero by entry. In addition, however, the model generates predictions about the approach paths, which may be asymptotic or oscillatory, depending upon (among other things) the relative speed of industry response to rent opportunities.

In both the Gordon and Smith models of a single patch, the number of vessels operating in the patch changes over time in a manner dependent upon bioeconomic conditions in the patch relative to *outside* opportunities. In generalizing to a multiple patch environment, it seems desirable to incorporate this same assumption that each vessel has opportunities outside the fishery in question, particularly in an open-access setting. In addition, however, in a multiple patch setting there is an opportunity cost to participating in any given patch, namely, the rents that could be earned in any other patch. Hence a fully specified model of behavior in a patchy environment needs to account for both external and internal profit opportunities. In this setting, movement to and from any particular patch ought to be related to relative profit opportunities both vis-à-vis opportunities outside the whole fishery as well as opportunities in other patches within the fishery.

These ideas can be operationalized as follows. Consider the system discussed in Section II in which there are n distinct biological patches represented by an integer valued index i. Let $E_i(t)$ be the effort level in patch i, let $x_i(t)$ be the biomass level in patch i in period t, and let $NR_i(E_i, x_i)$ be the instantaneous average net rents per vessel associated with harvesting in patch i. As in the single patch model in (3), net rents are assumed to be average gross operating profits per vessel, less an opportunity cost π per vessel. Gross operating profits are assumed to be a function of $E_i(t)$ and $x_i(t)$ via a harvesting function $H_i(E_i, x_i)$, a cost function $C_i(E_i, x_i)$, and a parametric output price p. Opportunity costs per vessel π are assumed to reflect alternative income earning opportunities outside of the fishery, which we assume to be constant per unit of vessel capacity and common across all patches.

In this n-patch system, we can model the forces operating to change the distribution of effort over space as

$$\dot{E}_i = \Phi_i(NR_i) + \sum_{\substack{j=1 \\ i \neq j}}^{n} \Delta_{ij}(NR_i, NR_j), \qquad i = 1, 2, \ldots, n. \tag{4}$$

This behavioral equation is a spatial generalization of the entry–exit part of equation system (3). Entry into and exit from a particular patch i is again hypothesized to depend upon own-patch rents vis-à-vis outside opportunities, depicted by the $\Phi_i(NR_i)$ function; when Φ_i is positive effort is entering the fishery into patch i from outside the fishery and when negative vessels are exiting to an outside pool. We continue to assume that Φ_i is increasing in NR_i so that patches with the highest net rents in a system are experiencing the highest rates of entry, other things equal. In addition, vessels move into and out of any given patch i to other patches depending upon relative opportunities within the fishery. The second component of (4) depicts this dispersal of effort to and from various other patches in the system. Total dispersal to and from other patches is designated by a summation of $j - 1$ pairwise dispersal functions. We choose the convention of adding dispersal relationships so that when Δ_{ij} is positive, vessels are assumed to be moving to patch i from j and vice versa when negative.[14,15]

The simplest explicit representation of the previous system, and one which we use in the remainder of this paper, is a linear specification where $\Phi_i(NR_i) = s_i NR_i$ and $\Delta_{ij} = s_{ij}(NR_i - NR_j)$ so that

$$\dot{E}_i = s_i NR_i + \sum_{\substack{j=1 \\ i \neq j}}^{n} s_{ij}(NR_i - NR_j), \qquad i = 1, 2, \ldots, n. \tag{5}$$

In this specification (qualitatively similar to the biological dispersal model discussed in Section I), the rate of change of effort in a particular patch is proportional to the level of rents in that patch relative to both outside opportunities (π), where the own-patch responsiveness rate is s_i, and is proportional to rents in different patches in the system. In this system, when rents are higher in patch i than j, there is dispersal from j into i by an amount proportional to rent arbitrage opportunities $(NR_i - NR_j)$, where the cross patch responsiveness rate is s_{ij}. The larger the difference in rents between patches i and j, the higher the rate of movement of vessels between the two patches, other things equal. Whether a given patch is a net receptor or net contributor to the rest of the system depends upon its interrelationships with all other patches under the summation. At any given point in time, patch i may be attracting vessels from patch j and contributing vessels to patch k in a manner that depends upon relative rents across the whole system. Other things equal, these spatial forces tend to redistribute effort in a manner that, in the long run, equalizes average net rents across all patches at the common opportunity cost. Note that this is not the *optimum* way to distribute vessels because this is an open-access system; there are too many vessels in the whole system and they are allocated across the fishing grounds to equalize average rather than marginal rents.

As we did with the spatially explicit biological model, we can stack Eq. (5) for all n patches and we can express the system as

$$\dot{\mathbf{E}} = \mathbf{S}\mathbf{r}(\mathbf{x}, \mathbf{E}) + \Delta\mathbf{r}(\mathbf{x}, \mathbf{E}). \tag{6}$$

The economic system in (6) is comprised of n patch-specific dynamic equations where $\dot{\mathbf{E}}$ is a $n \times 1$ vector whose components are the instantaneous rate of change of effort in each patch $(\dot{E}_i)$, $\mathbf{S}$ is a $n \times n$ diagonal matrix of entry–exit response rates $(S_{ii} = s_i)$,[16] Δ is an $n \times n$ matrix of dispersal response rates between patches i and j $(\Delta_{ii} = \sum_{j=1}^{n} s_{ij}$ and $\Delta_{ij} = -s_{ij})$, and $\mathbf{r}(\mathbf{x}, \mathbf{E})$ is a $n \times 1$ vector of net rents in patch i where net rents in patch i are a function of the biomass in patch i and the

---

[14] In order to avoid certain implausible results, we need to impose a symmetry condition on the dispersal functions such that $\Delta_{ij} = -\Delta_{ji}$ for all $i, j = 1, \ldots, n$. This condition states that whenever there is a gain recorded in patch i by the vessels attracted from patch j, there must be a corresponding loss associated with an exact symmetrical movement elsewhere in the system, namely, the number of vessels lost from patch j that are moving into patch i. This ensures that vessels are not lost during the process of dispersal.

[15] It is important to point out that the total effort operating in the fishery $(E^{total}(t) = \sum_{i=1}^{n} E_i(t))$ changes from one period to the next as a function of a weighted sum of net rents throughout the system $(\dot{E}^{total} = \sum_{i=1}^{n} \dot{E}_i = \sum_{i=1}^{n} \Phi_i(NR_i))$.

[16] If $s_i = s$ for all $i = 1, \ldots, n$, then $\mathbf{S} = s\mathbf{I}$ where $\mathbf{I}$ is an $n \times n$ identity matrix.

level of effort in patch i. The characteristics of the matrices S and Δ and the functional form of net rents will determine the steady-state levels and the stability of the system.

Equation system (6) is a simple representation of the process of effort distributing itself across different fishing grounds as first discussed by Gordon and it also generalizes Smith's ad hoc depiction of bioeconomic dynamics by adding spatial dispersal. The model is still ad hoc in the same ways that the Smith–Gordon models are. In particular, there is no explicit microtheory of behavior here; aggregate entry–exit behavior is simply assumed to depend in a myopic and proportional way on current rents. Several points can be made in defense of this structure however. First, with respect to the assumption that behavior is myopic, it should be expected that, under real world open access conditions, behavior is relatively myopic. This is because under open access it does not pay to be too foresightful since sacrifices made today for tomorrow are likely to be eroded away by open access competitive behavior. Second, even if we wished to include less myopic expectations behavior (see Berck and Perloff [4]), there really is no well developed microdynamic theory of open access behavior. Third, as it turns out, the simple model has been found to predict effort dynamics quite accurately. For example, Wilen [39] and Conrad and Bjorndal [12] and others have found that simple entry–exit functions fit the empirical data in open access settings quite well. Finally, from a modeling perspective, it makes sense to appeal to Occam's razor and to start with the simplest model which still captures fundamental forces. In this setting, we rely on only two premises: that individuals seek to arbitrage profit opportunities over time and space and that they do so sluggishly due to various intertemporal and spatial adjustment costs (the most important of which may be that it simply takes time to move across space).

IV. EQUILIBRIUM IN A SPATIAL BIOECONOMIC MODEL

In this section we bring together the two components discussed in Sections I and II and we discuss the implications of the spatially integrated bioeconomic model. First we discuss some general characteristics of the intertemporal–spatial dynamics and the nature of equilibrium in this system. Then, we solve for a closed form solution for a particular specification of the model and we illustrate how various biological systems give rise to corresponding bioeconomic equilibria. The spatial bioeconomic model developed here depicts a mobile metapopulation system and a mobile harvesting system operating over that population. The integrated system can be represented by

$$\dot{x} = F(x)x + Dx - h(x, E),$$
$$\dot{E} = Sr(x, E) + \Delta r(x, E). \tag{7}$$

Biomass levels in each patch change over time according to three additively separable forces: own-patch birth–death ($F(x)x$), between-patch dispersal (Dx), and harvest $h(x, E)$. Effort levels in each patch respond to own-patch rents relative to outside the fishery opportunities ($Sr(x, E)$), and between-patch rents which generate fleet dispersal ($\Delta r(x, E)$). With this integrated model, we can explore various features of bioeconomic heterogeneity associated with individual patches in the system, including possibilities of different biological productivity and differences in costs, prices, and/or catchability in each patch. In addition, the model is capable of depicting a range of potential linkages between patches that reflect biological and oceanographic features associated with the landscape ecology inhabited by the population. For example, a biological system along a narrow coastal upwelling environment might be best depicted as a system of patches located in a line, with each one linked mainly to those adjacent to it. In contrast, a system located in a large continental shelf area might best be viewed as a fully (or at least highly) integrated system, with each patch linked to most of the others via larval mixing and circulation. These two systems would be characterized by suitably specified restrictions on the biological dispersal matrix.[17]

In disequilibrium, both biological and economic forces are operating to influence convergence to the bioeconomic equilibrium. For example, in a relative density-dependent system, patches that are closer to their natural carrying capacities "feed" other patches via dispersal, thus speeding the process of system convergence to long run equilibrium. In addition, other things equal, patches with higher biomass levels attract vessels from other patches with lower levels, also reinforcing the ability of low biomass patches to catch up with higher biomass counterparts. Biological and economic dispersal thus both act as system averaging forces, smoothing out the influence of bioeconomic heterogeneity, and linking the inherently dynamic process of convergence across space. In the long run, the system equilibrium can be characterized by

$$\dot{x} = F(x)x + Dx - h(x, E) \overset{set}{=} 0 \quad \Rightarrow \quad h(x, E) = [F(x) + D]x,$$

$$\dot{E} = (S + \Delta)r(x, E) \overset{set}{=} 0 \quad \Rightarrow \quad r(x, E) = 0. \tag{8}$$

From inspection, several properties of the spatial equilibrium can be inferred. First, in an open access spatial equilibrium, net rents in each patch are zero as we would expect.[18] Second, the matrix of response parameters in the entry–exit–dispersal system does not affect the equilibrium levels of either effort or biomass but rather (as in Smith's model) affects the speed of adjustment and qualitative characteristics of the approach paths in each patch. High response rates generate oscillatory overshoot and undershoot in an dampened convergence and slow response rates correspondingly generate approach paths that are asymptotic. Third, the biological equilibrium is maintained where harvest equals yield in each patch, but patch yield is composed of both intrinsic growth and dispersal from and to other patches.

---

[17] With respect to economic dispersal, generally we would expect that the fleet would be relatively free to move from any patch to any other patch and hence the fleet dispersal matrix would be fully integrated. However, ecological landscape features or management policy might prevent vessel dispersal between some linkages and this would be modeled with restrictions on the fleet dispersal matrix. For example, a fishery operating over a system of peninsulas and fjords might find it possible to move from patch 1 to patch 2 to patch 3 and back, but not directly between patch 1 to patch 3, and this would be modeled with $s_{13} = s_{31} = 0$.

[18] It can easily be shown that the industry's equilibrium equation in (8) can be simplified to the following; $Ar = 0$ where $A = (S + \Delta)$. Furthermore, it can be shown that A is symmetric, quasi-dominant diagonal, and nonsingular. From the properties of A, a basic result from linear algebra is that the only solution to $Ar = 0$ is $r = 0$ (result (8)).

As discussed earlier, this bioeconomic system is capable of depicting a very broad range of ecological and economic circumstances. Many of the most interesting implications of this integrated model emerge out of various assumptions about linkages and dispersal summarized in the **D** matrix. The **D** matrix not only identifies exactly which patches are biologically linked to one another, but it also defines the mechanisms by which patches are linked and equilibrium is achieved. For example, biological dispersal might be driven by a density-dependent or unidirectional process but in either case, when the exploited biological system is in equilibrium, own-patch growth is balanced by harvesting and net migration. In contrast, the economic system is in equilibrium when net rents are equalized across space and entry–exit stops when these rents are zero. In general, the *combined* bioeconomic equilibrium is a nonhomogeneous or flux equilibrium in the sense that there is some movement across space in equilibrium, even though the population size and the fleet size in each patch is constant. The flux in the system arises out of the biological system and not the economic system, however, because each patch's population size is maintained by the balance between own-patch natural growth and harvesting, and system wide immigration or emigration to and from the other patches.

The kinds of issues that this type of system can address are several. For example, what does the equilibrium distribution of effort and population look like over space? In which patches are population levels (or harvest or effort levels) large or small? How is the spatial distribution affected by various configurations of patch-specific parameters? What role is played in this system by biological dispersal? Because the avenues for dispersal are captured in the properties of the **D** matrix, how do alternative configurations of these linkages affect the spatial distribution of the population and fleet? How are aggregate variables such as total effort or total population size affected by the types of linkages and dispersal rates? What types of biological systems equilibrate with higher or lower spatial variability in effort and populations? Do dispersal processes amplify or mute the inherent sources of heterogeneity in a patchy system?

In order to get a feel for the answers to some of these questions, we consider next a special case of the preceding general model which allows us to solve for a closed form solution to an n-patch system. As it turns out, it is easy to derive a closed form solution for the model in which net rents per patch are multiplicatively separable with respect to effort ($NR_i(x_i, E_i) = NR_i(x_i)E_i$). Here we consider the common functional representation where rents are linear in x and proportional to E_i, namely, $NR_i = pq_i x_i E_i - (c_i + \pi)E_i$. If we set this equation equal to zero and if we solve for x_i, we get a rent dissipating equilibrium biomass level for each patch which depends upon own-patch economic parameters only and which we designate w_i.[19] In this special case, the proportionality of net rents to effort generates a recursive solution because the equilibrium biomass levels can be solved independently from the equilibrium effort levels. The biomass levels can then be substituted into the effort equilibrium equations. The results are closed form solutions for biomass levels (x_i^{ss}) which are functions of the economic parameters and

---

[19] Thus for this rent specification $x_i = w_i \equiv (c_i + \pi)/pq_i$. Note we do not need to impose the assumption that net rents are linear in x_i to be able to isolate a solution for x_i. For example, if net rents in patch i are $NR_i = pq_i x_i E_i - (c_i/x_i)E_i - \pi E_i$ then the steady-state biomass level is $x_i^{ss} = (-\pi + \sqrt{\pi^2 + 4pq_i c_i})/(2pq_i)$.

TABLE II

Steady-State Levels for the Different Types of Spatial Biological Links

Spatial configuration	Biomass levels	Effort levels	Patches
Fully integrated	$x_i^{ss} = \dfrac{c_i + \pi}{pq_i} = w_i$	$E_i^{ss} = \dfrac{1}{q_i}\left[r_i\left(1 - \dfrac{w_i}{k_i}\right) + d_{ii} + \displaystyle\sum_{j=1}^{n} d_{ij}\dfrac{w_j}{w_i} \right]$	$\forall i$
Closed	$x_i^{ss} = \dfrac{c_i + \pi}{pq_i} = w_i$	$E_i^{ss} = \dfrac{1}{q_i}\left[r_i\left(1 - \dfrac{w_i}{k_i}\right) \right]$	$\forall i$
Sink-source	$x_i^{ss} = \dfrac{c_i + \pi}{pq_i} = w_i$	$E_i^{ss} = \dfrac{1}{q_i}\left[r_i\left(1 - \dfrac{w_i}{k_i}\right) + d_{ii} \right]$	i = source
		$E_j^{ss} = \dfrac{1}{q_j}\left[r_j\left(1 - \dfrac{w_j}{k_j}\right) + d_{ji}\dfrac{w_j}{w_i} \right]$	j = sink
Multiple-source	$x_i^{ss} = \dfrac{c_i + \pi}{pq_i} = w_i$	$E_i^{ss} = \dfrac{1}{q_i}\left[r_i\left(1 - \dfrac{w_i}{k_i}\right) + d_{ii} + \displaystyle\sum_{j=1}^{n} d_{ij}\dfrac{w_j}{w_i} \right]$	i = source area
		$E_j^{ss} = \dfrac{1}{q_j}\left[r_j\left(1 - \dfrac{w_j}{k_j}\right) + d_{jj} + d_{ji}\dfrac{w_i}{w_j} \right]$	j = sink
Spatially linear	$x_i^{ss} = \dfrac{c_i + \pi}{pq_i} = w_i$	$E_i^{ss} = \dfrac{1}{q_i}\left[r_i\left(1 - \dfrac{w_i}{k_i}\right) + d_{ii} + d_{ji}\dfrac{w_j}{w_i} + d_{ki}\dfrac{w_k}{w_i} \right]$ = center
			j, k = neighbors

solutions for the effort levels (E_i^{ss}) which are functions of both the biological and economic parameters. The solutions for this system under different biological dispersal conditions are given in Table II.

The implications of various configurations of biological dispersal and linkages in this particular system are relatively straightforward because they are essentially isolated in the equilibrium levels of effort in each patch. As can be seen in Table II, the equilibrium biomass levels for each patch in this system are only functions of that particular patch's own economic parameters. In contrast, the equilibrium levels of effort in each patch depend upon the spatial structure of the entire system's biology; specifically, the dispersal links embodied in the **D** matrix, as well as own-patch growth rates and carrying capacities. For example, the *closed* system, which is essentially n biologically independent patches, generates a bioeconomic equilibrium in which the equilibrium level of effort in each patch is inversely proportional to the equilibrium biomass level, which is itself dependent upon economic parameters only for the patch in question. In contrast, the equilibrium levels of effort in each patch in the *fully integrated* system depend upon not only own biomass but on the equilibrium biomass levels of **all** other patches (each dependent, in turn, on own-patch-specific economic parameters).[20]

To better visualize some of the implications of different ecological structures and biological dispersal mechanisms, consider a simple three patch example. The first

---

[20] Note that the presence of economic and/or biological heterogeneity is necessary to generate interesting conclusions; if the economic and biological parameters are equal across all patches in a fully integrated system, the bioeconomic equilibrium is identical to the closed system.

BIOECONOMICS OF SPATIAL EXPLOITATION 143

TABLE III

Steady-State Biomass Density and Effort Levels for the Different Types of
Spatial Biological Links in the Three Patch Example

Spatial configuration	Patch effort levels	Total effort level ($\sum_{i=1}^{3} E_i$)
Closed	$E_1^{ss} = E_2^{ss} = \frac{r}{q}(1 - \bar{x})$ $E_3^{ss} = \frac{r}{q}(1 - \lambda\bar{x})$	$\frac{r}{q}[2(1 - \bar{x}) + (1 - \lambda\bar{x})]$
Fully integrated	$E_1^{ss} = E_2^{ss} = \frac{r}{q}\left[(1 - \bar{x}) - \frac{b}{r}(1 - \lambda)\right]$ $E_3^{ss} = \frac{r}{q}\left[(1 - \lambda\bar{x}) - \frac{2b}{r}\left(1 - \frac{1}{\lambda}\right)\right]$	$\frac{r}{q}[2(1 - \bar{x}) + (1 - \lambda\bar{x})] - \frac{2b}{q}\left[2 - \lambda - \frac{1}{\lambda}\right]$
Spatially linear (two = center)	$E_1^{ss} = \frac{r}{q}(1 - \bar{x})$ $E_2^{ss} = \frac{r}{q}\left[(1 - \bar{x}) - \frac{b}{r}(1 - \lambda)\right]$ $E_3^{ss} = \frac{r}{q}\left[(1 - \lambda\bar{x}) - \frac{b}{r}\left(1 - \frac{1}{\lambda}\right)\right]$	$\frac{r}{q}[2(1 - \bar{x}) + (1 - \lambda\bar{x})] - \frac{b}{q}\left[2 - \lambda - \frac{1}{\lambda}\right]$
Sink-source (one = source)	$E_1^{ss} = \frac{r}{q}\left[(1 - \bar{x}) - \frac{2b}{r}\right]$ $E_2^{ss} = \frac{r}{q}\left[(1 - \bar{x}) + \frac{b}{r}\right]$ $E_3^{ss} = \frac{r}{q}\left[(1 - \lambda\bar{x}) + \frac{b}{r\lambda}\right]$	$\frac{r}{q}[2(1 - \bar{x}) + (1 - \lambda\bar{x})] - \frac{b}{q}\left[1 - \frac{1}{\lambda}\right]$

question of interest is: how does the spatial equilibrium pattern of effort depend upon the type of biological system the industry is exploiting? We choose parameters as follows: first, prices, catchability coefficients, intrinsic growth rates, and opportunity costs are assumed equal across all patches at values p, q, r, and π, respectively. Second, we normalize so that population is measured in density terms ($x_i = X_i/k_i$). Third, costs are (arbitrarily) assumed greater in patch 3 than in patches 1 and 2, perhaps due to strong currents, or other oceanographic conditions. In particular, we assume $(c_3 + \pi) = \lambda(c_2 + \pi) = \lambda(c_1 + \pi)$, where $\lambda > 1$. Let $\bar{x}$ be the population density satisfying the rent dissipation equation for the parameters associated with patches 1 and 2.[21] By our assumption regarding cost differences in patch 3, we know that the equilibrium density in patch 3 is $\lambda\bar{x}$ with $\lambda > 1$. Thus the equilibrium density of biomass will be higher in the high cost patch compared with the two lower cost patches.

Now we are in a position to examine how different types of biological dispersal–linkage mechanisms influence the spatial equilibrium determination of effort. Note first that the closed system is simply three separate and unconnected patches and hence it is a good benchmark with which to compare all of the other cases. Table III shows the equilibrium population and effort levels in each patch

---

[21] Thus, for example, if $pq_i x_i E_i - (c_i + \pi)E_i = 0$, we have $x_1 = (c_1 + \pi)/pq_1 = \bar{x} = x_2$ and $x_3 = \lambda(c_1 + \pi)/pq_1 = \lambda\bar{x}$.

for several types of biological systems.[22] Note next that the biomass levels do not change across system types with different linkages; this is because we are using a special case of the rent function which allows each patch's population density to be determined only by its own-patch economic variables.[23] Note also that equilibrium biomass levels reflect the inherent heterogeneity of the economic conditions; in particular, where costs are relatively high, biomass levels are correspondingly high.

Now, what happens as the closed system is opened up via biological dispersal linkages? Consider first the fully integrated system, with density-dependent dispersal flowing between all patches. From Table III, we can see that the spatial distribution of effort in equilibrium in the integrated system involves more effort in patches 1 and 2, and less effort in patch 3, relative to the closed case. Why does this happen? Because with biological dispersal, biomass flows out of the high density patch (patch 3), to the lower density patches, dispersing effort as a result. Interestingly, the total effort drawn into the integrated system is also greater than in the closed system. This occurs because the biological dispersal gradient operates in the direction of intrinsically more profitable areas. Thus, in this case at least, biological dispersal and linkages increase overall potential rents and skew the relative distribution of effort more than would be the case in a closed system.

Consider next the spatially linear case in which, for example, patches are located along a line. Suppose that there are three patches so that the center patch is connected to both outside patches whereas each outside patch is only connected to the center patch. This differs from the fully integrated case in an important way, because the two outside patches are not directly connected to each other. Thus this case illustrates the importance of so-called "edge effects" in a spatial system. Consider first a case where the economic parameters for each outside patch are identical and at levels to generate higher equilibrium population densities than the center patch. Then biomass would flow from the edges to the center, skewing both effort and harvest toward the center. Now consider, in contrast, the case considered here where patch 3 is high cost relative to patches 1 and 2. As we would expect, the higher population density in patch 3 generates dispersal from 3 to 2, but because patches 1 and 2 have identical population densities, there is no dispersal from patch 2 to patch 1. Table III shows equilibrium effort levels for patches 1–3, and we find levels equal to, greater than, and less than corresponding levels in the closed case, respectively. As we might expect, total effort is greater than the level in the closed case for reasons discussed earlier, but not as great as in the fully integrated case. This is a sort of spatial Le Chatelier effect in operation; the

---

[22] Note in this paper we only consider interior solutions. In an n-patch setting, the conditions for interior solutions are rather complex, but Table III shows the restrictions on the parameters that are required to guarantee them. For example, in the closed system, for effort to be positive in patches 1 and 2 we require that $\bar{x} < 1$ for $i = 1, 2$. For effort to be positive in patch 3, we require $\lambda \bar{x} < 1$. Thus, to ensure interior solutions, costs cannot be too high relative to price in any of the patches, but in the high cost patch in particular. In other linked systems, the conditions ensuring interior solutions are more complicated because they generally depend upon combinations of parameters in all (or subsets of) the patches. For example, in a two-patch sink-source system (with patch 1 the source) we would need $1 - w_1 - b/r_1 > 0$ and $1 - w_2 + b/r_2 > 0$ for interior solutions with E_1 and E_2 positive.

[23] This system is equivalent, in Smith's single patch model [35, 36], to one with a phase diagram in (E, X) space characterized by a horizontal $\dot{E}$ nullcline. A more general model, with rents nonlinear in effort, would integrate our decoupled special case, and would make it necessary to solve for the equilibrium values of E_i and x_i simultaneously (for example, if $C(E_i, x_i)$ is twice-differentiable and concave in E_i).

BIOECONOMICS OF SPATIAL EXPLOITATION   145

spatially linear system is more constrained by fewer between-patch linkages than the fully integrated case and as a result, potential rents are not as high.

Now consider the three patch sink-source case where we (arbitrarily) designate patch 1 as the source patch. Recall that with sink-source linkages, biomass is hypothesized to depend upon absolute population levels (or densities) in the source and not to depend upon relative densities between patches. Assume the total biomass outflow from patch 1 to be $2b\bar{x}$, distributed equally between patches 2 and 3. Again, the intuition behind what should happen here relative to the closed case is clear; biomass flows from patch 1 to patches 2 and 3, attracting effort into both receptor patches from patch 1. What happens to total system effort in this case? Interestingly, it falls relative to the closed system and a bit of reflection suggests why. In this case, the biological dispersal gradient is working against the spatial distribution of profitability and hence overall rent potential falls as dispersal occurs. Basically vessels are being repelled out of low cost area 1 into higher cost area 3, so that every unit of harvest associated with dispersal is caught with higher average costs. Note also that dispersal in this case acts to homogenize the inherent heterogeneity generated by the differences in fundamental cost–price ratios, averaging out these differences over space.

Figure 2 illustrates and compares these and an additional multiple source case. For comparison purposes, we compare each case with the reference closed system of biologically independent patches. Thus the shaded areas represent effort *differences* in each patch, *relative* to the closed system, induced by various structural biological linkages among patches. Each biological dispersal system generates a different spatial equilibrium distribution of effort. As Figure 2 shows, the qualitative variety of spatial equilibria possible is large and reflective of the dispersal characteristics in the biological system. In some cases biological integration exacerbates the skewness of effort relative to the closed system case. For example, when patches 1 and 2 are fully integrated to patch 3, biomass flows from patch 3 to the other patches, enhancing their attractiveness and drawing effort. On the other hand, in a simple sink–source system in which patch 1 is the source patch linked unilaterally to patches 2 and 3, the effect is to enhance the biomass levels in both sinks, causing effort flows out of one and into the other two receptors. This qualitative pattern is dramatically reversed in the multiple source case, where the addition of density-dependent flows results in effort dispersal from patch 3 into patch 1 and no change in patch 2.

This simple example hints at some of the richness that emerges when we explicitly consider spatial factors in a model of bioeconomic exploitation. While there are numerous other aspects of both the dynamics and equilibria we could consider, we have focused primarily on how effort and the population distribute over space in response to various bioeconomic phenomena and the nature of biological linkages. As we have shown here, the spatial pattern of effort is driven fundamentally by patch-specific cost–price ratios. In areas where the ratio $(c_i + \pi)/q_i p_i$ is high, effort is low and vice versa. Because effort governs exploitation, patches with low effort have relatively high biomass levels. These observations are illustrated by the closed structure, which effectively maps the heterogeneity in the economic parameters into corresponding heterogeneity in the spatial distribution of effort and biomass.

What role do linkages play in determining the spatial equilibrium? Basically biological and economic linkages serve either to smooth out or to exacerbate any

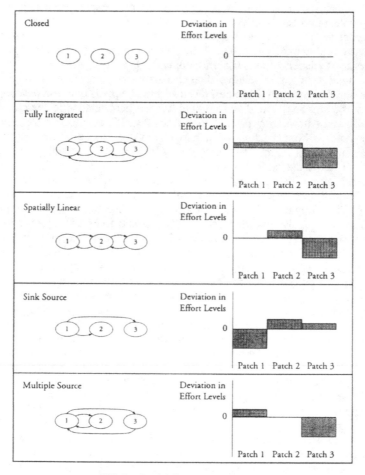

FIG. 2. Spatial distributions of effort.

bioeconomic imbalances via dispersal of both the population and effort. For example, suppose we begin with the closed system and then we imagine it transformed into one with a relative density based dispersal process. Then the initial (closed system) conditions, which involve a higher biomass level in the relatively low profit patch, generate biological dispersal from the high biomass patch to the other two lower biomass patches. This, in turn, generates higher rents there, attracting more effort until the exit from the low profit patch is matched by dispersal to the higher profit patches. In this case, dispersal in both the biological and economic components of the system exacerbates the initial heterogeneity due to fundamental differences in patch-specific profitability. In equilibrium, the integrated system has a **more** spatially skewed distribution of effort because dispersal acts to complement fundamental relative profit forces.

As we have seen with the sink-source case, however, linkages can also work against fundamental heterogeneities induced by cost–price differences across patches. In our sink-source case, biomass is (arbitrarily) assumed to flow unidirectionally from patch 1 to patches 2 and 3, with a flow based on patch 1's density. This enhances biomass levels in each of the receptor patches, making each relatively more attractive, compared with the closed system case. In equilibrium, the source ends up losing biomass and effort, and patches 2 and 3 gain biomass and effort. This works contrary to the fundamental incentives to move away from patch 3 toward patch 1 (and patch 2) based on economic parameters.

V. DISCUSSION

This paper generalizes the basic open access models of Gordon and Smith in the spatial dimension. The biological model embeds spatial features in a manner consistent with developments in biology focusing on patchiness, heterogeneity, and linkages between patches. The economic model of harvester behavior is simple, positing myopic open-access behavior responding sluggishly to potential arbitrage opportunities both outside the fishery and within the fishery across space. While the combined bioeconomic system is relatively simple, it captures the fundamental forces that we would expect to be operating in an open-access system exploited by a spatially mobile and economically responsive fleet.

As we have demonstrated, space matters in interesting and complicated ways, and there are predictable patterns that emerge in a spatially explicit bioeconomic system. These spatial patterns reflect both within-patch bioeconomic forces and between-patch biological and economic linkages. The range of possibilities of both patch-specific heterogeneity and types of dispersal linkages is large in the real world, of course. Our model is general and capable of depicting many of these scenarios. At the same time, the model suggests some important generalizations about how various within- and between-patch forces interact. For example, because harvesters are assumed to respond to relative as well as to absolute rents, there is a fundamental and endogenous "economic gradient" operating at each point in time, depicting incentives to reallocate effort over space toward patches that exhibit higher relative rents. At the same time, there is a "biological gradient" operating to reallocate biomass over space via migration to areas of lower density or that are otherwise natural sinks. The manner in which these two forces operate together determines the process of bioeconomic convergence over space and time. In some systems they operate in a complementary fashion and in other systems they work against each other.

Although it is reasonably straightforward to characterize the dynamics of convergence, we focus in this paper on the equilibrium properties of this system. As we have shown with a simple example, it is possible to predict how equilibrium spatial patterns of the two important state variables (effort and biomass) are affected by different configurations of patch-specific parameters and between-patch linkages. In our example, the pattern of biomass levels is determined by bioeconomic ratios specific to each patch. In patches where costs are high or catchability or prices are low, effort is driven away by lower rents toward high rent patches in a manner determined by the economic gradient. But as effort reallocates, it also affects biological dispersal, which is reacting according to the biological gradient. In a density dependent dispersal system, patches with relatively higher biomass levels will also generate biomass dispersal, driving away even more effort in second and subsequent rounds of adjustment. Hence in some cases, the ability of the fleet to respond to rent differentials across space reinforces any initial forces due to heterogeneity in cost–price ratios, skewing effort even more dramatically than would be the case in an unlinked system. Of course, it is also possible with other biological linkages to homogenize the fundamental heterogeneity driven by bioeconomic ratios. For example, in a sink-source system in which the source is a lower cost area relative to the sink, the fleet would skew toward the source without any biological dispersal. But with sink–source dispersal, in this example biomass will flow toward the sink, supporting more effort there. Thus in this case, spatial dispersal by the fleet smoothes the initial spatial distribution associated with patch-specific fundamentals.

An important implication of these results is that, in an exploited system, whether a particular patch is a source or sink in equilibrium (and in transition) may depend as much on economic parameters as on biological parameters. This seems to be an important point missed by biologists who have begun to promote closed zones and natural refugia as spatial management measures. A typical argument made to justify setting a particular area aside and protected from harvesting involves pointing to high productivity as a source area for juveniles, larvae, etc. In instances where evidence is used to support the case for a particular area, biologists typically attempt to measure indicators of dispersal by tagging juveniles or doing genetic analysis of adults at various locations. In an integrated bioeconomic system, however, whether a particular patch acts as a de facto source depends not only upon biological but also on economic factors as well. Any tagging or measurement of movement in this type of system might conclude that observed net dispersal is due to special biological characteristics, when in fact, it is due to special economic circumstances.

A question that might naturally be asked is: is adding the spatial dimension really worth all of the effort? This is a legitimate question because, even though the model discussed here is relatively simple, it requires some effort before patterns emerge from among the richness of all the cases that can be examined. Our answer to the fundamental question is an unequivocal yes. In the first place, we may simply be missing a great deal of important information by aggregating over space. An important part of what economists do is to try and understand economic behavior. Renewable natural resource exploitation is a somewhat unique activity because the "firm" often moves to the resource. This opens up the possibility of adding degrees of freedom in empirical work by taking advantage of the behavioral reaction to changes in incentives over space as well as time. Of course adding spatial complexity adds parameters to be estimated too, but as our model suggests, there are sensible specifications that might be utilized to increase estimation efficiency over the system as a whole. In addition, by ignoring spatial decisions and simply estimating aggregate time series models, we may be grossly biasing estimates of behavioral relationships by washing out all of the true responsiveness in the system. At the very least, then, this modeling system seems useful for framing empirical analysis that can take advantage of richer data sets incorporating both spatial and intertemporal features of behavior.

A second important reason why spatial analysis may be worth the effort is more normative than predictive. As discussed at the beginning of the paper, there is a virtual groundswell of interest among biologists in using spatially differentiated management techniques. The suggestions seem to be emerging from new ecological paradigms that move away from presumed obsolete views which treat populations as homogeneous and spatially uniform units. In the new biological models, populations are comprised of many spatially distinct patches, some of which may be linked by dispersal processes. This suggests a new suite of options involving managing the population by managing the patches. For example, one popular suggestion is to set aside specific areas from harvesting, so that they might act as sources to the other areas. Another related suggestion is to close an area for a period, and then open it while closing another, etc. This "rotating harvest zones" suggestion (or the permanent closed zone or refugia idea) is thought by proponents to give the system a chance to rejuvenate from commercial exploitation, to develop a richer and more diverse species portfolio, and to provide other often unspecified but implied beneficial effects.

Unfortunately, most of the analysis of these proposals focuses exclusively on biological implications of spatial management, paying virtually no attention to the harvesting sector. There are no analyses of the costs of establishing refugia or rotating harvest zones and there is no analysis of how the fleet might react to these policies. What happens to those parts of the fleet that are removed from an area? How will changes in fleet distribution affect the remaining open access part of the system? Under what circumstances will total system biomass increase? When will total harvest rise or fall? These are questions that must be answered before these new proposals are given any chance in a political system in which harvesters are stakeholders with a legitimate position at the table. In order to answer these kinds of questions, we must begin with conceptual structures of the sort developed here, which account for economic motivations and behavior over space and time, just as they account for biological growth and dispersal processes.

REFERENCES

1. H. J. Albers, Modeling ecological constraints on tropical forest management: Spatial interdependence, irreversibility, and uncertainty, *J. Environ. Econom. Management* 30(1), 73–94 (1996).
2. J. C. Allen, Mathematical models of species interactions in time and space, *Amer. Naturalist* 109, 319–342 (1975).
3. P. Berck, Open access and extinction, *Econometrica* 4, 877–882 (1979).
4. P. Berck, and J. M. Perloff, An open-access fishery with rational expectations, *Econometrica* 2, 489–506 (1984).
5. M. G. Bhat, R. G. Huffaker and S. M. Lenhart, Controlling forest damage by dispersive beaver populations: Centralized optimal management strategy, *Ecological Appl.* 3(3), 518–530 (1993).
6. M. G. Bhat, R. G. Huffaker and S. M. Lenhart, Controlling transboundary wildlife damage: modeling under alternative management scenarios, *Ecological Modeling* 92, 215–224 (1996).
7. N. E. Bockstael, Modeling economics and ecology: The importance of a spatial perspective, *Amer. J. Agricultural Econom.* 78(5), 1168–1180 (1996).
8. J. A. Bohnsack, Marine reserves they enhance fisheries, reduce conflicts, and protect resources, *Oceanus* 63–71 (Fall, 1993).
9. L. W. Botsford, J. F. Quinn, S. R. Wing, and J. G. Brittnacher. "Rotating Spatial Harvest of a Benthic Invertebrate, the Red Sea Urchin, Stronglyocentrotus franciscanus," International Symposium on Management Strategies for Exploited Fish Populations: Alaska Sea Grant College Program, 1993.
10. G. M. Brown, and J. Roughgarden, A metapopulation model with private property and a common pool, *Ecological Econom.* 22(1), 65–71 (1997).

11. M. H. Carr, and D. C. Reed, Conceptual issues relevant to marine harvest refuges: Examples from temperate reef fishes, *Canad. J. Fisheries Aquatic Sci.* **50**, 2019–2028 (1993).

12. J. Conrad and T. Bjorndal, A bioeconomic model of the Harp Seal in the Northwest Atlantic, *Land Econom.* **67**(2), 158–171 (1991).

13. G. E. Davis, Designated harvest refugia: The next stage of marine fishery management in California, *CalCoFl Rep.* **30**, 53–58 (1989).

14. J. E. Dugan and G. E. Davis, Applications of Marine Refugia to Coastal Fisheries Management, *Canad. J. Fisheries Aquatic Sci.* **50**, 2029–2042 (1993).

15. J. Geoghegan, L. A. Wainger, and N. E. Bockstael, Spatial landscape indices in a hedonic framework: An ecological economics analysis using GIS, *Ecological Econom.* **23**(3), 251–264 (1997).

16. H. S. Gordon, The economic theory of a common-property resource: The fishery, *J. Political Econom.* **62**, 124–142 (1954).

17. A. Hastings, Dynamics of a single species in a spatially varying habitat: The stabilizing role of high dispersal rates, *J. Math. Biol.* **16**, 49–55 (1982).

18. A. Hastings, Can spatial variation alone lead to selection for dispersal, *Theoret. Population Biol.* **24**, 244–251 (1983).

19. R. D. Holt, Population dynamics in two-patch environments: Some anomalous consequences of an optimal habitat distribution, *Theoret. Population Biol.* **28**, 181–208 (1985).

20. F. R. Homans and J. E. Wilen, A model of regulated open access resource use, *J. Environ. Econom. Management* 1–21 (1997).

21. R. G. Huffaker, M. G. Bhat, and S. M. Lenhart, Optimal trapping strategies for diffusing nuisance-beaver populations, *Natur. Resource Modeling* **6**(1), 71–97 (1992).

22. A. Leung, and A. Y. Wang, Analysis of models for commercial fishing: Mathematical and economical aspects, *Econometrica* **44**, 295–303 (1976).

23. S. A. Levin, Dispersion and population interactions, *Amer. Naturalist* **108**, 207–227 (1974).

24. S. A. Levin, Population dynamic models in heterogenous environments, *Ann. Rev. Ecology Syst.* **7**, 287–310 (1976).

25. A. Man, R. Law, and N. V. C. Polunin, Role of marine reserves in recruitment to reef fisheries: A metapopulation model, *Biol. Conservation* **71**, 197–204 (1995).

26. H. P. Possingham and J. Roughgarden, Spatial population dynamics of a marine organism with a complex life cycle, *Ecology* **71**(3), 973–985 (1986).

27. H. R. Pulliam, Sources, sinks, and population regulation, *Amer. Naturalist* **132**, 652–661 (1988).

28. C. M. Roberts and N. V. C. Polunin, Are marine reserves effective in management of reef fisheries? *Rev. Fisheries Biol.* **1**, 65–91 (1991).

29. J. N. Sanchirico, "The Bioeconomics of Spatial and Intertemporal Exploitation: Implications for Management," Ph.D. thesis, Department of Agricultural and Resource Economics, University of California at Davis (1998).

30. J. N. Sanchirico and J. E. Wilen, "The Bioeconomics of Marine Reserves," selected paper presented at the European Environmental and Resource Economics Conference, Lisbon, Portugal, 1996.

31. C. E. Schulz and A. Skonhoft, Wildlife management, land-use and conflicts, *Environ. Development Econom.* **1**, 265–280 (1996).

32. M. B. Shaefer, Some considerations of population dynamics and economics relation to the management of marine fisheries, *J. Fisheries Res. Board Canada* **14**, 669–681 (1957).

33. J. G. Skellam, Random dispersal in theoretical populations, *Biometrika* **38**, 196–218 (1951).

34. A. Skonhoft and J. T. Solstad, Wildlife management, illegal hunting and conflicts. A bioeconomic analysis, *Environ. Development Econom.* **1**, 165–181 (1996).

35. V. L. Smith, Economics of production from natural resources, *Amer. Econom. Rev.* 409–431 (1968).

36. V. L. Smith, On models of commercial fishing, *J. Political Econom.* **77**, 181–198 (1969).

37. G. N. Tuck, and H. P. Possingham, Optimal harvesting strategies for a metapopulation, *Bull. Math. Biol.* **56**(1), 107–127 (1994).

38. R. Vance, The effects of dispersal on population stability in one-species, discrete space population growth models, *Amer. Naturalist* **123**(2), 230–254 (1984).

39. J. E. Wilen, "Common Property Resources and the Dynamics of Overexploitation: The Case of the North Pacific Fur Seal," University of British Columbia, Vancouver, Department of Economics Working Paper No. 3, (1976).

Schooling Species

[33]

The dynamics of an open access fishery

TROND BJØRNDAL Norwegian School of Economics
JON M. CONRAD Cornell University

Abstract. A discrete time non-linear deterministic model for an open access fishery is developed and the equilibrium is characterized. The open access exploitation of North Sea herring during the period 1963–77 is analysed. Alternative production functions are considered and estimated for the Norwegian purse seine fishery. The bionomic equilibrium and approach dynamics are presented when prices and costs are changing. The results indicate that the resource stock was saved from possible extinction by the closure of the fishery at the end of the 1977 season.

Sur la dynamique d'une zone de pêches quand l'entrée est libre. Les auteurs développent un modèle déterministe non-linéaire en temps discret d'une zone de pêches où l'entrée est libre et définissent les caractéristiques de l'équilibre. L'exploitation du hareng de la Mer du Nord qui s'est faite sans entraves à l'entrée pendant la période 1963–1977 est analysée avec ce modèle. Des fonctions de production de rechange sont examinées et calibrées pour la pêche à l'essaugue par la flotte norvégienne. L'équilibre bionomique et la dynamique de l'approche à cet équilibre sont examinés dans un univers où les prix et les coûts sont changeants. Les résultats de l'analyse montrent que le stock de ressource a échappé à la disparition possible grâce à la fermeture de la zone de pêches à la fin de la saison de 1977.

INTRODUCTION

Open access exploitation of common property fish resources frequently causes severe stock depletion. Indeed, the question whether open access may cause stock extinction has been analysed by several authors (Smith, 1968 and 1975; Berck, 1979; Hartwick, 1982). Moreover, as Smith (1968) has pointed out, although stock equilibrium under open access may be positive, the stock may be driven to extinction along the path of adjustment. Stock equilibrium may also be stable and positive with fixed prices and technology and still drift towards extinction over time, since these fixed variables drift in the long run.

Canadian Journal of Economics Revue canadienne d'Economique, xx, No. 1
February février 1987. Printed in Canada Imprimé au Canada

With the exception of Wilen (1976) the work on the dynamics of open access or free entry fisheries is mainly theoretical. The purpose of this paper is to provide an empirical application, based on the North Sea herring fishery, with special reference to the question of stock extinction under open access. Herring is a schooling species. The schooling behaviour has permitted the development of very effective harvesting techniques. With modern fish-finding equipment, harvesting can remain profitable even at low stock levels. Open access exploitation of a number of schooling species has caused severe stock depletion (Murphy, 1977). The question of possible stock extinction thus takes on special importance for schooling species.

In the second section we shall develop a deterministic model for an open access fishery based on Smith (1968) and give a characterization of open access equilibrium. In the following section open access exploitation of North Sea herring during the period 1963–77 will be analysed. Alternative production functions are considered and estimated for the Norwegian purse seine fishery. The bionomic equilibrium and approach dynamics are presented when prices and costs are changing. Finally, the work is summarized and some policy implications are discussed.

THE OPEN ACCESS MODEL

In this section we construct a simple open access model to discuss steady state (equilibrium) conditions and system dynamics. The model will be specified in discrete time as a system of difference equations. Time is partitioned into annual increments, a procedure consistent with the data used to estimate production and growth functions and the equation for capital (vessel) dynamics. It is also consistent with the observation that vessel owners are reluctant to incur the cost of regearing once a decision has been made to enter the herring fishery which, in the North Sea, has a season running from May until September. While the steady state equilibria for differential equation systems and their difference-equation analogues are usually equivalent, the stability and thus approach dynamics can be qualitatively different. The distinction becomes more than a mathematical curiosity in resource systems, where discrete-time and possibly lagged adjustment to biological and economic conditions can lead to overshoot and greater potential for overharvest and possibly species extinction.

The model presumes an industry production function

$$Y_t = H(K_t, S_t), \tag{1}$$

where Y_t is yield (harvest) in year t, K_t are the number of vessels in the fishery during year t, and S_t is the fishable stock at the beginning of year t.

The number of vessels, K_t, may be a crude measure of actual fishing effort. In demersal fisheries the best measure might be the volume of water 'screened' by nets during the season (Clark, 1985). However, in a fishery on a schooling species like herring, search for schools of herring is of predominant importance. Accordingly, in such fisheries the number of participating vessels may be an appropriate measure of effort.

For schooling stocks, like herring, there is some question as to 'elasticity' of yield with respect to stock size, S_t. If as a population declines it continues to concentrate in (fewer) schools of the same approximate size, and if these schools can be located with relative ease by electronic search, then yield may be essentially determined by effort, independent of stock, until the population declines to a small number of schools. If this were the case, the production function $H(\cdot)$ might depend strictly on K_t, and catch per unit effort, often used to estimate stock, would not predict the collapse of the fishery (Clark and Mangel, 1979; Ulltang, 1980).

Assuming that vessel numbers are an appropriate measure of effort and that yield is stock dependent, the standard open access model proceeds by defining industry profit (net revenue) in year t as

$$\pi_t = pH(\cdot) - cK_t, \tag{2}$$

where p and c are the per unit price for yield and cost per vessel, respectively. Two additional assumptions are implicit in equation (2). First, the fishery must be one of several sources of the species in question; otherwise price would depend on yield, that is, $p_t = p(Y_t)$ where $p(\cdot)$ is an inverse demand function. Cost per vessel is also assumed given. Second, the unit prices and costs are assumed constant through time. Neither assumption is likely to hold in 'real world' fisheries, but their maintenance permits the estimation of an open access or bionomic equilibrium, which may give an indication of the extent of overfishing.

Vessels are assumed to enter a profitable fishery and exit an unprofitable fishery according to

$$K_{t+1} - K_t = n\pi_t \tag{3}$$

where $n > 0$ is an adjustment parameter (unit: vessels/$). With n positive, it will be the case that (a) $K_{t+1} > K_t$ if $\pi_t > 0$; (b) $K_{t+1} < K_t$ if $\pi_t < 0$; and (c) $K_{t+1} = K_t$ if $\pi_t = 0$. It is possible that the rates of entry and exit may differ, in which case n^+ might apply if $\pi_t > 0$ and n^- might apply if $\pi_t < 0$ where $n^+, n^- > 0, n^+ \neq n^-$.

Finally, the resource stock is assumed to adjust according to

$$S_{t+1} - S_t = F(S_t) - H(K_t, S_t), \tag{4}$$

where $F(S_t)$ is a net natural growth function. It is often assumed that there exist stock levels $\mathbf{S}$ and $\bar{S}$ where $F(\mathbf{S}) = F(\bar{S}) = 0$, $F(S) < 0$ for $0 < S < \mathbf{S}$, $F(S) > 0$ for $\mathbf{S} < S < \bar{S}$, and $F(S) < 0$ for $S > \bar{S}$.

Taken together equations (3) and (4) constitute a dynamical system (or an iterative map). More specifically, with given values for S_0 and K_0 the system

$$K_{t+1} = K_t + n[pH(K_t, S_t) - cK_t]$$
$$S_{t+1} = S_t + F(S_t) - H(K_t, S_t)$$

(5)

can be iterated forward in time. The trajectory (S_t, K_t) may be plotted in phase-space. A stationary point (S, K) is one for which $K_{t+1} = K_t = K$ and $S_{t+1} = S_t = S$ for all future t. Such a point must satisfy $K = pH(K, S)/c$ and $H(K, S) = F(S)$.

For the Gordon-Schaefer model (Clark, 1976) where $F(S_t) = rS_t(1 - S_t/L)$ and $H(K_t, S_t) = qK_tS_t$, the differential equation system takes the form

$$\dot{K} = n(pqKS - cK)$$
$$\dot{S} = rS(1 - S/L) - qKS,$$

(6)

where r is the intrinsic growth rate, L is the environmental carrying capacity, and q is the catchability coefficient. The system has an equilibrium at $S_\infty = c/(pq)$ and $K_\infty = r(1 - S_\infty/L)/q$, which is the focus of a stable spiral (see figure 1a). The difference equation analogue might be written

$$K_{t+1} = [1 + n(pqS_t - c)]K_t$$
$$S_{t+1} = [1 + r(1 - S_t/L) - qK_t]S_t,$$

(7)

and is capable of more complex behaviour, including limit cycles (see figure 1b).[1]

THE NORTH SEA HERRING FISHERY 1963–77

The North Sea herring fishery takes place in the central and northern North Sea, with the main season in the months May to September. In the present case study data for the Norwegian purse seine fleet will be used to estimate production functions and vessel dynamics. The fishery, utilizing this technology, started in 1963. In the middle of the 1970s, however, the stock was severely depleted under an open access regime and the fishery was closed at the end of 1977. Severe regulations have been in effect ever since to allow the stock to recover.

Table 1 contains data on stock size, Norwegian purse seine harvest and the number of Norwegian purse seiners for the period 1963–77. Other countries (Denmark, the Netherlands, Germany, and the United Kingdom) were also harvesting the herring stock using a variety of gear, including single and pair

---

1 For system (6) with $S_\infty = c/(pq) > 0$ and $K_\infty = r(1 - S_\infty/L)/q > 0$ the open access equilibrium is stable (a node or spiral) and limit cycles are precluded by the Bendixon-du Lac test (see Clark, 1976, 203–4). For the difference equations in system (7), simulation for $p = 1,000$, $c = 3,000$, $q = 3.8 \times 10^{-5}$, $n = 0.0001$, $r = 0.5$, and $L = 250,000$ from $S_0 = 250,000$ and $K_0 = 1,000$, results in a convergent spiral. Changing n to 0.000175 leads to a limit cycle and with $n = 0.000175$ and $r = 2.6$, ceteris paribus, an invariant circle is obtained. The difference equation system, with its inherent lag, is capable of much more complex, possibly 'chaotic' behaviour.

78   Trond Bjørndal and Jon M. Conrad

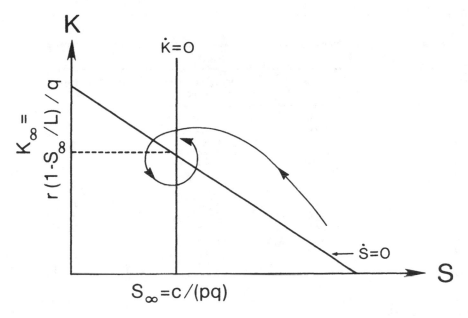

FIGURE 1a: Phase plane analysis of system (6). The point (S_∞, K_∞) is the focus of a stable spiral.

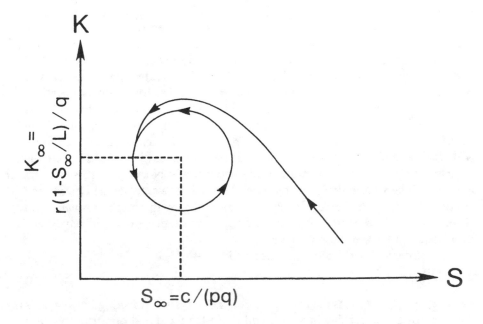

FIGURE 1b: Phase plane analysis of system (7). The point (S_∞, K_∞) is the focus of a limit cycle.

TABLE 1

North Sea herring stock, Norwegian purse seine harvest, and the number of Norwegian purse seiners

Year	Stock size S_t (tonnes)	Norwegian harvest Y_t (tonnes)	Number of participating purse seiners K_t
1963	2,325,000	3,454	8
1964	2,529,000	147,933	121
1965	2,348,000	586,318	209
1966	1,871,000	448,511	298
1967	1,434,000	334,449	319
1968	1,056,000	286,198	352
1969	696,000	134,886	253
1970	717,000	220,854	201
1971	501,000	210,733	230
1972	509,000	136,969	203
1973	521,000	135,338	153
1974	345,000	66,236	165
1975	259,000	34,221	102
1976	276,000	33,057	92
1977	166,000	3,911	24

SOURCE: Bjørndal (1984)

trawl and drift nets. After 1963, however, the purse seine technology became the dominant gear, and lacking data on the number and harvest of other gear types, we used the Norwegian purse seine data to estimate parameters for several alternative production forms. The stock estimates (S_t) were obtained by virtual population analysis (Ricker, 1975). Unrestricted OLS regressions were run, and table 2 shows four estimating equations (a)(i)–(d)(i) and four associated production functions (a)(ii)–(d)(ii). The exponents on K_t in (a)(ii) and (b)(ii) would indicate a yield/vessel elasticity greater than one. This is presumably the result of economies of scale in searching for schools of herring, since information about locations of schools tends to be shared between boats in this fishery. The yield-stock elasticity in (b)(ii) and (d)(ii) are both significantly positive and less than one. Thus, as stock declines, catch per vessel will decline and there will be a stock-dependent incentive to exit from the industry, as indicated by the rather rapid departure of Norwegian purse seiners from the fishery after 1968. The remaining vessels, however, seemed more than adequate to continue harvest in excess of natural growth and recruitment, and from inspection of table 1 it is still not clear whether exit would have been rapid enough for the stock to increase.

The expression for profits was specified as

$$\pi_t = p_t H(K_t, S_t) - c_t K_t,$$ (8)

where $c_t = e_t \tilde{c}_t + f_t$; e_t is the average number of days spent fishing herring, $\tilde{c}_t$ is the operating cost per day in year t, and f_t are the fixed and opportunity costs incurred during the herring season.

80   Trond Bjørndal and Jon M. Conrad

TABLE 2

Estimates of production function parameters for the Norwegian purse seine fleet; all regressions OLS with *t*-statistics in parentheses[a]

(a)(i) $\ln Y_t = 4.5408 + 1.4099 \ln K_t$ (adjusted $R^2 = 0.85$)
                  (5.86)   (9.11)

  (ii) $Y_t = 93.769 \, K_t^{1.41}$

(b)(i) $\ln Y_t = -2.7876 + 1.3556 \ln K_t + 0.5621 \ln S_t$ (adjusted $R^2 = 0.96$)
                  (2.11)   (16.39)   (5.84)

  (ii) $Y_t = 0.06157 \, K_t^{1.356} S_t^{0.562}$

(c)(i) $\ln(S_t - Y_t) = -0.5683^b + 1.0398^c \ln S_t - 0.0011 \, K_t$ (adjusted $R^2 = 0.99$)
                           (1.29)   (30.81)   (3.74)

  (ii) $Y_t = S_t(1 - e^{-0.0011 K_t})$

(d)(i) $\ln(Y_t/K_t) = -1.6718^b + 0.6086 \ln S_t$ (adjusted $R^2 = 0.54$)
                          (0.84)   (4.16)

  (ii) $Y_t = S_t^{0.609} K_t$

---

[a] Autocorrelation was indicated only in equations (a) and (d). First-order correction did not significantly alter the magnitude of the estimated coefficients. Two-stage least squares did not indicate the presence of simultaneous equations bias which can occur if estimates of S_t are based on current period harvest. This is less of a problem when stock estimates are obtained by virtual population analysis.

[b] Not significantly different from zero; parameter assumed to be zero in the associated production function.

[c] Not significantly different from 1.00; parameter set equal to one in the associated production function.

Vessel dynamics were assumed to occur according to

$$K_{t+1} - K_t = n\pi_t/(p_t K_t). \tag{9}$$

Equation (9) assumes that entry or exit will depend on the sign of normalized profit per boat. This form was employed to take advantage of previous analysis by Bjørndal and Conrad (1985). Estimates of n ranged between 0.08 and 0.10.

A discrete-time analogue to the logistic growth function might be written as

$$S_{t+1} - S_t = rS_t(1 - S_t/L), \tag{10}$$

where estimates of r and L were 0.8 and 3.2×10^6 metric tonnes. Equation (10) is an approximation to a more complex delay-difference equation discussed in Bjørndal (1984).

Of the four production models the Cobb-Douglas form $Y_t = aK_t^b S_t^g$, resulted in the most plausible values for the bionomic equilibrium and open access dynamics. The open access system may be written as

TABLE 3

Costs (per season per vessel) and herring price (per tonne): figures in Norwegian kroner[a]

Year	c_t	p_t
1963	190,380	232
1964	195,840	203
1965	198,760	205
1966	201,060	214
1967	204,880	141
1968	206,800	128
1969	215,200	185
1970	277,820	262
1971	382,880	244
1972	455,340	214
1973	565,780	384
1974	686,240	498
1975	556,580	735
1976	721,640	853
1977	857,000	1,415

[a] Price figures have been adjusted by a factor of 0.6, which represents the boat owner's share of income. Costs cover only costs incurred by the boatowner.
SOURCES: p_t: The Directorate of Fisheries, Norway
           c_t: The Budget Committee for the Fishing Industry, Norway

$$K_{t+1} = K_t + n(aK_t^{b-1}S_t^g - c_t/p_t)$$
$$S_{t+1} = S_t + rS_t(1 - S_t/L) - aK_t^b S_t^g. \tag{11}$$

If $c_t = c$ and $p_t = p$, then one obtains the following equations for the bionomic equilibrium

$$S_\infty = [c/(paK_\infty^{b-1})]^{1/g}$$
$$K_\infty = [rS_\infty(1 - S_\infty/L)/(aS_\infty^g)]^{1/b} \tag{12}$$

While it is not possible to solve for explicit expressions for S_∞ and K_∞, it is possible to solve for S_∞ and K_∞ numerically. By making an initial guess for K_∞, the first equation in (12) provides a value for S_∞. Substituting this value into the second equation one obtains a value for K_∞, consistent with growth and yield. Calling the initial guess Z_∞, one can evaluate $|Z_\infty - K_\infty|$. If this is not within an arbitrary ϵ, readjust the guess according to $Z_\infty = (Z_\infty + K_\infty)/2$. This process will converge to the bionomic equilibrium from above or below K_∞.

During the period 1963–77 prices and costs were changing as indicated in table 3. If the 1975 values of $c = 556,580$ and $p = 735$ (both in Norwegian kroner) were somehow fixed into the future and all other parameters remained unchanged, then the bionomic equilibrium is calculated at $S_\infty = 430,191$ (tonnes), $K_\infty = 393$ (boats), and $Y_\infty = 297,887$ (tonnes). When c_t and p_t are allowed to vary as per table 3, the time paths for S_t and K_t are given in table 4

82   Trond Bjørndal and Jon M. Conrad

TABLE 4
Bionomic equilibrium and open access dynamics

System

$$K_{t+1} = K_t + n(aK_t^{b-1}S_t^g - c_t/p_t)$$

$$S_{t+1} = S_t + rS_t(1 - S_t/L) - aK_t^h S_t^g.$$

Parameter values

$a = 0.06157$,	$b = 1.356$,	$c = 556,580$,	$g = 0.562$
$L = 3,200,000$,	$n = 0.1$,	$p = 735$,	$r = 0.8$

Bionomic equilibrium

$$S_\infty = 430,191, \qquad K_\infty = 393, \qquad Y_\infty = 297,887$$

Open access dynamics

With c_t and p_t as given in table 3, $S_0 = 2,325,000$, $K_0 = 120$, then

Year	S_t	K_t	Y_t
1963	2,325,000	120	153,698
1964	2,679,895	166	258,531
1965	2,769,820	225	398,376
1966	2,669,323	305	588,874
1967	2,434,586	404	818,564
1968	2,081,887	461	897,077
1969	1,766,756	494	898,025
1970	1,501,779	559	970,003
1971	1,169,363	626	983,051
1972	779,950	626	782,754[a]
1973	469,075	538	479,232
1974	310,096	480	325,061
1975	209,071	410	210,287
1976	155,113	385	163,580
1977	109,609	343	115,016

[a] After 1972 harvest exceeds S_t but *not* S_t plus growth. This is possible, since growth to the resource occurs before harvesting (see equation for S_{t+1}, above).

and plotted in phase-space in figure 2. The values for K_t might be interpreted as an estimate of 'purse seine equivalents' fishing herring in the *entire* North Sea. Thus K_t is larger than the number of Norwegian purse seiners that participated in the fishery during the period. The stock actually increases until 1965 and then decreases monotonically. The estimates of the herring stock in table 1 are a bit more ragged, lower than the simulated estimates until 1973 and higher thereafter. Of particular interest is the overshoot 'past' the 1975-based bionomic equilibrium and the continued decline in stock. In contrast to the results of Wilen, there is no increase in the stock and the 'first loop' of a convergent spiral has *not* been completed.

In 1977 Norway and the EEC agreed to close the fishery. There are no official

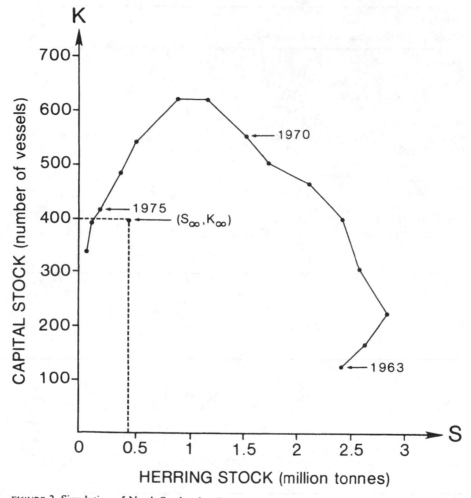

FIGURE 2. Simulation of North Sea herring fishery.

prices nor data to estimate costs after this year. One can only speculate what the future evolution of stock and vessel numbers would have been. It seems entirely plausible that with declining harvest, relative price increases would have exceeded relative cost increases with species extinction the result. If the price in 1978 were increased to 2,000 NoK/metric tonne and costs held steady, the species 'simulates' to extinction in 1983. Under the moratorium which lasted until 1981 the stock was allowed to recover, and fisheries scientists estimated the 1983 stock level at 600,000 metric tonnes.

CONCLUSIONS AND POLICY IMPLICATIONS

In the empirical analysis of open access systems it is important to note that

84   Trond Bjørndal and Jon M. Conrad

non-linear difference equations, with or without longer lags, are capable of more complex dynamic behaviour than their continuous-time (differential equation) analogues. The lag in adjustment by both the exploited species and the harvesters themselves is often a more accurate depiction of dynamics, and the differential equation systems are best viewed as theoretical approximations.

If adjustment in an open access system is discrete, there is a greater likelihood of overshoot, severe depletion, and possibly extinction. When discrete adjustment takes place in a system where the species exhibits schooling, declining stocks may fail to reduce profits rapidly enough to turn the critical 'first corner' in an approach to bionomic equilibrium. The fact that the economic and natural environments are subject to fluctuations places greater importance on modelling the dynamics of non-autonomous systems as opposed to the calculation of equilibria based on long-run or average values.

The analysis of the North Sea herring fishery would seem to support many of the above points. During the 1963–77 period the resource (1) was subject to open access exploitation by Norway and members of the EEC; (2) exhibited a weak yield-stock elasticity (because of schooling) which failed to encourage a rapid enough exit of vessels from the fishery; and (3) was saved from more severe overfishing and possibly extinction by the closure of the fishery at the end of the 1977 season.

Recent analysis by Bjørndal (1985) indicates that the optimal stock level is likely to be in the range 1.0–1.4 million tonnes, supporting a harvest of 550,000–600,000 tonnes. With the recovery of the resource, the stock might be initially managed through a system of *inter*nationally assigned but *intra*nationally transferable quotas. In the longer run a system allowing fisheries managers from one country to purchase or lease the quota rights of another should permit the total allowable catch (TAC) to be harvested at least cost. The theory and institutions for the management of transboundary resources are still at an early stage of development, but likely to be of critical importance if the value of fisheries resources are to be maximized among coastal countries.

REFERENCES

Berck, P. (1979) 'Open access and extinction.' *Econometrica* 47, 877–82
Bjørndal, T. (1984) 'The optimal management of an ocean fishery.' Unpublished PH D dissertation, Department of Economics, University of British Columbia
— (1985), 'The optimal management of North Sea herring.' Working Paper No. 2/1985, Centre for Applied Research, Norwegian School of Economics and Business Administration, Bergen
Bjørndal, T. and J.M. Conrad (1985) 'Capital dynamics in the North Sea herring fishery.' Working Paper No. 01/85, Institute of Fisheries Economics, Norwegian School of Economics and Business Administration, Bergen
Clark, C.W. (1976) *Mathematical Bioeconomics: The Optimal Management of Renewable Resources* (New York: Wiley)

— (1982) 'Concentration profiles and the production and management of marine fisheries.' In W. Eichhorn, et al. eds, *Economic Theory of Natural Resources* (Würzburg: Physica Verlag)

— (1985) *Bioeconomic Modelling and Fisheries Management* (New York: Wiley)

Clark, C.W. and M. Mangel (1979) 'Aggregation and fishery dynamics: a theoretical study of schooling and the purse seine tuna fisheries.' *Fishery Bulletin* 77, 317–37

Hartwick, J.M. (1982) 'Free access and the dynamics of the fishery.' In L.J. Mirman and D.F. Spulber, eds, *Essays in the Economics of Renewable Resources* (Amsterdam, New York, Oxford: North-Holland)

Murphy, G.I. (1977) 'Clupeids' In J.A. Gulland, ed., *Fish Population Dynamics* (New York: Wiley)

Ricker, W.E. (1975) *Computation and Interpretation of Biological Statistics of Fish Populations* (Ottawa: Environment Canada)

Smith, V.L. (1968) 'Economics of production from natural resources.' *American Economic Review* 58, 409–31

— (1975) 'The primitive hunter culture, Pleistocene extinction, and the rise of agriculture.' *Journal of Political Economy* 83, 727–55

Ulltang, O. (1980) 'Factors affecting the reactions of pelagic fish stocks to exploitation and requiring a new approach to assessment and management.' *Rapp. P.-V. Réun. Cons. Int. Explor. Mer.* 177, 489–504

Wilen, J.E. (1976) 'Common property resources and the dynamics of over-exploitation: the case of the North Pacific ful seal.' Paper No. 3 in the Programme in Resource Economics, Department of Economics, University of British Columbia

[34]

JOURNAL OF ENVIRONMENTAL ECONOMICS AND MANAGEMENT 15, 9–29 (1988)

The Optimal Management of North Sea Herring[1]

TROND BJØRNDAL

Institute of Fisheries Economics, The Norwegian School of Economics and Business Administration, Helleveien 30, N-5000 Bergen, Norway

Received January 13, 1983; February 1986

A discrete time dynamic bioeconomic model for a fish resource is developed. The objective is maximization of discounted net revenues subject to changes in stock size. The model of population dynamics is described by a delay-difference equation. Natural growth and recruitment are related to stock size, with recruitment taking place with a time lag. Conditions characterizing the optimal stock level are derived. The model is applied to North Sea herring. Estimates of the optimal stock level are given, and optimal trajectories derived. Due to the schooling behaviour of herring, it is shown that open access may cause stock extinction. © 1988 Academic Press, Inc.

1. INTRODUCTION

The objective of this paper is to study the optimal management of North Sea autumn spawning herring. This fish stock was severely depleted in the 1960s and early 1970s due to overfishing under an open access regime, but has since been permitted to recover. This leads to questions about the target level for the rebuilding program and the approach to this level. These questions, which constitute the essence of a management plan, will be analyzed by means of a dynamic bioeconomic model for North Sea herring.

On of the difficulties of implementing a management plan is the transboundary nature of the fishery. This is because North Sea herring is harvested by several European nations. The North-East Atlantic Fisheries Commission was responsible for making policy recommendations, but had no power to implement these over the wishes of its member countries. Because of the serious stock depletion, a total ban on the herring fishery was introduced in 1977. Severe regulations have been in effect ever since so as to allow the stock to recover.[2] After the introduction of extended fisheries jurisdiction, North Sea herring has been considered a common resource between Norway and the European Economic Community (EEC). Therefore, management decisions, e.g., determining annual catch quotas, are decided upon jointly by Norway and the EEC.

In view of the difficulties that have occurred in the past with regard to the management of North Sea herring, it is of great interest to study what an appropriate management policy for this resource would be. Gains that can be obtained

---

[1] I would like to thank G. R. Munro, R. S. Uhler, C. W. Clark, R. Hannesson, R. Hilborn, Ø. Ulltang, N. J. Wilimovsky, C. M. Wernerheim, and two anonymous referees for comments on various versions of this paper. The research has been supported by a grant from the Norwegian Fisheries Research Council.
[2] The fishery was reopened in the southern North Sea in October 1981 and in the central and northern North Sea in 1983. The fishery is *inter alia* regulated by total allowable catch quotas.

10   TROND BJØRNDAL

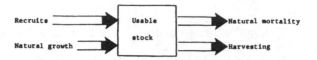

FIG. 1. Stock dynamics.

through cooperation ought to be an impetus to improved cooperation. Therefore, we feel it is of importance to devise such a plan, which must consist of the following elements:

(1) The target level of the stock, i.e., the optimal stock level in a bioeconomic context, the corresponding effort level and harvest rate.

(2) The policy to achieve the target level, i.e., the time path of exploitation of the resource.

However, the plan must analyze the sensitivity of the optimal stock level to changes in important parameter values.

In Section 2 a deterministic bioeconomic model is developed. Although the model is stock specific, it should prove useful for other species as well. In Section 3 the model is applied to North Sea herring. Estimates of the optimal stock level are presented, and the optimal approach path is derived. Finally, in Section 4 we briefly summarize the work and point out areas where future research is warranted.

2. THE BIOECONOMIC MODEL

In this section we will develop a model of population dynamics. This will be combined with a net revenue function in order to formulate a dynamic bioeconomic model. An implicit function for the optimal stock level will be derived.

2.1. *The Model of Population Dynamics*

In its most simple form, changes in the biomass of a fish stock over time will come from additions to the stock due to recruitment and natural growth and deductions from the stock due to natural mortality and harvesting as illustrated in Fig. 1 (adapted from Ricker [22, p. 25]).

A fish population can generally be divided into several subpopulations. In a bioeconomic as opposed to a purely biological context, the primary concern is to identify the harvestable population. Since for North Sea herring this coincides with the spawning stock, the model will be formulated in terms of this variable.

The interactions between recruitment, natural growth, natural mortality and harvesting have been fundamental in the development of the model of population dynamics. The following delay-difference equation[3] will be used to explain changes in the biomass over time:

$$S_{t+1} = (S_t - H_t)e^{G-M} + G(S_{t-\gamma}) \tag{1}$$

---

[3]Other examples of delay-difference models are given by Clark [6] and Deriso [12].

where

S_t = spawning stock (biomass) in period t
H_t = harvest quantity in period t
G = mean instantaneous natural growth rate
M = mean instantaneous natural mortality rate
$G(S_{t-\gamma})$ = recruitment, to the stock, taking place with a delay of γ periods.[4]

The reasoning behind the model requires some amplification. The first part of the right-hand side of Eq. (1) denotes stock changes due to natural growth, natural mortality, and harvesting. In the model, it is assumed that harvesting occurs in a short season at the beginning of the period. The escapement, $S_t - H_t$, is left to grow at the net instantaneous growth rate $G - M$. Therefore, in the absence of recruitment, changes in the spawning stock over time are given by

$$S_{t+1} = (S_t - H_t)e^{G-M}. \qquad (2)$$

Alternatively, we could assume that the fishery took place at the end of the period, without affecting the qualitative nature of the model.[5]

The second part of the right-hand side of Eq. (1) represents addition to the stock due to recruitment, which is assumed to occur at discrete time intervals. Moreover, recruits will normally join the parent population several years after spawning. We postulate that

$$R_{t+1} = g(S_t), \qquad (3)$$

where R_{t+1} is the number of recruits to the juvenile population as a function of the previous period's spawning biomass. A certain fraction, λ, will survive the juvenile stage[6] and join the spawning stock, so that

$$g(S_{t-\gamma})\lambda \qquad (4)$$

is the number of recruits joining the spawning stock with a delay of γ periods. The delay occurs while the juveniles mature to spawning age. Letting w denote the weight of new recruits, we get

$$G(S_{t-\gamma}) = g(S_{t-\gamma})\lambda w, \qquad (5)$$

where $G(S_{t-\gamma})$ denotes recruitment in weight to the spawning stock.

Herring spawn in September every year. The following year a number of recruits, called zero-group herring, join the juvenile population as indicated by Eq. (3). After another two years the survivors (Eq. (4)) become sexually mature and joint the spawning or adult population (Eq. (5)). Thus for this species, $\gamma = 2$, and the delay

---

[4] The model can be formulated in terms of total biomass X, by replacing S with X in Eq. (1). Recruitment will then generally occur with a time lag of one period, giving a γ value of zero.

[5] In that case, Eq. (2) would become: $S_{t+1} = S_t e^{G-M} - H_t$. In the case of continuous fishery throughout the period, we would have $S_{t+1} = S_t e^{G-M-F}$ where F is fishing mortality. A seasonal fishery [4, 13] can be formulated as a further development of this model. Since F in a herring fishery may depend on stock size (cf. Section 3.2), it is appropriate to focus on harvest quantity rather than stock size.

[6] The model may be generalized by letting λ depend on juvenile stock density.

between spawning and recruitment to the spawning stock is three years. The time periods are defined such that the beginning of one period coincides with the point of time when a year-class of new recruits enters the spawning biomass.

Before proceeding to the production function, the behaviour of the model under natural conditions will be considered. In the absence of fishing, Eq. (1) is reduced to

$$S_{t+1} = S_t e^{G-M} + G(S_{t-\gamma}) = S_t e^{\delta} + G(S_{t-\gamma}),$$

$$\delta \equiv G - M, \tag{1'}$$

where δ is the mean net natural growth rate. Under natural conditions, a fish population will grow towards its carrying capacity, which is the upper limit of the stock size as determined by environmental conditions. However, in Eq. (1'), it is assumed that G and M are constants. In reality, natural growth will be density-dependent because *ceteris paribus*, there will be relatively more food available to a small stock than to a large one. Natural mortality may also be density dependent, e.g., if the effectiveness of predation depends on stock size[7] or if cannibalism occurs or becomes more frequent at high stock densities. Hence one expects net natural growth to depend on stock density, with

$$\delta = \delta(S) \quad \text{and} \quad \delta(\bar{S}) \leqslant 0,$$

where $\bar{S}$ is the carrying capacity of the stock; otherwise an upper limit to stock size would not exist under natural conditions. Equation (1) can now be restated as

$$S_{t+1} = (S_t - H_t)e^{\delta(S_t)} + G(S_{t-\gamma}). \tag{6}$$

For the stock to remain at its carrying capacity, one of the following relationships between recruitment and net natural growth is required to hold:

(i) $G(\bar{S}) = 0 \rightarrow \delta(\bar{S}) = 0$
(ii) $G(\bar{S}) > 0 \rightarrow \delta(\bar{S}) < 0$ such that

$$\bar{S} = G(\bar{S})/(1 - e^{\delta(\bar{S})}).$$

The second relationship is probably the more realistic one.

2.2. *The Bioeconomic Model*

In the economics of fisheries, a fish resource is considered a capital stock. Changes in stock size over time are viewed as positive or negative investments in the stock. These ideas are essential in the paper by Clark and Munro [9], where they formulate a dynamic bioeconomic model and derive a modified golden rule equation, which is an implicit expression for the optimal stock level. The authors also derive the optimal path to the optimal stock level. This approach may appear particularly relevant in the case of an overexploited resource. For North Sea herring, there seems to be general agreement that the stock should be allowed to rebuild to a

---

[7]This is believed to be the case for schooling fish like herring (cf. Section 3.1).

OPTIMAL MANAGEMENT OF HERRING 13

higher level than the present. However, there is considerable disagreement about the speed of the stock recovery. Decision makers must thus make a choice with regard to both the desired stock level and the harvest path over time.

For the purpose of this analysis, it is assumed that the resource in question is managed by a sole owner whose objective is to maximize the present value of net revenues from the fishery. The net revenue function is given by

$$\Pi(H_t, S_t). \tag{7}$$

We assume the manager of the fishery maximizes the present value of (7) subject to changes in the population level given by Eq. (6). In addition, we have the feasibility constraint

$$0 \leqslant H_t \leqslant H_{\max}. \tag{8}$$

This gives a discrete time dynamic bioeconomic model, with H_t and S_t as control and state variable, respectively.

The method of Lagrange multipliers can be used to derive equilibrium conditions for an optimum. We define the Lagrangean expression

$$L = \sum_{t=0}^{\infty} \left\{ \alpha^t \Pi(H_t S_t) - q_t \left[S_{t+1} - (S_t - H_t) e^{\delta(S_t)} - G(S_{t-\gamma}) \right] \right\}, \tag{9}$$

where $\alpha = 1/(1 + r)$ is the discount factor, r the rate of discount, and q_t the discounted value of the shadow price of the resource. Performing the dynamic optimization (see Appendix 1), the implicit expression for the optimal spawning stock S^* is derived as

$$e^{\delta(S^*)} \left[\frac{\Pi_S + \Pi_H}{\Pi_H} \right] + \delta'(S^*)[S^* - G(S^*)] + \alpha^\gamma G'(S^*) = 1 + r. \tag{10}$$

The term $(\Pi_S + \Pi_H)/\Pi_H$ is the marginal stock effect (MSE) in a discrete time nonlinear model (Clark [5]). The MSE represents the impact of stock density on harvesting costs. This effect will cause an increase in the optimal stock level. Intuitively, it can be understood by considering that an increase in stock size will increase catch per unit effort and hence reduce unit harvesting costs. The MSE is analogous to the wealth effect in modern capital theory (Kurz [15]).

In the following section we will estimate the optimal stock size S^* under different assumptions and derive the optimal path to S^*.

3. AN EMPIRICAL ANALYSIS OF NORTH SEA HERRING

The bioeconomic model, as outlined in the previous section, will now be applied to data for the North Sea herring fishery. We will first present empirical estimates of the parameters of the model of population dynamics, i.e., Eq. (6). We will then estimate the optimal spawning stock and the corresponding harvest quantity, and derive the optimal approach path to this level.

14   TROND BJØRNDAL

3.1. The Model of Population Dynamics

North Sea autumn-spawning herring (*Clupea harengus L*) consists of three spawning stocks with spawning grounds east of Scotland, east of England and in the English Channel. However, the three stocks mix on the feeding grounds in the central and northern North Sea rendering it impossible to distinguish between catches from the three stocks. It is therefore customary to treat the three stocks as one, which is also the approach in this study. Herring become sexually mature in agegroup two. Thus, the juvenile stock consists of agegroups zero and one, while the adult or spawning stock consists of agegroup two and older. On the biology of North Sea herring, see Saville and Bailey [23].

The model of population dynamics (Eq. (6)), defined in terms of the spawning stock, consists of two parts: the stock-recruitment function and the net growth function. We will first look at these two parts separately.

In the model, *knife-edge* recruitment is assumed. This implies that all fish of a given age become vulnerable to fishing at a particular time in a given year, and their vulnerability remains constant throughout their lives.[8] In many instances the available data presuppose this type of model, as is the situation with North Sea herring. The following alternative functional forms will be used to estimate the relationship between recruits in period $t + 1$ (R_{t+1}) and spawning stock in period t (S_t):

(1) $R_{t+1} = S_t e^{a(1-S_t/b)}$,

(2) $R_{t+1} = aS_t/(b + S_t)$,

(3) $R_{t+1} = aS_t^b$,

(4) $R_{t+1} = aS_t - bS_t^2$.

The first three functional forms are known as the Ricker. Beverton-Holt, and Cushing stock recruitment functions, respectively (Cushing [11]), while the fourth specification is a quadratic function.[9]

It should be noted that while recruitment here is modelled in a deterministic manner, it is in practice often a highly stochastic process. Ricker notes that "year-to-year differences in environmental characteristics cause fluctuations in reproduction at least as great as those associated with variation in stock density over the range observed—sometimes much greater" [22, p. 274]. For this reason, it is often difficult to detect any clear relationship between the number of recruits and parent stock size from observed data.

---

[8]In addition to knife-edge recruitment, two other types of recruitment can be distinguished (Ricker [22]): (i) platoon recruitment, where the vulnerability of a yearclass increases gradually, but during any year each individual fish is either fully catchable or not catchable; (ii) continuous recruitment, where there is a gradual increase in vulnerability of members of a yearclass over a period of two or more years. Continuous recruitment is probably most common while knife-edge recruitment is least common. However, both platoon and continuous recruitment are often approximated by knife-edge recruitment in analytical models.

[9]For the Ricker function, (i) $R = S$ for $S = b$ and (ii) $(\partial R/\partial S)|_{S=0} = e^a$. For the Beverton-Holt function, $\lim_{S \to \infty} R = a$. The quadratic function is only well behaved for $0 \leq S \leq a/b$.

OPTIMAL MANAGEMENT OF HERRING 15

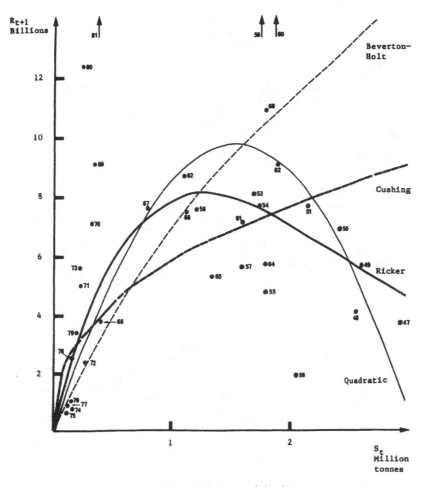

FIG. 2. The stock recruitment relationship.

Appendix 2 gives the data on which these regressions are based: annual spawning stock and number of recruits for the period 1947–1981. The data, estimated by virtual population analysis, are given in research reports published by the International Council for the Exploration of the Sea. Figure 2 is a scatter diagram relating the number of recruits—zero-groups herring—in year $t + 1$ (R_{t+1}) to the spawning biomass in the previous year (S_t). We note the high number of recruits in the years 1956, 1960, and 1981, which presumably is due to exceptionally favourable environmental conditions. This is a phenomenon that is known to occur for clupeids at certain intervals.

Ulltang [27] suggests that the stock-recruitment relation exhibits depensation at low stock levels. This means that the rate of recruitment (recruits/parents) is

16   TROND BJØRNDAL

TABLE I

Estimated Stock-Recruitment Functions[a]

Functional form	Regression	R^2	F^b	DW
Ricker	$\ln(R/S) = 2.87^c - 0.81^c \times 10^{-3} S$ $(16.68)(-6.90)$	0.59	47.62	1.64
Beverton–Holt	$(1/R) = 0.32 \times 10^{-4} + 0.11^c \times (1/S)$ $(0.43) \qquad (5.17)$	0.65	26.77	—
Cushing	$\ln R = 5.89^c + 0.40^c \ln S$ $(5.99) \quad (2.75)$	0.36	7.54	—
Quadratic[d]	$R = 12.74^c S - 0.004^c S^2$ $(6.53) \quad (-4.57)$	0.12	—	1.78

[a] Time series: 1947–1981 ($n = 35$). R is measured in millions while S is measured in 1,000 tonnes. The Beverton-Holt and Cushing models have been estimateed with first order autocorrelation, using an iterative Cochrane–Orcutt procedure. In the ordinary least squares regressions of these two models, the Durbin–Watson statistics were 1.25 and 1.32, respectively. t statistics in parentheses.

[b] To estimate the F statistic, the first observation had to be excluded in the Beverton–Holt and Cushing functions.

[c] Significant at the 95% confidence level.

[d] For the quadratic function, the regression is forced through the origin. Therefore, the F test is invalid and the R-square may be incorrect. The function is well defined for $0 \leqslant S \leqslant 3.07$ million tonnes.

TABLE II

Maximum Recruitment

Functional form		S_{max} (1,000 tonnes)	R_{max} (millions)
Ricker	$R = Se^{2.87(1 - S/3,551)}$	1,240	8,040
Beverton–Holt	$R = 31,405\,S/(3,408 + S)$	∞	31,450
Cushing	$R = 361.6e^{0.4 S}$	∞	∞
Quadratic	$R = 12.74\,S - 0.004\,S^2$	1,530	9,800

increasing over a range of stock densities.[10] This phenomenon is not uncommon for schooling fish. Schooling provides among other things protection against predation. Predation may become less effective at higher stock levels, which means that the relative mortality rate can be decreasing in stock size. This phenomenon may give rise to a depensatory stock-recruitment curve (Clark [5, Chap. 7]). If there is depensation, this will influence the speed of recovery if the actual stock level is in this range. Indeed, it may cause slow recovery of overexploited clupeids.[11]

The estimated regressions for the four stock-recruitment functions postulated above are given in Table I. The relationships between R_{t+1} and S_t are drawn in Fig. 2. Table II gives maximum recruitment for the $R_{t+1} = g(S_t)$ regressions.

[10] This implies that there are increasing returns to recruitment (or growth, in a growth model) at low stock levels. If average recruitment is increasing for $0 \leqslant S \leqslant S_0$, marginal recruitment will reach a maximum at some stock level below S_0.

[11] An example is probably given by Norwegian spring spawning herring.

None of the estimated stock-recruitment functions exhibits depensation. The same result was obtained when recruitment was estimated as third and fourth deg polynominals. Moreover, for North Sea herring, the present stock level is probably outside the range where there may be depensation. Thus, we need not be concerned with this phenomenon in our analysis.

The Ricker and Beverton-Holt functions explain 59% and 65%, respectively of the total variance of the dependant variable. Considering the time span and taking possible changes in the environmental variables into account, the degree of explanation is fairly good. However, this is not so much the case for the Cushing function.

One of the parameters in the Beverton-Holt function is insignificant. The inverse of this parameter gives maximum recruitment (Table II), which is considerably higher than what has actually been observed during the data period. For this reason, we must be cautious in using this function for predictions. The Cushing function has been estimated for a number of Atlantic and Pacific herrings with point estimates of b in the range 0.2 to 0.7 [10]. The present estimate is also in this range.

The above regressions relate zero-group herring in year $t + 1$ to the previous year's spawning stock. However, not all recruits survive to agegroup two. The survival rate of Eq. (4) is defined as

$$\lambda = e^{-(M_0 + M_1)},$$

where M_0 and M_1 are the natural mortality rates of yearclass zero and one, respectively. Estimates of M_0 and M_1 are 0.4 and 0.3, respectively.[12] The mean weight of agegroup two herring is estimated to be 126 g [1]. We then have estimates of all parameters of the stock-recruitment Eq. (5),

$$G(S_{t-2}) = g(S_{t-2})e^{-(M_0 + M_1)}w, \tag{5'}$$

where w is the mean weight of new recruits. If the Ricker function is used, this becomes

$$G(S_{t-2}) = S_{t-2}e^{2.87(1 - S_{t-2}/3,551)}e^{-0.7}0.126$$

$$= 1.10S_{t-2}e^{-0.81 \times 10^{-3}S_{t-2}},$$

where $G(S_{t-2})$ is measured in million tonnes.

The stock-recruitment functions for North Sea herring have been estimated on the basis of data for 1947–1981. This period is probably not long enough for any major long-run changes in the aquatic environment to have taken place.[13] Moreover, data that can be used to test the effects of possible changes in the aquatic environment on recruitment are not available.

The basic postulate about the net natural growth rate, δ, is that it is related to the size of the biomass. The value of δ has been estimated annually for the time period

---

[12] These estimates include fishing mortality for juvenile herring. M_0 and M_1 are the sums of average fishing mortality for the period 1976–1980 [2] and natural mortality, estimated to be 0.10 [23]. Fishing mortality is mainly due to bycatches of juvenile herring. If these could be reduced, the yield from the spawning stock would increase.

[13] Posthuma [21] shows that temperature conditions on the spawning grounds during the incubation period seem to affect yearclass strength, possibly through differential egg mortality at different temperatures.

TABLE III
Estimation of Net Natural Growth Functions[a]

Functional form	R^2	F	DW
$\delta(S) = 0.15^b - 0.43^b \times 10^{-4} S$ (7.96) (−3.63)	0.37	20.31	1.44
$\delta(S) = 0.11^b + 0.46 \times 10^{-4} S - 0.27^b \times 10^{-7} S^2$ (4.52) (1.25) (−2.51)	0.47	14.87	1.82

[a] Time series: 1947–1982 ($n = 36$). Stock size as of January 1 has been used. The Durbin–Watson statistic for the first regression is in the indeterminate range, but autocorrelation has not been corrected for. The estimate of the first order rho is only significant at the 90% confidence level. In addition, correcting for autocorrelation causes only negligible changes in parameter estimates. t statistics in parentheses.

[b] Significant at the 95% confidence level.

1947–1982 (Appendix 2). Two functional forms for the relationship between δ and S are specified:

(i) $\delta(S) = a + bS$,

(ii) $\delta(S) = a + bS + cS^2$.

The results of the regressions are given in Table III.

The linear regression of δ on S explains 37% of the total variance of the dependent variable. Moreover, $\delta'(S) < 0$ for all stock levels. The quadratic regression on S explains 47% of the total variance of the dependent variable. The degree of explanation is relatively good considering the time-span and effects of possible changes in the aquatic environment. Furthermore, the net natural growth rate while positive at low stock levels, is continuously decreasing with increases in stock size.

In the absence of recruitment and harvesting, changes in the stock level are given by (cf. Eq. (2))

$$S_{t+1} = S_t e^{\delta(S_t)}.$$

Some of the characteristics of this function are given in Table IV. There are two main differences between the two functional forms. First, although maximum net growth occurs at roughly the same stock levels, net growth is considerably higher in the quadratic model. The second difference is that net growth remains positive until a much higher stock level in the linear than in the quadratic model.

For the quadratic net growth function, an interesting feature of the function $S_t e^{\delta(S_t)}$ is that it exhibits depensation at stock levels lower than 0.85 million tonnes.

TABLE IV
Characteristics of Function $S_t e^{\delta(S_t)}$

Functional form	Maximum net growth		Level of zero growth (million tonnes)
	Stock level (million tonnes)	Net growth (million tonnes)	
$S_t e^{a - bS_t}$	1.71	0.14	3.49
$S_t e^{a + bS_t - cS_t^2}$	1.84	0.20	3.05

This effect is not present when the linear net growth function is used. Thus, there is a qualitative difference between the two models.

Both parts of the model of population dynamics have now been estimated and we turn to the full model. Various functional forms both for the stock recruitment and the net growth function have been estimated. The different alternatives may be combined in the model of population dynamics. The model of population dynamics that will be used in the estimation of the optimal stock level (Section 4) will be the combination of the Ricker stock-recruitment function and the quadratic net growth function. There are several reasons for this choice. First, from an empirical point of view, these two functions perform as well as or better than their alternatives. Second, the predictions of this model appear to be quite reasonable. The model of population dynamics is thus:

$$S_{t+1} = \left(S_t - H_t \right) e^{0.11 + 0.46 \times 10^{-4} S_t - 0.27 \times 10^{-7} S_t^2}$$

$$+ 1.10 S_{t-2} e^{-0.81 \times 10^{-3} S_{t-2}}.$$

For this model

  (i) $\bar{S} = 3.55$ million tonnes[14]

  (ii) $S_{msy} = 1.57$ million tonnes with

  (iii) $MSY = 0.61$ million tonnes.

S_{msy} is the stock level corresponding to Maximum Sustainable Yield (MSY). Moreover, simulation results show that for $H_t = 0$, $\forall t$,

$$\lim_{t \to \infty} S_t = \bar{S}.$$

In other words, the model is stable under natural conditions.

The model of population dynamics is illustrated in Fig. 3. To facilitate graphical representation of a function involving a time lag, a steady-state stock level is assumed. The harvest quantity is represented by the difference between S_{t+1} and the 45° line. It is noteworthy that the steady-state harvest quantity is fairly constant over a wide range of stock values.

It should be stressed that the bioeconomic model will not be qualitatively affected by the choice of a particular model of population dynamics.[15] There could be some quantitative differences, however, depending on the combination of stock-recruitment and net growth function in question.

3.2. Estimation of the Optimal Stock Level

We assume that North Sea herring is managed by a sole owner whose objective is to maximize the present value of net revenues from the fishery. As pointed out in the introduction North Sea herring is a common resource of Norway and the

---

[14] The stock of North Sea herring may in recent times have been closest to its carrying capacity in the early post World War II years, due to low fishing pressure during the war. Stock estimates for this period range between three and four million tonnes.

[15] In all combinations, net growth is nonpositive at the carrying capacity, while recruitment is nonnegative.

20   TROND BJØRNDAL

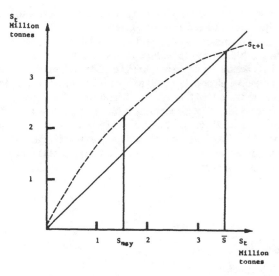

FIG. 3.   Stock dynamics (steady state stock levels).

European Economic Community. In other words, cooperative management is assumed. This means that problems associated with the transboundary character of the herring fishery are disregarded (see [18]). Thus, there is scope for further research in this field.

In this section, estimates of the optimal stock level (S^*) will be presented. However, we must first specify the net revenue function (7). For a constant price of herring,[16] we will give the specification of the net revenue function

$$\Pi(H_t S_t) = pH(E_t, S_t) - cE_t, \tag{7}$$

where E_t is the amount of effort in period t and c its unit cost. It is commonly assumed that harvesting costs are decreasing in stock size (cf. Section 2.2, see also [5, 9]). However, evidence indicates that harvesting costs for herring are not density dependant. If that is the case costs will not affect the maximization problem.

As noted above, a very important behavioural characteristic of North Sea herring, as of other clupeids, is the schooling behaviour. Schooling takes place both to make the search for food more effective and to reduce the effectiveness of predators [20]. Moreover, schooling fish contract their feeding and spawning range as the stock is reduced, with the size of schools often remaining unchanged. The schooling behaviour has also permitted the development of very effective means of harvesting, especially the purse seine. With modern fish-finding equipment, this

---

[16] The assumption of constant price is not inconsistent with the assumption of a sole resource manager. This is because the major part of the herring deliveries will be used for reduction into fish meal and fish oil. In this market, the supply of herring will be a negligible part of total supply and is thus unlikely to affect price.

means that harvesting can remain profitable even at declining stock levels. Hence, there are no economic brakes on harvesting until the fishery collapses (Murphy [19, p. 274]).

Clark [7] argues that the marginal concentration of schooling fish like herring tends to remain constant, except for very low stock levels. The consequence of this is that harvesting costs are not density-dependant, except at very low stock levels. Clark also gives the following analytical results for schooling fish[17]

(1) $r = 0 \rightarrow S^* = S_{msy}$
(2) $r \rightarrow \infty \rightarrow S^* = 0$.

Ulltang [26] provides some empirical evidence supporting these theoretical results. It is commonly postulated that

$$F = qE, \tag{11}$$

where F is fishing mortality, q is the catchability coefficient and E fishing effort. It is often assumed that q is constant. The harvest in numbers from the stock is

$$H = FN = qEN, \tag{12}$$

where N is the number of fish in the stock. For Norwegian spring spawning herring. Ulltang found evidence of changes in q with changes in the abundance of the stock (N). The relationship was postulated as

$$q = dN^{-e}, \tag{13}$$

where d and e are constants. Inserting (13) into (12), we obtain

$$H = dEN^{1-e}. \tag{12'}$$

The regression between q and N (Eq. (13)) gave an e-value slightly less than one.[18] We note that with e equal to one catch is proportional to E, and catch per unit effort is constant and independent of stock size.

In a subsequent paper, Ulltang observes that his findings for Norwegian spring spawning herring "indicate that a constant fishing effort would generate a more or less constant catch instead of a constant fishing mortality, thus creating rapidly increasing fishing mortalities as the stock size decreases" [27, p. 491]. There are apparently no similar empirical studies for North Sea herring. However, the behavior of this species should presumably not differ significantly from Norwegian spring spawning herring.

---

[17] These results are made transparent in the context of the Clark–Munro model [9], where $F(X)$ is the natural production function. When costs are non-density dependent the modified Golden Rule equation is given by $F'(X^*) = r$. In other words, the optimal stock level is determined by the equality of the marginal physical product of the stock and the discount rate. See Appendix 1 for the counterpart of this expression in terms of the present model.

[18] This result must be viewed with caution, as it was estimated on the assumption that effort was constant during the period of investigation (1950–1960).

22   TROND BJØRNDAL

TABLE V
Optimal Stock Level (S^*) and Corresponding Harvest (H^*)

Discount rate	S^* (million tonnes)	H^* (million tonnes)
0	1.57	0.61
0.02	1.51	0.61
0.04	1.46	0.60
0.06	1.40	0.60
0.08	1.35	0.60
0.10	1.29	0.59
0.12	1.23	0.59
0.20	1.00	0.55
0.52	0	0

We now return to the net revenue function, which using Eq. (12) can be rewritten as

$$\Pi(H_t, S_t) = pH_t - cE_t = pqE_tS_t - cE_t = \left[p - \frac{c}{qS_t} \right] H_t. \qquad (7')$$

The theoretical and empirical evidence described above suggest that q is inversely related to stock size, and we will assume this is the case for North Sea herring. With $q = d/S_t$, d constant, Eq. (7') becomes

$$\Pi(H_t, S_t) = \left[p - \frac{c}{d} \right] H_t. \qquad (7'')$$

The consequence of this assumption is that costs are not density dependent and will thus not affect the maximization problem[19] (see Appendix 1).

Thus, the present case represents what may be called a pure schooling fishery, where harvesting costs are not density dependent.[20] Even if this assumption is not entirely correct, it is still an interesting case study. This is because optimal stock levels under this assumption always are less than (or equal to) optimal stock levels with density-dependent costs. In other words, the results presented here represent lower bounds for the optimal stock level.

Having specified the net revenue function, we can now proceed to estimate the optimal spawning stock. As the implicit expression for the optimal stock level (Eq. (10)) cannot be solved explicitly for S^*, the solution is found by numerical methods (simulation). Moreover, the corresponding steady state harvest quantity has been estimated. The results are given in Table V.

The results in Table V permit us to draw a number of conclusions.

(1) For $r = 0$, $S^* = S_{msy}$. This case represents maximization of physical yield from the stock. S_{msy} can be estimated in the following manner. Assuming a steady-state stock level, the corresponding harvest can be expressed as a function of S by rewriting Eq. (6),

$$H(S) = S - \frac{S - G(S)}{e^{\delta(S)}}.$$

---

[19] The maximization problem is actually independent of both price and costs provided harvesting is profitable, i.e., $p > c/d$. If $p < c/d$, the fishery is unviable.

[20] For a more elaborate analysis of the consequences of non-density dependent costs on optimal harvesting, see [16].

S_{msy} is then the stock level at which the derivative of H with respect to S is equal to zero; i.e.,

$$H'(S_{msy}) = 1 - \frac{1 - G'(S_{msy}) - \delta'(S_{msy})[S_{msy} - G(S_{msy})]}{e^{\delta(S_{msy})}} = 0$$

or alternatively,

$$G'(S_{msy}) + e^{\delta(S_{msy})} - 1 + \delta'(S_{msy})[S_{msy} - G(S_{msy})] = 0.$$

Thus S_{msy} corresponds to the stock level where surplus production is at its maximum. For the model of population dynamics in question, $S_{msy} = 1.57$ million tonnes. It can be noted that if $\delta = 0$, S_{msy} is equal to S_{msy} of the stock-recruitment function. Similarly, if δ is independent of S and a positive (negative) constant, S_{msy} is greater (less) than S_{msy} of the stock-recruitment function.

(2) Clark [7] shows that stock extinction is "optimal" at an infinite discount rate. The numerical results show that this is the case even at a finite discount rate, as S^* goes to zero for $r = 0.52$. This corresponds to the own rate of return on the resource at zero stock size.

(3) S^* changes 2–3% for a one percentage point change in the discount rate ($r \leqslant 0.20$). H^*, on the other hand, changes only marginally for moderate changes in the discount rate. The latter result is caused by the properties of the estimated model of population dynamics.

(4) For "reasonable" discount rates ($0 \leqslant r \leqslant 0.12$), the optimal spawning stock is in the range 1.2–1.6 million tonnes with an annual harvest of about 0.60 million tonnes.

Gallastegui [14], in a study of a sardine fishery in the Gulf of Valencia, reported S^* to be not very sensitive to changes in the discount rate. The consequences of discounting in the case of baleen whales are illustrated by Clark [6]. An increase in the discount rate from zero to five percent reduces the optimal stock level by almost 50%. The reason for this severe effect of discounting as compared to the results presented in this paper and Gallastegui's is the low growth rate in the whale population.

It is commonly asserted that an open access fishery is characterized by an infinite discount rate, because fishermen will not be concerned with the effect of today's harvest on tomorrow's stock size (the stock externality). The presence of density dependent harvesting costs serves as a brake on stock depletion. However, we have argued above that harvesting costs in a pure schooling fishery are not density dependent. This implies that the open access equilibrium stock level—bionomic equilibrium—is zero. This result, combined with the numerical results in Table V, puts the management of this resource in perspective: while an open access fishery could cause stock extinction, a sole owner would—for finite discount rates—aim at achieving a stable stock level with a sustained harvest flow ad infinitum. The development of the stock in the post World War II period gives credibility to these

24   TROND BJØRNDAL

predictions. Under an open access regime, stock size was reduced from about 3–4 million tonnes in the years 1945–1950 to about 0.12 million tonnes in 1977, when the fishery was closed (cf. Appendix 2). Had the fishing moratorium not been imposed then, one would have expected further stock depletion. Open access exploitation of a number of other clupeids tells a similar story [19].

As illustrated by both theoretical results and empirical evidence, unrestricted entry to schooling fisheries may have irreversible consequences. Therefore, proper management of these resources becomes particularly important.

3.3. The Optimal Approach to the Steady State

In the previous section, the steady-state stock level was estimated under different assumptions. The optimal approach to the steady state is of interest both from a theoretical and from a practical point of view.

Clark and Munro [9] have shown that if the objective function is linear in the control variable and capital is perfectly malleable, the optimal path to equilibrium is of the bang-bang type, i.e., the most rapid approach. In the Clark–Munro model, linearity is obtained by assuming that the demand for fish is infinitely elastic and that costs are linear in the harvest rate. As is evident from Section 3.2, these assumptions are met in our model. We can therefore conclude that in the case of our model for North Sea herring, the optimal approach to S^* is of the bang–bang type.

The assumptions underlying the model ensure a most rapid approach to S^*. It is therefore of interest to investigate whether changing them would affect the nature of the optimal approach. Clark and Munro [9] have shown that if the linearity in the control variable is violated, the optimal approach to S^* is a gradual, asymptotic one. The control variable in our model is the harvest quantity.

We have assumed that the demand for herring is infinitely elastic. While this may be appropriate for deliveries for reduction into fish meal and fish oil, it may be more appropriate to assume a finite-demand elasticity for deliveries for human consumption. This would give a gradual asymptotic approach to S^*. Similarly, the presence of a fixed-factor-like capital in a linear model would also give rise to an asymptotic approach [8]. Thus changing any of the basic underlying assumptions would affect the nature of the approach to the steady state stock level.

The task of finding the optimal approach path in a nonlinear empirical problem would, however, be nontrivial and especially so in a model with a time lag. From a policy point of view, the appropriate approach may therefore be to simulate various paths and compare them according to criteria that are deemed important. Examples of two such paths are given in Table VI.

The first case simulates the development of the stock under natural conditions, which corresponds to the bang-bang approach. This path may be both inoptimal and inconceivable, as a complete fishing moratorium would be very difficult to achieve over a prolonged period time. The case is given to allow consideration of the delay in attaining the steady state if an asymptotic approach is used rather than the most rapid approach. Accordingly, an an example, in the second simulation it is assumed that there is an annual harvest quantity of 0.20 million tonnes.

Assuming a 6% discount rate, the optimal stock level is 1.40 million tonnes with a harvest quantity of 0.60 million tonnes (cf. Table V). With a most rapid approach, this stock level would be attained in 1984, after a two-year fishing moratorium. An

TABLE VI

Simulation of Spawning Stock 1983–1987[a]

Year	$H_t = 0$ spawning stock (million tonnes)	$H_t = 0.20$ (million tonnes) spawning stock (million tonnes)
1983	1.05	0.82
1984	1.50	1.01
1985	2.05	1.29
1986	2.74	1.70
1987	3.33	2.19

[a]$S_{82} = 0.50$ million tonnes. The rapid increase in stock size from 1982 to 1983 is influenced by the good 1981 yearclass of recruits (cf. Appendix 2).

annual harvest quota of 0.20 million tonnes would cause a delay in reaching the equilibrium of two years. The two policies can be compared in present value terms. Assume that once the steady state has been reached, the fishery continues on a sustained basis, i.e., with an annual harvest quantity of 0.60 million tonnes ad infinitum. Calculating the present values of catches or gross revenues[21] for the two policies—still assuming $r = 0.06$—gives a difference of about 2.7 percent.[22] One would expect that if the present values of net revenues were calculated, the differences would be of the same magnitude. This result suggests that the loss—measured in terms of present value—of deviating from the optimal approach path may not be great. This observation is supported by Clark [5, p. 323] and Ludwig [17].[23] The practical consequence of this result is that the resource manager may have some degree of freedom in determining the approach to the steady state stock level.

4. SUMMARY

This paper introduces and estimates a model of population dynamics. This has been combined with a net revenue function to formulate a dynamic bioeconomic model, which is used to analyze the optimal management of North Sea herring. Although the model is stock specific, it should be useful for other fisheries as well. It can also be noted that most studies in fisheries economics are based on the Schaefer model [24]. The present analysis shows that the basic theory works also when other biological models are used.

The results show that the size of the optimal spawning stock is fairly sensitive to changes in the discount rate. The harvest quantity, on the other hand, is considerably less sensitive to changes in the discount rate. This result is caused by the

[21] The optimal stock level would typically be "overshot" in the year it was reached (cf. Table V), giving a harvest quantity for that year which is higher than the sustained one. This effect has been neglected in the following calculation.

[22] The bang-bang approach has the highest present value.

[23] Although Ludwig's results are in terms of a stochastic model, they are likely to carry over to the deterministic case.

26   TROND BJØRNDAL

properties of the estimated model of population dynamics. Altogether, the model appears to be robust.[24]

An important result of the paper is to illustrate that an open access fishery could cause stock extinction. This is a result of the schooling behavior of the herring. In other words, failure to manage schooling fish resources could have irreversible consequences.

This analysis can be extended in a number of ways. On the production side, an empirical study of the relationship between the catchability coefficient and the stock size should be undertaken. Depending on the outcome of this study, the costs of the industry might need to be estimated. One interesting extension of the analysis would be to formulate a stochastic model. The first step would presumably be to formulate a stochastic stock-recruitment function, as a consequence of the great uncertainty that is implicit in this relationship. From a practical point of view, the optimal trajectories—taking into account other objectives in fisheries management than rent maximization—should also be analyzed more closely. The same may be said for the transboundary nature of the fishery.

APPENDIX I: DERIVATION OF AN IMPLICIT EXPRESSION FOR THE OPTIMAL STOCK LEVEL

The Lagrangean for the maximization problem is

$$L = \sum_{t=0}^{\infty} \left\{ \alpha^t \Pi(H_t, S_t) - q_t \left[S_{t+1} - (S_t - H_t) e^{\delta(S_t)} - G(S_{t-\gamma}) \right] \right\}. \quad (1)$$

In the following, constant prices will be assumed. First order necessary conditions for an optimum are [5]

(i) $\dfrac{\partial L}{\partial H_t} = 0, \qquad t = 0, 1, 2,$

(ii) $\dfrac{\partial L}{\partial S_t} = 0, \qquad t = 1, 2, 3.$

We want to solve for the steady-state equilibrium biomass level and take derivatives of (1) for general t ($t \geqslant 1$):

$$\frac{\partial L}{\partial H_t} = \alpha^t \Pi_H - q_t e^{\delta(S_t)} = 0, \quad (2)$$

$$\frac{\partial L}{\partial S_t} = \alpha^t \Pi_S + q_t \delta'(S_t)(S_t - H_t) e^{\delta(S_t)} + q_t e^{\delta(S_t)} - q_{t-1}$$
$$+ q_{t+\gamma} G'(S_t) = 0, \quad (3)$$

$$\alpha^t \Pi_S + q_t \delta'(S_t)(S_t - H_t) e^{\delta(S_t)} + q_t e^{\delta(S_t)} + q_{t+\gamma} G'(S_t) = q_{t-1}. \quad (3')$$

Equation (2) can be solved for q_t, the shadow price of the resource

$$q_t = \alpha^t \frac{\Pi_H}{e^{\delta(S_t)}}. \quad (4)$$

---

[24] Using any other model of population dynamics, i.e., combination of the estimated stock-recruitment and net growth functions, gives the same qualitative results as those represented in this paper. In addition, quantitative results do not differ much, except for the Beverton–Holt function.

(4) is then inserted in (3'),

$$\alpha' \Pi_S + \alpha^t \frac{\Pi_H}{e^{\delta(S_t)}} \delta'(S_t)(S_t - H_t) e^{\delta(S_t)} + \alpha^t \Pi_H + \alpha^{t+\gamma} \frac{\Pi_H}{e^{\delta(S_t)}} G'(S_t) = \alpha^{t-1} \frac{\Pi_H}{e^{\delta(S)}}.$$

This equation can be simplified to

$$e^{\delta(S_t)} \left[\frac{\Pi_S}{\Pi_H} + 1 \right] + \delta'(S_t)(S_t - H_t) e^{\delta(S_t)} + \alpha^\gamma G'(S_t) = \frac{1}{\alpha}. \tag{5}$$

For the stock to be in steady-state equilibrium, we require

$$S_{t+1} - S_t = (S_t - H_t) e^{\delta(S_t)} + G(S_{t-\gamma}) - S_t = 0. \tag{6}$$

Equation (6) can be solved for the steady-state harvest quantity

$$H(S) = S - \frac{S - G(S)}{e^{\delta(S)}}. \tag{7}$$

Moreover, we get

$$S - H(S) = \frac{S - G(S)}{e^{\delta(S)}}. \tag{8}$$

Equation (8) can be inserted in (5) to yield the following implicit expression for the optimal stock level S^*,

$$e^{\delta(S^*)} \left[\frac{\Pi_S}{\Pi_H} + 1 \right] + \delta'(S^*)[S - G(S^*)] + \alpha^\gamma G'(S^*) = \frac{1}{\alpha}. \tag{9}$$

For North Sea herring, $\gamma = 2$. The net revenue function is

$$\Pi(H_t, S_t) = \left(p - \frac{c}{qS_t} \right) H_t = \left(p - \frac{c}{d} \right) H_t, \qquad q = \frac{d}{S_t}. \tag{10}$$

Taking derivatives, we get

$$\frac{\Pi_S}{\Pi_H} = 0. \tag{11}$$

Inserting (11) and $\gamma = 2$ in Eq. (9), it reduces to

$$e^{\delta(S^*)} + \delta'(S^*)[S^* - G(S^*)] + \alpha^2 G'(S^*) = \frac{1}{\alpha} = 1 + r. \tag{9'}$$

Equation (9') is an implicit expression for the optimal stock level of North Sea herring (cf. footnote 17). The only variables appearing are S and r. Once the δ and G functions are specified, Eq. (9') can be solved for S^* for given values of r. The corresponding harvest quantity can be found from Eq. (7).

APPENDIX 2: BIOLOGICAL DATA

Year	Spawning stock per September 1 (1,000 tonnes)	Recruits the following year (millions)	Harvest quantity[a] (1,000 tonnes)	Instantaneous net natural growth rate[b]
1947	2,945	4,720	587	−0.1159
1948	2,581	4,100	502	−0.0380
1949	2,618	5,680	509	−0.0531
1950	2,428	6,900	492	0.0053
1951	2,169	7,690	600	−0.0307
1952	1,908	9,100	664	−0.0174
1953	1,707	8,070	699	0.0807
1954	1,745	7,700	763	0.0216
1955	1,821	4,768	806	0.1193
1956	1,741	21,429	675	0.0770
1957	1,593	5,641	683	0.0051
1958	1,236	7,555	671	0.1763
1959	2,063	1,954	785	0.1454
1960	1,871	16,686	696	0.0230
1961	1,601	7,085	697	0.0926
1962	1,132	8,740	628	−0.0252
1963	1,800	10,907	716	0.1397
1964	1,829	5,709	871	0.0903
1965	1,340	5,289	1,169	0.1000
1966	1,116	7,581	896	0.0893
1967	817	7,623	696	0.0669
1968	390	3,820	718	0.1754
1969	359	9,081	547	0.1470
1970	318	7,146	564	0.1306
1971	219	4,975	520	0.1514
1972	269	2,398	498	0.18
1973	228	5,583	484	0.1593
1974	166	773	275	0.1404
1975	117	720	313	0.1382
1976	141	1,064	175	0.2265
1977	123	899	46	0.1607
1978	154	2,582	11	0.0237
1979	208	3,423	25	0.0385
1980	238	12,414	61	0.0891
1981	368	14,958	95	0.0589
1982	498	—	—	0.1137

[a] Harvest of juvenile herring is included.
[b] The net growth is estimated according to the formula

$$S_{t+1} = S_t e^{\delta_t - F_t} \rightarrow \delta_t = F_t + \ln(S_{t+1}/S_t); \qquad F_t = \text{fishing mortality}$$

Note. For these estimates the stock size as of January 1 has been used. Sources: S_t: Anon. [1] for 1947–1954, Anon. [2] for 1955–1974, Anon. [3] for 1975–1982. R_t and F_t: Anon. [1] for 1947–1954, Anon. [2] for 1955–1972, Anon. [3] for 1973–1981. H_t: Schumacher [25] for 1947–1976, Anon. [2] for 1977–1981.

REFERENCES

1. Anon., "Assessment of Herring Stock South of 62°N 1973–1975, Cooperative Research Report of the International Council for the Exploration of the Sea," Vol. 60 (1977).

2. Anon., "Report of the Herring Assessment Working Group for the Area South of 62°N," International Council for the Exploration of the Sea, C.M. 1983/Assess: 7 (1982).

3. Anon., "Report of the Herring Assessment Working Group for the Area South of 62°N," International Council for the Exploration of the Sea. C.M. 1983/Assess: 9 (1983).

4. P. G. Bradley, Some seasonal models of the fishing industry, *in* "Economics of Fisheries Management," (A. D. Scott, Ed.), Institute of Animal Resource Ecology, University of British Columbia, Vancouver (1970).

5. C. W. Clark, "Mathematical Bioeconomics," Wiley, New York (1976).

6. C. W. Clark, A delayed-recruitment model of population dynamics, with an application to Baleen whale populations, *J. Math. Biol.* **3**, 381–391 (1976).

7. C. W. Clark, Concentration Profiles and the Prodution and Management of Marin Fisheries, *in* "Economic Theory of Natural Resources" (W. Eichhorn, R. Henn, K. Neumann, and R. W. Shephard, Eds.), Physica-Verlag, Würzburg (1982).

8. C. W. Clark, F. H. Clarke, and G. R. Munro, The optimal exploitation of renewable resource stocks: Problems of irreversible investment, *Econometrica* **47**, 25–47 (1979).

9. C. W. Clark and G. R. Munro, The economics of fishing and modern capital theory: A simplified approach, *J. Environ. Econom. Management* **2**, 92–106 (1975).

10. D. H. Cushing, The dependence of recruitment on parent stock in different groups of fishes, *J. Cons. int. Expl. Mer* **33**, 340–362 (1971).

11. D. H. Cushing, The problems of stock and recruitment, *in* "Fish Population Dynamics" (J. A. Gulland, Ed.), Wiley, New York (1977).

12. R. B. Deriso, Harvesting strategies and parameter estimation for an age-structured model, *Canad. J. Fish. Aquat. Sci.* **37**, 268–282 (1980).

13. O. Flaaten, The optimal harvesting of a natural resource with seasonal growth, *Canad. J. Econom.* **16**, 447–462 (1983).

14. C. Gallastegui, An economic analysis of sardine fishing in the Gulf of Valencia (Spain), *J. Environ. Econom. Management* **10**, 138–150 (1983).

15. M. Kurz, Optimal economic growth and welfare effects, *Internat. Econom. Rev.* **9**, 348–357 (1968).

16. D. Levhari, R. Michener, and L. J. Mirman, Dynamic programming models of fisheries, *Amer. Econom. Rev.* **7**, 649–661 (1981).

17. D. Ludwig, Harvesting strategies for a randomly fluctuating population, *J. Cons. int. Explor. Mer* **39**, 168–172 (1980).

18. G. R. Munro, The optimal managementof transboundary renewable resources, *Canad. J. Econom.* **12**, 355–376 (1979).

19. G. I. Murphy, Clupeids, *in* "Fish Population Dynamics" (J. A. Gulland, Ed.), Wiley, New York (1977).

20. B. L. Partridge, The structure and function of fish schools, *Scientific Amer.* **246**, 90–99 (1982).

21. K. H. Posthuma, The effect of temperature in the spawning and nursery areas on recruitment of autumn-spawning herring in the North Sea, *Rapp. P.-v. Réun. Cons. int. Explor. Mer* **160**, 175–183 (1971).

22. W. E. Ricker, "Computation and Interpretation of Biological Statistics of Fish Populations," Environment Canada, Ottawa (1975).

23. A. Saville and R. S. Bailey, The assessment and management of the herring stocks in the North Sea and to the West of Scotland, *Rapp. P.-v. Réun. Cons. int. Explor. Mer* **177**, 112–142 (1980).

24. M. B. Schaefer, Some considerations of population dynamics and economics in relation to the management of marine fishes, *J. Fish. Res. Board Canad.* **14**, 669–681 (1957).

25. A. Schumacher, Review of North Atlantic Catch Statistics, *Rapp. P.-v. Réun. Cons. int. Explor. Mer* **177**, 8–22 (1980).

26. Ø. Ulltang, "Catch per Unit Effort in the Norwegian Purse Seine Fishery for Atlanto-Scandian (Norwegian Spring Spawning) Herring," FAO Fish. Tech. Paper 115, pp. 91–101 (1976).

27. Ø. Ulltang, Factors affecting the reaction of pelagic fish stocks to exploitation and requiring a new approach to assessment and management, *Rapp. P.-v. Réun. Cons. int. Explor. Mer* **177**, 489–504 (1980).

The Share System

[35]

The Share System in Open-Access and Optimally Regulated Fisheries

Lee G. Anderson

INTRODUCTION

Although the share system is almost universal in fisheries throughout the world, it has received little treatment in the literature, the exceptions being papers by Zoeteweij (1956), Sutinen (1979), and Griffin, Lacewell, and Nichols (1976). The latter shows that a share system can divide the rent from the fishery between vessel owners and crew members. See Anderson (1982), however. This paper introduces the share system into the traditional static deterministic model of an exploited fishery, as described in Gordon (1954), Christy and Scott (1965), and Anderson (1977). The model will then be used to answer the following questions: (1) How is the share rate determined and how will it affect open-access fishing and the rents earned by boat owners and crew? (2) How will imperfect competition in the determination of the share rate affect open-access fishing and rent distribution? (3) What are the implications of the share system for management, and do they differ depending on how the share rate is determined?

The next section will present the basic elements of this new model and use them to derive the results of the traditional model. The third section shows that if fishing effort can be produced using variable proportions and if the crew has some control over the input mix, then the least cost combination of inputs will not be used unless variable costs and revenues

are shared at the same rate. The new model will be derived in the fourth section. The next few sections use the model to answer the above questions. A concluding section summarizes the main results and makes several generalizations from the conclusions. While it is generally agreed that a main purpose of the share system is to spread risks between boat owners and crews, the model presented here is deterministic. A stochastic model was not used because it would unnecessarily complicate the analysis without adding much to the essential points to be made.

ASSUMPTIONS AND DEFINITIONS

A fishery model where revenues and costs are shared by owner and crew is presented below. Because of the nature of the share system, it is necessary to modify the traditional model by introducing vessels and the level of vessel output. Although vessels are often owner-operated, the financial considerations of the owner as a crew member (he receives a crew share) and as an owner

The author is associate professor at the University of Delaware. This article was written while he was Visiting Professor at the Institute for Economics, University of Bergen, Norway. Research support from the Norwegian Fishery Research Council is gratefully acknowledged. Helpful comments were provided by Wade Griffin, Røgnvaldur Hannesson, Kenneth Roberts, an anonymous reviewer, and participants in seminars at the universities of Bergen and Delaware.

(he receives a boat share) are distinguishable. In what follows the two roles will be treated separately except where otherwise noted. Let:

$y(Ne)$ = the sustainable yield function for the exploited fish stock

where
$y' > 0$ for $Ne < \bar{N}e$
$y' < 0$ for $Ne > \bar{N}e$
$y'' < 0$

N = The number of vessels operating in the fishery. This is the basic control variable of the model.

e = The fixed annual amount of effort produced by each boat. (See final section for discussion of variable effort per boat.)

P = The constant price of fish.

C_1 = The opportunity cost of labor used to produce one unit of effort.

C_2 = Opportunity costs of nonlabor inputs used to produce one unit of effort.

C_3 = Opportunity costs of inputs used to produce the onboard processing and handling services associated with one unit of output.

K = The annual fixed cost of each boat.

S_r = Percentage share of revenues received by crews.

S_c = Percentage share of variable costs paid by crews.

As Zoeteweij (1956, p. 19) reports, there is enormous variety in the share systems of the world, and one of the major differences is the degree to which the crew pays the variable costs. In most instances, however, $S_r = S_c$. (See for example, Huppert et al. 1973, p. 13). To simplify the analysis here, it will be assumed that the two rates are equal. When this is not true, as in the case of the Gulf of Mexico shrimp fishery (Sass and Roberts 1979, p. 14), there are serious

efficiency effects. This will be discussed in a separate section.

The implicit assumptions behind the above definitions are: (1) Effort is produced with nonvariable proportions; (2) there is perfect competition elsewhere in the economy; (3) the output of this particular fishery is small relative to total output and so price can be considered a parameter; and (4) sustainable yield is a deterministic function of effort.

As a basis of comparison, the traditional model, which assumes that labor is paid an opportunity cost wage, has an open-access equilibrium (i.e., profit is equal to zero), which can be expressed as:

$$\pi^* = (P - C_3)\,y - (C_1 + C_2)\,Ne - NK = 0 \quad [1]$$

or

$$(P - C_3)\frac{y}{N} - (C_1 + C_2)\,e - K = 0 \quad [2]$$

That is, open-access equilibrium will occur where average revenue equals average cost for each vessel. For ease of notation let the solution to this equation be N_0, and the level of output which results be y_0.

The economically efficient point occurs where the difference between the value of output and the opportunity cost of producing it is maximized. With a constant price of output, this is where profit to the fishery is maximized. From [1] this occurs when:

$$\frac{\partial \pi^*}{\partial N} = (P - C_3)\,y'e - (C_1 + C_2)\,e - K = 0 \quad [3]$$

Efficiency occurs where the marginal net value of vessel catch equals marginal cost. Let the solution to equation [3] be N^* and the level of output it produces be

y^*. Since $y'' < 0$ then $y'e < y/N$ and so $N^* < N_o$.

THE SHARE SYSTEM AND EFFICIENT PRODUCTION OF EFFORT

This section will be a short diversion to show that if effort can be produced using variable proportions and if a nonowner captain determines the usage of some inputs, then for optimal input utilization, S_r must equal S_c. When there are absentee owners, as for example in many boats in the Gulf shrimp fishery (Roberts 1981), the captain often has latitude in determining where in a particular area the vessel will fish and when to change areas during a particular trip. Hence, he can control use of variable inputs such as fuel and lubrication. Therefore the question is of more than academic interest.

It will be necessary to modify the model introduced above to allow for variable proportions in the production of effort. For ease of exposition, however, the constant proportion model will be used in all other sections of the paper. For the moment, however, assume that effort per boat is a function of labor and another input a. That is:

$$E = E(a, L)$$

Also assume that there are diminishing returns to all inputs. The long-run cost-minimizing combination of inputs for any level of effort will be when the value of the marginal products are equal to the input prices.

An absentee owner can select the amount of labor but has only loose control over other inputs. The captain will select that level of input a which maximizes crew share constrained only by the fact that the boat must cover its

cost. Since any one vessel in a large fleet can not affect catch per unit of effort, the revenue received for each unit of effort, call it R_E, is the product of the price of fish and the existing catch per unit of effort. The operational maximand for the captain is therefore:

$$L = S_r R_E E - S_c (P_a a) + \lambda [(1 - S_r) R_E E - (1 - S_c)(P_a a) - K] \qquad [4]$$

The first order condition for a maximum is:

$$R_E E_1 = P_a \frac{S_c(1 - \lambda) + \lambda}{S_r(1 - \lambda) + \lambda} \qquad [5]$$

Except in the special case where λ equals one, the value of the marginal product will equal P_a only when $S_r = S_c$. If $S_r > S_c$, the variable input will be overutilized because the absentee owner will bear the larger share of its cost. The reverse will be true if $S_c > S_r$. In both instances the cost of producing effort will be higher than need be, because, assuming the owner ultimately chooses that amount of labor where its VMP equals the effective wage, the ratios of VMP and input prices will not be equal.[1]

The policy implications are that, where effort can be produced in variable proportions and the captain can determine the use of some variable inputs, owners and crew should share both variable costs and revenue and further, they should share them at the same rate. Otherwise, the cost of producing effort at every level of output will be higher than necessary.

---

[1] During final editing of this paper, I came across an unpublished manuscript by Ola Flaaten (1981) which discusses the efficiency aspects of share systems in some detail.

THE BASIC MODEL

The optimal operation point in the share system model has been described above, see equation [2]. The open-access point and the process of achieving it will now be described assuming that owners and crew share both revenues and variable costs using the same percentage rate, i.e., $S_r = S_c = S$. To do this the operational behavior of both owners and crew members must be examined. Open-access equilibrium will occur at that level of effort where, given the share rate, the number of boats that owners are willing to use equals the amount that can be operated by workers who are willing to fish as opposed to alternative employment. However, if the share rate is fixed, the open-access equilibrium occurs at the number of boats that owners are willing to use or at the number for which they are actually able to find crews, whichever is smaller.

Open-access equilibrium for the owners occurs where total profits to boats are zero and is expressed as follows:

$$\pi_B = (1 - S)[(P - C_3) y - NeC_2] - NK = 0 \quad [6]$$

It is obvious that the number of boats that owners are willing to use depends upon the share rate. Solving the above for S generates a boat equilibrium curve (*BEC*).

$$S = 1 - \frac{NK}{(P - C_3) y - NeC_2} \quad [6']$$

$$= \frac{(P - C_3) \frac{y}{N} - eC_2 - K}{(P - C_3) \frac{y}{N} - eC_2}$$

As can easily be shown and as depicted in Figure 1, where S and N are plotted in

the vertical and horizontal axes respectively, the *BEC* curve intersects the vertical axis at 1 and is downward sloping. That is, the lower the share rate, the more vessels can be assimilated into the fishery before the rent earned by *vessels* is driven to zero.

On the other hand, workers are willing to supply their services as long as fishing income is at least equal to foregone wages, which is assumed to be the opportunity cost of labor.[2] That is, workers will continue to enter until their total "profit" is equal to zero, i.e.:

$$\pi_c = S [(P - C_3) y - NeC_2] - NeC_1 = 0 \quad [7]$$

The first term is the total income earned by workers, and the second term is their total opportunity cost. Notice that the number of vessels that workers are willing to crew is also a function of S. Solving [7] for S generates a crew equilibrium curve (*CEC*).

$$S = \frac{NeC_1}{(P - C_3) y - NeC_2} \quad [7']$$

$$= \frac{eC_1}{(P - C_3) \frac{y}{N} - eC_2}$$

This function is positively sloped (Figure 1) and approaches zero as N approaches zero from above. The higher the share rate, the more vessels can be assimilated before the rent earned *by the crew* is driven to zero.

If the share rate is market-determined, the intersection of these curves will determine the equilibrium share rate and open-access level of effort. At a rate

[2] To be precise, this is true only if there are no non-cancelling nonpecuniary benefits related to either type of work, see Anderson (1980).

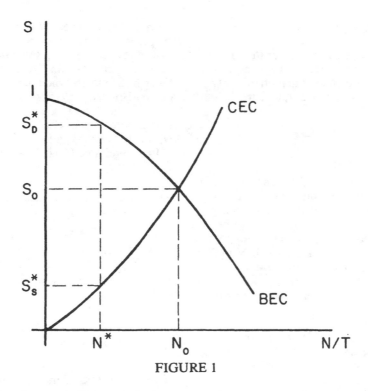

FIGURE 1

above S_0, the number of boats that are willing to work is less than that which is necessary to offer positions to all those individuals desiring to fish, but working crew members would be earning rents. However, the excess supply of labor to the fishery will bid the share rate down until all rents to workers are dissipated.

Alternatively, if S is below S_0, the number of boats that can be crewed is less than that which owners wish to use, but boat owners will be earning rents. However, the excess demand for labor will bid the share rate up until all rents to boat owners are dissipated.

By equating expressions [6'] and [7'], the open-access equilibrium condition can be derived.

$$\frac{(P - C_3)\frac{y}{N} - eC_2 - K}{(P - C_3)\frac{y}{N} - eC_2} = \frac{eC_1}{(P - C_3)\frac{y}{n} - eC_2} \qquad [8]$$

Equation [8] can be alternatively expressed in a manner identical to [2] above. Therefore, if the share rate is determined by market forces, the basic results of the traditional model and the share rate model are the same. The share rate has nothing to do with the determination of the economic efficiency point, and market forces in an open-access fishery will guarantee that the share rate that evolves will result in an opportunity cost wage rate for labor. The share model, however, does contribute a fuller description of the process of achieving the open-access equilibrium.

By solving equation [2] for the value of y/N at the open-access point and substituting it into either equation [6'] or [7'], the equilibrium share rate can be expressed as:

$$S_n = \frac{eC_1}{eC_1 + K} \qquad [9]$$

This "natural" share rate is a function of the costs that are not shared. It is the ratio of the opportunity cost of labor per boat to the sum of the opportunity cost of labor per boat and annual fixed opportunity costs per boat. The logic behind this is that at open access the revenues must be shared such that both owners and crew members cover their opportunity costs.

If the share rate is not determined by the market, the results of the share system and the traditional models are not the same. For example, if, by government fiat, union contracts, etc., the share rate is fixed above S_o, then the *BEC* is operational in determining the open-access level of effort. If S were set at S_D^* (see Figure 1), then N^*, the optimal number of boats, would be employed. At this point more workers would want to join the fishery, but at the mandated share rate, boat owners will not hire them. The fishery will be operating at the economically efficient point and the rent will accrue to those crew members who are able to secure positions. For fixed share rates between S_D^* and S_o the "constrained" open-access level of effort will be between N^* and N_o according to the *BEC* curve.

On the other hand, if S is set below S_o, then the *CEC* constrains the fishery. If the rate were fixed at S_s^*, the optimal amount of vessels would again be

utilized. Boat owners would want to use more, but no more workers would be available at the existing share rate. In this case the rent from the fishery accrues to those boat owners who are able to secure crews. For fixed share rates between S_o and S_s^* the "constrained" open-access equilibrium level of effort will be between N_o and N^* according to the *CEC* curve.

From equation [6'] it can be seen that

$$S_D^* = \frac{(P - C_3)\dfrac{y^*}{N^*} - eC_2 - K}{(P - C_3)\dfrac{y^*}{N^*} - eC_2} \qquad [10]$$

This is the ratio of net returns to labor per boat to net returns to labor and capital per boat evaluated at N^*. Similarly from [7']

$$S_s^* = \frac{eC_1}{(P - C_3)\dfrac{y^*}{N^*} - eC_2} \qquad [11]$$

This is the ratio of the average opportunity cost of labor per boat to the net returns to labor and capital per boat evaluated at N^*. Equations [10] and [11] have the same denominator but in S_D^* (which gives all rent to labor) the numerator is the greatest possible amount that each boat can pay to labor and still pay for other inputs. In S_s^* (which gives all rent to boat owners) the numerator is the lowest dollar amount that can be paid to the workers necessary to operate a boat and still have them work.

From this, one might conclude that a fishery can be regulated by controlling the share rate; i.e., managers should be able to achieve a fleet size of N^* by fixing the share rate at either S_D^* or S_s^* depending on whether it chooses to give the rent to crew members or vessel owners. This will be true, however, only if the fixed nonequilibrium share rate can be maintained and if no side payments are permitted. If side payments are possible, the rents will be dissipated in the normal manner.

Note that if fixed share rates can be maintained and side payments can be prevented and if there are alternative means to achieve a fleet size of N^* without rent dissipation, the agency can divide the rents to the resource between the two sectors by selecting a rate somewhere between S_D^* and S_s^*. The higher the share rate, the larger the proportion of the rents that will go to crews and vice versa. The alternative means of regulation is necessary because at fixed rates other than S_D^* and S_s^*, the constrained open-access level of effort is different from N^*.

THE SHARE SYSTEM AND MARKET POWER

The purpose of this section is to consider the effects of[1] market power by either boat owners or crew members on the open-access utilization of the fishery. Assume that there is a single owner or that boat owners collude to form a monopsony in the purchase of labor. This single entity will consider S as a function of the number of boats, and hence labor it uses, rather than as a parameter determined by the market. Assuming that the fleet owner desires to maximize profits, the operational maximizing equation can be obtained by substituting the "supply" equation for S [7'] into the boat profit equation [6].

$$\pi_B = \left[1 - \frac{NeC_1}{(P - C_3) y - NeC_2} \right] \times \quad [12]$$

$$[(P - C_3) y - NeC_2] - NK$$

With some manipulation this can be simplified to the total profit function for the fishery as a whole. See equation [1]. Therefore, the single fleet owner will use the optimal sized fleet and the minimum share rate necessary to achieve it, S_s^*. This is the familiar result, that a sole owner will optimally utilize a fishery if the price of output is fixed. The existence of a share system does not alter this result.

Consider now the case of the central labor organization that maximizes total returns less the opportunity costs of labor. Due to their position, the actual share rate they can get depends upon the number of workers they actually place, i.e., the number of boats they allow to fish. Therefore their operational maximizing equation can be found by substituting the "demand" equation for the share rate [6'] into the profit equation for the labor sector [7]. This is

$$\pi_c = \left[\frac{(P - C_3) y - NeC_2 - NK}{(P - C_3) y - NeC_2} \right] \times \quad [13]$$

$$[(P - C_3)y - NeC_2] - NeC_1$$

This also reduces to the total industry profit equation. Therefore, a sole authority for labor trying to maximize labor's total profits will also choose N^* units of effort. It will, however, choose S_D^*, the share rate that gives all the returns to labor.

Labor organizations are also concerned with total employment and the distribution of rent to individuals. Therefore, unless the rent or the work opportunities can be easily distributed, some combination of total rent earned and level of employment will be the operational objective. Therefore, effort will vary between N^* and N_0 and the share rate will vary between S_D^* and S_0, depending on the relative weights placed on each.

THE SHARE SYSTEM AND REGULATION

Introduction

The purpose of this section is to discuss regulation in terms of the share system model and to describe how various types of regulation will affect the share rate and the distribution of rents. It will be useful to discuss limited entry and non-limited entry regulations separately. Limited entry regulations include license programs to restrict the number of operating units, properly designed taxes, and individual quotas. Non-limited entry regulations include total quotas, closed seasons or areas, and gear restrictions. The difference between the two is that the latter directly or indirectly adversely affect the efficiency of producing effort, and in the long run they are often ineffective in reducing the total amount of effort.[3] Most of the discussion will be in terms of optimal tax policies. They are not necessarily the best policies, but they do provide a pedagogical device to explain the workings of the model and a framework in which to describe other types of regulation.

In terms of Figure 2, the regulation required to achieve economic efficiency can be summarized as follows. If the share rate is market-determined, regulation must shift the BEC and the CEC curves such that they intersect above N^*, and do so such that the social opportunity cost function for effort is not affected. The latter constraint is necessary because optimal utilization entails an optimal number of vessels, each operating efficiently. On the other hand, if the

share rate is fixed, either the BEC or the CEC needs to be shifted such that one of them intersects the vertical line above N^* at the fixed share rate, and such that the opportunity cost of producing effort is not affected.

Limited Entry Regulations

Given the assumption of constant effort per vessel, license limitation programs and individual quotas are identical. The optimal strategy would be to permit only N^* vessels to operate or to issue individual quotas, such that the total amount of allowable effort is equal to what can be produced by N^* vessels. If the rights to fish are given to specified boat owners, then the effective BEC will become a vertical line above N^* as is depicted in Figure 3. With a market-determined share rate, the new equilibrium rate will be S_s^* and all the rents will accrue to boat owners who have the right to fish. With a fixed share rate between S_D^* and S_s^*, the fleet will still be N^*, but the rents will be shared by owners and crew members, with the owners' share varying inversely with the share rate. If the rights are given to specified crews, the effective CEC will become vertical above N^*. With a market-determined share rate, the equilibrium rate will be S_D^* and these crews will receive the rents. With a fixed share rate, the distribution of rents will depend upon the actual rate chosen.

The assumption of fixed vessel capacity is very limiting here because experience has shown (Pearse 1979; Rettig and Ginter 1980) that with variable capacity

---

[3] See Pearse (1979) and Rettig and Ginter (1980) for a detailed discussion of the economic efficiency and other aspects of limited entry programs, and see Crutchfield (1961), Scott (1962), and Anderson (1977) for a complete analysis of the weaknesses of nonlimited entry regulations.

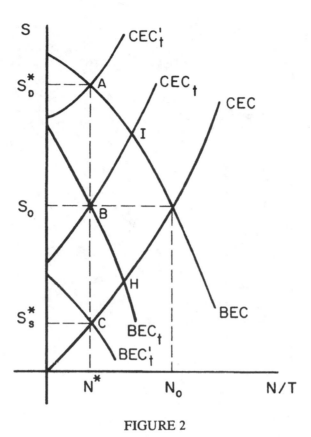

FIGURE 2

and substitutability of inputs, license programs lead to inefficiency in both the amount of effort used and the cost with which it is produced. Individual quotas do not appear to suffer the same weaknesses.

The analysis of optimal taxes is also based on appropriate changes in the *CEC* and the *BEC*. The open-access and optimal conditions of a fishery with opportunity cost wages are described by equations [2] and [3] above, respectively. The tax per unit of effort that causes an otherwise open-access fishery to operate the optimal number of boats (and

incidentally, causes each boat to be operated in an efficient manner) is

$$T_E = (P - C_3)\left(\frac{y}{Ne} - y_1\right) \qquad [14]$$

This tax can be determined by putting eT_E, the total tax bill per boat, into equation [2], equating it to [3] and solving for T_E. See Anderson (1977, ch. 4). Since the optimal tax is the rent earned per unit of effort, it takes away the incentive to overutilize the fishery.

The derivation of optimal tax programs with a share system is analogous, although there are more types of taxes to

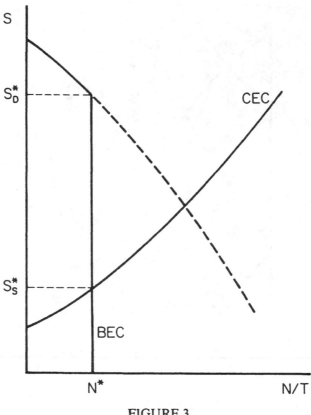

FIGURE 3

be considered. Equations [15] and [16], which are modified forms of [6] and [7] above, are the equilibria conditions for boat owners and crew members, respectively. The term t_s represents a tax on the nonlabor inputs that must be shared by owners and crew. The terms t_B and t_C represent taxes on boats and labor, respectively.

$$\pi_B = (1 - S) [(P - C_3) y - Ne (C_2 + t_s)] \qquad [15]$$
$$- N(K + t_B) = 0$$

$$\pi_C = S [(P - C_3) y - Ne (C_2 + t_s)] \qquad [16]$$
$$- Ne(C_1 + t_C) = 0$$

A tax program will achieve economic efficiency if, in the simultaneous solution of [15] and [16] for N and S, N equals N^*. Consider first a shared tax on nonlabor inputs. If t_B and t_C equal zero, N will equal N^* if $t_s = T_E$ as defined in [14]. At this point the share rate will equal S_o as defined in [9]. In terms of Figure 2, the optimal shared tax will shift the equilibrium curves to BEC_t and CEC_t.

Consider now the possibility of separate taxes on owners and crew members. If t_s equals zero and t_B and t_C are set as follows, N will always equal N^*.

$$t_B = T_E - \left[S \left[(P - C_3) \frac{y}{Ne} - C_2 \right] - C_1 \right] \quad [17]$$

$$t_C = T_E - \left[(1 - S) \left[(P - C_3) \frac{y}{Ne} - C_2 \right] - \frac{K}{e} \right] \quad [18]$$

The second term in equation [17] is the rent per unit of effort earned by labor. Therefore since T_E is the total rent per unit of effort, t_B is the rent per unit of effort earned by owners. Similarly, since the second term in [18] is rent per unit effort for boat owners, t_C is the rent per unit of effort earned by crew members.

These taxes shift the appropriate equilibrium curves such that they intersect the vertical line above N^* at whatever S rate is used in determining the taxes. In terms of Figure 2, imposition of t_B and t_C each computed with S equal to S_o will shift the two curves to BEC_t and CEC_t. A new equilibrium will occur at N^* and S_o.

These separate taxes are different than the shared tax in that there is no one share rate associated with the optimal fleet size. Given a specific share rate, there exists a tax policy that can obtain N^*. Another way to view this is that the share rate must differ according to the relative tax burden placed on owners and crew members, the higher the tax burden on the crew, the higher the share rate must be and vice versa. If all the burden is placed on the crew (i.e., $t_B = o$), then S must equal S_D^* and $t_c = T_E$. Similarly, if all taxes are paid by owners (i.e., $t_c = o$), then S must equal S_s^* and $t_B = T_E$. It follows that if the share rate is market-determined, the results of the traditional model, which indicate a per unit effort tax of T_E on boat owners, still hold. Such a tax will, in the presence of a market-determined share rate, lead to the opti-

mal level of effort and a share rate equal to S_s^* by shifting the BEC to BEC_t'.

The policy conclusions are that if economic efficiency is to be achieved when the share rate is market-determined, all rents must be taxed away from both crew members and boat owners. This can be done either by a shared tax or by a set of separate taxes. By using separate taxes the management authority can determine the share rate as well as the optimal fleet size. The ability to determine the share rate is not very important, however, unless the industry suffers from a "share rate illusion," because in the presence of the taxes, no matter what the share rate, boat owners and crew members will be covering their opportunity costs and nothing more.

When the share rate is fixed, the analysis is somewhat different. Consider Figure 4, and assume S is fixed at S_F. The BEC constrains the fishery and hence N_F will be the open-access fleet size. At this point rents will be earned by crews. To achieve economic efficiency either the BEC or the CEC must shift until it intersects the vertical line above N^* at S_F. This means that either t_B or t_C, with $S = S_F$, will be appropriate. If the proper t_B is used, the BEC will shift down as indicated. This will reduce effort to N^* and since the BEC' curve will be the binding constraint, rents will still be earned by crews. In fact, with the tax and the resulting reduction in effort, the amount of rent earned will increase, although employment will fall.

A tax on crews with t_C determined using S_F will shift the CEC to CEC'', which will also result in a fleet size of N^*. Here the CEC'' will be the effective constraint, and hence, rents will be earned by owners. If both taxes are used, the curves will both shift until they intersect at S_F above N^* and all of the rent will be

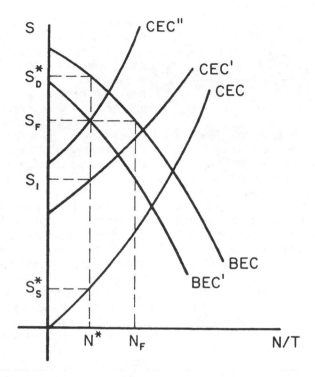

FIGURE 4

appropriated. However, with a fixed share it is only necessary to appropriate the rents from one sector.

Note that when S_F is above S_o, and the *BEC* is the initial constraint, taxes on crews will in some sense be "inefficient," because unless they are higher than some minimum level, they will extract rent but will not change the level of effort. For instance if t_C is calculated using S_1 in Figure 4, the *CEC* will shift to *CEC'*. However, given that the share rate is fixed at S_F, the *BEC* is still the binding constraint, and the level of effort will not be affected. The same analysis applies to the taxes on boats and shifts in the *BEC* if the fixed share rate is

below S_o, and hence, the *CEC* is the initial binding constraint.

Non-Limited Entry

The above results are directly applicable to closed seasons and areas, gear restrictions, and total quotas. Directly or indirectly, to the degree that they are effective in reducing effort, each will increase costs. Closed seasons and areas, and gear restrictions do so directly by affecting the way effort is produced. Total quotas indirectly increase costs because fishermen are motivated to increase their rate of harvest so as to get more before the total quota is reached. In terms of this analysis, these increases in costs cause either or both of the equilibrium curves to shift. The exact nature of the shift depends upon what types of costs are affected.

For example, if only boat operation costs are affected by the regulations, (i.e., because of specific restrictions or because in response to total quotas or area closures, private profit-seeking behavior dictates a change in boat operations), only the *BEC* will change. Assuming that the goal of the regulation is to reduce fleet size to N^* (see Figure 2), these regulations will have to increase the cost of operating a vessel by an amount equal to the optimal t_B when t_s and t_C are zero as described above. This means that the share rate will fall to S_s^*. On the other hand if only real labor costs are affected (i.e., the crew size must be increased), only the *CEC* will be affected. The total increase in costs will have to equal the tax bill generated by using the optimal t_C when t_s and t_B are zero and the share rate will rise to S_D^*.

If the regulations affect only the shared nonlabor costs, both of the curves will be affected. These unit costs will have to increase by an amount equal to T_E in order to force the fishery to N^*. There will be no change in the share rate in this instance.

To reiterate, nonlimited entry regulations reduce effort because they increase the cost of producing it, and as such, they are not capable of achieving maximum economic yield. Such regulations will reduce fleet size, but the effect they have on the share rate depends upon which types of costs are changed. Increased boat costs will tend to decrease the share rate, increased costs borne by the crew will tend to increase

it, and increased shared costs will have no effect.

When the share rate is fixed, the analysis is also analogous to that of taxes in the same situation. Refer again to Figure 4. If the share rate is fixed at S_F, it is only necessary for one of the curves to shift until it intersects the vertical line above N^* at S_F. For instance a gear restriction program that increased annual boat costs could shift the *BEC* to *BEC'*. With a flexible share rate, an increase in boat costs high enough to cause the *BEC* to intersect the *CEC* at S_D^* would be necessary. Therefore, the necessary cost increase to achieve N^* will be less with $S = S_F$. At this new regulated equilibrium, the *BEC'* is the binding constraint and crew members will be earning rents.

The same position could be achieved by an increase in the shared costs. The necessary cost increase would be less than with a flexible share rate because it is not necessary to shift the curves until they intersect at N^*. It is only necessary to shift them until one intersects the vertical line above N^* at the fixed share rate. In this case since $S_F > S_o$, the *BEC* will be the first curve to do so as shared costs are progressively increased. Both the *BEC* and the *CEC* would move in this case but the *BEC* would still be the binding constraint. Crew members will still earn rents but they will be smaller than in the case of increases in vessel costs. If the fixed share rate were below S_o, the above argument would be reversed in that the *CEC* would be the first curve to intersect the N^* line at S_F and that boat owners would earn rents before and after regulation.

A fleet size of N^* could also be achieved by increases in labor costs. Again the necessary increase is less than would be required with a flexible share rate because the *CEC* only has to shift up until it intersects the vertical line above N^* at S_F rather than at S_D^*. Note, however, that until a certain point is reached, increases in cost are not effective in reducing effort. As in the discussion of taxes, a shift from *CEC* to *CEC'* will have no effect on fleet size. The same argument holds in reverse if the fixed share rate is below S_o. In that case the *CEC* is the binding constraint and regulations that only affect boat costs will have no effect on fleet size unless they are larger than a certain amount depending on the difference between the fixed share rate and S_o.

SUMMARY AND CONCLUSIONS

The main conclusions of this paper may be summarized as follows.

1. If effort is produced with variable proportions and if a nonowner captain can control input usage, cost inefficiencies will result with a share system unless revenues and cost are shared at the same rate.

2. The existence of a share system will have no effect on the economically efficient level of output.

3. If the share rate is market-determined, the open-access equilibrium that occurs will be the same one that results from a simple opportunity-cost wage system. If the share rate is fixed, it will depend upon the level of the share rate but will always be less than or equal to the traditional open-access point. Maintaining a fixed rate not equal to the equilibrium rate could be quite difficult.

4. If either owners or crew have market power, they will operate the fishery efficiently (under the assumption of a fixed price); they will, however, select different share rates.

5. If the share rate is market-determined, optimal regulation policy drawn from the traditional model is applicable, although modifications based on the fact that both crew members and boat owners can earn rents can also be used. If the share rate is fixed, only one sector need be regulated to achieve economic efficiency. Nonlimited entry type regulations can affect both fleet size and the share rate depending upon how much and which types of costs are affected.

The following generalizations of the above model can be made:

1. If S_r does not equal S_c, the model must be modified by using S_c as a shift parameter for the *BEC* and the *CEC* (this will follow directly from appropriate modification of equations [6] and [7], respectively, to include both S_r and S_c). The system then becomes indeterminate; an equilibrium level of S_r can only be achieved for a given level of S_c, and vice versa. As long as one share rate is fixed however, only minor modifications are necessary and the same general conclusions apply.

2. The basic conclusions also follow if the assumption of constant output per boat is dropped. It would be necessary to specify a separable vessel cost function for effort which distinguishes between labor and nonlabor inputs (otherwise the share system can not be explicitly treated). The nonlabor part must increase with output so that an optimal output per boat may be specified. Then it is necessary to set up a profit function per boat so as to determine output per boat, as well as an industry profit function so as to determine the number of vessels. Under these more general assumptions, output per boat increases as the number of boats decreases. At the intersection of the *BEC* and the *CEC*, each would be operating at the minimum of its average cost of effort

curve. These are the same conclusions that follow from the traditional model, see Anderson (1976).

3. In the case where there are many fisheries, all with market-determined share rates, it is entirely possible that different equilibrium share rates could exist side by side, since each is dependent upon the biological productivity of the stocks and the price and cost structure. Although each fishery would have a different share rate, the real wage and the real return to capital would be the same in all.

References

Anderson, Lee G. 1976. "The Relationship between Firm and Fishery in Common Property Fisheries." *Land Economics* 52: 180–91.

———. 1977. *The Economics of Fisheries Management*. Baltimore: .The Johns Hopkins University Press.

———. 1980. "The Necessary Components of Economic Surplus in Fisheries Economics." *Canadian Journal of Fisheries and Aquatic Sciences* 37: 858–70.

———. 1982. "Optimum Effort and Rent Distribution in the Gulf of Mexico Shrimp Fishery: Comment." *American Journal of Agricultural Economics* 64: 157–59.

Christy, Francis T., and Scott, Anthony. 1965. *The Common Wealth in Ocean Fisheries*. Baltimore: The Johns Hopkins University Press.

Crutchfield, J. A. 1961. "An Economic Evaluation of Alternative Methods of Fishery Regulations." *Journal of Law and Economics* 4 (Oct.): 131–43.

Flaaten, Ola. 1981. "Resource Allocation and Share Systems in Fish Harvesting Firms." Resources Paper No. 72, Department of Economics, University of British Columbia.

Gordon, H. Scott. 1954. "The Economic Theory of a Common Property Resource." *Journal of Political Economy* 62 (Apr.): 124–42.

Griffin, Wade L.; Lacewell, Ronald D.; and Nichols, John. 1976. "Optimum Effort and Rent Distribution in the Gulf of Mexico Shrimp Fishery." *American Journal of Agricultural Economics* 58: 644–52.

Huppert, Daniel, et al. 1973. "NORSIM II: An Expanded Simulation Model for the Study of Multi-species Exploitation." NORFISH Technical Report No. 40. Manuscript, University of Washington.

Pearse, P. H., ed. 1979. "Symposium on Policies for Economic Rationalization of Commercial Fisheries." *Journal of the Fisheries Research Board of Canada* 36: 711–866.

Rettig, R. Bruce, and Ginter, J. C., eds. 1980. *Limited Entry as a Fishery Management Tool*. Seattle: University of Washington Press.

Roberts, Kenneth J. 1981. Personal communication.

Sass, M. E., and Roberts, Kenneth J. 1979. "Characteristics of the Louisiana Shrimp Fleet, 1978." Sea Grant Publication No. LSU-TL-79-006, Louisiana State University.

Scott, A. D. 1962. "The Economics of Regulating Fisheries." In *The Economic Effects of Fishery Regulation*, ed. R. Hamlisch. FAO Fisheries Report No. 5 (Fle/R5). Rome: FAO.

Sutinen, Jon G. 1979. "Fishermen's Remuneration Systems and Implications for Fisheries Development." *Scottish Journal of Political Economy* 26: 147–62.

Zoeteweij, H. 1956. "Fishermen's Remuneration." In *The Economics of Fisheries*, eds. R. Turvey and J. Wiseman. Rome: FAO.

Recreational Fisheries

[36]

JOURNAL OF ENVIRONMENTAL ECONOMICS AND MANAGEMENT 6, 127–139 (1979)

Bioeconomic Models of Marine Recreational Fishing [1]

KENNETH E. McCONNELL AND JON G. SUTINEN

*Department of Resource Economics, University of Rhode Island
Kingston, Rhode Island 02881*

Received January 20, 1978; revised July 29, 1978

The theory of recreational fishing is developed and conditions are derived for optimal management policy, with special attention given to functional relationships that must be empirically verified. Determinants of the optimal allocation between commercial and recreational fishing effort are derived. The theory is extended to include selected peculiar features of recreational fishing: Some anglers sell their catch; a small proportion of the fishing population accounts for a large proportion of the catch; and anglers throw back a fraction of what they catch. Optimal policies are derived under these more realistic conditions.

I. INTRODUCTION

This paper is concerned with the development of bioeconomic models for recreational fishing. Most bioeconomic models assume that production, or catch, is a function of the marine fish stock, which is not only scarce but also growing according to the laws of population dynamics. Perhaps most importantly, bioeconomic models explicitly account for the intertemporal interdependence between present catch and future fish stock levels. In this paper we discuss some implications of this interdependence for research and policy in marine recreational fishing.[2]

The most compelling reason for developing such models is the recent enactment of the Fishery Conservation and Management Act of 1976, which gives the U.S. Department of Commerce the authority to manage fishery stocks optimally for their recreational as well as commercial yields (90 Stat. 331, Sections 2.6.3 and 2.6.4). Though not anticipated, the regional management councils set up by the Act have already had to deal with the impact of recreational fishing. For example, in New England, the competition between sport and commercial fishermen for cod led to the recommendation for a daily bag limit on cod. In the Middle Atlantic area, a program for allocating mackerel between recreational

---

[1] Contribution No. 1786 of the Rhode Island Agricultural Experiment Station. The authors are members of the Department of Resource Economics at the University of Rhode Island. Partial support for the research was provided by Sea Grant and the International Center for Marine Resource Development, University of Rhode Island. Senior authorship is not designated. An earlier version of this paper was presented at the 1976 summer meetings of the AAEA, University Park, Pennsylvania.

[2] While we address primarily marine recreational issues, there is no reason why this analysis would not be of value for freshwater fisheries as well.

128   McCONNELL AND SUTINEN

and foreign commercial harvesting was developed. Recreational fishing has
evoked other controversies emanating from the scarcity of the resource stock:
For example, there is a bluefin tuna allocation program in the Northeast.

The size of the recreational fishery in terms of catch is striking. In 1970, the
most recent year for which data for recreational fishing is available, the weight
of the recreational catch is estimated to be 1.58 billion pounds [22, p. 9, Table 1].
During the same year, the weight of commercial finfish catch is estimated to be
3.87 billion pounds [21, p. 11].[3] These very rough figures indicate that about 29%
of the weight of the total fisheries landing is recreational. The total weight figures
give only rough magnitudes of the relative volume of activity in sport and
commercial fishing. More important are the net benefits generated by sport and
commercial fishing. Total volume figures give no indication of the relative size
of the net benefits generated by sport and commercial fishing. Hence, a substantial proportion of the total fisheries catch may come from recreational
fishing, and there is no reason to suppose that recreational benefits are small in
comparison with the net benefits generated by commercial fishing.

Hence the purpose of this paper is threefold:

(i) to develop the notion of scarcity as it applies to recreational fish stocks;
(ii) to demonstrate that unregulated use of the oceans by recreational fishers
may induce an economically inefficient allocation of resources; and (iii) to
develop an analytical model which can be used for guidance in research directions
and policy decisions.

Although it is difficult to determine the relative net benefits generated by
sport and commercial fishing, researchers generally have favored commercial
fisheries with their attention. There are many attempts to develop bioeconomic
models of commercial fisheries and to define management policies in terms of
these models. Commercial fisheries research includes theoretical models such as
those by Clark and Munro [8], Clark *et al.* [7], Lewis [15], Quirk and Smith
[18], Clark [6], and Brown [3]. The state of the theoretical art is summarized
in Clark [5]. Examples of empirical studies are Gates and Norton [11], Bell [1],
Crutchfield and Zellner [10], and Lewis [15]. These works develop management
strategies by maximizing the economic rents from commercial fisheries. The
desirability of managing commercial species is justified by the technological
external diseconomies vessels impose upon each other by exploiting the same
fishery stock.

In contrast, the standard economic model of a recreational fishery has been
the static demand model which attempts to measure the economic value of
sport fishing (for example, see the excellent work by Brown *et al.* [4], Gordon
et al. [12], and Talhelm [20]). These models traditionally have used indirect
methods of valuation originally associated with Clawson [9]. (In fact, Brown
provided one of the earliest applications of the Clawson method.) McConnell
and Norton [16] survey the various approaches to estimating net benefits of
recreational fishing. As part of the research on the demand side of recreational
fisheries, there has been some work relating the sport fisher's catch rate (success
ratio) to the net benefits of recreational fishing. Empirical research by Stevens

---

[3] The figure of 3.87 million pounds excludes catch from Hawaii, the Mississippi River, and the
Great Lakes in order to make it comparable with sport fisheries statistics.

[19], Talhelm [20], and Goodreau [13] has identified the importance of the catch rate as a determinant of demand. The work in recreational fishing, including the use of the catch rate, has focused on proper specification and estimation of the demand function. In recreational fisheries research, particularly marine, it has been implicitly assumed that open access is optimal, and that the marginal social cost of each sport fishing trip is zero. This approach ignores the scarce fishery stock, implicitly setting the marginal economic value of sport-caught fish at zero. There have been no attempts to incorporate the impact of the sport fishing catch on the population dynamics of the exploited species or to determine the optimal level of marine recreational fishing.[4]

The increasing demand for marine recreational fishing makes the management of predominantly recreational stocks important for its own sake. However, the dynamics of the interaction of commercial and recreational fishing greatly exacerbates the problems of regulating only the commercial sector. If the regulation of the commercial sector successfully increases stock abundance, it would draw recreational fishers who respond to higher catch rates, which are caused by greater stock abundance. In effect, unless the recreational sector is included in the management efforts, the more successful the commercial scheme is in stock rebuilding, the more likely that the benefits from such a scheme will be distributed to the recreational sector.

The charge to manage fishery stocks offers the opportunity to improve the efficiency with which we utilize our fishery resources. However, unless we recognize that recreational fishery stocks may be subject to scarcity and develop models to reflect that scarcity, we are inviting potential inefficiencies in recreational fishing. The attempt to devise policies with respect to the exploitation of sport fishery stocks is likely to be futile until economists and biologists develop analytical frameworks which reflect not only the impact of additional catches on the demand for recreational fishing, but also the impact of additional catches on the growth of the fishery stock.

In this paper, we are concerned with the management of fishery resources from the national point of view, and hence the appropriate criterion for management is the direct benefits of the fishery to the user. We assume that net benefits per user take the form of the area under the user's demand curve, and further that this demand curve has a zero income effect. Thus we can ignore the difference between willingness to pay and willingness to be compensated for marginal adjustments in the resource stock.

II. THE THEORY OF RECREATIONAL FISHING

Recreational fishers are assumed to fish primarily due to recreational motives rather than for the pecuniary purposes usually ascribed to commercial fishers. Specifically, we follow McConnell and Norton [16] and assume that each individual sports fisher derives utility from both the quantity and quality of fishing outings. The quantity is measured by x, the number of outings, and the quality by catch per outing, $h(w)$.[5] Catch per outing increases as the fish stock, w,

---

[4] The models developed in this paper are closely related to the bioeconomic models found in the commercial fisheries literature cited above.

[5] Another endogenous quality argument could be the size of the individual fish caught, where larger fish yield greater benefits. In such a case the natural growth relationship, Eq. (1) below,

increases, i.e., $h'(w) > 0$.[6] The individual's marginal benefit function, or inverse demand function for outings, is given by $p[x, h(w)]$, $\partial p/\partial x < 0$, and $\partial p/\partial h > 0$.[7] To simplify the analysis we assume there are n fishers with identical preferences and that benefits accrue only to those who fish. Therefore, total benefits from recreational fishing are given by

$$\int_0^{x^*} np[x, h(w)]dx,$$

where x^* is the number of outings per fisher in the period.

Recreational fishing involves the use of scarce resources, e.g., the individual's time and the appropriate equipment. We assume a constant marginal cost of an outing, c, identical for every fisherman.[8] The expression for net benefits per period can then be written as

$$\int_0^{x^*} n\{p[x, h(w)] - c\}dx.$$

The natural growth rate of the fish stock is assumed given by the differential equation

$$\frac{dw}{dt} = g[w(t)], \tag{1}$$

where $w(t)$ is the biomass of the single fish stock at time t. The function $g(\cdot)$ is everywhere concave and reaches a maximum at $w = \bar{w}$. The total recreational harvest in the period is $nxh(w)$, and when it is introduced into Eq. (1) the growth of the exploited fish stock becomes

$$\frac{dw}{dt} = g(w) - nxh(w). \tag{2}$$

(A) Central Management and User Cost

Suppose a central authority is endowed with the powers to manage this recreational fishery, the object being to maximize the discounted value of the stream

---

would have to explicitly account for the age structure of the fish population. This could be accomplished by using the population dynamics model developed by Beverton and Holt [2]. And since selective gear likely would be used in such situations, the specification of the technical production relationships would involve multiple heterogeneous inputs and outputs. Also, quality may be better represented by total catch per year than by catch per outing. While this specification is important for empirical research, it does not alter the theoretical results derived here.

[6] Catch also is a function of other factors such as skill, weather, season, and the type and quantity of fishing gear used. For simplicity we assume these factors are exogenously determined and constant across individuals and over time.

[7] Some empirical evidence for the relationship $p[x, h(w)]$ is given by Stevens [19], who estimates a long-run unitary elasticity between catch and number of fishing outings for salmon. Goodreau's [13] study of striped bass fishermen from cross-sectional data indicates an elasticity of 0.1 of outings with respect to the number of striped bass caught per trip.

[8] The analysis that follows can be modified easily to allow for fishers who travel different distances to fish. However, the results are not significantly changed by such a modification. Results are available from the authors.

MODELS OF RECREATIONAL FISHING 131

of net benefits derived from the recreational harvest. The discounted value of net benefits is given by

$$\int_0^\infty e^{-\delta t} \int_0^{x^*} n\{p[x, h(w)] - c\} dx dt,$$ (3)

where δ = the instantaneous social discount rate. Therefore, the task of the authority is to maximize Eq. (3) subject to Eq. (2) by choosing an optimal x^* for each period. For simplicity we assume the number of fishers, n, is exogenously determined, although the determination of the optimal n could be treated easily in this model.[9] The current value Hamiltonian for this problem is

$$H = e^{-\delta t} \int_0^{x^*} n\{p[x, h(w)] - c\} dx + \lambda(t)[g(w) - nxh(w)].$$

The necessary conditions for an interior maximum include

$$p[x^*, h(w)] - c - \lambda(t)h(w) = 0,$$ (4)

$$\frac{d\lambda}{dt} - \delta\lambda = -\left\{ \frac{\partial}{\partial w} \int_0^{x^*} n\{p[x, h(w)] - c\} dx + \lambda(t)[g'(w) - nx^*h'(w)] \right\},$$ (5)

$$\frac{dw}{dt} = g(w) - nx^*h(w).$$ (6)

Conditions similar to these have been interpreted and analyzed at length by others (e.g., see [3, 12, 13]). Briefly, Eq. (4) is the condition determining the optimal number of outings per fisherman in the period, where the adjoint variable, $\lambda(t)$, is the imputed marginal social value of the fish stock.

Equations (4), (5), and (6) form an autonomous system which attains a stationary equilibrium where $dw/dt = 0$ and $d\lambda/dt = 0$. From Eq. (5), when $d\lambda/dt = 0$,

$$\lambda = nh'(w) \int_0^{x^*} \frac{\partial p[x, h(w)]}{\partial h} dx / [\delta - g'(w) + nx^*h'(w)].$$ (7)

In a stationary state, λ measures the marginal user cost of the scarce fishery stock. By inspection, the stock is not scarce and the user cost is zero if catch per outing does not affect marginal benefits ($\partial p/\partial h = 0$) or if the stock is so abundant that changes in stock size do not affect catch per outing [$h'(w) = 0$]. Naturally, λ increases as the impact of w on net benefits increases. The impact of changes in x and n on resource scarcity is ambiguous: Increases in x and n increase benefits, causing λ to increase; however, for a given catch rate, increasing x and n may increase or decrease the natural growth of the biomass, depending on the sign and magnitude of $g'(w)$, the change in the biological growth rate.

(B) Open Access and Decentral Management

We now turn to the case of an open access recreational fishery with competition in all markets. The identical fishers are assumed to maximize their net benefits from the recreational harvest of the fishery. Since no private property rights

---

[9] Another issue not treated here is crowding, where the number of other fishermen present directly detracts from the benefits received by an individual fisherman.

exist, the fishery resource is shared by all and each fisher does not consider the full social cost of more fishing trips and catching more fish. The optimization problem for the representative fisher is as follows:

$$\underset{\{x^*\}}{\text{maximize}} \int_0^{x^*} \{p[x, h(w)] - c\} dx,$$

where all terms are as defined above.

The necessary conditions for an interior maximum include

$$p[x^*, h(w)] - c = 0. \tag{8}$$

Equation (8) implies that individuals adjust their outings until the net marginal value is zero. Comparing condition (8) of the open access fishery condition (4) of the fishery under control of a central authority, we easily can see that the open access solution is not socially optimal because private fishers, seeking to maximize their own well-being, ignore the impact of their catches on the welfare of others. The efficient allocation condition (4) is not satisfied by the open access model unless $\lambda(t) = 0$. We may also think of the difference as reflected in the time preference of individuals operating on open access resources. They are unwilling to refrain from current catches and invest in the "ocean bank" under open access, because they have no assurance that they will benefit from their abstention. Equivalently, they have an infinitely high discount rate for future fish in the ocean bank. In (7), λ goes to zero as the discount rate goes to infinity, implying that the open access and central management plans are the same.

For expository purposes we consider an alternative to centralized control where the authority simply assesses a user fee equal to $\lambda(t)$ on each unit of catch per outing. Assuming no significant income effects of such a measure, the resulting number of outings in an open access fishery would be optimal. An individual fisher would seek to maximize

$$\int_0^{x^*} \{p[x, h(w)] - c - \lambda h(w)\} dx.$$

The necessary condition for an interior maximum is

$$p[x^*, h(w)] - c - \lambda h(w) = 0,$$

which is the same as condition (4) in the stationary state if the correct λ is chosen.

TABLE I

Commercial and Recreational Landings, New England and New York, 1965 and 1970

Species	Sport[a]		Commerical[a]	
	1965	1970	1965	1970
Cod	28,978	35,688	35,831	53,029
Flounder	40,966	36,295	120,450	111,241
Haddock	21,390	2,528	133,892	26,887
Pollock	9,348	5,584	11,709	8,790
Porgy	10,150	2,296	19,004	4,670

Source. U.S. Department of Commerce, National Fisheries Service, "Fisheries of the United States," various years.
[a] In thousands of pounds.

MODELS OF RECREATIONAL FISHING 133

III. ALLOCATION BETWEEN COMMERCIAL AND RECREATIONAL FISHERS

Many fisheries have both significant commercial and recreational use (see Table I). Often, serious conflict has developed over the allocation of the fish stock between the two interest groups, as, for example, with the striped bass fishery in Chesapeake Bay. Despite some rather well-known conflicts between the sport and commercial sectors, no theoretical framework has been developed to handle this important problem.

To examine the problem of a fishery exploited for both commercial and recreational purposes, the model in Section II is modified as follows. Let $F(E, w)$ be the inverse derived demand function for the commercial fishing effort, E, and let the total commercial catch be given by $Eq(w)$, where $q(w)$ is catch per unit effort and $q'(w) > 0$. Following the argument of Just and Hueth [14], consumer and producer surplus for the commercial fishery is given by

$$\int_0^{E^*} [F(E, w) - k]dE,$$

where k is the cost of a unit of effort, assumed to be constant, and E^* is the optimal quantity of effort.

Assuming the central authority wishes to maximize the consumer and producer

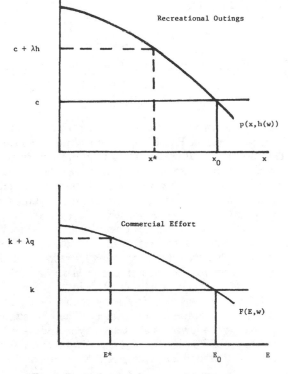

FIG. 1. Recreational and commercial allocations.

surplus from both recreational and commercial fishing, the problem is to

$$\underset{\{x^*, E^*\}}{\text{maximize}} \int_0^\infty e^{-\delta t} \left\{ \int_0^{x^*} n\{p[x, h(w)] - c\} dx + \int_0^{E^*} [F(E, w) - k] dE \right\} dt,$$

subject to $dw/dt = g(w) - x^* n h(w) - E^* q(w)$.

Following the same procedure as above, we derive the following expression for the stationary-state marginal value of the fish stock:

$$\lambda = \frac{n \int_0^{x^*} \dfrac{\partial p[x, h(w)]}{\partial h} h'(w) dx + \int_0^{E^*} \dfrac{\partial F(E, w)}{\partial w} dE}{[\delta + nx^* h'(w) + E^* q'(w) - g'(w)]}. \tag{9}$$

As in the case of the simple recreational fishery, to ensure an optimal harvest in the steady state the authority could assess a user fee given by Eq. (9) on each unit of catch by both sport and commercial fishers.[10]

An issue of some concern is how the participation of recreational and commercial fishery will change if an open access regime is replaced by optimal management. In Fig. 1 recreational and commercial participation rates under open access are at x_0 and E_0, respectively (determined, of course by the conditions $p[x, h(w)] = c$ and $F(E, w) = k$). Optimal management requires a reduction in x and E to where $p(\cdot) = c + \lambda h$ and $F(\cdot) = k + \lambda q$ hold. For given values of λ, $h(w)$, and $q(w)$, the extent to which x and E are reduced depends upon the elasticities of $p(\cdot)$ and $F(\cdot)$. Define

$$\eta_x = - \frac{\partial x}{\partial p} \frac{p}{x},$$

where $p = p[x, h(w)]$, the derived demand price for recreational outings, and

$$\eta_E = - \frac{\partial E}{\partial F} \frac{F}{E},$$

where $F = F(E, w)$, the derived demand price for commercial effort. If η_E is large relative to η_x (as shown in Fig. 1), then optimal management would require a larger reduction in commercial participation than in recreational participation. Therefore, while the open access mix may have substantial involvement by commercial fishers, the optimal mix may allow little commercial participation relative to recreational participation. Alternatively, if η_x is large and η_E small the optimal mix would favor commercial fishers. This suggests that prescribed allocations between commercial and recreational fishers based on historical catch shares can result in allocations significantly different from the efficient allocation.

When should a fishery be exclusively exploited by recreational fishers or, vice versa, by commercial fishers? The answer lies in the intercept values of $p(\cdot)$ and $F(\cdot)$ at $x = 0$ and $E = 0$, respectively. In Fig. 2 the intercept of F_1 lies below $k + \lambda q$ while the intercept of p_1 is greater than $c + \lambda h$, allowing no commercial participation under optimal management. The opposite case, where no recreational participation is allowed, is illustrated by p_2 and F_2 in Fig. 2.

Since user fees are prohibited by the Fishery Conservation and Management Act other means must be found to manage a fishery. Regardless of the means

---

[10] As Brown [3] has shown, a tax to limit entry in the commercial fishery also would be necessary.

MODELS OF RECREATIONAL FISHING   135

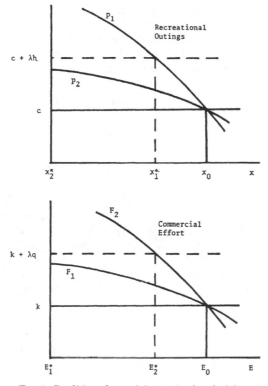

FIG. 2. Conditions determining optimal exclusivity.

employed (quotas, closures, etc.), the estimation of $p[x, h(w)]$ and $F(E, w)$ and calculation of λ can be valuable for fisheries management.[11]

IV. ADDITIONAL ISSUES

The above model, while useful for several purposes, ignores some significant features of recreational fishing. For example, recreational catch is often sold, adding significantly to some fishers' incomes; it is not uncommon for a small proportion of the recreational fishing population to account for a large proportion of the total catch; and fishers often throw back a fraction of what they catch. In this section we modify our simple model to account for these features and develop the implied policy ramifications.

(A) Consumption and Sale of Recreational Catch

The motivation of fishers is an important issue. Some claim it is only the fish caught that matter, while others maintain it is the number kept that matter. For fishers who keep their catch and consume or sell it, their money income is

---

[11] Wang [23] provides one of the few estimates of $F(E, w)$. His study yields $\eta_E = -2.95$ and $(\partial F / \partial w)(w/F) = 1.4$ for the Canadian sockeye fishery.

supplemented by the market value of the catch. What implications do the consumption and sale of recreational catch have for management?

Let b represent the market price of food fish (for simplicity we assume no difference between the wholesale and retail prices of fish). Since a fisher's income is supplemented by an amount bxh per period, net benefits now are given by

$$\int_0^{x^*} n\{p[x, h(w)] - c + bh(w)\}dx. \tag{10}$$

Under central management the user cost in the stationary state for this case is

$$\lambda = nh'(w)\int_0^{x^*} \left\{ \frac{\partial p[x, h(w)]}{\partial h} + b \right\} dx/[\delta - g'(w) + nx^*h'(w)]. \tag{11}$$

Because of the addition of b to the marginal value of catch per trip, and because x^* is likely to be larger for those who consume or sell their catch, the marginal user cost in this case will be larger (compare (11) with (7)). Also note that here, unlike the simple case, λ is not zero when catch per outing does not affect marginal recreational benefits (i.e., $\partial p/\partial h = 0$).

Under open access in this case, the individual fisher maximizes (10) with respect to x; and the necessary condition for the maximum is

$$p[x^*, h(w)] - c + bh(w) = 0. \tag{12}$$

The addition of $bh(w)$ to the marginal value of a fishing outing means the representative fisher likely will be more responsive to changes in catch rates and, hence, in stock abundance.

(B) Distribution of Catch

In many recreational fisheries a small proportion of the fishing population accounts for a large proportion of the catch. The better fishers generally sell their catch, while the others do not, their catch being too small to justify the costs of transacting a sale. To examine the implications of these features we subscript the variables so that 1 represents the better fishers and 2 represents the others, and all variables remain as defined above. Net benefits per period become

$$\int_0^{x^*_1} n_1\{p_1[x_1, h_1(w)] - c_1 + bh_1(w)\}dx_1 + \int_0^{x^*_2} n_2\{p_2[x_2, h_2(w)] - c_2\}dx_2. \tag{13}$$

Under central management the allocation of outings between the better and other fishers is determined by the conditions

$$p_1(x^*_1, \cdot) - c_1 + bh_1 - \lambda h_1 = 0, \tag{14}$$
$$p_2(x^*_2, \cdot) - c_2 - \lambda h_2 = 0.$$

Under open access and with differences in skill levels only—that is, the two groups have the same preferences and face the same cost per outing—the better fishers clearly fish more often per period. Under central management, however, the better fishers may be allocated fewer fishing outings. Both $x^*_1 > x^*_2$ and $x^*_1 < x^*_2$ are possible.

MODELS OF RECREATIONAL FISHING 137

The marginal user cost in the stationary state is now

$$\lambda = \left\{ n_1 h'_1(w) \int_0^{x^*_1} \left[\frac{\partial p_1}{\partial h_1} + b \right] dx_1 + n_2 h'_2(w) \int_0^{x^*_2} \frac{\partial p_2}{\partial h_2} dx_2 \right\} \bigg/$$

$$\{\delta + n_1 x^*_1 h'_1 + n_2 x^*_2 h'_2 - g'(w)\}. \quad (15)$$

By making several fairly reasonable simplifying assumptions, we can get a good appreciation for the expression for λ in (15). First define $\theta_i = n_i/(n_1 + n_2)$, where θ_i is the proportion of better fishers ($i = 1$) or other fishers ($i = 2$). Then suppose that their catch rates respond identically to changing stock sizes, i.e., $h'_1(w) = h'_2(w) = h'(w)$. (This assumption does not require that they have the same catch rates.) Then we can write

$$\lambda = h'(w) \left(\theta_1 \int_0^{x^*_1} \left[\frac{\partial p_1}{\partial h_1} + b \right] dx_1 + \theta_2 \int_0^{x^*_2} \frac{\partial p_2}{\partial h_2} dx_2 \right) \bigg/$$

$$\{[\delta - g'(w)]/(n_1 + n_2) + h'(w)(\theta_1 x^*_1 + \theta_2 x^*_2)\}. \quad (16)$$

Additionally, if $(\delta - g'(w))/(n_1 + n_2)$ is close to zero,

$$\lambda = \left(\theta_1 \int_0^{x^*_1} \left[\frac{\partial p_1}{\partial h_1} + b \right] dx + \theta_2 \int_0^{x^*_2} \frac{\partial p_2}{\partial h_2} dx \right) \bigg/ (\theta_1 x^*_1 + \theta_2 x^*_2). \quad (17)$$

Expression (17) tells us that this approximation of the marginal user cost equals the weighted average of the marginal value of changing catch rates to each group (where the weights are the number of individuals in the groups) divided by the mean number of trips per person (mean trips $= \theta_1 x^*_1 + \theta_2 x^*_2$). We may interpret λ as the increase in net benefits per day from changing the catch rate, averaged over both groups. The importance of (17) is not only that θ_1 and θ_2 are important to measure, but the estimation of the marginal benefit functions is also crucial. It is not unreasonable to assert that $\partial p/\partial h = 0$ for a fairly large proportion of recreational fishermen. If so, data gathering and regulation could take place only for the better fishermen (who are relatively easy to locate), because other fishermen have so little interaction with the stock. The equity of such a strategy might be subject to debate.

(C) Catch and Kill Rate Differences

Not all fish caught are kept for consumption or sale. For various reasons a fraction of fish caught is thrown back. In fact, a "pure" sport fisher may release all fish caught. We modify the simple model to treat this case; i.e., all fishers are alike and sale is not considered.

Let f be the proportion of caught fish that are thrown back, and of the fh fish thrown back per outing, assume s is the proportion that survive. Therefore, for an outing's catch of $h(w)$ a smaller number, $h(w)(1 - sf)$, are actually killed. For this situation the growth relationship (2) is modified to become

$$\frac{dw}{dt} = g(w) - nxh(w)(1 - sf). \quad (18)$$

138   McCONNELL AND SUTINEN

The marginal user cost is now

$$\lambda = nh'(w) \int_0^{x^*} \frac{\partial p}{\partial h} \, dx / [\delta - g'(w) + nx^*h'(w)(1 - sf)], \qquad (19)$$

and the necessary condition determining the optimal number of outings is

$$p[x, h(w)] - c - \lambda h(w)(1 - sf) = 0. \qquad (20)$$

In Section 2B we argued that a user fee equal to λ on each unit of catch would yield an optimal solution. This is not the case here. Instead, the optimal user fee will have to be equal to $\lambda \cdot (1 - sf)$ on each unit of fish actually caught. This strategy likely will induce fishers to conceal the fish thrown back in order to minimize their fees. Therefore, we ask, What would the optimal fee be if applied only to those fish kept?

The quantity of fish kept is $h(w)(1 - f)$. The problem is to determine some fee Λ where

$$\Lambda h(w)(1 - f) = \lambda h(w)(1 - sf).$$

Therefore the optimal fee per unit of fish kept is

$$\Lambda = \lambda(1 - sf)/(1 - f). \qquad (21)$$

Clearly, if, as is true for some fisheries, s and f are significantly greater than zero, Λ can be significantly different from λ. Note too that $\Lambda > \lambda$ always. That is, the optimal fee on fish kept is greater than the optimal fee on fish caught. While this result may seem to some intuitively obvious, Eqs. (19) and (21) provide the means to calculate the optimal fees in this case.

V. SUMMARY AND CONCLUSIONS

The bioeconomic model of a recreational fishery developed in this paper has shown that, under plausible assumptions, recreational fisheries under open access result in too much fishing. The conditions determining the optimal number of fishing outings show the functional relationships which must be empirically estimated.

In a fishery exploited by both commercial and recreational fishers under open access, we derive explicit conditions determining the optimal levels of commercial and recreational effort. A critical parameter for allocating between commercial and recreational fishing effort is the own-price elasticity of effort in each.

Also developed are conditions determining optimal recreational effort when recreational fishers supplement their income with their catch, and when the catch and kill rates differ—two prevalent features of recreational fisheries. When a small number of skilled fishers account for a large proportion of the recreational catch, we show how optimal management may be approximated by regulating only the few skilled fishers.

REFERENCES

1. W. Bell, Technological externalities and common-property resources: An empirical study of the U.S. Northern Lobster Fishery, *J. Political Econ.* 70, 148–158 (1972).
2. R. J. H. Beverton and S. J. Holt, On the dynamics of exploited fish populations, in *Fish. Invest. Ser. II*, Vol. 19 Min. Agric., Fish. and Food, Her Majesty's Stationery Office, London (1957).

MODELS OF RECREATIONAL FISHING 139

3. G. M. Brown, An optimal program for managing common property resources with congestion externalities, *J. Polit. Econ.* 82, 163–174 (1974).
4. W. G. Brown, D. M. Larson, R. S. Johnston, and R. T. Wahle, "Improved Economic Evaluation of Commercial and Sport-Caught Salmon and Steelhead of the Columbia River," Final Report for the Pacific Northwest Regional Commission (1975).
5. C. W. Clark, "Mathematical Bioeconomics: The Optimal Management of Renewable Resources," Wiley–Interscience, New York, (1976).
6. C. W. Clark, Profit maximization and the extinction of animal species, *J. Polit. Econ.* 81, 950–961 (1974).
7. C. W. Clark, F. H. Clarke, and G. R. Munro, The optimal exploitation of renewable resource stocks: Problems of irreversible investment, *Econometrica*, in press.
8. C. W. Clark and G. R. Munro. The economics of fishing and modern capital theory: A simplified approach, *J. Environ. Econ. Manag.* 2, 92–106 (1975).
9. M. Clawson, "Methods of Measuring the Demand for and Value of Outdoor Recreation," Resources for the Future, Washington, D.C. (1959).
10. J. A. Crutchfield and A. Zellner, "Economic Aspects of the Pacific Halibut Fishery," U.S. Department of Interior, Washington, D.C. (1962).
11. J. M. Gates and V. J. Norton, "The Benefits of Fisheries Regulation: A Case Study of the New England Yellowtail Flounder Fishery," Marine Technical Report No. 21, University of Rhode Island, Kingston (1974).
12. D. Gordon, D. W. Chapman, and T. C. Bjornn, Economic evaluation of sport fisheries—What do they mean? *Trans. Amer. Fish. Soc.* 102, 293–311 (1973).
13. L. J. Goodreau, "Willingness to Pay for Striped Bass Sportsfishing in Rhode Island," Master's thesis, University of Rhode Island, 1977.
14. R. E. Just and D. L. Hueth, Welfare measures in a multimarket framework, *Amer. Econ. Rev.*, in press.
15. T. R. Lewis, "Optimal Resource Management under Conditions of Uncertainty: The Care of an Ocean Fishery," Ph.D. thesis, University of California, San Diego (1975).
16. K. E. McConnell and V. J. Norton, An economic evaluation of marine recreational fishing, in "Marine Recreational Fisheries" (R. Stroud, Ed.), Sport Fishing Institute, Washington, D.C. (1976).
17. C. G. Plourde, A simple model of replenishable natural resource exploitation, *Amer. Econ. Rev.* 60, 518–521 (1970).
18. J. P. Quirk and V. L. Smith, Dynamic models of fishing, in "Economics of Fishery Management: A Symposium" (A. Scott, Ed.), Univ. of British Columbia Press, Vancouver (1970).
19. J. B. Stevens, Angular success as a quality determinant of sport fishery recreational values, *Trans. Amer. Fish Soc.* 95, 357–362 (1966).
20. D. R. Talhelm, "Evaluation of the Demands for Michigan's Salmon and Steelhead Sport Fishery of 1970," Fisheries Research Report No. 1797, Michigan Department of Natural Resources, Lansing (1973).
21. U.S. Department of Commerce, National Marine Fisheries Service, "Fisheries Statistics of the United States 1970," Statistical Digest No. 64, Washington, D.C. (1970).
22. U.S. Department of Commerce, National Marine Fisheries Service, "1970 Salt-Water Angling Survey," Current Fishery Statistics No. 6200, Washington, D.C. (1973).
23. D.-H. Wang, "An Economic Study of the Canadian Sockeye Salmon Market," Ph.D. dissertation, Oregon State University (June 1976).

[37]

JOURNAL OF ENVIRONMENTAL ECONOMICS AND MANAGEMENT 24, 272–295 (1993)

Toward a Complete Economic Theory of the Utilization and Management of Recreational Fisheries*

LEE G. ANDERSON

College of Marine Studies and Department of Economics, University of Delaware, Newark, Delaware 19716-3501

Received July 19, 1992; revised November 22, 1992

A model of individual behavior for recreational fisheries which considers both the participation decision and the activity level decision is developed. The model also distinguishes between the catch rate, which is a biologically determined parameter, and the landings rate, which is a control variable. Individual and fisherywide equilibria under open access are described for both homogeneous and heterogeneous participants. Optimal utilization is also described. Optimal utilization differs from open access in terms of activity levels of participants and number and type of participants. Regulations to achieve optimal utilization are described. © 1993 Academic Press, Inc.

INTRODUCTION

The theoretical economic analysis of open-access and optimally utilized fisheries has, for the most part, been developed in the context of commercial fisheries. The analysis of recreational fisheries has received comparatively short shrift. The exceptions are McConnell and Sutinen [20, 21]; Bishop and Samples [3], Anderson [1, 2], and McConnell and Strand [19]. Although the main conclusions of these papers are the same as those on commercial fishing (open access in a recreational fishery does not maximize potential rents because individuals do not consider the costs they impose on other participants) there are specific reasons for studying recreational fisheries separately. Bishop and Samples [3, p. 221] state that theoretical work on recreational fishing is needed to provide "an adequate theoretical basis for planning empirical research" and "a conceptual framework consistent with sound economic principles to guide public debates about fisheries management polices." McConnell and Sutinen [20, p. 129] point out:

> The attempt to devise policies with respect to the exploitation of sport fishery stocks is likely to be futile until economists and biologists develop analytical frameworks which reflect not only the impact of additional catches on the demand for recreational fishing, but also the impact of additional catches on the growth of the fishery stock.

While the analysis of optimal utilization of recreational fisheries is lean, there is extensive literature on valuation in recreational fisheries (see the references). While some of the constructs of the valuation literature are used here, the emphasis is on extending the conceptual framework of fisheries utilization. The object is to understand the behavior of individuals as they pursue recreational fishing activities. The problem of measuring the values that are generated is

*This research was sponsored by NOAA Office of Sea Grant, Department of Commerce, under Grant NA86AA-D-SG040 (Project SG 89 R/E-8). The author is appreciative of helpful comments from Nancy Bockstael, George Parsons, V. Kerry Smith, and several anonymous reviewers, but the usual disclaimer applies.

272

secondary. Also, because a better understanding of individual behavior will help to formulate regulation policies, considerable emphasis is placed on the two-way relationship between stock size and the benefits derived from recreational fishing emphasized by McConnell and Sutinen. Other influences on participation such as prices of other goods and services and the income, tastes, and other demographic parameters of participants are assumed constant.

Previous works have not fully modeled the complex decision making framework of the individual recreational fisheries participant, but, as is shown, this complexity can have a major impact upon both open-access and optimal utilization. The improved model generates some interesting policy implications. For instance, a stock enhancement program which increases social welfare may actually lead to a decrease in the number of days fished by some participants. As exemplified by McConnell and Sutinen, the existing models assume that benefits are a function of the number of days fished and a biologically determined catch rate per unit of effort and that there are a fixed number of homogeneous participants. The present work introduces a more general net benefit function which distinguishes between landings and the catch rate.[1] This distinction is important because it captures another facet of individual behavior and it also allows for the analysis of catch and release programs which will likely play an increasingly important role in recreational fisheries policy if sports fishing pressure continues to intensify. Another difference is that the number of participants is a variable and participation is analyzed in a manner analogous to the entry and exit of firms in an industry. While the analogy is not perfect, it follows directly from utility theory. An individual will fish if the utility obtained is greater than or equal to the full opportunity cost in terms of lost utility.

The discussion focuses on the behavior of recreational fishermen as both individuals and a group. The first section describes the benefit functions and some implications of the distinction between landings and the catch rate. Subsequent sections describe the open-access and optimal utilization of a recreational fishery with homogeneous and with heterogeneous participants. A final section provides examples of how the model can be used.

THE BASIC MODEL

The benefit function. The received theory on recreational fishing assumes that participants fish primarily for pleasure and that the relationship between utility and fish consumed (is called landings in this paper) is not as direct as it is with commercial fishers. Recreational utility is assumed to be a function of both the quantity of fishing outings and their quality. Quality is measured by the catch rate per outing (or sometimes fish size). Since the catch rate and average fish size are functions of the fish stock, the requirement to reflect attributes of the stock on demand is satisfied.

While the quantity–quality formulation in terms of days fished is useful, and is retained here, it slights the fact that fish also provide utility as food. In addition, while the received theory considers the effects of fishing on stock growth, a richer

[1]McConnell and Sutinen [20, p. 137] do briefly discuss situations where landings are less than the catch rate but they do not include landings in the benefit function.

LEE G. ANDERSON

and more complete analysis is possible by considering that landings can be, and often are, different than potential catch. It is assumed that the total net benefit derived from recreational fishing is a function of days fished, d, the landings per day, l, and the catch per day (CPD), $H(X, D, Dl)$:

$$B = B[d, l, H(X, D, Dl)]. \tag{1}$$

The catch rate per day, $H(X, D, Dl)$, is an increasing function of stock size, X, a decreasing function of aggregate effort, D, and a decreasing function of aggregate landings.[2] While aggregate effort may be thought of as causing a congestion externality, a more straightforward interpretation would be diminishing returns to a variable input, D, in the presence of a fixed input, X.[3] Aggregate landings affect catch rate because the more fish that are landed (i.e., not returned to the water after capture) the smaller the number of fish available for harvest for the remainder of the period.

The total net benefit from recreational fishing is the difference between the utility obtained from participation and the opportunity cost in terms of utility lost as a result of the income and time spent engaging in fishing. Rigorous descriptions of such constrained maximizing individual behavior can be found in the literature (see, for example, McConnell [16]) and is not repeated here. While the current analysis focuses on consumption decisions regarding recreational fishing, such decisions are not made in a vacuum. Income, time constraints, flexibility of employment, fixed and variable costs of fishing, tastes, and prices of other goods and services will affect the net benefit derived from recreational fishing. These items are implicit arguments in the total net benefit function.

This benefit function is different from the received theory in two ways. First, there is a distinction between landings per day and the catch rate per day, CPD. Both are parameters in the benefit function but only l is a decision variable. Landings are obviously constrained by the catch rate.[4] Previously, landings and the catch rate were assumed to be identical. When participants fished they were assumed to keep whatever catch the stock allowed them to make. One rationale for making a distinction is that participants sometimes choose to stop fishing or to stop keeping fish after a certain number have been brought on board. For example, when mackerel are "hitting," it is common to take 50 fish in a morning. Diminishing marginal utility for fish as a food may set in and the rest of the fish may be released as they are brought to the boat or the afternoon may be spent resting or boating. The point is that 50 fish caught in a frenzied morning may provide more utility than 50 fish which come in slowly over a full day, even though both provide the same food. Landings and the catch rate are different attributes of utility. A second rationale is that this formulation allows for a theoretical analysis

---

[2] For a large part of what follows, it is assumed that all participants are homogeneous and D can be represented as nd, where n is the number of participating individuals. The terms D and nd are used interchangeably depending on the issue at hand.

[3] The CPD function is derived from a total yield function. As long as total yield is not linear in days fished, CPD will be a function of aggregate effort.

[4] In actuality, there is no reason for landings per day to be the same for every day. The utility of the nth fish on the 10th day could well be less than the utility of the nth on the 1st day. Therefore, a more complete model would have landings on each day as a separate variable. However, introducing this nicety would make notation significantly more complicated without adding substantially to the results.

of the pure sports fisher who throws all fish back and of bag limits and catch and release programs.

The second difference in this benefit function is the diminishing returns to aggregate effort through the introduction of D as a variable in the CPD function. This makes the model analogous to commercial fishing theory where there are both intertemporal and intratemporal externalities. The additional reality added by this expanded assumption shows clearly some of the management problems that must be faced with heterogeneous participants.

It is important to note that in order to make clear the relationship between stock size and participation variables, days fished is measured in a manner which is consistent with measuring fishing effort in the H function. This abstracts from important issues concerning skill levels and fishing behavior over the course of a particular fishing outing.

While an explanation of all aspects of the total net benefit function is possible only in the context of the discussions to follow, several aspects can be summarized here. (1) Total net benefits will be less than zero at any positive level of d and l for some minimum level of H. If the catch rate is too low, it just is not worth it to fish. This specification is necessary if the participation decision is to be endogenous. (2) Total net benefits can be less than zero at low levels of d and l if fixed costs of fishing are high. Just as a commercial fisherman will not operate to the left of where MC intersects AVC, a fisherman will have negative benefits if fixed costs are not spread over a large enough number of fishing days. (3) B_1, B_2, and B_3 will each be positive over some range, but for the decision variables, days fished and landings, the marginal net benefit must equal zero at some point if there is to be an optimal activity level. B_3 will always be positive if there is no satiety for the catch rate. (4) B_{11}, B_{22}, and B_{33} will be negative after some point but there may be increasing returns initially. (5) $B_{13} > 0$. The higher the rate at which fish are caught, the larger the marginal utility of a fishing day. (6) It is not possible to specify the signs of the other cross-partials, a priori. There are equally appealing arguments why a certain one can be positive, negative, zero, or change signs over the relevant range. However, in order to make the discussion below more tractable, it is assumed that $B_{12} < 0$ and $B_{23} = 0$. $B_{12} < 0$ implies that d and l are net substitutes, which will be the case if total landings are an important part of total utility. It is seen that the signs on these cross-partials are important in predicting behavior. Reference is made to the sensitivity of the results to these specific assumptions where appropriate.

Implications of the distinction between landings and the catch rate. It is useful to discuss the implications of this distinction. Remember that $H(\ )$ represents the catch rate (the rate at which fish can be brought on board) while l represents landings (the fish that are kept). $H(\ ) - l$ equals fish returned to the sea. Aggregate catch is

$$Y = ndH(X, nd, ndl) = DH(X, D, Dl) \tag{2}$$

when n is the number of homogenous participants. The change in catch due to a change in aggregate effort is

$$\frac{dY}{dD} = H + DH_2 + DlH_3. \tag{3}$$

As aggregate effort goes up, total catch will go up by the current catch per day, but will fall because of decreases in CPD. Further,

$$\frac{\partial Y}{\partial n} = d \frac{dY}{dD} \tag{4}$$

$$\frac{\partial Y}{\partial d} = n \frac{dY}{dD} \tag{5}$$

$$\frac{\partial Y}{\partial l} = (nd)^2 H_3. \tag{6}$$

For comparison purposes, the catch of a single participant is

$$Y_i = d_i H(X, D, Dl). \tag{7}$$

The marginal catch due to an extra day fished by one individual is

$$\frac{\partial Y_i}{\partial d_i} = H + d_i H_2 \frac{\partial D}{\partial d_i} + d i H_3 l \frac{\partial D}{\partial d_i}. \tag{7.1}$$

If $\partial D / \partial d_i$ is close to zero, or if individual fishermen perceive it to be so, operationally the second and third terms will drop out.

Given that l can be less than H, landings can be less than fishing mortality. However, the latter is a necessary component of user cost and must be used in the economic model. Since total landings will be ndl, then if α is the survival rate for fish which are returned to the sea, aggregate fishing mortality is $ndl + (1 - \alpha)nd[H(X, nd, ndl) - l]$, or

$$M = \alpha ndl + (1 - \alpha)ndH(X, nd, ndl). \tag{8}$$

The first term is the mortality from landings and the second is the mortality from catch. This expression shows that when the CPD constraint is not binding, a unit increase in landings will have a less than one unit effect on mortality because some of the fish that are kept would have died anyway if they were released. Note that when the CPD constraint holds, $l = H$ and the mortality Eq. (8) collapses to landings Eq. (2). If $\alpha = 0$, the problem is equivalent to the case where the CPD constraint is binding. If $\alpha = 1$, there is no difference between landings and mortality.

The partial derivatives of mortality when $l < H$ are

$$\frac{\partial M}{\partial n} = d \left[\alpha l + (1 - \alpha) \frac{dY}{dD} \right] \tag{9}$$

$$\frac{\partial M}{\partial d} = n \left[\alpha l + (1 - \alpha) \frac{dY}{dD} \right] \tag{10}$$

$$\frac{\partial M}{\partial l} = \alpha nd + (1 - \alpha)(nd)^2 H_3. \tag{11}$$

The terms in parentheses are the same in both (9) and (10). The first is the effect on mortality of the increased landings for all days fished for the marginal participant in (9) and for the marginal day fished for all participants in (10). Only a portion of increased landings results in new mortality; some would have died anyway. The second is the effect of the increased catch on mortality in the same situations. Only a portion of the increased catch results in mortality; some survive being released.

AN ANALYSIS OF INDIVIDUAL BEHAVIOR

Individuals face three interrelated decisions in recreational fishing. First there is the participation decision, which is a discrete choice of whether to fish. The decision will be in the affirmative if the net benefits of doing so (the gains from fishing less all opportunity costs) are nonnegative. The participant must also decide how many days to fish (the frequency decision) and how many fish to land per day (the landing decision) if he or she chooses to participate. In certain instances the landing decision may be constrained by the CPD. The level of activity will determine the size of the net benefits.

In formal terms, the problem the potential participant faces is to maximize (1) subject to $l \leqslant H(X, D, Dl)$. The required Lagrangian is

$$L = B[d, l, H] + \lambda_1(H - l).$$

The optimal levels of d and l can be found using the Kuhn–Tucker conditions although all the required conditions are not listed here:

$$\frac{\partial L}{\partial d} = B_1[d, l, H] = 0 \tag{12}$$

$$\frac{\partial L}{\partial l} = B_2[d, l, H] = \lambda_1. \tag{13}$$

In Eq. (12), it is assumed that $\partial D / \partial d_1$ is effectively zero.[5]

When the CPD constraint is not binding, λ_1 is equal to zero. In that case conditions (12) and (13) state that the individual should increase days fished and landings per day as long as the net benefit of doing so is positive. When the CPD constraint is binding, λ_1 is nonnegative, landings per day equal CPD, and the marginal fish landed has a value equal to λ_1.

Assuming the second-order conditions are met, the simultaneous solution of (12) and (13) is the net benefit maximizing combination of d and l for any level of H. If net benefits are nonnegative at the optimal levels of d and l, the individual will participate. There are many combinations of stock size, aggregate effort, and aggregate landings that will result in any given level of H. While this has no bearing on individual decision making, it adds a considerable complexity to aggregate analysis (see below).

A more complete analysis is possible by studying graphical representations of Eqs. (12) and (13) in (l, d) space. Assume for the moment that the CPD constraint

[5]Note that since the catch rate is a function of actions of other current participants, individuals may adopt strategic behavior based on their beliefs about what others will do [11]. Space prevents further consideration of these issues, but further research is justified.

278   LEE G. ANDERSON

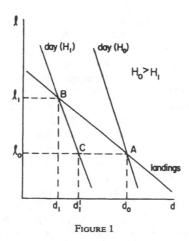

FIGURE 1

does *not* hold. In Fig. 1, the curve labeled "day (H_0)" represents Eq. (12) when the catch rate is H_0. It represents the optimal level of d for each level of l when H equals H_0. It is negatively sloped when B_{12} is assumed negative. As the number of fish landed increase, the optimal number of days fished decreases. Because B_{13} is assumed positive, increases in the catch rate will shift the day curve to the right. At each level of l, a higher catch rate will mean that a higher d will be optimal.

The curve for Eq. (13) is labeled "landings." It represents the optimal level of landings for each level of days fished assuming that the constraint is not binding and so the right-hand side is equal to zero, not λ_1. It is also negatively sloped because of the assumption that B_{12} is negative. As the number of days fished increase, the optimal landings decrease. The assumption that $B_{23} = 0$ means the catch rate does not affect the position of the landings curve. The intersection of the day curve for the existing catch rate and the landings curve represents the optimal combination of d and l for the given catch rate. The day curves for Eq. (12) must be steeper than the landings curves for Eq. (13) if the second-order conditions for utility maximization are to hold. The qualitative results which follow depend on these assumptions and they should be interpreted accordingly.

Figure 1 can be used to derive what can be called the catch rate utilization path. Note that as the catch rate decreases, the day curve moves to the left, which, given our assumptions, shifts the equilibrium point upward and to the left. When catch rate decreases, the active participant will decrease the number of days fished but increase landings per day. Remember at this point it is assumed that the CPD constraint is nonbinding.

Let H_0 be the catch rate of the virgin biomass when there are no other participants and let H_1 be the catch rate where the maximum net benefits are equal to zero. The landings curve over the range from A to B traces out the path of d and l utilization for the relevant catch rates. As the catch rate decreases from H_0 due to reductions in stock size or increases in D or Dl, the individual's optimal combination of l and d will move up the landings curve toward B. If catch rate is less than H_1, the optimal combination of d and l will produce negative benefits and the individual will not fish.

To summarize, a catch rate utilization path is a useful construct for understanding individual recreational fishing behavior. As drawn here it shows that as the stock is fished down or aggregate effort or landings go up, the participant will react by fishing fewer days but keeping more fish on those days that they do fish. Alternatively, government stock enhancement programs which increase the stock size will cause the individual to increase days fished but decrease landings per day. The exact types of behavioral changes will depend upon the cross-partials of the net benefit function and the results described here follow from the specific assumptions described above. The point is that the stock size and the way it is utilized by aggregate effort and aggregate landings will affect the choice of d and l of the individual participant and may affect the participation decision. To understand recreational fisheries utilization and the potential for management, more research on the exact nature of these cross-partials will be required.

The catch rate utilization path will be different when the CPD constraint is binding. Using Fig. 1, assume that the catch rate is H_1 where $H_1 = l_0$. The individual will want to operate at point B, but under the circumstances the highest possible utility will be achieved at point C. The operational curves are day (H_1) and the constraint which is a horizontal line at l_0. Faced with the constraint, the individual will increase d from the desired level of d_1 to d_1' to partially compensate for the restriction on l.

When the CPD constraint is binding, the analysis can be summarized in Fig. 2, which has four day curves, each for a different catch rate. The larger the subscript, the lower the catch rate. Landings equal to these catch rates are plotted on the vertical axis. Using this information, the constrained catch rate utilization path can be derived. As the catch rate decreases from H_0 toward H_2, the utility-maximizing combination of d and l will move along the l curve, d will decrease, and l will increase. At catch rates below H_2, CPD becomes a binding constraint. The utility-maximizing point will occur on the day curve for a given catch where l equals the catch rate. As the curves are drawn, a decrease from H_2 to H_3 will lead to a decrease in d with the reduction in l, but a reduction from H_3 to H_4 will lead to an increase in d. The constrained catch rate utilization path is drawn in heavy

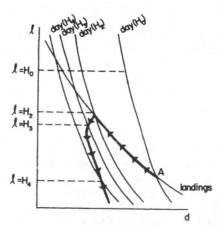

FIGURE 2

lines with arrows representing the direction of change as catch rate decreases. As drawn, a stock enhancement program which increases catch rate marginally above H_4, while it will always increase utility, will lead to a decrease in days fished. This contradicts the conventional wisdom.

This analysis can also be used to examine the effects of changes in other parameters. When the CPD constraint does not hold, an increase in the costs associated with d will shift the d curve to the left. This will result in a decrease in d but an increase in l. This makes intuitive sense given d and l are substitutes, i.e., that B_{12} is negative. The participant will shift out of days fished, the relatively expensive method of obtaining extra fish, and increase l, the number of fish kept each day. An increase in the costs associated with l will shift the l curve to the left, but since this will also affect the total cost associated with a fishing day, the d curve will also shift to the left. Depending on the relative size of the two shifts, l and d could both decrease or one could increase while the other decreases. When the CPD constraint is binding, a change in the cost of l or d will always cause a reduction in d, but will obviously have no effect on l.

HOMOGENEOUS PARTICIPANTS

Open access. An open-access bioeconomic equilibrium occurs when there is no tendency for individuals to begin or cease recreational fishing, and, at the same time, when the aggregate effort and landings generated by participating individuals is such that fishing mortality is equal to the growth of the existing stock. Assume there are k potential homogeneous participants. This assumption allows for a concise statement of principles which also applies to more realistic cases. Heterogeneous individuals are discussed below.

Assuming the CPD constraint does not hold, there are four variables that must be determined to obtain an open-access equilibrium: the number of actual participants, n; the days fished and the landings per day for each participant, d and l; and the stock size, X. The equations necessary to solve for these four variables are Eqs. (12) and (13) and

$$B[d, l, H(X, nd, ndl)] = 0 \qquad (14)$$

$$F(X) = \alpha ndl + (1 - \alpha)ndH(X, nd, ndl). \qquad (15)$$

All terms are as defined above and $F(X)$ is the standard Schaefer growth function. All individuals must be choosing that combination of d and l which maximizes utility given the catch rate [Eqs. (12) and (13)], but the net utility gain from fishing must be pushed to zero [Eq. (14)]. This will be an economic equilibrium. Further, the number of participants must be such that total fishing mortality equals natural growth at existing stock size as stated in Eq. (15). This will be a biological equilibrium.

The nature of the open-access equilibrium and the process of obtaining one can vary depending upon the structure of the equations. For example, if H is a function of stock size only, which is to say there are no intraperiod effects on H, the simultaneous solution of the four equations is relatively easy. In that case Eqs. (12), (13), and (14) can be solved independently for d, l, and X. Let X_1 be the stock size for which the catch rate is such that the maximum net benefit is equal to zero. If X is greater than X_1, the net benefit of fishing is positive; for nonparticipants, individual utility will be increased by switching from other activities to fishing. Therefore participation will increase, causing the stock to decrease. The reverse will be true when X is less than X_1. The number of participants will decrease, allowing the stock size to grow. At X_1 each participant will use a given level of d and l, called d_1 and l_1. Using these levels, Eq. (15) can be solved for the number of participants that can be supported by stock size X_1 such that total mortality equals growth.

The situation is more complex if H is a function of stock size, aggregate effort, and aggregate landings as is assumed in this paper. All four equations must be solved simultaneously to obtain the solution. Because catch rate is not dependent solely on stock size, at nonequilibrium points large stock sizes for which growth is greater than catch will not always lead to a case where participants and mortality will have a tendency to increase. Similarly, small stock sizes for which growth is less than catch will not always lead to a case where participants and mortality decline.

The open-access equilibrium solution is constrained because n must be less than or equal to k. If the solution to the above four equations yields an n greater than k, then the relevant solution for d, l, and X can be found by solving Eqs. (12), (13), and (15) when k is substituted for n.

In this case, all benefits from recreational fishing are not dissipated. The net benefits per person at the equilibrium can be bound by substituting k and the equilibrium levels of d, l, and X into Eq. (14). The analogous situation does not occur in the normal firm and industry analysis. If economic profits are greater than zero, it is assumed that entrepreneurs will organize new firms. However, the number of recreational participants is limited by the population size and the structure of individual tastes. Therefore, when the number of potential participants is relatively small, all rents may not be lost under open access. However, the rents will not be maximized because d and l will not be optimally determined (see below).

When the CPD constraint does hold, λ_1 in Eq. (13) becomes a fifth variable and the CPD constraint becomes a fifth equation. The basic results are similar to the above discussion and are left as an exercise.

Optimal utilization.   The use pattern which maximizes the present value of the sum of net benefits can be determined from the model. There are three control variables: the number of participants and the levels of d and l used by each. The size of the fish stock is the stock variable. The constrained, current value Hamiltonian is

$$\mathcal{H} = nB[d, l, H(X, nd, ndl)] + \phi[F(X) - M] + \lambda_2(H(X, nd, ndl) - l).$$

$$(16)$$

The ϕ term is the value of a unit of fish stock in place. The expression for M, and hence the form of the first-order conditions, will vary depending upon whether the CPD constraint holds. See Eqs. (2) and (8) above. When the CPD constraint holds, the λ_2 term is the social marginal value of an extra fish caught per day by *all* participating individuals.

The necessary conditions for an interior solution with respect to the control variables (other necessary conditions are discussed below) are

$$\frac{\partial \mathcal{H}}{\partial n}\bigg| \quad \frac{B}{d} = -nB_3[H_2 + H_3l] + \phi\frac{\partial M/\partial n}{d} - \lambda_2[H_2 + H_3l] \qquad (17)$$

$$\frac{\partial \mathcal{H}}{\partial d}\bigg| \quad B_1 = -nB_3[H_2 + H_3l] + \phi\frac{\partial M/\partial d}{n} - \lambda_2[H_2 + H_3l] \qquad (18)$$

$$\frac{\partial \mathcal{H}}{\partial l}\bigg| \quad B_2 = -nB_3(H_3d) + \phi\frac{\partial M/\partial l}{n} - \frac{\lambda_2}{n}[H_3D - 1]. \qquad (19)$$

Assume for the moment that the CPD constraint is not binding. This means that mortality changes are represented by Eqs. (9), (10), and (11) and that the λ_2 term drops out of all three equations. The above become

$$\frac{B}{d} = -nB_3[H_2 + H_3l] + \phi\left[\alpha l + (1 - \alpha)\frac{dY}{dD}\right] \qquad (17.1)$$

$$B_1 = -nB_3[H_2 + H_3l] + \phi\left[\alpha l + (1 - \alpha)\frac{dY}{dD}\right] \qquad (18.1)$$

$$B_2 = -nB_3(H_3d) + \phi d[\alpha + (1 - \alpha)DH_3]. \qquad (19.1)$$

These conditions can be interpreted in the context of the three ways a fish can be allocated at any point in time. There can be more participants, there can be more days fished per participant, and there can be more landings per day. The rhs of (17.1) and (18.1) are the same and represent the marginal user cost of aggregate effort. The first term is the intraperiod user cost. An extra unit of effort will decrease CPD (H_2 and H_3 are negative), which will decrease the benefits for all participants. The second term is the interperiod user cost, which is value of a unit of stock in place at that point in time multiplied by the mortality of a single unit of effort either at the extensive or at the intensive margin. Together, Eqs. (17.1) and (18.1) state that the benefits of a unit of aggregate effort, whether it comes from the extensive margin (B/d, adding an additional participant at the existing number of days fished) or the intensive margin (B_1, increasing days fished for all individuals) should equal the marginal user cost.

Equation (19.1) means that landings per day should be increased until the marginal benefit equals the marginal social cost. The first term on the rhs is an intraperiod user cost. Increased landings decrease CPD, which will decrease net benefits for all participants. The second term is an interperiod user cost. It is the value of a unit of stock in place times the number of days that landings will be increased times the actual amount of new mortality which results because an extra fish is kept rather than released. Recall that α is the survival rate of fish that are released.

As a way of further elucidating the optimal conditions as well as initiating a discussion of regulation, consider the tax policy that will achieve the optimal conditions. When the CPD constraint does not hold, two separate taxes will be necessary, one on each fish landed and one on days fished:

$$t_l = \phi[\alpha + (1 - \alpha)ndH_3] - nB_3H_3 \tag{20}$$

$$t_d = (1 - \alpha)(H + ndH_2)\phi - nB_3H_2. \tag{21}$$

The tax on landings forces the participant to consider the social cost due to the mortality caused by increased landings and due to the loss of total utility because of the reduction in the catch rate. Similarly, the tax on days fished forces the participant to consider the social cost due to the mortality caused due to increased catch and due to the loss of utility of other participants because of the reduction in catch rate. Application of this tax program will require information on marginal catch rates, changes in the CPD, and the discard mortality rate as well as the shadow price of a unit of fish in place at that period in time. In addition it will be necessary to have information on the benefit functions and the number of participants.

Given these taxes the net benefit function of the individual at equilibrium becomes

$$B(d, l, H(X, nd, ndl)) - dlt_l - dt_d = 0. \tag{22}$$

This is equivalent to Eq. (17.1). Also the first derivatives of the new net benefit function with respect to d and l are equivalent to (18.1) and (19.1), respectively. The fact that two types of controls will be necessary is of general policy interest. A daily license fee combined with a daily catch limit has the potential of improving fisheries utilization. It is interesting to note that both these techniques are currently used by state agencies in managing recreational fisheries.[6] This is in contradiction to traditional commercial fishing regulations such as closed seasons and gear restrictions, which do not address the economic efficiency aspects of management. That is, the achievement of efficient utilization in recreational fisheries will not require the introduction of new types of controls.

When the CPD constraint is binding, the mortality terms are represented by Eqs. (4), (5), and (6), and λ_2 is positive. The above equations become

$$\frac{B}{d} = -[H_2 + H_3l][nB_3 + \lambda_2] + \phi\frac{dY}{dD} \tag{17.2}$$

$$B_1 = -[H_2 + H_3l][nB_3 + \lambda_2] + \phi\frac{dY}{dD} \tag{18.2}$$

$$B_2 = -H_3d[nB_3 + \lambda_2] + \phi H_3\,dD + \frac{\lambda_2}{n}. \tag{19.2}$$

The first term in the three equations is the intraperiod user cost. These terms are the same in (17.2) and (18.2). They represent the change in CPD caused by an increase in effort (on either the intensive or the extensive margin) times the value

---

[6]While seasonal fishing licenses are most common, many states have a daily license which can be purchased by out-of-state users for a fraction of the seasonal rate. These licenses can be purchased at sporting goods stores. There may be problems extending such a program to all users, but it is not a completely new concept.

effect this will have on participants. Net benefit will decrease as before due to the decrease in the catch rate, but in addition, since landings equal the catch rate, a decrease in the catch rate will also decrease landings. Recall that λ_2 is the shadow value of an extra unit of landings for all participants. The intraperiod user cost term in (19.2) is analogous except the change in the catch rate in the first factor is due to the change in landings. The second term in all the equations is the interperiod user cost, which is the value of a unit of fish stock in place times the mortality caused by a change in the relevant control variable.

The optimal tax structure when the constraint holds is

$$t'_l = \phi[ndH_3] - H_3(nB_3 + \lambda_2) \qquad (21.1)$$

$$t'_d = \Phi[H + ndH_2] - H_2(nB_3 + \lambda_2). \qquad (22.1)$$

They differ from those in the nonconstrained case by the absence of the α factors in the first term and the presence of λ_2 in the second.

The remaining conditions for a full interior solution of (16) are

$$\frac{dX}{dt} = F(X) - \alpha ndl - (1 - \alpha)ndH(X, nd, ndl) \qquad (23)$$

$$\frac{d\phi}{dt} = r\phi - \frac{\partial H}{\partial X}. \qquad (24)$$

When there is a steady-state equilibrium, growth will equal mortality, and

$$\phi = \frac{nB_3H_1}{r - F'(X) + (1 - \alpha)ndH_1}. \qquad (25)$$

Comparisons of open access and optimal utilization. The conditions for an open-access equilibrium are (12), (13), (14), and (15) while the analogous conditions for optimal utilization are (18), (19), (17), and (23). In addition, optimization has a new condition, which is (24). Analysis of the open-access conditions demonstrates the well-known conclusion that uncontrolled utilization will result in inefficiencies. Participants will continue to increase days fished and landings per day until the marginal value of either is zero. Further, because of the effects that each participant has on the others, net individual gains will be fully dissipated except when the number of potential participants is below a critical level.

When the CPD constraint holds, the last fish landed in open access will still have a positive value. However, fish will not be allocated over participants and days fished, such that welfare is maximized. The positive value of the marginal fish landed is due to the biological constraint, not socially optimal individual decision making.

Comparisons of open access and optimal utilization at the individual participant level are possible using Fig. 3. Assume that the CPD constraint does not hold and point B is the solution of Eqs. (12) and (13) at H_1, the open-access equilibrium catch rate. This is an equilibrium for an individual. The marginal benefits just equal the marginal opportunity costs. Let the curves labeled d^* and l^* represent

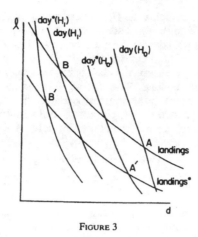

FIGURE 3

Eqs. (18.1) and (19.1) when catch rate is equal to H_1. These curves represent the optimal conditions when the stock is at the open-access level and they will be to the left of the open-access curves for the same catch rate because of the user cost.

As drawn here, the optimal levels of d and l at the open-access stock size (point B') will be a reduction from the open-access levels. However, one could increase and the other could decrease depending upon the relative shifts of the curves. As the stock size increases due to reduced fishing, the d^* curve will shift to the right. The furthest it can go will be $d^*(H_0)$ because H_0 is the catch rate at the unexploited virgin stock. If an interior steady-state solution exists, it will be somewhere between B' and A' on l^*, assuming that the l^* curve does not shift with changes in stock size. The extension of the above analysis when the constraint holds produces the same general results and is left as an exercise for the reader. A comparison of how n, d, and l in the aggregate will change with a shift to optimal utilization is provided in the discussion of heterogeneous participants.

HETEROGENEOUS PARTICIPANTS

Open access: A preliminary analysis. In the analysis of firms in an industry, heterogeneous firms with homogeneous products can be classified by the minimum point on their average cost curves. The marginal firm is the one whose minimum point equals the market-determined equilibrium price. But how can heterogeneous participants in a recreational fishery be classified? Because there is nothing analogous to the fixed (to the firm) price, cost curves do not provide a useful method of classification.

In order to address this problem, the discussion of heterogeneous participants is split. In this and the following subsections, it is assumed that there are no intraperiod effects on catch rate. H is assumed to be a function of stock size only, $H = H(X)$. This will provide a way to order participants so as to allow a graphical analysis which illustrates many of the important economic concepts involved. The

LEE G. ANDERSON

final subsection describes the nature of the open-access and optimal utilization situations when $H = H(X, D, Dl)$ and this ordering is not possible. The contrast highlights some of the problems of recreational fisheries management.

Given that CPD is a function of stock size only, potential participants can be ordered by the stock size at which their net benefits equal zero. The marginal participant is the one whose net benefit function equals zero at the stock size where fishing mortality equals natural growth; that is, each potential participant will have a utilization path as pictured in Fig. 2 and at some point on the curve net benefits will equal zero. With different tastes, skills, and opportunity costs, the stock size at which net benefits are zero will differ for different individuals.

The open-access equilibrium can be described in terms of Fig. 4, which contains a graph of the natural growth curve, $F(X)$. At stock size X_0, the virgin biomass, and the CPD that is associated with it, there are a certain number of individuals who have net benefits greater than zero at their optimum combination of d and l. Let the mortality represented by this combination of participants, days fished, and fish landed be M_0. At lower stock sizes the number of individuals who have positive net benefits (and hence will participate) will fall and the optimal combination of d and l for those who remain will change. Therefore, each stock size will generate a unique level of fishing mortality given the tastes of potential participants and the prices and opportunity costs they face. Let the curve labeled M_{OA} in Fig. 4 represent the relationship between stock size and fishing mortality under open access. (The other curve is discussed below.) As drawn it is monotonically increasing, and while this is likely, it need not always be the case. The open-access equilibrium will occur at stock size X_1. At that point the marginal participant is achieving no net benefits but the intramarginal participants are receiving positive net benefits. The CPD constraint may be operative for some participants but not for all.

The conditions for an open-access equilibrium with heterogeneous participants are analogous to those for the homogeneous case. Equations (12) and (13) must hold for all participants but Eq. (14) holds only for the marginal participant. The expression for fishing mortality in Eq. (15) must be modified to take into account

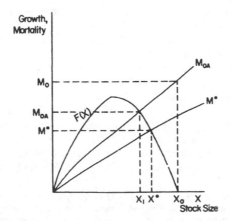

FIGURE 4

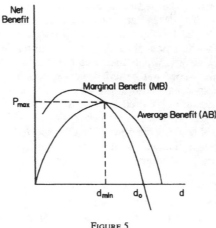

FIGURE 5

the number of participants and the order they enter or leave the fishery. That is, it is necessary to use a mathematical expression of the M_{OA} curve in Fig. 4.

This equilibrium and the process of obtaining it can be depicted in a manner that is analogous to standard firm industry analysis. Figure 5 depicts a possible relationship between average and marginal net benefits and days fished for a hypothetical participant given a specified stock size. The total benefit curve from which they are derived can be derived from Fig. 1 in the following manner. Assuming that CPD equals H_0, net benefits are maximized when days fished and landings are d_0 and l_0, respectively. Given that the CPD constraint does not hold, maximum net benefits for any given number of days fished occur along the landings curve. Therefore, total possible net benefits as a function of days fished can be found by tracing out net benefits at various points on the landing curve. When the CPD constraint does hold, they can be measured along a horizontal line equal to the constraint.

The average and marginal curves as pictured here would occur if there are increasing marginal returns at low levels of effort. If there are high fixed costs, the curves would have a similar shape except that the average curve would intersect the horizontal axis to the right of the origin. In that case, unless the number of days is high enough, the total benefit received will not cover the fixed opportunity costs.[7]

Note that just as the portion of the marginal cost curve above the average variable cost curve is the supply curve for the firm, the portion of the marginal benefit curve below the average benefit curve is the demand curve for days fished when price is expressed in terms of full opportunity cost. This is a special demand curve because the normal ceteris paribus condition does not hold since l can vary with d when the CPD constraint does not hold. Formally the curve shows the

[7]If the total curve does not emanate to the right of the origin and if there are decreasing returns throughout, the average and marginal curves will emanate from the same point on the vertical axis and the MB curve will always be below the AB curve. In that case, therefore, average benefit approaches marginal benefit only as the number of days fished approaches zero.

288   LEE G. ANDERSON

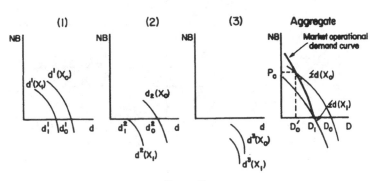

FIGURE 6

maximum net benefit achieved for the marginal fishing day given that l is not constrained. At the stock size for which these curves are defined, the individual will never pay more than P_{max} as a user fee because at higher prices total net benefits from participation would be negative. Also, as drawn here, the demand curve does not intersect the vertical axis. There is a maximum price, above which participation falls to zero, and a minimum amount of fishing days, d_{min}, which is greater than zero.

There will be a different demand curve for each stock size. As stock size decreases, the marginal benefit curve will shift downward and to the left. What is not clear, however, is how the minimum number of fishing days will change with changes in stock size. If the point at which average benefits are maximized does not change, then the d_0 point will not change.

As stock size decreases, the demand curve shifts down. When the stock size reaches a certain point, the demand curve will be reduced to a single point at d_{min}. The participant is just "breaking even"; the gains just cover the lost utility from foregone expenditures of time and income. If there is a positive user fee, the individual will no longer engage in fishing. At lower stock sizes, participation will cease.

The thrust of this discussion is that with heterogeneous participants, the curves for different individuals can be different for the same stock size. Therefore, participants can be ordered according to the height of their maximum point on the average net benefit curve. Consider Fig. 6; the first three graphs represent demand curves for two stock sizes of three potential participants arranged in decreasing order. Assume that there are other potential participants whose curves fall on either side of those pictured. When stock size is X_0, these participants will operate on the $d^i(X_0)$ demand curves. With no user fee, individuals 1 and 2 will fish d_0^1 and d_0^2 days per year, respectively, and individual 3 will not participate. The fourth graph in Fig. 6 represents the sum of the demand curves for all potential participants.[8] Any point on this aggregate demand curve provides information on the number of participants and aggregate number of days fished.

---

[8]These analyses are similar to earlier constructs of aggregate demand curves for recreational fishing days (see Anderson [2] and the references cited therein).

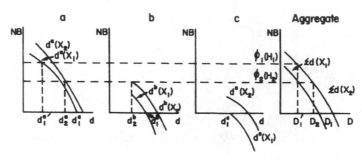

FIGURE 7

Given stock size X_0, the aggregate days fished is D_0. This will be a biological equilibrium only if the total mortality associated with D_0 equals growth. Assume that with stock size of X_0, growth equals mortality at D_0^1. Then at D_0, stock size will decrease and the demand curves of all participants for the next period will shift down. Say that stock size in the next period is X_1. The individual and aggregate demand curves shift down as indicated. Individual 3 will still not participate and individuals 1 and 2 will cut back on days fished. Individual 2 is just on the margin of participating. If D_1 produces a mortality equal to growth at X_1, this will be an open-access bioeconomic equilibrium. People with average benefit curves lower than that of individual 2 will not participate. People with higher curves will and they will earn positive net benefits. Individual 2 is the marginal participant.

As a sidelight, note that if stock size is X_0, a bioeconomic equilibrium is possible if there is a daily user fee of P_0. This is so because given $\Sigma d(X_0)$, the user fee of P_0 will bring forth an aggregate level of D_0^1, the point where mortality is assumed to be equal to growth. A similar price will be associated with the aggregate demand curve associated with each stock size. A zero user fee gives a bioeconomic equilibrium at stock size X_1. The collection of these points traces out an operational market demand curve for recreational fishing days (see [2]).

Optimal utilization: A preliminary analysis.   Optimal utilization and the nature of the path to it with heterogeneous participants can be described using Fig. 7. As before the curves in the first three graphs are for representative participants and the fourth curve is the aggregate demand curve. Let D_1 in the right-hand graph, where $\Sigma d(X_1)$ intersects the horizontal axis, represent an open-access equilibrium. It can be seen that participant c is the open-access marginal participant and he or she fishes d_1^c days per season. Participants a and b fish d_1^a and d_1^b days each year, respectively.

Assume that the CPD constraint is binding so that an optimal tax program, given the temporary assumption about H, is a tax on d equal to $\phi H(X)$. $H(X)$ is the catch per day and ϕ. is the shadow price of a unit of fish in place during this period. Given the stock size at open access, the optimal tax the first period should be $\phi_1 H(X_1)$. Given this tax, participants b and c cease fishing and participant a decreases days fished to $d_1^{a'}$.

Given the reduction in aggregate days fished, the stock size will grow, which will shift the demand curves up. The change in the stock size will also change $H(X)$ and ϕ. To make a long story short, assume that $\Sigma d(X_2)$ is the new aggregate demand curve, that $\phi_2 H(X_2)$ is the optimal tax for the next period, and that D_2, the level of aggregate days fished which is generated, produces a total mortality equal to growth at the new stock size. Participant b will fish d_2^b days per year and is the marginal participant under optimal utilization. There are no net benefits after the optimal user fee is paid. By construction his or her marginal benefits are equal to average benefits. Participant a fishes d_2^a days per year. The marginal benefit of his or her last fishing day is equal to that of participant b and all other active participants. However, like other nonmarginal participants, average benefits will be greater than marginal benefits and so net benefits after paying the user fee will be positive.

This discussion has shown the path to optimal utilization in one step and as a result many interesting aspects have been neglected. For example, note that if the optimal tax in the second year were low enough, d_2^a could be higher than d_1^a. That is, it may be optimal to increase days fished for remaining participants. Similarly, if the optimal tax in the last year were low enough, the optimal number of days fished for individual c could become positive even though it was zero under open access.

From the above it can be seen that there is an optimal amount of mortality at every stock size. It is the mortality generated by the aggregate d and l provided by the number of individuals who choose to participate given an optimal tax structure. Let the curve labeled M^* in Fig. 4 represent this relationship. As drawn, the change from open access to optimal utilization will increase stock size and decrease the amount of aggregate fishing mortality. However, if the M_{OA} curve intersects the upward-sloping segment of the $F(X)$ curve, the move to optimal utilization could allow for an ultimate increase in aggregate fishing mortality.

Therefore, it is not possible to state a priori whether n, d, or l will increase or decrease when the fishery changes from open access to optimum utilization. Since the utility of recreational fishing is tied only partially to the number of fish landed, it may be that optimization will require that the landings be spread as widely as possible. This may mean an increase in n and d but a decrease in l.

Open access and optimal utilization: A complete analysis. The above describes the operation of individual participants and the nature of, and the process of obtaining, the open-access equilibrium and the optimal utilization point with heterogeneous participants. However, the relative simplicity of the graphics was possible because of the assumption that H was a function of stock size only. This means that the height of the average benefit curve is a function only of stock size. If H is a function of stock size, aggregate effort, and aggregate landings, things become much more complicated. The height of the average benefit curve of any potential participant is a function of the number and the exact composition of other participants. As such, a single ordering of participants may not be possible.

However, following a comparable analysis of heterogeneous firms in a commercial fishery by Clark [8, pp. 246–254], it is possible to provide a more rigorous analysis of heterogeneous participants. Let there be k potential participants, each trying to independently maximize utility subject to a possible CPD constraint. The superscript i refers to the $i = 1$ to k potential participants. Utility maximization for each of the k potential participants implies that

$$\left.\begin{array}{l} B_1^i = 0 \\ B_2^i = \lambda_1^i \end{array}\right] \quad \text{if } B^i \geqslant 0 \tag{26}$$

$$\left.\begin{array}{l} d^i = 0 \\ l^i = 0 \end{array}\right] \quad \text{otherwise.} \tag{27}$$

If the CPD constraint holds for any participant j, then $l^j = H$, and if it does not hold, then $\lambda_1^j = 0$. Biological equilibrium requires that

$$F(X) = \alpha L_T + (1 - \alpha)D_T H(X, D_T, L_T), \tag{28}$$

where

$$D_T = \sum^k d^i \tag{29}$$

$$L_T = \sum^k d^i l^i. \tag{30}$$

Solution of this system is possible only by numerical methods. The actual participants will be those for whom $B^i \geqslant 0$, and this could hold true for all k potential participants or only a subset of them.

The conditions for open-access equilibrium can shed light on economic valuation studies for recreational fishing. When the CPD constraint does not hold at the equilibrium point, the marginal value of an extra fish to all users is zero. Fish will have only a transient positive marginal value as the fishery moves toward the equilibrium. Empirical work to measure the value of extra fish should be designed with this in mind and the results should be interpreted accordingly. When the CPD constraint does hold, a marginal fish landed by a participant can have a positive value. However, the value could be vastly different for different individuals.

It is possible to have meaningful discussions of marginal values in general only for market goods and services. While individuals can have different tastes, incomes, and consumption bundles, if they purchase a hamburger they will only do so if the increased utility is equal to the opportunity cost. The market insures that hamburgers will be allocated to the highest value uses at the margin, so an extra unit will have the same value wherever it is consumed. This is not the case with recreational fish. While heterogeneous individuals will all operate where the equivalent of Eq. (26) holds, there is no guarantee that the value of λ_1 will be the same for all of them. In fact, given that H can be specific to particular individuals according to their natural abilities, level of experience, and types and amounts of fishing gear, the CPD constraint can be binding on some, but not on others, at the same stock size.

Two points about empirical valuation studies on open-access fisheries seem obvious. First, the results will depend critically on the way the study sample is chosen. Second, at best these studies can only measure the average of the marginal values to different individuals.

Optimal utilization requires the maximization of the net present value of net benefits over all potential participants. The current value Hamiltonian is

$$\mathscr{H} = \sum_{}^{k} B^i\big[d^i, l^i, H(X, D_T, L_T)\big]$$
$$+ \phi\big[F(X) - \alpha L_T - (1 - \alpha)D_T H(\ )\big] \tag{31}$$
$$+ - \sum_{}^{k} \lambda_2^i(H - l^i).$$

The necessary conditions for the control variables for each potential participant are[9]

$$B_i^i = - \sum_{}^{k} B_3^i(H_2 + H_3 l_i)$$
$$+ \phi\big[(1 - \alpha)H + D_T(H_2 + H_3 l_i)\big] \tag{32}$$
$$+ \lambda_i(H_2 + H_3 l_i)$$

$$B_2^i = - \sum_{}^{k} B_3^i H_3 d_i + \phi(\alpha + (1 - \alpha)D_T H_3 d_i) - \lambda_i(H_3 d_i - 1). \tag{33}$$

The above equations will determine d_i and l_i if $B^i \geqslant 0$; otherwise they will equal zero. What makes this problem difficult (but interesting from a policy point of view) is that as this set of equations is used to solve for optimal utilization through time, there is no way to know which particular individuals will have nonnegative net benefits in any given period. Whether conditions (32) and (33) for a particular i will be relevant for the general solution (i.e., whether that individual will be a participant) depends upon the partial derivatives of the B^i and H functions.

The other necessary conditions are that growth equals mortality [see Eq. (28) above] and that

$$\frac{d\phi}{dt} = r\phi - \frac{\partial H}{\partial X}. \tag{34}$$

In the unconstrained case, this means that, in equilibrium,

$$\phi = \frac{\sum_{}^{k} B_3^i H_1}{r - F'(X) - (1 - \alpha)D_T H_1}. \tag{35}$$

This formulation sheds some light on the complexity of optimal management. The relationship between the marginal benefit to each consumer of the last days fished and average benefit per day fished of the marginal consumer does not fall out of this formulation, but it does follow from the graphical analysis and from basic economic maximization principles.

An important conclusion is that there is no special way to order the participants. There are so many things which determine whether the net benefits of a particular individual are positive, that along an optimizing trajectory, it is not possible to say

---

[9]Solution of this formulation requires that $B^i[0, 0, H(\ )] = 0$ and that all cross-partials be zero at this point.

that once it is optimal for a person to participate, it will always be optimal to do so. For example, in an over exploited fishery, the optimal combination of participants may include those to whom catch rate is not too important. They get their satisfaction mostly from participation, and given the status of the stock, these are the types of people that can be satisfied. However, as the stock continues to grow and catch rates go up, it may be optimal to remove these participants and replace them with a smaller number of individuals who place high values on catch rate (see [17]).

The policy implications of the heterogeneous participant model are really quite staggering. Optimal management will require knowledge of the net benefit function of potential participants and how CPD varies with stock size and aggregate effort. One can make the standard conclusion that a tax on days fished equal to the rhs of (32) and tax on landings equal to the rhs of (33) will change the open-access situation to the optimal one. However, the difficulty of deriving such taxes should not be underestimated.

APPLICATIONS

The model presented here can be used to analyze a range of issues. To demonstrate, consider the implications of a daily catch limit imposed on an otherwise open-access fishery with heterogenous participants. There are three potential categories of affected individuals. First, there are those for whom the daily catch limit is a binding constraint because their open-access equilibrium l is higher than the limit. These participants will increase days fished in response to the bag limit. If the optimizing combination of d and l with the limit is such that net benefits become negative, the participant will cease fishing. To the extent that the regulation is successful in reducing aggregate mortality, X will be increased in the long run. The individuals in this group will react to this by further increasing days fished, and some of those who ceased fishing may find it beneficial to start again.

A second group of participants would be those who are not affected by the CPD constraint. Initially they will not change their fishing activities. However, as the stock increases, given the above assumptions, they will increase the number of days fished but will decrease l if the limit remains a nonbinding constraint. This means that the long-term effects of a daily bag limit on those people who are initially not constrained by it may be to decrease their landings per day.

A third group of affected people would be the marginal nonparticipants whose optimizing level of l is less than the bag limit. As the stock increases, some of them may find it beneficial to start fishing.

The application of a particular regulation policy will affect participants differently. The overall affect on welfare is impossible to predict, however. Some individuals may cease fishing, others may begin, and all who participate will change their combination of l and d. Welfare could go up or down. For example, if the bag limit restricts those who get high benefits from participation and the resultant increase in stock size induces the entry of less-enthused participants, total welfare could go down.

The above provides an idea of the types of analysis that are possible using this model. Further research is necessary to fill out the details and to analyze potential

294   LEE G. ANDERSON

effects of other regulation measures such as catch and release programs, closed seasons, gear restrictions, licenses, gear limitations, and various forms of privatization. In addition, the model could be expanded to consider utilization and management implications of issues such as the use of fishing guides, learning activities to increase skill levels, the purchase and use of fishing gear, and the simultaneous use of many species with different utility-producing attributes such as relative locations and catch rates.

REFERENCES

1. L. G. Anderson, An economic analysis of joint recreational and commercial fisheries, *in* "Allocation of Fishery Resources, Proceedings of the Technical Consultation, Vichy, France, 1980" (J. H. Grover, Ed.), pp. 16–26, FAO, Rome (1980).
2. L. G. Anderson, The demand curve for recreational fishing with an application to stock enhancement activities, *Land Econom.* 59(3), 279–286 (1983).
3. R. C. Bishop and K. C. Samples, Sport and commercial fisheries conflicts: A theoretical analysis, *J. Environ. Econom. Management* 7, 220–233 (1980).
4. N. E. Bockstael and K. E. McConnell, Theory and estimation of the household production function for wildlife recreation, *J. Environ. Econom. Management* 8(3), 199–214 (1981).
5. N. E. Bockstael, W. M. Hanemann, and I. E. Strand, Jr., "Measuring the Benefits of Water Quality Improvements Using Recreation Demand Models," Vol. III, Report to U.S. Environmental Protection Agency, Department of Agricultural and Resource Economics, University of Maryland (1987).
6. N. E. Bockstael, W. M. Hanemann, and I. E. Strand, Jr., Time and the recreational demand model, *Amer. J. Agric. Econom.* 69(May), 292–302 (1987).
7. F. J. Cesario, Value of time in recreation benefit studies, *Land Econom.* 52(Feb.), 32–41 (1976).
8. C. W. Clark, "Mathematical Bioeconomics: The Optimal Management of Renewable Resources," Wiley, New York (1990).
9. M. Clawson, "Methods of Measuring the Demand for the Value of Outdoor Recreation," Reprint 10, Resources for the Future, Washington, DC (1959).
10. M. Clawson, and J. L. Knetsch, "Economics of Outdoor Recreation," Johns Hopkins Univ. Press, Baltimore (1966).
11. R. Cornes, and T. Sandler, "The Theory of Externalities, Public Goods, and Club Goods," Cambridge Univ. Press, Cambridge (1986).
12. W. M. Hanemann, Discrete/continuous models of consumer demand, *Econometrica* 52(May), 541–561 (1984).
13. W. M. Hanemann, Welfare evaluations in contingent valuation experiments with discrete responses, *Amer. J. Agric. Econom.* 66(Aug.), 332–341 (1984).
14. M. J. Kealy and R. C. Bishop, Theoretical and empirical specification issues in travel cost demand studies, *Amer. J. Agric. Econom.* 69(Aug.), (1986).
15. C. L. Kling, Comparing welfare estimates of environmental quality changes from recreation demand models, *J. Environ. Econom. Management* 15(Sept.), 331–340 (1988).
16. K. E. McConnell, The economics of outdoor recreation, *in* "Handbook of Natural Resource and Energy Economics" (A. V. Kneese and J. L. Sweeney, Eds.), Vol. II, Elsevier Science, Amsterdam/New York (1985).
17. K. E. McConnell, Heterogeneous preferences for congestion, *J. Environ. Econom. Management* 15, 251–258 (1988).
18. K. E. McConnell and N. E. Bockstael, Aggregation in recreation economics: Issues of estimation and benefit measurement, *Northeastern J. Agric. Resource Econom.* Oct., 181–186 (1984).
19. K. E. McConnell and I. E. Strand, Jr., Some economic aspects of managing marine recreational fishing, *in* "Economic Analysis for Fisheries Management Plans" (L. G. Anderson, Ed.), Ann Arbor Science, Ann Arbor (1981).
20. K. E. McConnell and J. G. Sutinen, Bioeconomic models of marine recreational fishing, *J. Environ. Econom. Management* 6(2), 127–139 (1979).
21. K. E. McConnell and J. G. Sutinen, An analysis of congested recreation facilities, *in* "Advances in Applied Micro-Economics" (V. K. Smith, Ed.), Vol. 3, pp. 9–36, Jai Press, Greenwich, CT (1984).

22. E. R. Morey, The demand for site-specific recreational activities: A characteristic approach, *J. Environ. Econom. Management* **8**(Dec.), 345–371 (1981).

23. V. K. Smith, Selection and recreation demand, *Amer. J. Agric. Econom.* **70**(Feb.), 29–36 (1988).

24. V. K. Smith, W. H. Desvousges, and M. P. McGivney, The opportunity cost of travel time in recreation demand models, *Land Econom.* **59**(Aug.), 259–278 (1983).

25. V. K. Smith and W. H. Desvousges, "Measuring Water Quality Benefits," Kluwer–Nyhoff, Boston (1986).

26. V. K. Smith and Y. Kaoru, The hedonic travel cost model: A view from the trenches, *Land Econom.* **63**(May), 179–192 (1987).

27. V. K. Smith and R. J. Kopp, The spatial limits of the travel cost recreation demand model, *Land Econom.* **56**(Feb.), 64–72 (1980).

Part V
Analysis of Management Agencies

[38]

The Economics of Fisheries Law Enforcement

Jon G. Sutinen and Peder Andersen

In his analysis of the fishery, Cheung (1970) shows that externalities arise when exclusive property rights are absent. Cheung and others (see, for example, Demsetz 1967) argue that the presence or absence of exclusive rights depends on, among other things, the costs of defining and enforcing exclusivity. This paper explores this issue in greater depth with a formal model of fisheries law enforcement to show how fishing firms behave and fishery policies are affected by costly, imperfect enforcement of fisheries law.[1] This is achieved by combining standard bioeconomic theory and the economic theory of crime and punishment (Becker 1968).[2]

The enforcement of exclusive property rights has become increasingly important as the new ocean regime tends towards removing nonexclusive rights to marine fish resources. We argue that the cost of defining and securing exclusive property rights under the new extended jurisdiction regime is a principal determinant of management measures that eventually will emerge. Thus, the presence of enforcement costs may have a significant impact on fish production and allocation in the future as coastal states move to regulate fishing activity in their waters.

The paper is organized as follows. Section I illustrates the empirical significance of the fisheries law enforcement problem. Sections II and III briefly present the formal model of fisheries law enforcement and examine how optimal management policies are affected by costly, imperfect enforcement. The final section contains a summary of the main results and comments on some unresolved issues.

I. BACKGROUND

History reveals changes in property rights to ocean resources have been significantly influenced by enforcement costs.[3] According to Clarkson (1974), the common property doctrine for ocean resources was articulated by the Romans as early as the second century, and later codified by Justinian in the sixth century. While feudal law in medieval Europe transferred to the state all property that previously had been common, only "utilized" fisheries were given legal status since "feudal law ignored resources whose definition or enforcement were prohibitively costly" (120). By the fifteenth century Scotland claimed exclusive rights to fishing within fourteen miles of its shores. "These exclusionary policies reflect several forces, including Scotland's comparative advantage in maritime activities

Respectively, Department of Resource Economics, University of Rhode Island, and Institute of Economics, University of Aarhus, Denmark. We are indebted to Lee Anderson, Rögnvaldur Hannesson, Marc Mangel, James Opaluch, Jørgen Søndergaard, and two anonymous reviewers for their valuable comments. We assume responsibility for remaining errors. Financial support is gratefully acknowledged from the Research Foundation of Aarhus University, the Carlsberg Foundation, Denmark, and from the NOAA office of Sea Grant, U. S. Department of Commerce (under grant NA81AA-D-00096) and the Agriculture Experiment Station, University of Rhode Island. RIAES Contribution No. 2180.

[1]For an excellent survey of the economics literature on fisheries regulation, see Scott (1979) who briefly discusses the information and enforcement costs of tax and quota systems. More formal analyses of fisheries regulation include Clark (1980) and Andersen (1982).

[2]Bradley (1974) incorporates Becker's theory in an analysis of enforcing regulations directed at reducing oil spills. Viscusi and Zeckhauser (1979) study the problem of setting pollution standards where enforcement is costly and imperfect.

[3]For an excellent related study of changes in property rights for land, livestock, and water in the American West, see Anderson and Hill (1975).

Land Economics, Vol. 61, No. 4, November 1985
0023-7639/85/004-0387 $1.50/0
©1985 by the Board of Regents
of the University of Wisconsin System

and lower costs of policing their coastal fisheries. . ." (120).

By the seventeenth century, an extensive treaty network recognized national claims to territorial seas. Enforcement of these claims involved "substantial naval forces to provide escorts for fleets, evict trespassers, confiscate catches or ships, and other similar activities" (Clarkson 1974, 121). Changes in technology and economic conditions during this period stimulated political and legal debates about ocean property rights. Grotius's doctrine of open access to the seas was based on "the assertion that the rewards of exclusive rights were not sufficient to offset the costs of obtaining and holding those rights" (122). Application of Grotius's doctrine eventually restricted national claims to narrow bands of coastal waters.

When in the nineteenth century important fisheries were threatened with depletion, multilateral agreements, such as the North Sea Convention of 1882, were formed to establish and enforce rights on the open seas. Such efforts were only partially successful, however, largely because of the high costs of definition and enforcement of property rights (Clarkson 1974). Thus the open-access doctrine prevailed for most of the oceans' resources well into the twentieth century.

After World War II the United States significantly altered the structure of property rights for ocean resources by asserting jurisdiction and control over the natural resources of the subsoil and seabed of its continental shelf. This action was a catalyst for similar and more extensive unilateral claims to ocean resources by other coastal nations. In 1952, Chile, Ecuador, and Peru asserted jurisdiction over ocean resources out to 200 nautical miles from their shores.

In the early 1970s coastal nations became increasingly concerned about the rapid depletion of fish stocks off their coasts by foreign distant-water fleets. This concern was a major force leading to the Third United Nations Conference on the Law of the Sea in 1973. Since progress at the conference was slow, several coastal nations took unilateral action, and by the end of 1977—the pivotal year—a majority of coastal nations had declared extended jurisdiction over their fishery resources (Copes 1983). The zones of extended jurisdiction most commonly extend 200 nautical miles from shore.

According to Eckert (1979, 354), this surge in extended jurisdiction was induced by, among other things, "new technologies which have lowered the costs of monitoring and enforcement (making) it economical for coastal nations to expand their areas of jurisdiction."

The claims of extended jurisdiction have allowed coastal nations to displace foreign exploitation with domestic exploitation of their fishery resources. Copes (1983) reports that for the period 1976 to 1979 catches by the principal distant-water fishing nations declined significantly. Except for Japan, declines ranged from 10% (USSR) to 30% (Portugal). Japan's catch declined by less than 1%. During the same period, nations with large coastal fishery resources experienced significant increases in their catches. Catch in the United States increased by 15%, in Canada by 21%, and in Iceland by 67%.

In addition to excluding or controlling foreign exploitation, many coastal nations also control domestic exploitation of the fishery resources in their zones of extended jurisdiction. By the end of 1981, 26 major fisheries in the U.S. were being governed by fisheries management plans and twelve of these plans include regulation of domestic fishing vessels (U.S. Department of Commerce 1982).

The costs of controlling fisheries exploitation appear high relative to potential benefits. Following the United States's extension of jurisdiction over marine fisheries (the Magnuson Fishery Conservation and Management Act of 1976), annual federal expenditures on fisheries law enforcement more than doubled and, in 1983, exceeded $100 million (Sutinen and Hennessey forthcoming).[4] Additional transactions costs (which include the costs of administration, data collection, and research)

---

[4]Bell and Surdi (1979) estimate the U. S. government spent nearly $280 million during the fiscal years 1977 and 1978 implementing the Fisheries Conservation and Management Act of 1976. Federal expenditures on fisheries law enforcement are projected to reach $125 million (in 1982 dollars) by the late 1980s (Sutinen and Hennessey, forthcoming).

may approach $200 million annually (Norton 1983). Potential benefits from fisheries, in the form of economic rent, may range from $200 million to $500 million annually.[5]

Few enforcement programs result in perfect compliance and the U.S. fisheries law enforcement program is no exception. In the U.S. fishery conservation zone, approximately 20% of the foreign and 4% of the domestic vessels boarded have been cited for violating fisheries law.[6] For the years 1979 through 1981 there were averages of 440 civil penalty actions, 16 seizures of (foreign) vessels, and fines collected of $2.5 million per year (Peterson 1982). There also is reason to believe that significant violations are going undetected (e.g., in one of the largest U.S. fisheries, effective monitoring of Japanese pollock catches is not possible). For these and other reasons, the economic rent being realized likely is nowhere near its potential.

Given the high costs and modest benefits, it is reasonable to ask how much fisheries law enforcement is desirable. An even more fundamental question is how fisheries law and regulations should be modified to reflect costly, imperfect enforcement. The following sections develop a framework for examining these and related issues.

II. A MODEL OF FISHERIES LAW ENFORCEMENT

In this section we construct a formal model to explain how firm and aggregate catch rates are determined when enforcement of regulations are costly and imperfect. The fundamental problem in fisheries management, of course, is to curb the tendency under open access to overexploit the resource stock. Starting at any initial stock size, a means must be found to reduce open-access catch rates and, as the stocks recover over time, to let catch rates increase (but by less than they would under open access at each stock size). The most commonly used methods of curbing the tendency to overexploit fish stocks include quotas (aggregate and individual), gear restrictions, area and seasonal closures, and other forms of limiting effective effort applied to the fishery.

A. Supply Conditions

We begin our analysis by assuming that whatever means are applied to reduce catch rates, they can be interpreted to mean that any catch above the level $\bar{q}$ is illegal. Perhaps the simplest regulation to imagine here is that of nontransferable individual quotas in a single species fishery. In such a case, the amount of the individual firm's catch above its quota, $q_i - \bar{q}_i$, is illegal.

If detected and convicted, a penalty fee[7] is imposed on the firm in an amount given by $f = f(q_i - \bar{q}_i)$, where

$$f \begin{cases} > 0, & \text{if } q_i > \bar{q}_i \\ = 0, & \text{otherwise,} \end{cases}$$

and

$$\frac{\partial f}{\partial q} = f_q > 0, \qquad \frac{\partial^2 f}{\partial q^2} = f_{qq} > 0, \qquad \forall q_i > \bar{q}_i.$$

We allow for discontinuity at $q_i = \bar{q}_i$, but assume $f(\cdot)$ is continuous and differentiable for all $q_i > \bar{q}_i$.[8] The penalty fee has a finite upper bound, f_i, no greater than the assets of the firm (restricted by bankruptcy laws). Each firm is assumed to face the same penalty fee schedule, except the upper bound on the fee can vary across firms.

An individual firm's profits before penalties are given by

$$\pi^i(q_i, x) = pq_i - c^i(q_i, x),$$

---

[5]Estimated by Robert R. Nathan Associates, cited in Eckert (1979, 51). These estimates are for 1985 in 1972 dollars and, therefore, not strictly comparable to the cost estimates. We use them regardless since only the orders of magnitude are essential to the discussion.

[6]These are approximate averages for six-month periods from March 1977, through February 1980 (U.S. Coast Guard 1977–80).

[7]A referee observes that Smith (1969) was the first to incorporate a penalty function in a formal analysis of fishery regulation. While unique in that respect, Smith's analysis is deterministic and implicitly assumes perfect, costless enforcement.

[8]In practice, $f(\cdot)$ may have several discontinuities and even have a negative slope in places. Over large ranges of q_i, however, actual assessed penalties rise with more serious infractions. Therefore, $f(\cdot)$ is at least an approximation of penalty fee schedules employed in practice.

where $\pi_q^i > 0$ in the relevant range

$$\pi_{qq}^i < 0, \pi_x^i > 0, \pi_{qx}^i > 0$$

(subscripts other than i and j denote partial derivatives), p is the price of fish, x is the size of fish stock and $c^i(\cdot)$ is the cost function. We assume firms are price takers.

In an imperfect law enforcement system, not every violator is detected and convicted. Let the probability of detection and conviction be given by θ. To simplify the analysis, we assume all firms face the same θ.[9] If detected and convicted of a violation, a firm's profits will be $\pi^i(q_i, x) - f(q_i - \bar{q})$; and if not, $\pi^i(q_i, x)$. Therefore, expected profits are

$$\theta\left[\pi^i(q_i, x) - f(q_i - \bar{q}_i)\right] + (1 - \theta)\pi^i(q_i, x).$$

Assuming firms are risk neutral and maximize expected profits, each q_i is determined by the first-order condition

$$\pi_q^i(q_i, x) \geq \theta f_q(q_i - \bar{q}_i). \qquad [1]$$

The inequality is due to the discontinuity in $f(\cdot)$ at $q_i = \bar{q}_i$.

The solution to [1] for one form of the marginal penalty schedule, f_q, is shown in Figure 1, panel (a). For a given stock size, x, the firm sets its catch rate at q'_i, in excess of its quota, where marginal profits equal the expected marginal penalty. If there were no penalty for fishing beyond $\bar{q}_i$, or if there were no chance of being detected and convicted (i.e., either $f = 0$ or $\theta = 0$), the firm would set its catch at q_i^o, the open-access catch rate. If the expected marginal penalty schedule lies above the marginal profit schedule for all $q_i > \bar{q}_i$, as illustrated in panel (b), the firm's optimum catch rate equals its quota.

Firms with no quota have an expected marginal penalty schedule that emanates from the vertical axis. If their expected marginal penalty schedule lies everywhere above their marginal profit schedule, they have no incentive to enter this fishery. If, however, their expected marginal penalty schedule begins below their marginal profit schedule, there is an expected net gain to illegally entering the fishery.

The first-order condition, [1], can be solved for a firm's catch rate as

$$q_i = q_i(\theta, x, \bar{q}_i). \qquad [2]$$

The catch rate also depends on price, production cost parameters, and parameters of the penalty fee schedule. These other arguments in [2] are suppressed for notational simplicity. The properties of [2] important for the present discussion are

$$\partial q_i/\partial \theta < 0, \partial q_i/\partial x > 0 \text{ and } \partial q_i/\partial \bar{q}_i \geq 0.[10]$$

In other words, an increase in the probability of detection and conviction decreases (or leaves unchanged) a firm's catch rate as the expected marginal penalty schedule, θf_q, becomes steeper (refer to Figure 1). An increase in the stock size shifts up the marginal profit schedule, π_q, and increases (or leaves unchanged) a firm's catch rate. An increase in the quota shifts the expected marginal penalty schedule to the right and increases a firm's catch rate (so long as the initial

$$q_i < q_i^o).$$

Aggregating [2] for all firms in the fishery yields the aggregate catch function

$$q = q(\theta, x, \bar{q}). \qquad [3]$$

where q is aggregate catch and $\bar{q}$ is the total quota (i.e.,

$$\bar{q} = \sum_{i=1}^{N} \bar{q}_i$$

---

[9]In practice, some firms stand a greater or lesser chance than others of being detected and convicted for a given offense. Relaxing our simplifying assumption, while making the analysis more cumbersome, would not alter our principal results.

[10]To derive these signs, three cases must be considered. The first case is where the equality in [1] holds and $\hat{q} < q_i^o$ (panel (a) in Figure 1). Differentiating [1] yields $\partial q_i/\partial \theta < 0, \partial q_i/\partial x > 0$ and $\partial q_i/\partial \hat{q}_i > 0$. The second case is where the inequality in [1] holds and $\bar{q}_i < q_i^o$ (panel (b) in Figure 1). In this case $q_i = \bar{q}_i$ and $\partial q_i/\partial \theta = 0, \partial q_i/\partial x = 0, \partial q_i/\partial \bar{q}_i > 0$. The third case is where either the equality or the inequality in [1] holds and $\hat{q}i > q_i^o$. Since $q_i = q_i^o, \partial q_i/\partial \theta = 0, \partial q_i/\partial x > 0$ and $\partial q_i/\alpha \bar{q}_i = 0$.

for an exogenously determined set of the $\bar{q}_i$, where N is the number of firms allocated quotas). We assume there is sufficient heterogeneity across firms in the fishery to allow [3] to be everywhere continuous. Therefore inverse forms of [3] exist and

$$\partial q/\partial\theta < 0, \ \partial q/\partial x > 0, \ \partial q/\partial\bar{q} > 0.\text{[11]}$$

There also is a (nearly) continuous distribution of firm catch rates for each $(\theta,\bar{q})$ combination, with firms producing below, at, and above their individual quota levels.

B. Enforcement Costs

Detecting and convicting firms violating their quotas require costly inputs (e.g., aircraft, patrol boats, on-board and on-shore observers, judicial personnel). Let the quantities of such inputs be represented by a vector k which has an associated vector of unit prices w. The probability of detecting and convicting violators is assumed to depend on the inputs k, i.e., $\theta = \theta(k)$ where

$$\partial\theta/\partial k_j > 0, \ \partial^2\theta/\partial k_j^2 < 0.\text{[12]}$$

Assuming the least-cost combination of k is chosen for each level of θ, there is an enforcement cost function, $e(\theta)$, where

$$\partial e/\partial\theta > 0, \ \partial^2 e/\partial\theta^2 > 0.\text{[13]}$$

Using an inverse form of

[3], $\theta = q^{-1}(q,x,\bar{q})$,

enforcement costs can be represented by

$$e\big(q^{-1}(q,x,\bar{q})\big) = E(q,x,\bar{q}), \tag{4}$$

where

$$\partial E/\partial q < 0, \ \partial E/\partial x > 0 \text{ and } \partial E/\partial\bar{q} > 0.\text{[14]}$$

That is, a reduction in the catch level (below the open-access level and for a given stock size and quota) requires an increase in enforcement costs. And an increase in the fish stock or quota requires greater enforcement costs to achieve a given catch level.[15]

Before proceeding we should note that without an upper bound on the penalty fee, f, the least-cost θ is arbitrarily close to zero.[16] A θ near zero implies enforcement costs also near zero. We assume above, however, an upper bound on f which, if low enough, results in a θ not near zero and significantly large enforcement costs.

The size of the quota, $\bar{q}$, also affects enforcement costs, and the least cost quota depends on the form of the marginal penalty schedule, f_q. If f_q takes the form shown in panel *(a)* of Figure 1, the least-cost quota is zero (effectively a tax on all catches). If f_q

---

[11]To derive these signs an aggregate form of [1] is used to allow for price effects: $p(q)-c_q(q,x) = \theta f_q(q-\bar{q})$. The results follow immediately.

[12]Other factors, such as the number of fishing vessels would also affect the level of θ, but we assume all other factors constant.

[13]In practice, fisheries law enforcement programs are operated by bureaucratic entities which may not be minimizing costs in the manner specified here. Since our analysis is primarily normative, however, it is appropriate to use this specification.

[14]Using the properties of [3], the derivation of these properties is straightforward:

$$\frac{\partial E}{\partial q} = \frac{\partial e}{\partial q}\cdot\frac{\partial\theta}{\partial q}, \quad \text{where} \quad \frac{\partial\theta}{\partial q} = \frac{1}{(\partial q/\partial\theta)} < 0;$$

$$\frac{\partial E}{\partial x} = \frac{\partial e}{\partial\theta}\cdot\frac{\partial\theta}{\partial x}, \quad \text{where} \quad \frac{\partial\theta}{\partial x} = -\frac{(\partial q/\partial x)}{(\partial q/\partial\theta)} > 0$$

from $dq = \left(\dfrac{\partial q}{\partial\theta}\right)d\theta + \left(\dfrac{\partial q}{\partial x}\right)dx = 0$;

$$\frac{\partial E}{\partial\bar{q}} = \frac{\partial e}{\partial\theta}\cdot\frac{\partial\theta}{\partial\bar{q}}, \quad \text{where} \quad \frac{\partial\theta}{\partial\bar{q}} = -\frac{(\partial q/\partial x)}{(\partial q/\partial\theta)} > 0$$

from $dq = \left(\dfrac{\partial q}{\partial\theta}\right)d\theta + \left(\dfrac{\partial q}{\partial\bar{q}}\right)d\bar{q} = 0$.

[15]This specification of the enforcement cost function ignores some possibly important aspects of enforcement practices. For example, it is conceivable that complete closure of the fishery (i.e., at the lower bound on $q-\bar{q}$) costs less to enforce than permitting some positive amount of fishing. There also may be economies of scale in enforcement. That is, the marginal enforcement costs may decrease over some range of $q-\bar{q}$. The implications of these alternative specifications are not considered here.

[16]That is, to minimize costs, θ is set arbitrarily close to zero and the penalty fee is set sufficiently high so that the expected fee, θf, induces the desired catch rate.

takes the form shown in panel *(b)* of Figure 1, the least-cost quota is at or below the desired catch rate. The present model does not explicitly recognize the several costly steps involved from detection of a violation through conviction. Therefore, we must assume the form of f_q and the value of $\bar{q}$ are exogenously determined. We leave for later analysis the determination of the least-cost form of f_q. However, our conjecture is that when multiple costly steps are involved, the least-cost form of f_q is similar to that shown in panel *(b)*,

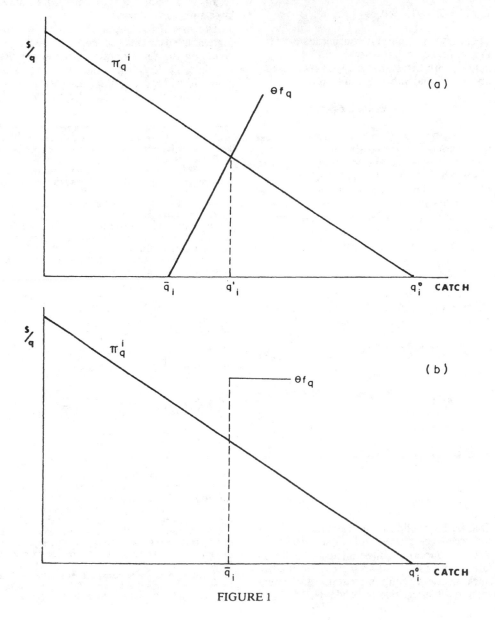

FIGURE 1

Figure 1, since by setting the quota equal to the desired catch rate the legal and other related costs are minimized.

III. OPTIMAL POLICY

We examine in this section how optimal management policies are affected by costly, imperfect enforcement. Optimal policies are based on the usual criterion of maximizing the discounted sum of net social benefits. Net social benefits in each period are given by

$$\int_0^q p(s)\,ds - c(q,x) - E(q,x),$$

where $p(q)$ is the inverse demand function, $c(q,x)$ is the aggregate catch cost function (with $c_q > 0$ and $c_x < 0$) and $E(q,x)$ is the enforcement cost function. We assume that social benefits can be appropriately measured by the area under the ordinary demand curve. The aggregate cost function depends on the fixed set of quotas, and does not include penalty fees. Penalty fees are excluded from the criterion since they are transfers from fishing firms to the general treasury. Since the quota allocation is assumed exogenously determined, $\bar{q}$ is suppressed as an argument in the enforcement cost function.

The stock dynamics are assumed given by the standard differential equation

$$\dot{x} = h(x) - q, \qquad [5]$$

where $x = x(t)$, $q = (t)$, and $h(x)$ is the natural growth rate.[17]

Optimal policies are found by maximizing

$$\int_0^\infty \left[\int_0^q p(s)\,ds - c(q,x) - E(q,x) \right] e^{-\delta t}\,dt$$

subject to [5], where the social discount rate is represented by δ. The first-order conditions for this problem (assuming an interior solution) are

$$p - c_q - E_q - \lambda = 0, \qquad [6]$$

$$\lambda = c_x + E_x + \lambda(\delta - h_x). \qquad [7]$$

where $\lambda = \lambda(t)$ is the dynamic multiplier.

Setting $\lambda = 0$ and solving [6] and [7] yields

$$\delta - h_{x^{**}} = \frac{-(c_{x^{**}} + E_{x^{**}})}{p^{**} - (c_{q^{**}} + E_{q^{**}})} \qquad [8]$$

which, together with [5] when $\dot{x} = 0$, determines the steady-state optimal stock size, x^{**}, and the optimal catch rate, q^{**}, and resulting price, p^{**}.

Assuming costless and perfect enforcement, i.e., where catch rates are perfectly controlled at zero cost, the condition for optimality is

$$\delta - h_{x^*} = \frac{-c_{x^*}}{p^* - c_{q^*}} \qquad [9]$$

where x^* is the optimal stock size, q^* the optimal catch rate, and p^* the resulting price. By comparing [8] and [9] we show in the appendix that the presence of costly, imperfect enforcement results in a smaller optimal stock size than otherwise, i.e., $x^{**} < x^*$. Similarly, higher enforcement costs result in a lower optimal stock size.

The economic reasoning behind this result is as follows. At the open access equilibrium, i.e., with no enforcement, enforcement cost is nil. Moving the fishery away from the open-access equilibrium towards a larger stock size increases enforcement costs and management benefits (net consumers and producers surplus). For the interior solution assumed here, marginal enforcement costs increase and marginal management benefits decrease as the steady-state stock size is increased. The optimal stock size, x^{**}, is where marginal management benefits equal marginal enforcement costs.[18] With costless, perfect en-

---

[17] $h(\underline{x})$ is strictly concave with $h(\underline{x}) = h(\bar{x}) = 0$, $\underline{x} \geq 0$, $x < \bar{x}$, where $\underline{x}$ and $\bar{x}$ are lower and upper bounds on x, respectively.

[18] To see this result combine [6] and [7], for $\lambda = 0$, and rearrange to become

$$ph_x - c_q h_x - c_x) - \lambda\delta = E_q h_x + E_x,$$

which is equivalent to

$$d\left[\int_0^{h(x)} p(s)\,ds - c(h(x),x) - \delta V(x) \right]$$

$$/dx = dE(h(x),x)/dx, \text{[i]}$$

where $V(x) = \max \int_{\underline{x}}^{\bar{x}} e^{-\delta t} [\int_0^q p(s)\,ds - c(q,x) - E(q,$

forcement, the optimal stock size, x^*, is where marginal management benefits equal zero. Hence, the result $x^{**} < x^*$.

To compare catch rates for costless, perfect enforcement with costly, imperfect enforcement, we must specify whether the stock sizes are above or below the maximum sustainable yield (MSY) level.[19] The results on catch rates can be summarized as follows:

$$x^{**} < x^* < x_{MSY} \Rightarrow q^{**} < q^*$$

$$x_{MSY} < x^{**} < x^* \Rightarrow q^{**} > q^*$$

$$x^{**} < x_{MSY} < x^* \Rightarrow q^{**} \doteq q^*$$

In addition, for an interior solution, the optimal stock sizes for both optimal cases are clearly greater than the open-access stock size.

These results are illustrated in Figure 2. The demand curve, D, intersects the backward-bending, open-access supply curve, S_{OA}, at "a", resulting in a low equilibrium catch rate, q^o, and stock size, x^o. With costless, perfect enforcement, the social marginal cost schedule, SMC^*, intersects the demand curve at "b", resulting in a higher catch rate, q^*, and stock size, x^*. Costly, imperfect enforcement results in a lower social marginal cost schedule, SMC^{**}, lying everywhere below SMC^* for corresponding values of q and x.[20] The lower SMC^{**} intersects the demand curve at "c", resulting in an optimal catch rate $q^{**} < q^*$ and stock size $x^{**} < x^*$. Were the demand schedule to intersect the SMC schedules below where they cross, at "d", then $q^{**} > q^*$. Therefore, $q^{**} \gtreqless q^*$ as the demand schedule intersects the two SMC schedules below/at/above where they cross. Regardless of the height of the demand schedule, however, $x^{**} < x^*$ always.

IV. CONCLUDING REMARKS

Both historical evidence and logical reasoning demonstrate that enforcement costs are a major determinant of regulatory policy for nonexclusive resources. We have shown that costly, imperfect enforcement of fisheries law significantly affects the behavior of fishing firms and optimal fisheries manage-

ment policy. For the case of individual, nontransferable quotas, we show how firm and aggregate catch rates (and the distribution of violations) are determined. The principal result is that costly, imperfect enforcement results in an optimal, steady-state stock size that lies between the smaller open-access stock size and the larger stock size where enforcement is assumed costless and perfect. The optimal catch rate for costly, imperfect enforcement can be greater or less than the costless, perfect enforcement optimum, depending on the relative height of the demand schedule.

The analytical framework in this paper contains only the basic principles of fisheries management and useful extensions can follow several directions. For example, our model examines only nontransferable quotas. A preliminary analysis of transferable quotas reveals that the equivalence of transferable quotas and taxes (a result obtained by Maloney and Pearce [1979] and Clark [1980 and 1982]) may not hold when enforcement is imperfect. The analysis of taxes, where collection is both costly and imperfect, appears to more closely parallel the case of nontransferable quotas treated in this paper. The more common forms of regulation—simple aggregate quotas, gear restrictions, and area and seasonal closures—also require further study. Since gear restrictions can increase the costs of production, they are typically viewed as inefficient regulatory measures (Crutchfield 1961). However, since gear restrictions may be less costly to enforce than other measures, gear restrictions could turn out to be the most efficient method of regulation when enforcement costs are taken into account.

There also remains the issue of how optimal policies leading to the steady state are af-

---

$x)]\,dt$ subject to [5], and $q = h(x)$ in the steady-state. The LHS of expression [i] represents marginal management benefits, i.e., the change in the sum of consumers and producers surplus, net of the opportunity cost of the resource stock, due to a change in the steady-state stock size. The RHS of expression [i] similarly represents marginal enforcement costs.

[19]The maximum sustainable yield occurs at the stock size x_{MSY} where $h_x(x_{MSY}) = 0$.

[20]The SMC^{**} is lower, of course, because $E_q < 0$.

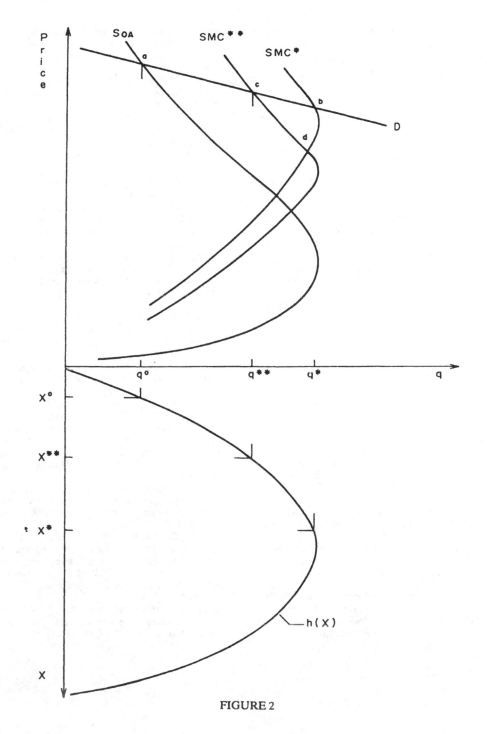

FIGURE 2

fected by costly, imperfect enforcement, and how enforcement costs behave along the optimal policy path. These and the issue of stability are left for future work.

References

Andersen, P. 1982. "Commercial Fisheries under Price Uncertainty." *Journal of Environmental Economics and Management* 9 (Mar.):11–28.

Anderson, T., and Hill, P. 1975. "The Evolution of Property Rights: A Study of the American West." *Journal of Law and Economics* 12 (Apr.):163–79.

Becker, G. S. 1968. "Crime and Punishment: An Economic Approach." *Journal of Political Economy* 86 (Mar.):169–217.

Bell, F., and Surdi, R. 1979. "An Economic Benefit-Cost Analysis of the Fishery Conservation and Management Act of 1976." *Mimeo*. Economics Department, Florida State University.

Bradley, P. G. 1974. "Marine Oil Spills: A Problem in Environmental Management." *Natural Resource Journal* 17 (July):337–59.

Cheung, S. 1970. "The Structure of a Contract and the Theory of a Non-Exclusive Resource." *Journal of Law and Economics* 13 (Apr.):49–70.

Clark, C. W. 1980. "Towards a Predictive Model for the Economic Regulation of Commercial Fisheries." *Canadian Journal of Fisheries and Aquatic Sciences* 37 (Nov.):1111–29.

Clark, C. W. 1982. "Models of Fishery Regulation." In *Essays in the Economics of Renewable Resources*, ed. L. J. Mirman and D. F. Spulber. Amsterdam: North Holland.

Clarkson, K. W. 1974. "International Law, Seabeds Policy and Ocean Resource Development." *Journal of Law and Economics* 17 (Apr.):117–42.

Copes, P. 1983. "Extended Jurisdiction and Its Effect on World Trade in Seafood." In *Proceedings of the International Seafood Trade Conference*. Alaska Sea Grant Report No. 83–2.

Crutchfield, J. A. 1961. "An Economic Evaluation of Alternative Methods of Fishery Regulation." *Journal of Law and Economics* 4 (Apr.):131–43.

Demsetz, H. 1967. "Toward a Theory of Property Rights." *American Economic Review* 57 (May):347–59.

Eckert, R. D. 1979. *The Enclosure of Ocean Resources: Economics and the Law of the Sea.* Stanford: Hoover Institution Press.

Maloney, D. G., and Pearse, P. H. 1979. "Quantitative Rights as an Instrument for Regulating Commercial Fisheries." *Journal of the Fisheries Research Board of Canada* 36 (July):859–66.

Norton, V. 1983. "Problems and Opportunities." In *The U.S. Fishing Industry and Regulatory Reform,* ed. T. M. Hennessey. Wakefield, R.I.: Times Press.

Peterson, A. E., Jr. 1982. "Testimony before the Subcommittee on Coast Guard and Navigation Committee on Merchant Marine and Fisheries of the U.S. House of Representatives." Washington, D.C.

Scott, A. 1979. "Development of Economic Theory on Fisheries Regulation." *Journal of the Fisheries Research Board of Canada* 36 (July):725–41.

Smith, V. L. 1969. "On Models of Commercial Fishing." *Journal of Political Economy* 77 (Mar./Apr.):181–98.

Sutinen, J. G., and Hennessey, T. M. Forthcoming. "Enforcement: The Neglected Element in Fisheries Management." In *Natural Resource Policy and Management: Essays in Honor of James A. Crutchfield,* ed. E. Miles, R. Pealy, and R. Stokes. Seattle: University of Washington Press.

U.S. Department of Commerce. 1982. "Calendar Year 1981 Report on the Implementation of the Magnuson Fishery Conservation and Management Act of 1976." Washington, D.C.

U.S. Coast Guard. 1977–1980. "Semiannual Reports to Congress on Degree and Extent of Known Compliance with the Fishery Conservation and Management Act of 1976." *Unpublished*. Department of Transportation, Washington, D.C.

Viscusi, W. K., and Zeckhauser, R. J. 1979. "Optimal Standards with Incomplete Enforcement." *Public Policy* 27(4) (Fall):437–56.

APPENDIX

In this appendix we formally derive the effects enforcement costs have on the fishery system. For this purpose we rewrite the Hamiltonian with a multiplicative shift parameter, ϕ, before the enforcement cost function:

$$H = \int_0^q p(s)\,ds - c(q,x)$$

$$- \phi E(q,x) + \lambda[h(x) - q]$$

where $\phi = 0$ in the case of no enforcement costs and $\phi = 1$ when enforcement costs are accounted for. The first-order conditions in the steady-state equilibrium are:

$$p - c_q - \phi E_q - \lambda = 0 \qquad [A.1]$$

$$\dot{\lambda} = \delta\lambda + c_x + \phi E_x - \lambda h_x = 0 \qquad [A.2]$$

$$\dot{x} = h(x) - q = 0 \qquad [A.3]$$

totally differentiating [A.1–3] with respect to ϕ yields the system of equations:

$$[D]\begin{bmatrix} dq/d\phi \\ dx/d\phi \\ d\lambda/d\phi \end{bmatrix} = \begin{bmatrix} E_q \\ -E_x \\ 0 \end{bmatrix}$$

where

$$[D] = \begin{bmatrix} (p_q - c_{qq} - \phi E_{qq}) & (c_{qx} + \phi E_{qx}) & (-1) \\ (c_{qx} + \phi E_{qx}) & (c_{xx} + \phi E_{xx} - h_{xx}) & (\delta - h_x) \\ (-1) & (h_x) & 0 \end{bmatrix}$$

The sufficiency condition that H be concave requires $|D| < 0$. Solving this system then yields:

$$\frac{dx}{d\phi} < 0 \quad \text{and} \quad \frac{dq}{d\phi} - 0 \quad \text{as} \quad h_x - 0.$$

The results hold for all values of ϕ. Therefore, the effects of both the presence and an increase in enforcement costs are the same.

[39]

Optimal Governing Instrument, Operation Level, and Enforcement in Natural Resource Regulation: The Case of the Fishery

Lee G. Anderson and Dwight R. Lee

Most regulation studies have used industry output or inputs as the control variable(s), but these are only indirectly controlled by government action through its choice of governing instrument, enforcement procedure, and penalty structure and the operational level of each. A model is developed which demonstrates how profit-maximizing firms will react to these control variables taking into account the benefits (extra production) and costs (possible penalties) of noncompliance and the ability to avoid detection of noncompliance. The optimal operation level for two sets of control variables is derived and discussed.

Key words: common property, enforcement, fisheries, natural resources, regulation.

There is extensive literature on the optimal regulation of open-access resources, such as airsheds, watersheds, and fisheries. The dominant themes are (*a*) unregulated market behavior results in socially excessive exploitation of open-access resources, and (*b*) government regulations can adjust market incentives in ways that will generate socially efficient outcomes. The first theme comes from the most developed and accepted part of the literature. That markets fail under a variety of circumstances, including open-access settings, is acknowledged without serious dissent by economists.

However, when attention turns to regulatory policy designed to correct open-access market failure, the theory is less developed. There are at least three areas where it can be improved. First, the actual regulation process has not been completely modeled. The question normally asked concerns the optimal output from a fishery or the optimal amount of

pollutant released into an air or watershed. Most models have been built with these outputs as the control variables. In actuality, the regulatory control variables are the choice of governing instrument (i.e., taxes, direct controls, etc.), the enforcement procedure, and penalty structure for noncompliance and the operation level of each. For example, if tax is the governing instrument, the operation level is the size of the tax; if incarceration is the penalty, the jail term is the operation level. The output of the regulated industry is only indirectly controlled through firms' reactions to these regulatory control variables. While the choice of control variable may be of little consequence as far as pure economic theory is concerned, it does have a significant impact on the ability to draw policy-relevant implications.

A second incomplete area of theory development is evasion activities by those being regulated. This is important because of the actual resources used, and hence the economic waste which occurs, and also because of the way in which industry output behavior is actually modified with evasion. The final weakness of existing theory is the failure to deal adequately with implementation and enforcement activities. Regulation programs will have little or no effect if they are not adequately imple-

Lee G. Anderson is a professor of economics and marine studies, University of Delaware; Dwight R. Lee is Ramsey Professor of Economics, University of Georgia.

Research was sponsored by the National Science Foundation Grant No. SES 8308968, Division of Social and Economic Sciences, Applied Research on Public Regulation.

Helpful comments from Adi Ben-Israel, Richard C. Bishop, and Scott Milliman are appreciated, but the usual disclaimer applies.

Review was coordinated by Peter Berck, associate editor.

mented and enforced, but since doing so uses resources and affects the degree to which desired behavior modifications actually occur and the amount of resources devoted to evasion, it is not possible to define properly an optimal policy without considering these direct and indirect opportunity costs. The final two areas are of more than mere theoretical interest. They are at the very heart of both fisheries and environmental regulations.

The purpose of this paper is to add to the theory of regulatory policy in these three less-developed areas. The discussion will be in terms of a fishery, but many of the results will be general. From the analysis, it can be shown that enforcement is far more important than might be evidenced by the treatment it receives in the literature, and that given certain types of enforcement, the distinction between what have been called market and nonmarket controls becomes somewhat blurred. Although there are many stochastic elements in this problem (i.e., the nature of the biological growth function, the chances of being apprehended for noncompliance, etc.) the analysis will be deterministic in order to keep the model from being unduly complex.

A Model of Government Regulation

After a brief discussion which provides a frame of reference for the analysis, this section contains a detailed model of government regulation. The discussion focuses on the nature of an equilibrium in a regulated fishery, the determination of the optimum regulated equilibrium, and the dynamics of achieving the optimum.

The Problem

To set the stage, a brief static analysis of the open-access fisheries problem will be presented. Figure 1a contains the now familiar total sustainable revenue and cost curves of an open-access fishery where the former has been drawn assuming a fixed price of fish and a pure compensation model of biological growth. (See Gordon, Clark 1976, and Anderson 1986). Fishing effort, which for purposes here is best thought of as nominal days fished by standardized vessels, is measured on the horizontal axis. The actual measurement of effort, however, is a nontrivial problem on both theoretical and applied bases and is a cru-

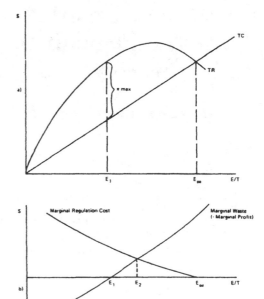

Figure 1. Open-access versus optimal fisheries utilization

cial part of any management problem (see Rothschild; Clark 1985, chap. 2). In order to focus on the issue at hand, however, we will abstract from these problems in the theoretical construction of the model. Under open access, individuals will enter the fishery up to E_{oa}, where the total revenue equals total cost. At lower levels of effort, individual boats will be making economic profits, and with no entry controls this incentive will cause the fleet to expand until all rent is dissipated. The static economically efficient operation point is E_1, where the marginal value of production equals marginal opportunity cost of effort.

This information is displayed in figure 1b in a manner more suitable to discussions of regulation. The curve labeled marginal waste is the negative of marginal profit as effort expands. Looked at in reverse, it shows the marginal benefit from reducing effort from the open-

680 *August 1986*

Amer. J. Agr. Econ.

access level. The marginal benefit of effort reduction is positive for reductions over the range between E_{oa} and E_1. Therefore, with costless regulation it would pay to decrease effort to E_1. However, with a marginal regulation cost curve as depicted in the diagram, the optimal amount of effort is E_2, which is greater than the standard static maximum economic yield level, E_1 (see Andersen and Sutinen, and Sutinen and Andersen). If the biological growth function exhibits depensation, the sustainable yield curve can bend forward on itself toward the origin, in which case the open-access level of effort can be less than the optimal amount (see Clark 1976, p. 34). The path toward the optimum, however, requires a temporary decrease in effort to allow the fish stock to grow. Only at the larger stock size is the higher level of effort optimal. The dynamic interpretation of the static model pictured in figure 1, applies for both compensation and dispensation, however. At any point in time, the optimal amount of regulated effort will be higher when enforcement costs are considered. For the remainder of this discussion, it will prove useful to use the optimal level of E with and without enforcement costs, E_2 and E_1, respectively, as benchmarks for comparison purposes.

The intricacies of optimal regulation are more complex than this, however. As indicated above, it is necessary to select the governing instrument (i.e., taxes, input specifications, etc.) the enforcement or monitoring procedure (dockside checks, at sea boardings, etc.), and the penalty structure (fines, gear confiscation, jail terms) by which the fishery will be regulated and then to determine the operation level of each. The governing instrument, enforcement procedure, and the penalty structure may be thought of as the fixed regulatory inputs, and operation level of each as variable inputs. The remainder of this section will discuss regulation equilibria, optimum regulation equilibrium, and optimal dynamic regulation assuming that effort production rights are the governing instrument, that there is only one practical enforcement procedure, and that fines are used to penalize noncompliance. Effort production rights fall into that classification of policy instruments that Baumol and Oates (p. 218) call market processes as opposed to direct controls. In the fisheries literature these two classes are usually called limited entry and nonlimited entry, and the former are praised for their ability to

achieve economic efficiency. In the next section the analysis will be repeated for a non-limited entry governing instrument. In order to keep the models tractable, it will be necessary to use stylized one-dimensional versions of each regulation class. For a more complete discussion of the various types of fisheries regulation, see Crutchfield, Pearce, Ginter and Rettig, and Beddington and Rettig.

Regulated Equilibria

Assume that a fishery is controlled by issuing a limited number of effort production rights, $\bar{E}$, which allow the owner to produce a specified amount of fishing effort each year; and there is a fine, K, for all units of E that are produced without such a right and are detected. In these circumstances, individuals may be motivated to produce unauthorized effort depending upon the profitability of fishing, the amount of the fine, and the chances of being detected. While it is rather cynical to assert that all individuals will base their compliance on the pure net revenue effects of their actions, it would be naive to assume that none would. However, to focus on these issues, the first extreme will be used.

The amount of effort detected by the authorities, E^D, will be a function of what the fishermen do (i.e., produce effort, E, and engage in detection avoidance activities, A), and what the government does (i.e., issue effort production rights and engage in monitoring activities, m). This can be expressed as an aggregate detected effort function:

$$(1) \qquad E^D = E^D[\overset{(+)}{E}, \overset{(-)}{A}, \overset{(-)}{\bar{E}}, \overset{(+)}{m}].$$

The likely signs of the first derivatives are as indicated. Assume also that $E^D(E, A, \bar{E}, 0) = 0$ for all values of E, A, and $\bar{E}$.

To ensure the clarity of what follows, it will be useful to describe what is meant by detection avoidance activities. They make it more difficult for agencies to detect the actual amount of fishing. They can be anything from underreporting to subterfuges such as fishing or landing fish at night or the use of remote ports or fishing grounds. For simplicity, assume the activity can be measured in discrete units and produced at a constant cost. Although this assumption is contrary to these examples, it is innocuous and will not change the results (see Lee).

Modifying the analysis of an open-access

equilibrium (see Smith, Anderson 1976), the profit function for one of the assumed identical boats in the fishery can be represented as follows:

$$P\bar{f}(E, X)e - c(e) - K(e^D - \bar{e}) - C^a a,$$

where X is fish stock size; the lower case letters e, e^D, $\bar{e}$, and a represent actual effort, detected effort, production rights, and avoidance activities, respectively, at the vessel rather than an aggregated level; $c(e)$ is the vessel cost function for effort with the normal assumptions; and C^a is the unit cost of avoidance. Harvest is a function of effort and stock size, $f(E, X)$, and $\bar{f}(E, X)$ is the catch per unit of effort, $CPUE$, which at any point in time is considered a constant by individual industry participants. It is a function of E (i.e., Σe), and each vessel operator is a "$CPUE$ taker" in the same way that pure competitors are price takers. If the number of boats is sufficiently large, the amount of effort per individual boat has no effect on $CPUE$ for the fishery as a whole. See Anderson (1976, p. 183).

The first-order profit-maximizing conditions for the individual vessel are

$$(2) \quad P\bar{f}(E, X) - c' - K \frac{\partial e^D}{\partial e} = 0 \text{ and}$$

$$(3) \quad -K \frac{\partial e^D}{\partial a} - C^a \leq 0.$$

The solution to these equations will be the optimal effort and avoidance levels for each boat. Equation (2) states that each participant will produce effort as long as the marginal returns are greater than the sum of marginal harvesting costs and marginal fine payments, while equation (3) states that they will engage in avoidance activities as long as the marginal reduction in fine payments is larger than marginal avoidance costs. If the inequality in (3) holds (i.e., the former is less than the latter), the optimal amount of avoidance is zero.

These equations show the optimizing conditions for the "$CPUE$ taker" firms. To have a complete equilibrium, it is necessary that catch equal stock growth so that X remains constant and that Σe for all firms equals the E each uses in the $\bar{f}(E, X)$ expression. This is analogous to the normal competitive industry model, where the total output produced by each firm operating where $P_e = MC$ must be such that the market demand curve generates a price equal to P_e.

Note that such an equilibrium is different from the simple open-access equilibrium depicted in figure 1. Some of the rent is dissipated through avoidance and enforcement costs, but at the same time some rents are captured through fine revenues; and since the limited entry program and its penalties puts a constraint on entry, the operating vessels may earn rents as well.

Assuming that it is possible to derive the aggregate equations analogous to (2) and (3), the solution to those equations for any combination of $\bar{E}$, m, K, and X will yield the aggregate regulatory equilibrium levels of E and A. That is, for policy analysis, the levels of effort and avoidance activity will be functions of the size of the fish stock and of the variable regulatory inputs. Mathematically, this can be expressed as

$$(4) \quad E = E(\overset{(+)}{\bar{E}}, \overset{(-)}{m}, \overset{(-)}{K}, \overset{(+)}{X}) \text{ and}$$

$$(5) \quad A = A(\overset{(-)}{\bar{E}}, \overset{(\pm)}{m}, \overset{(+)}{K}, \overset{(+)}{X}).$$

While the exact nature of these equations will depend upon the types of functions in equations (2) and (3), it is likely that the signs of the first derivatives are as indicated. The presence of X in these equations is interesting because it implies that the effectiveness of the actual control variables depends upon the stock. For any combination of $\bar{E}$, m, and K, the higher the stock, the higher the returns to fishing and hence the higher the actual effort and the level of avoidance activity will be. Further it is likely that

$$E(\bar{E}, m, 0, X) = E(E, 0, K, X) = E_{oa};$$
$$E(\bar{E}, \infty, K, X) = \bar{E} \text{ if } K \geq K_1.$$

$$A(\bar{E}, 0, K, X) = A(\bar{E}, \infty, K, X)$$
$$= A(\bar{E}, m, 0, X) = 0.$$

That is, whatever $\bar{E}$ is selected, with a zero fine, effort will equal E_{oa}. Similarly, with no enforcement the fishery will operate at the open-access level, but if enforcement is high enough effort will equal the specified $\bar{E}$ when the fine, K, is higher than K_1, that level necessary to deter violations if they are sure to be detected. On the other hand, at both extremes of enforcement, avoidance will be zero. More generally, detection avoidance will be zero with no enforcement (noncompliance may be high, but there will be no incentive to distort the perceived amount of effort produced because there is no monitoring) and will initially

682   *August 1986*

Amer. J. Agr. Econ.

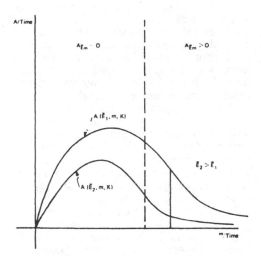

Figure 2.   Avoidance costs as a function of monitoring

increase with m (monitoring increases the chance of detection, and it may be productive to reduce the detected portion of produced effort) but will ultimately fall back to zero as monitoring increases (as monitoring continues to increase the chance of being caught becomes so high that the productivity of detection avoidance decreases). See figure 2.

Because of its importance in the discussion to follow, a small diversion to the cross-partial derivative of the A function is in order. As indicated above, A will decrease with increases in $\bar{E}$. It is plausible, but by no means necessary, that the total effect of an increase in $\bar{E}$ will be as indicated in figure 2. The A curve will shift down with increases in $\bar{E}$ and, in addition, E will reach its maximum and equal zero at lower levels of m. This means that for low levels of m the slope of $A(\bar{E}_2, m, K)$ will be less than that of $A(\bar{E}_1, m, K)$ or that $A_{\bar{E}m} < 0$. But at some point after A is declining with m, the slope of $A(\bar{E}_2, m, K)$ is greater (less negative) which means that $A_{\bar{E}m} > 0$. Note the slopes of the two curves at the points where they are intercepted by the solid line. The dotted line is meant to represent the level of m at which point both curves have the same slope. To the left, the slope of the bottom curve is lower (less positive or more negative) than the top curve, while to the right it is greater (less negative).

Optimum Regulated Equilibrium

In order to regulate for maximum economic efficiency, the government should choose that combination of $\bar{E}$, m, and K which maximizes the net present value of

$$Pf[E(\bar{E}, m, K, X), X] - C(E) \\ - A(\bar{E}, m, K, X)C^a - C^m m$$

subject to $\dot{X} = g(X) - f(E, X)$, $\bar{E} \geq 0$, $m \geq 0$, and $K \geq 0$, where $f(E, X)$ is the harvest function as opposed to the *CPUE* function used by the firms in their individual decision making. In addition, X is the fish stock size, $C(E)$ is the aggregate cost function, C^m is the unit cost of monitoring, and $g(X)$ is the biological growth function of the fish stock.

The current value Hamiltonian for this problem is

$$(6) \quad H = Pf(E, X) - C(E) - C^a A - C^m m \\ + \mu[g(X) - f(E, X)].$$

The adjoint variable μ represents the shadow price of fish in place. The necessary conditions for a maximum to the Hamiltonian are

$$(7) \quad H_{\bar{E}} = (MNB_E)E_{\bar{E}} - C^a A_{\bar{E}} \leq 0,$$

$$(8) \quad H_m = (MNB_E)E_m - C^a A_m - C^m \leq 0,$$

$$(9) \quad H_K = (MNB_E)E_K - C^a A_K \leq 0,$$

$$(10) \quad \frac{d\mu}{dt} = \mu(r - g_x) - (P - \mu)f_x \\ - [(MNB_E)E_X - C^a A_X].$$

The term MNB_E in equations (7), (8), and (9) is $[(P - \mu)f_E - C_E]$, which is the marginal net benefit of effort. To elaborate, P is the dockside price of fish and μ is the marginal value of a unit of stock in place. Alternatively, μ can be thought of as the user cost of removing a unit of stock. In any event, this means that the first term in MNB_E is the marginal net social value which results from producing a unit of effort, and the second term is the marginal cost of producing it. Note that MNB_E is the dynamic equivalent of the negative of the curve labeled marginal waste in figure 1.

The interpretation of these equations is that $\bar{E}$, m, and K should be selected each period such that marginal net contribution to efficiency of each is pushed to zero subject to the constraints on the control variables. Note that increases in m and K will decrease E, while increases in $\bar{E}$ will increase it. More on this point below. Equation (10) stipulates the rate at which the shadow price of fish must grow. The economic interpretation of this condition is deferred to the discussion of equation (10a) below.

To begin the discussion of the nature of the optimum regulation program, it will be useful to make several simplifying assumptions and then gradually relax them. First, consider the implications of the maximizing equations when a stationary solution exists and assume that government ignores the costs of avoidance activities in selecting the optimal combination of control variables and that the level of K is set exogenously. In this case, equation (9) drops out of the maximizing conditions and the A terms drop out of equations (7) and (8).

Since there is no compliance with no enforcement, (8) will hold as an inequality only in the special case where no regulation is optimal and m should equal zero, i.e., when the rewards for monitoring are never greater than the costs and it is best to leave the open-access fishery alone. However, if there is a policy-relevant solution, (8) will have to hold as an equality. Therefore, given that $E_m < 0$, it follows that when there is no A term, MNB_E must also be negative. Therefore, since $E_{\bar{E}} > 0$, it follows that (7) must hold as a strict inequality, which is to say that optimum $\bar{E}$ is zero. Reductions would technically further reduce the negative MNB_E, but negative rights to produce effort are not possible.

Therefore, with no avoidance costs and a fixed positive fine higher than K_1, (i.e., high enough to deter violations if they are sure to be detected), the optimal solution is to issue no rights but to be so lax in enforcement that the optimal amount of fishing occurs. The optimal fishing effort is greater than that conventionally considered to be efficient (that which equates $(P - \mu)f_E$ with C) since it has been shown from equation (8) that MNB_E is negative. In terms of the simple model in figure 1, this means that the optimal output of effort must be greater than E_1. See Lee for similar discussions with respect to pollution taxes. The intuition behind issuing no rights is explained by the fact that the direct cost of reducing rights is zero, and therefore substituting reductions in these rights for costly monitoring allows the same outcome to be realized at less cost. This reasoning is similar to that which supports some of the results in

Becker's seminal article on the economics of crime and punishment.[1]

Obviously, such an extreme policy will not be a relevant consideration for most governments, but the general principle may be valid when avoidance costs are small or are not considered relevant by the government. Because enforcement is the operational "variable" control for a regulatory agency, it may pay to be as harsh as is politically feasible with the level of the governing instrument. The point may be of little relevance, however, because things are more complex, and the nature of the solution can change drastically when the full social implications of avoidance are considered.

To continue with analysis under different assumptions, consider the case where K is a policy variable but avoidance costs are still ignored. In these circumstances, K should be set at infinity because this will further reduce the need for costly monitoring to achieve the actual optimal level of effort. In fact, if E approaches $\bar{E}$ when K approaches infinity, then the penalty will become dominant control variable and the optimal solution will have K approach infinity, m approach zero, and $\bar{E}$ approach E_1. If there is a socially acceptable upper limit on K, then the policy trade-off between monitoring and the number of rights issued remains relevant.

When considering the reverse case where avoidance costs are considered but where K is still fixed, it is helpful to examine first the case where $C^m = 0$. The efficient solution in this case requires that avoidance cost equal zero regardless of the size of $\bar{E}$. It cost nothing to increase monitoring, and so whatever level is necessary to completely eliminate socially wasteful avoidance is worthwhile. Therefore, $A(\bar{E}, m)$ will equal zero in the efficient solution. This equality allows m to be considered a function of $\bar{E}$, and so the objective function can be expressed as a function of only $\bar{E}$. Without redoing all the mathematics, expressions (7) and (8) collapse into a single equation in E which may be expressed as

$$(11) \qquad MNB_E \cdot E_{\bar{E}} = 0,$$

---

[1] Becker discusses the substitution of harsh, but socially costless, penalties for the costly activities that lead to capture and conviction. Given the assumptions thus far, however, the per unit penalty for unauthorized effort, K, is assumed to be beyond the control of the regulatory authority.

where the constraint $\bar{E} \geq 0$ is ignored for reasons that will become obvious below. Since $\bar{E}_E > 0$, the solution of (11), and hence optimum E, must be where MNB_E is zero. That is, with zero monitoring costs, the optimum industry output is that level of fishing effort that is conventionally considered efficient (i.e., the one period analog of E_1 in fig. 1). This will be achieved by setting $\bar{E} = E_1$ and increasing m until MNB_E equals zero. To see why $\bar{E}$ must equal E_1 in the optimum solution, note that actual effort, E, can be equal to E_1 with any level of effort rights, $\bar{E}$, less than E_1 if enforcement activities are set accordingly. That is, with lax enforcement actual E can be greater than $\bar{E}$. However, in those instances fishermen will be engaging in avoidance activities. Recall that for large levels of m, avoidance costs go down (see fig. 3) and E approaches $\bar{E}$ for increases in m. If $\bar{E}$ is not set appropriately, avoidance costs can still be positive when E is reduced to E_1. Therefore, optimal $\bar{E}$ must equal E_1 when $C^m = 0$.

It is possible to investigate the optimal $\bar{E}$ when $C^m > 0$ (but when the level of the fine is still fixed) by using equations (7) and (8) to consider the effect of marginal changes in C^m when $C^m = 0$. These are the equations in the necessary conditions for maximization of the Hamiltonian which state that the function is maximized at each point along the optimal trajectory by choice of the control variables, in this case $\bar{E}$ and m. If the maximization is actually to occur, then the second-order conditions associated with these equations must also hold. If they do, these equations can be used to define optimal $\bar{E}$ and m as functions of C^m. Both will hold as equalities: the first because C^m is initially equal to zero, and the second because otherwise there is no reason for a regulation policy because monitoring is not beneficial at any level. Differentiating through (7) and (8) with respect to C^m, under the assumptions that there will be interior solutions for all relevant values of C^m and that (10) always holds, and solving the system of equations for

$$\frac{\partial \bar{E}}{\partial C^m} \text{ and } \frac{\partial m}{\partial C^m}$$

allows for the determination of the optimal policy response at a point in time to an increase in C^m.

This exercise yields

$$(12) \quad \frac{\partial \bar{E}}{\partial C^m} = -\frac{H_{\bar{E}M}}{|\Delta|}$$

$$= \frac{-\left[(MNB_E)E_{\bar{E}m} + \dfrac{dMNB_E}{dE} E_{\bar{E}}E_m - C^a A_{\bar{E}m}\right]}{|\Delta|},$$

and

$$(13) \quad \frac{\partial m}{\partial C^m} = \frac{H_{\bar{E}\bar{E}}}{|\Delta|}$$

$$= \frac{(MNB_E)E_{\bar{E}\bar{E}} + \dfrac{dMNB_E}{dE} E_{\bar{E}}^2 - C^a A_{\bar{E}\bar{E}}}{|\Delta|},$$

where $H_{\bar{E}M}$ and $H_{\bar{E}\bar{E}}$ are the derivatives of (7) with respect to m and $\bar{E}$, respectively. The determinant Δ is composed of derivatives of equations (7) and (8)

$$|\Delta| = \begin{vmatrix} H_{\bar{E}\bar{E}} & H_{\bar{E}m} \\ H_{m\bar{E}} & H_{mm} \end{vmatrix}.$$

It must be positive if the Hamiltonian is to be maximized at all points along the trajectory. The numerator in (13) is negative for the same

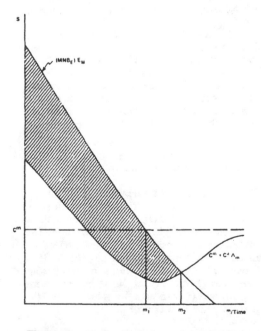

Figure 3. Optimal amount of monitoring activity

reason. Not surprisingly, then, an increase in C^m causes a decrease in the efficient level of monitoring at any point in time all else equal.

To sign (12) note from above that $E = E_1$ when $C^m = 0$, and so at that point $MNB_E = 0$. Also, when $C^m = 0$ monitoring is fully effective (i.e., compliance is complete), hence, $E_m = 0$. Therefore, it follows that

(13a)
$$\left. \frac{\partial \bar{E}}{\partial C^m} \right|_{C^m = 0} = \frac{C^a A_{\bar{E}m}}{|\Delta|}.$$

From figure 2 and the associated discussion, it is likely that $A_{\bar{E}m} > 0$ when M is large as it will be when $C^m = 0$. Therefore, it follows that (13a) will be positive in these instances. When $C^m = 0$ the efficient number of rights to be issued at a point in time is increasing in C^m. Since $\bar{E} = E_1$ when $C^m = 0$, it follows that over some range of positive C^m, the efficient number of fishing rights is likely greater than E_1.

The point of all this is that the consideration of avoidance costs changes the optimal policy such that the efficient number of fishing rights goes from zero to a positive amount which may exceed the number that is conventionally seen as efficient. This has some basis in intuition. Avoidance cost can be reduced by either increasing the number of fishing rights or, when monitoring is heavy, as it will be when C^m is low, increasing monitoring. An increase in C^m will increase the relative advantage of controlling avoidance cost by increasing the number of fishing rights.

Other issues dealing with avoidance costs can be summarized in terms of figure 3 where the first and the second and third terms of equation (8) are plotted separately for a given level of $\bar{E}$. The former is the marginal benefit of monitoring curve, while the latter is the marginal cost of monitoring curve. Note the peculiar shape of the marginal cost curve for m when avoidance costs are included. Marginal monitoring costs exceed C^m as long as avoidance costs are increasing in m. Beyond the maximum point on the relevant avoidance cost function, marginal monitoring cost lies below C^m until monitoring becomes so pervasive that all avoidance is eliminated. As constructed in figure 3, monitoring should be greater when attention is given to avoidance costs than otherwise would be the case. Specifically, m_2 is optimal if avoidance costs are considered, while m_1 would be if they are not. Monitoring not only helps eliminate the

inefficiencies of overexploiting the fishery, it also helps reduce avoidance costs, another wasteful use of resources. A different situation would occur if the marginal benefit curve were lower, such that it intersected the marginal cost curve (still from above) to the left of the point where the complete marginal cost curve intersects the C^m line (not pictured). In that instance, monitoring should be less when attention is given to avoidance costs (i.e., the points analogous to m_1 and m_2 should be reversed). With marginal regulation benefits relatively low, the optimum point will occur in the upward-sloping portion of the avoidance cost curve. While monitoring still reduces overexploitation inefficiencies, it now increases avoidance costs in the neighborhood of the optimum.

The shaded area represents the maximum net benefit that can be obtained for the given level of $\bar{E}$. The optimum policy is to select that combination of $\bar{E}$ and m such that this area is maximized. Note also that the marginal benefit of monitoring is positive at the optimal point. Since E_m is negative, this means that MNB_E must also be negative at this point. Therefore, the optimal industry operation point, regardless of the efficient number of fishing rights, is analogous to E_2 in figure 1 above.

In the case where K is a policy variable, the presence of avoidance costs changes the above only in that the optimum fine would be less than infinity or even less than the maximum feasible limit because avoidance activity will increase with the level of the fine. The social cost associated with increases in the level of the fine would no longer be zero.

Optimum Dynamic Regulation

The discussion thus far deals with optimal regulation at any point in time. The purpose of this section is to extend the analysis to show how policy should be changed over time. The analyses will be primarily intuitive, based on an informed interpretation of modifications of above equations and previous interpretations (see Clark 1976). To set the stage, equations (7), (8), and (9) can each be solved for μ, the shadow price of a unit of fish in the sea:

(7a)
$$\mu = P - \frac{C^e E_E + C^a A_{\bar{E}}}{f_E E_E};$$

686 *August 1986*

Amer. J. Agr. Econ.

(8a) $\quad \mu = P - \dfrac{C^e E_m + C^a A_m + C^m}{f_E E_m}$;

(9a) $\quad \mu = P - \dfrac{C^e E_K + C^a A_K}{f_E E_K}$.

The numerator of the fraction in the three equations is the change in all costs associated with the variable under consideration, $d\ cost/ d\ V_i$ (i.e., in equation (7a) it is the change in all costs associated with a change in fishing rights, while in equation (8a) it is the change in all costs associated with a change in enforcement, etc.). The denominator in each represents the change in catch brought about by a change in the control variables, $d\ catch/d\ V_i$. Therefore, in all cases, the expression simplifies to price minus the marginal cost of fish, $P - MC_F$. Therefore, the shadow price of fish is simply equal to the marginal net social value which is obtained from harvesting the marginal unit of fish. Over time, μ must change according to equation (10). A steady-state interior solution (i.e., when the values in each period are the same as those in the one preceding) will occur if X (the time rate of growth of the fish stock) and equation (10) equal zero simultaneously. Setting (10) equal to zero and solving for r obtains

(10a) $\quad r = g_x + \dfrac{(P - \mu)f_x}{\mu}$

$\qquad\qquad + \dfrac{[(MNB_E)E_X - C^a A_X]}{\mu}$.

The right-hand side is the composite own-rate-of-return to the fish stock. The first term is the return in terms of growth rate from investing in stock. The second is the percentage return in terms of increased productivity. The denominator, μ, is the decrease in net social value from cutting back on harvest (i.e., it is the value of the investment), while the numerator is the value of the extra annual production made possible by an increase in stock size. See Brown; Clark 1976, p. 40, 1985, p. 23; and Levhari, Michener, and Mirman for a more detailed discussion of this marginal stock effect term.

The third term is an addition to the marginal stock effect term under the conditions of this model. All else equal, as stock size increases, there will be increase in actual effort and avoidance activity. Therefore the net returns from this must be considered. Since MNB_E is negative for unregulated fisheries, the third

term will also be negative over this range. Therefore, this effect will lower the net benefits obtained from regulations which invest in the stock by restricting effort. Brown (p. 168–69) points out that the normal marginal stock effect means that the stationary optimal stock size, if one exists, will be larger than where the percentage rate of stock growth is equal to the discount rate because of the added productivity of larger stocks. This analysis shows that the incentives created by stock increases to produce extra effort (even when it is socially unproductive to do so) and to engage in socially wasteful avoidance activities will have a tendency to reverse this trend. In fact, if the third term overpowers the second, the stationary optimal stock size may be smaller than where the percentage rate of stock growth is equal to the discount rate.

The non-steady state can be described with reference to (10a). When the own-rate-of-return of the fish stock is greater than r, it pays to "invest" in the fish stock and the biomass should be increasing. The reverse is true when the own-rate-of-return is less than r. The likely policy implications are as follows. If avoidance costs are ignored and if K is fixed, then when the own-rate-of-return of the stock is greater than r so that the stock should be increasing, $\bar{E}$ should equal 0 as before, but enforcement should be relatively high to ensure that actual E is kept low enough to allow stock growth. Through time the optimal $\bar{E}$ will not change, but as the stock increases the optimal amount of enforcement will fall. When the own-rate-of-return is less than r and the stock should be decreasing, the optimal $\bar{E}$ should start out greater than 0, perhaps even as high as the open-access level of E. Optimal enforcement should start out very low and perhaps equal 0. Through time as the stock is fished down toward the steady-state size, $\bar{E}$ should be reduced to zero and the level of enforcement should be increased.

These conclusions will be slightly modified when avoidance costs are considered. If the stock should be reduced, E should be positive but decreasing through time and there will be even more incentive to keep enforcement low in order to discourage any avoidance costs. With an overexploited stock, in the linear case, the optimal $\bar{E}$ will be 0, at least temporarily, in order to achieve a "bang-bang" solution. At the same time there will be an incentive to have fairly high enforcement costs, again to keep the avoidance costs low.

Over time the optimal $\bar{E}$ will increase and the optimal monitoring will decrease.

Note that avoidance costs affect monitoring differently depending upon the size of the fish stock. When the stock should be decreased, low monitoring keeps avoidance costs down but it is still compatible with optimal stock size change because effort greater than $\bar{E}$ is optimal to reduce the stock. In the opposite case, however, where the stock should be allowed to grow, it is important to keep actual E close to $\bar{E}$ in order actually to achieve that growth, and so a high level of monitoring is necessary to prevent cheating. This will also keep avoidance cost low as well.

When avoidance costs are not considered but with a variable K, the above results would be modified as follows. When it is optimal to allow the stock to grow, the optimum $\bar{E}$ will be zero, while K should be infinite (or its maximum level) and m should be large enough to drive actual E to zero. As the stock grows $\bar{E}$ should increase to E_2 and m should decrease. When the stock should be reduced, $\bar{E}$ should be larger than E_2 and K and m should be low and perhaps even zero. As the stock is reduced to its optimum size, $\bar{E}$ should be reduced to E_2, K should be increased to infinity or its upper limit, and m should be increased to where its marginal benefits equal its marginal costs.

Finally, when avoidance costs are considered and K is variable, the dynamic policies will differ only in that the effects of changes in the policy variables on avoidance costs will have to be considered during the transition period. For example, with a depleted stock, since high fines increase avoidance costs, the optimal level of the fine will be lower than it would otherwise be, and so actual effort will be higher and the transition to the optimum stock will take longer. Over the range where monitoring increases avoidance, the same argument holds for the optimal m, but this is not so over the range where there is an inverse relationship between avoidance and monitoring.

Choice of Governing Instrument

The previous discussion has been in terms of effort production rights. However, taxes, total quotas, gear restrictions, etc., are also possible governing instruments. The overall optimum policy can be determined by finding the best combination of operation level, enforcement activities, and level of fine for each of the possible alternative governing instruments. The one with the highest net present value of returns when used properly is the optimum policy. It follows that to ensure an optimal overall policy, a wide range of governing instruments must be considered. Based on the results of more traditional regulation theory, economists have traditionally favored governing instruments that explicitly emulate market processes. However, using the above model, it can be shown that the distinction of explicit market emulation becomes blurred, and therefore some governing instruments may be dismissed too readily.

For example, consider a fishery regulation program whereby gear restrictions are imposed to reduce the potential power of the existing fleet. Such programs are commonly used in fisheries management the world over. An often-cited example is that oyster draggers on the Chesapeake are required to use sail power on certain days. If enforcement and avoidance activities are ignored, such policies involve net economic wastes because of the increase in the cost-of-producing effort. Net benefits from regulation will be less than they might otherwise have been and in some cases they may be zero. See Anderson (1977, p. 169, and 1985) for more details. These results do not necessarily follow, however, when the effects of enforcement and avoidance are considered.

The most simple way to introduce regulated inefficiencies into the model, and the way which allows the most straightforward comparisons with the above analysis, is exclusively via the cost of effort function. It is conceivable, however, that they could also affect the production function as well. Let the cost function of the fishery be

$$(14) \qquad C = C(\overset{+}{E}, \overset{+}{I}),$$

where I represents an "inefficiency" parameter representing the types of gear available to the fishermen. Increases in I represent changes to more costly ways of producing effort. Because increases in I increase the cost of producing effort but does affect its productivity, left to its own devices, the fleet will choose the lowest level, say zero.

If the government wishes to use a gear restriction policy, its variable policy instruments are the required level of I, call it $\bar{I}$, and the size

688   *August 1986*   *Amer. J. Agr. Econ.*

of the fine for deviance and the level of monitoring. Analogous to the detected effort function above, let the detected deviance from the required level of I be

(15) $$I^d = I^d(\overset{(-)(-)(+)(+)}{I, A, \bar{I}, m}),$$

where the first two parameters represent fleet variables and the second government variables. The likely signs of the first derivatives are as indicated.

Since there are no controls on entry per se, entry will continue, subject to the gear restriction, until all rent is dissipated; and hence the regulation equilibrium in this instance can be defined in aggregate terms from the outset. If the fleet is rational about cheating (i.e., deviating from I) and using avoidance activities (i.e., hiding the deviations from I), the regulation equilibrium will occur when the following three equations hold:

(16) $$Pf(E, X) - C(E, I) - AC^a \\ - KI^d(I, A, \bar{I}, m) = 0;$$

(17) $$-KI_I{}^d - C_I \le 0;$$

(18) $$-KI_A{}^d - C^a \le 0.$$

Equation (16) is the condition for rent dissipation, while (17) and (18) are the conditions for optimal cheating and avoidance. If the inequality in (17) or (18) holds, then the optimum cheating is zero and the optimum I is $\bar{I}$. Cheating or avoidance activities just do not pay.

From the above it is possible to derive equations from which to derive the regulatory equilibrium levels of E, I, and A as functions of the government's regulatory control variables:

(19) $$E = \overset{(-)(-)(-)}{E(\bar{I}, m, K)};$$

(20) $$I = \overset{(+)(+)(+)}{I(\bar{I}, m, K)};$$

(21) $$A = \overset{(-)(\pm)(+)}{A(\bar{I}, m, K)}.$$

The optimal levels of the variable control instruments can be found by maximizing the present value of net returns to the fishery. The net returns are

(22) $$\pi = [Pf(E, X) - C(E, I) - AC^a] \\ - mC^m.$$

The maximand is the net rent to the fishery after enforcement costs. But from (16) it can be seen that the term in brackets is the revenue from fines received by the government. Therefore, although the gear restriction policy is not an explicit market emulation program, when fines are used to insure compliance, market emulation is implicit. The maximand is still net rents to the fishery as captured by the government, although the revenue is from fines rather than from a tax as a price for using the fishery. See below. Note, however, that if jail, vessel confiscation, or other such measures are used as penalties, the potential net rents would be dissipated through these wastes of resources and not captured by the government, and there would be no market emulation.

The first three first-order conditions for the appropriate Hamiltonian are

(23) $$(MNB_E)E_{\bar{I}} - C_I I_{\bar{I}} - A_{\bar{I}}C^a \le 0;$$

(24) $$(MNB_E)E_m - C_I I_m - A_m C^a - C^m \le 0;$$

(25) $$(MNB_E)E_K - C_I I_K - A_K C^a \le 0.$$

Notice that these equations are similar to equations (7), (8), and (9) except for the presence of the I terms. Therefore, whether or not avoidance costs are considered does not matter here; even with the A terms omitted, the presence of the I terms rules out the necessity of (23) always holding as an inequality.

To return to the main argument, while increases in $\bar{I}$, m, and K all have positive gross benefits [i.e., $(MNB_E)E_{\bar{I}}$ will be positive over the relevant range for any of the control variables], each will have social costs in terms of inefficiency of production and avoidance costs and m will also involve actual government resource costs. Therefore, one of the concerns in determining the optimal combination of $\bar{I}$, m, and K will be the social costs involved. But by the very nature of gear restriction regulation, success in achieving the specified $\bar{I}$ will result in inefficiencies. Therefore, from an efficiency point of view, programmatic success may not be beneficial. More formally, the solution to these equations will maximize fine collections but will not necessarily achieve actual adherence to the specific policy.

Although it is difficult to specify exactly the nature of the optimum policy bundle, especially with respect to the optimum $\bar{I}$, it is likely that the optimal fine will be relatively low. In the previous analysis of effort rights, it was shown that the tax level had to be equal to or above that level which would guarantee compliance when detection was assured. This constraint will not hold in this case. The only way to prevent overcapitalization of the fishery is to capture the rent which leads to this dissipation. In order to do this, fines must be such that the rational fishermen will pay them rather than adhere to the gear restriction or engage in avoidance activities.

The optimal level of monitoring activities will be such that participants will pay their fines, but this must be balanced against its effect on avoidance. However, the simultaneous selection of the three variables may well determine a fine level that reduces the incentive to engage in costly avoidance activities regardless of the level of enforcement.

Many types of solutions are possible. In fact, given the possibility of inequalities in (17) and (18), the optimum gear restriction policy may be where absolutely no adherence and no avoidance take place. At the optimum it will be cheaper for the fleet to "cheat" and pay the fine when they are caught.

There are obviously some political reality constraints on a government policy where the desired end result is noncompliance. However, the general conclusion is valid. Removing the rent from the fishery eliminates the incentive for overcapitalization. Programs may not require a socially unacceptable level of noncompliance in order to collect these rents. And considering monitoring and avoidance costs, such a program may yield higher net benefits than alternative programs. Again, the main point is that, from an economic efficiency point of view, the goal of regulation is to maximize the net returns. This analysis has shown that when enforcement and avoidance costs are considered, the distinction between what may be called market processes and direct control regulation may not be so distinct, and the one with the largest bottom line figure may not be of the former classification.

Conclusions and Applications

It will prove useful to summarize important points raised in different parts of the paper and to make explicit other points that were ignored or glossed over, especially as both apply to policy formulation.

(*a*) Optimal regulatory policy involves the choice of an appropriate fixed input (the governing instrument) and a choice of appropriate variable control inputs. The choice of an op-

timum overall policy will therefore involve a comparison of the net benefits of a range of programs each optimally implemented.

(*b*) The social costs of avoidance can be important and should be considered when developing management programs. Further, their existence can affect the choice of governing instrument and the optimal combination of variable control inputs for any particular instrument.

(*c*) Since the behavior of the industry can only be indirectly affected by regulation, it is critical to understand the relationships between compliance, industry output, and avoidance and the regulatory control variables. The discussions here in terms of nominal effort in general and effort production rights and a general inefficiency parameter in particular have masked the complexity of these relationships. More research on deriving the meaningful economic measures of effort and how its components change with various regulation types and levels of enforcement are required.

(*d*) When enforcement policy includes payments of fines, the distinction between market and nonmarket regulatory programs blurs; and although the latter may involve some inefficiencies, they may still be optimal if they involve lower enforcement or avoidance costs.

(*e*) It is important to study enforcement activity, not only because of the costs involved but because compliance is directly related to it. In fact, the same level of industry operation is possible at different levels of other regulatory control variables depending on the amount of enforcement. This puts a different perspective on practical policy because, speaking in terms of the fishery, most of the arguments on policy deal with the allowable level of annual catch. This analysis shows that the level of enforcement is just as important. From a practical political point of view, it may be much easier to concentrate on the level of monitoring which, depending on the institutions, may be kept to a "back room" decision rather than opening up or paying more than lip service to the highly sensitive issue of allowable catch each year. As long as allowable catch or the number of fishing rights is set fairly low, and if changes in monitoring can be perceived rapidly by the fishermen and they change their activities accordingly, the amount of fish actually harvested can be changed fairly significantly by changing moni-

690   *August 1986*   *Amer. J. Agr. Econ.*

toring only. Because there is normally a well-structured grapevine in most fisheries, a heavy or lax annual enforcement policy should become known throughout the fleet fairly easily. One potential drawback to practical application of this in U.S. fisheries is that the management councils determine allowable catch and the Coast Guard and the National Marine Fisheries Service enforce it. Therefore, there may be potential gains for closer cooperation between these agencies.

(*f*) In fisheries especially, but also in environmental regulation, there has been a tendency to avoid market process regulatory controls for various political and social reasons. For example, pollution taxes are considered selling the "clean" environment, and instituting rights to fish is considered un-American because it takes away others' basic rights to use the oceans. In view of the conclusion that direct control governing instruments become implicit market process controls when noncompliance is penalized by fines, the political liabilities of market process instruments may not be such a problem. While people may be opposed to taxing fishermen, they may well be in favor of taxing deviance from gear restrictions. However, given the actual gear restriction, enforcement, and tax policy, the practical difference between the two may be quite small.

(*g*) Although they have been ignored in the analysis, there are many issues of moral values and ethics involved. First, all fishermen may not cheat even though it is financially profitable to do so. Also, the suggestion that policies be implemented assuming that people will not comply with them has the potential for eroding social capital which depends on respect for law. See Milliman.

[*Received September 1984; final revision received July 1985.*]

References

Andersen, Peder, and Jon G. Sutinen. "Fisheries Law Enforcement and International Trade in Seafood." *Proceedings of the International Seafood Trade Conference*, Alaska Sea Grant Rep. No. 83-2, University of Alaska, 1983.

Anderson, Lee G. "Potential Economic Benefits from Gear Restrictions and License Limitation in Fisheries Regulation." *Land Econ.* 61(1985):409–18.

———. *The Economics of Fisheries Management.* Baltimore MD: Johns Hopkins University Press, 1986, rev. ed.

———. "The Relationship Between Firm and Industry in Common Property Fisheries." *Land Econ.* 52 (1976):179–91.

Baumol, William J., and Wallace E. Oates. *Economics, Environmental Policy and the Quality of Life.* Englewood Cliffs NJ: Prentice-Hall, 1979.

Becker, Gary. "Crime and Punishment: An Economic Approach." *J. Polit. Econ.* 86(1968):169–217.

Beddington, J. R., and R. B. Rettig. *Approaches to the Regulation of Fishing Effort.* Rome: United Nations, FAO Fisheries Tech. Pap. No. 243, 1984.

Brown, Gardner M. "An Optimal Program for Managing Common Property Resources with Congestion Externalities." *J. Polit. Econ.* 82(1974):163–74.

Clark, Colin W. *Bioeconomic Modeling and Fisheries Management.* New York: John Wiley & Sons, 1985.

———. *Mathematical Bioeconomics.* New York: John Wiley & Sons, 1976.

Crutchfield, J. A. "An Economic Evaluation of Alternative Methods of Fishery Regulations." *J. Law and Econ.* 4(1961):131–43.

Ginter, J. J. C., and R. B. Rettig. "Limited Entry Revisited." *Limited Entry as a Fishery Management Tool.* Seattle, Washington Sea Grant, 1981.

Gordon, H. S. "The Economic Theory of a Common Property Resource." *J. Polit. Econ.* 62(1954):124–42.

Lee, Dwight. "The Economics of Enforcing Pollution Taxation." *J. Environ. Econ. and Manage.* 11 (1984):147–60.

Levhari, D., R. Michener, and L. J. Mirman. "Dynamic Programming Model in Fishing." *Amer. Econ. Rev.* 71(1981):649–61.

Milliman, Scott R. "Optimal Fishery Management in the Presence of Illegal Activity." *J. Environ. Econ. and Manage.*, in press.

Pearse, P. H. "Symposium on Policies for Economic Rationalization of Commercial Fisheries." *J. Fisheries Res. Board of Canada* 36(1979):711–866.

Rothschild, Brian J. "Fishing Effort." *Fish Population Dynamics*, ed. J. A. Gulland. New York: John Wiley & Sons, 1977.

Smith, V. L. "On Models of Commercial Fishing." *J. Polit. Econ.* 77(1969):181–98.

Sutinen, Jon G., and Peder Andersen. "The Economics of Fisheries Law Enforcement." *Land Econ.* 64 (1985):387–97.

[40]

JOURNAL OF ENVIRONMENTAL ECONOMICS AND MANAGEMENT **32**, 1–21 (1997)
ARTICLE NO. EE960947

A Model of Regulated Open Access Resource Use*

FRANCES R. HOMANS

Department of Applied Economics, University of Minnesota, St. Paul, Minnesota 55108

AND

JAMES E. WILEN

*Department of Agricultural and Resource Economics, University of California at Davis, Davis,
California 95616, and Giannini Foundation*

Received October 5, 1994; revised August 24, 1995

This paper develops a model of regulated open access resource exploitation. The regulatory model assumes that regulators are goal oriented, choosing target harvest levels according
to a safe stock concept. These harvest quotas are implemented by setting season lengths,
conditioned on the industry fishing capacity. The industry enters until rents are dissipated,
conditioned on season length regulations. Harvest levels, fishing capacity, season length, and
biomass are determined jointly. Using parameter estimates from the long-regulated North
Pacific Halibut fishery, predictions of these variables from the regulated open access model
are compared to predictions that arise from the Gordon model. © 1997 Academic Press

1. INTRODUCTION

A survey of the literature in fisheries economics would leave one with the
impression that fisheries fall into one of two institutional configurations: pure open
access or optimized rent maximizing. This impression would be conveyed because
these two paradigms totally dominate the published literature in fisheries economics. At the same time, if one were to examine the institutional configurations
under which real world fisheries operate, one would find virtually no fisheries
operating under either pure open access or rent maximizing conditions. Instead,
most of the world's most important fisheries operate under what might best be
termed *regulated open access*. In regulated open access fisheries, participants are
free to enter but subject to certain regulations imposed by a management agency.
These typically take the form of gear restrictions, area closures, and importantly,
season length restrictions.[1]

Economists have not, of course, ignored regulations in discussions of fisheries.
As early as the fifties, economists were hypothesizing about the economic impacts
of regulations on fisheries, including output and input taxes, gear restrictions, and
time and area closures, among others.[2] These somewhat speculative discussions
arose in a particular institutional context in which most of the world's important

---

*The authors acknowledge helpful comments from three anonymous referees. This is Minnesota
Agricultural Experiment Station Paper 21, 494.
[1] Some fisheries also operate under a more stringent form of regulation, in which regulations are
imposed and enforced in a restricted access setting that uses a license limitation scheme or another
form of closure to entry. These are perhaps best viewed as regulated restricted access fisheries.
[2] Cf. Turvey and Wiseman [22], Scott [18], and Crutchfield and Zellner [8].

1

2   HOMANS AND WILEN

fisheries were virtually unregulated and operating under open access conditions. H. S. Gordon's [10] important paper had just appeared, pointing out how open access fisheries inevitably dissipate potential rents and as a result, economists' early discussions about regulations focused on normative issues such as: how can regulations be designed in order to coax an open access fishery closer to a rent maximizing ideal? The extension of jurisdiction by coastal nations in 1976 changed the institutional context of fisheries dramatically since virtually all of the most important fisheries suddenly came under the authority of adjacent coastal nations. This important change set the stage for a new era of regulated fisheries use by legitimizing coastal nations' authority to explicitly manage the use of their fisheries.

As coastal nations have begun to exert more regulatory control over their adjacent fisheries, economists have continued to focus primarily on normative issues associated with regulatory design. There have been several important policy contributions emerging from this work. Economists were instrumental, for example, in framing management legislation (such as the Magnuson Act) in a manner that incorporated socioeconomic as well as biological goals and in promoting rationalization schemes including limited entry and transferable quota schemes. Interestingly, however, in their descriptive analytical and empirical work, economists have continued to depict fisheries as if they are still operating under pure open access conditions. This is in spite of the fact that most modern fisheries are heavily influenced by regulatory structures which proliferated especially after the extension of coastal jurisdictions at the end of the seventies.

A question which might be asked is, why not simply incorporate regulations as technological constraints and treat them as a minor modification of the basic open access model? The answer to that question is that the regulatory structure is fundamentally more important than this, particularly as a force affecting the character of fisheries in the long run. Moreover, regulatory constraints are not simply exogenous but instead are fluid and an outcome of purposeful behavior by institutions with goals and objectives. The menu of regulatory instruments is often extensive and regulators actively choose both the suite of instruments and their respective levels in a manner that reflects current and expected future conditions in each fishery. When external factors change, regulations also change in response and hence regulations are fundamentally endogenous and dynamic.

The implications of these observations are at least three. First, the technological and behavioral character of modern fisheries is intimately bound up with the nature and operation of the regulatory structure. Not only will technology and behavior be affected by the regulatory structure, but so also will the health and attributes of the harvested species. Second, discussions of policy alternatives need to measure the gains from rationalization relative to the appropriate status quo. Regulated fisheries may look considerably different from what would emerge under pure open access and they may, in fact, even be generating substantial economic rents. Finally, modeling the evolution of fisheries and particularly forecasting future conditions must account for not only the dynamics of the biomass and the technology and behavior of the industry but also the dynamics of the behavior and technology of the regulatory apparatus. This places an even greater burden on modelers to know and understand the particular features of the fisheries they are modeling.

In this paper we present a model of a regulated open access fishery which we believe more accurately captures some of the features of modern fisheries. It is important to highlight that this is a predictive rather than a normative model; we are interested in modeling and understanding the implications of structures which exist in real world setting rather than addressing questions about whether they are efficient or which institutions might be optimal. The model developed elevates the role of the regulatory structure to one on par with consideration of industry behavior and biological dynamics. We treat the regulatory sector as rational and purposeful (although not necessarily efficient), so that regulatory behavior and industry behavior jointly and endogenously determine the character of the fishery in question. We also view the process of regulator/industry interaction as dynamic so that when internal and external conditions change, the regulatory sector responds rather than being considered simply exogenous. In the next section the conceptual model is discussed and in Section 3 this model is used to generate some hypotheses about regulated open access fisheries. Section 4 discusses results of an empirical application of the model and the concluding section summarizes the differences between our model of regulated open access use and the more widely used pure open access model.

2. A MODEL OF A REGULATED OPEN ACCESS FISHERY

The model developed here has three fundamental components. First, in the industry component, it is assumed that the fishing industry commits a given amount of fishing capacity each season, based upon anticipated prices, costs, biomass level, and (importantly) the regulations set by the regulatory agency. Second, in the regulatory component, it is assumed that the regulatory agency selects regulations, based upon specific biologically oriented goals and the anticipated fishing capacity level of the industry. Thus there is a joint equilibrium established between the industry and the regulatory sector. Finally, in the biological component, we assume that the biomass evolves between seasons in a manner dependent upon how much has been harvested each season and the initial biomass level. The fishery is characterized by an equilibrium consisting of fishing capacity and regulation levels determined endogenously, and the biomass level.[3] The motivation for the specific characterizations of these components are as follows.

2.1. *Fishermen's Behavior*

H. S. Gordon's model of rent dissipation is a useful point of departure for considering industry behavior. We assume that fishermen behave as Gordon suggested, that is, they enter in response to rents and entry proceeds until effort is earning its opportunity cost. Rents will be assumed to be the difference between industry revenues and industry costs, defined over a given fishing season. Revenues are defined as total seasonal harvest multiplied by an exvessel price P per pound. Assume that there is an instantaneous Schaefer type harvest function defined by

$$h(t) = qEX(t), \tag{1}$$

---

[3] For a more complete exposition of this model, see Homans [13]. An earlier attempt to model regulation in fisheries as endogenous is in Wilen [24].

4   HOMANS AND WILEN

where h is the harvest rate, q is the catchability parameter, E is a measure of fishing capacity or power, and X is the biomass level in period t of a given season. Assume also that the industry commits an amount of capacity E each season so that E can be considered variable between seasons but fixed within a season.

Let the level of biomass at the beginning of a particular season be designated X_0. Assume also that within the fishing season, the biomass declines by the fishing rate so that:

$$\dot{X}(t) = -qEX(t). \tag{2}$$

Between fishing seasons we assume that the biomass grows in a density dependent fashion so that the beginning biomass level in one season is a function of the biomass level in the previous season. Natural mortality during the season is ignored here for analytical convenience. In Section 2.3 below we discuss the between-season dynamics. Under these assumptions, total cumulative harvest for the industry over a season of length T will be

$$H(T) = X_0 - X(T) = X_0(1 - e^{-qET}), \tag{3}$$

which is determined by integrating (2) over the season length T, assuming that E is constant and X_0 is given.

With respect to costs we assume simple linear costs related to both the level of fixed fishing capacity and to the cumulative level of variable effort expended over the season. Consider capacity costs first. We assume fixed costs of f per season per unit of capacity E must be incurred to participate in the fishery. These costs may be assumed to be outfitting, repair, and preparation costs associated with gear, the opportunity cost associated with the investment, and other implicit rents associated with other inputs. We also assume that there are variable costs v per unit of capacity used per unit time associated with the (assumed constant) rate of input use over the season.

With both cost and revenue formulations as described, we may write total industry rents anticipated for a season of length T as:

$$\text{Rents} = \left[PX_0(1 - e^{-qET}) \right] - \left[vET + fE \right]. \tag{4}$$

Total revenues are given by the left hand term and are determined by multiplying total seasonal harvest in (3) by the exvessel price P. Total costs are given by integrating (ignoring the discount factor) variable flow costs vE over the season length T and adding fixed capacity costs of fE. Setting rents equal to zero yields an implicit equation for capacity E as a function of T, P, X_0, v, f, and q. This gives the rent dissipating level of capacity that would be expected to be attracted into a fishery in a given season. We are thus assuming that capacity enters each season until rent dissipation occurs.

The implicit equation $J(E, T) = 0$ derived by setting rents in (4) equal to zero describes industry behavior associated with capacity levels that dissipate rents, given a season length T. The functional forms used here, while standard practice and sensible representations of production and cost functions, create some analytical complexities because it is not possible to explicitly isolate the rent dissipating capacity E as a function of the other variables and parameters. The function $E = E(T; X_0, P, v, f, q)$ may be analyzed by indirect methods and Fig. 1 depicts the

REGULATED OPEN ACCESS RESOURCE USE

5

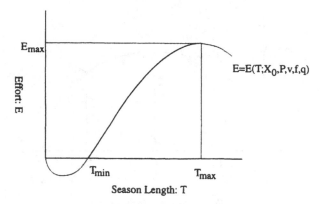

FIG. 1. Rent dissipating capacity.

shape of the function identified by setting (4) equal to zero (see Appendix A for details).

The function generally describes a monotonic relationship between E and T over the relevant range. Other things equal, longer season lengths will require larger amounts of fishing capacity to dissipate rents. Note that there is a minimum season length T_{min} below which no effort will be attracted, however. This is because there is a fixed cost f per unit of capacity and the season length must be long enough to generate sufficient variable profits to cover these costs for the first unit of positive capacity. In addition, there is a maximum capacity E_{max} associated with season length T_{max} which is the longest season length which the industry would voluntarily choose to use under any circumstance (see Appendix A).

2.2. *Regulator Behavior*

As discussed in the Introduction, economists have essentially ignored the fact that regulations are endogenous in modern fisheries. What hypotheses might we entertain to describe the motivations of a regulatory agency?[4] In this paper we assume a simplified goal structure for the regulatory body which captures the biological orientation of most real world fisheries regulatory bodies. In particular, we assume a two stage regulatory process where in the first stage, a target harvest quota is chosen to ensure stock safety. While there are several quota rules which regulators actually use, a common and analytically simple one is to assume that the

---

[4] One possibility is a rent seeking model, in which constituents are assumed to lobby regulators for actions that generate rents (Bhagwati [1], Buchanan *et al.* [2] and Rowley *et al.* [17]). A related possibility is a regulatory capture model in which regulatees "capture" the regulators through political or voting processes and manipulate outcomes (Karpoff [14]). These are plausible but at a level in the policy process once removed from that we are interested in. Fisheries in the United States and elsewhere are generally managed within broadly defined enabling legislation such as the Magnuson Fisheries Conservation and Management Act. Fisheries management acts such as this and related laws specify broad goals for management but not generally how to achieve these goals. Day to day implementation of fisheries regulations is usually left in the hands of local, regional, and fishery-familiar managers housed within biologically oriented field level agencies.

6   HOMANS AND WILEN

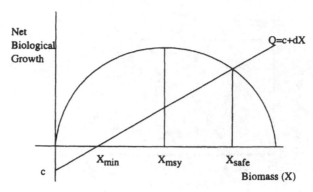

Fig. 2.   Regulator quota rule.

allowable quota or exploitation rate is a linear function of the biomass over some range.[5] The quota would be set, then, according to the rule

$$Q = c + dX_0. \tag{5}$$

In Fig. 2, for example, we show a quadratic yield function with a linear harvest quota rule superimposed. Whenever the biomass is below the desired long run "safe" stock level, the quota is set below yield so that the biomass grows and conversely if the biomass is above X_{safe}. This quota rule allows a gradual adjustment toward the equilibrium stock level whenever biomass is above or below it. Note that if $c < 0$, this rule would also call for a moratorium when the biomass falls to a level like X_{min}, hence a rule like $Q = \max(0, c + dX)$ better describes this situation. Note also that this simple rule encompasses the possibility that the safe stock is the maximum sustainable yield stock as well as others such as the $F_{0.1}$ strategy.[6] Obviously other variants of these types of rules are possible.

In the second stage of the regulatory process, we assume that regulatory instruments are selected to achieve the quota target determined from the quota rule. This distinction between targets and instruments is important and often ignored in descriptive analysis. In real fisheries, it is not enough to simply select an aggregate harvest level since there is nothing to ensure that the industry won't exceed the target. Typical regulatory instruments are the levels of various constraints on fishing technology and practices allowed by regulators, including mesh size restrictions, area and season length closures, engine size, and gear dimensions. By far the most commonly used instrument to ensure harvest quota adherence is the season length restriction. For the model discussed here, we assume that the

---

[5] Cf. the discussion in Hilborn and Walters [12], p. 22–43.
[6] The $F_{0.1}$ strategy prescribes a constant exploitation rate which leaves exploitation slightly less than the one which maximizes yield per recruit, cf. Hilborn and Walters [12].

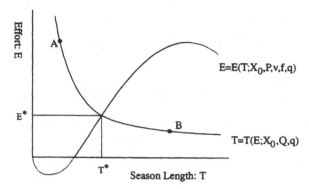

FIG. 3. Regulator season length choice and joint regulated equilibrium.

relevant instrument is the season length T, although it would be straightforward to generalize.[7]

Suppose that the biomass measured at the beginning of the fishing season is X_0. Then if the quota target is applied to the initial biomass, we have an allowable harvest quota of $Q = c + dX_0$, which, if realized, will leave a final end of season biomass of $X_T = X_0 - (c + dX_0)$. In the second stage, we thus assume that regulators select the season length instrument level which ensures that the target will be achieved. This is found by substituting the quota into Eq. (3), so that we have

$$Q = X_0(1 - e^{-qET}).\qquad(6)$$

This equation can be solved for the season length T as a function of capacity, beginning biomass, the harvest quota, and the catchability coefficient. In particular, the regulators are assumed to behave by choosing T so that

$$T = \frac{1}{qE} \ln\left[\frac{X_0}{X_0 - Q}\right].\qquad(7)$$

Note that, once X_0 is given, this describes a rectangular hyperbola in T, E space as shown in Fig. 3. When fishing capacity is large as at point A, regulators will only allow fishing over a short season but when capacity is low as at point B, a longer season can be permitted.

A jointly determined regulated open access equilibrium occurs at a capacity level and season length as depicted at the intersection of the two curves. The industry

---

[7] For example, since aggregate fishing capacity is E in our model, we could disaggregate to incorporate input controls by defining the capacity function to be effective capacity (on the grounds) which is a function of various regulatable inputs. If effective capacity is assumed to be a function of a vector of inputs, regulators could further be assumed to be able to choose the attributes of those inputs or to regulate their levels directly. For example, fishing capacity is certainly a function of gear dimensions and other specifications and these are almost always directly regulated. The complication introduced by multiple inputs is that one must also develop a theory about instrument choice which answers questions like: if both season length and mesh size restrictions can be used to curtail aggregate effective fishing pressure, how would the mix of the two be chosen by regulators?

equilibrium depicts the industry's "choice" of a rent dissipating capacity, given the season length, and regulatory behavior is depicted as a choice of season length, given fishing capacity, the initial biomass, and the quota target established in the first stage.

2.3. *Between-Season Dynamics and the Harvest Quota Target*

The discussions above characterize an industry and regulator equilibrium during a single fishing season in which the initial biomass and hence quota are predetermined. If the quota is an equilibrium quota, the biomass will remain constant and the regulated open access fishery will be in a full steady state. If the biomass is changing, however, both the industry and regulator equilibrium curves will shift from season to season, since they are both functions of a changing X_0.[8] We can then think of the intersection of the two curves as determining a sequence of temporary equilibria, in which the industry and regulators attain successive within-season equilibria, each associated with the existing biomass level each period. Over time we would observe the entire system moving through a sequence of temporary seasonal equilibria as the biomass approached its long run level. In order to close the model and describe this long run equilibrium, we thus need to specify the dynamic relationships which determine the evolution of biomass levels between seasons.

The simplest way to specify this process is to utilize a hybrid Schaefer/Beverton-Holt model in which within season dynamics are simply governed by Eqs. (1) and (2) above and between season dynamics are governed by a density dependent mechanism. That is, during the fishing season, total biomass is assumed to evolve to reflect the harvest rate, falling from its initial level X_0 to X_T by the end of the season. Then, between seasons, it is assumed that the net biological growth depends in some density dependent way on previous biomass. Let $X_{0,t+1}$ be the biomass at the opening date of next year's season and let $X_{T,t}$ and $X_{0,t}$ be the ending and beginning biomass levels this season, respectively. Here we are indexing seasons by t and denoting the beginning and ending dates of any particular season with 0 and T.

Assume that additions to the biomass can be defined as $G(X_{0,t}) = aX_{0,t} - bX_{0,t}^2$.[9] Then between season dynamics can be represented by

$$X_{0,t+1} = X_{T,t} + G(X_{0,t}) = X_{T,t} + aX_{0,t} - bX_{0,t}^2. \tag{8}$$

---

[8] In particular, the industry equilibrium function shifts up with increases in biomass. The regulatory equilibrium function may shift in or out, depending upon the form of the quota rule. It can be shown that the regulatory equation will shift out (in) if the regulatory rule is elastic (inelastic) with respect to biomass. Thus for a linear rule, the rectangular hyperbola will shift out as biomass increases if c is positive and in if c is negative.

[9] Our use of the biomass growth function deserves some comment. Since halibut reach reproductive age in 7 to 8 years, it would be more realistic to employ an age-structured cohort model to capture biomass dynamics rather than a simple annual model. While the introduction of a cohort model may improve the accuracy of our representation of biomass dynamics and would be tractable for that purpose, it would introduce considerable analytical complexities into the complete model of season length and effort determination. Using a lumped parameter annual model for halibut has several precedents, see Cook [4], Stollery [21], and Conklin and Kolberg [5].

That is, beginning biomass next season is equal to ending biomass this season plus net growth added to reproductive activities taking place or near the beginning of this season.[10]

With between season dynamics operating, under a linear exploitation rate rule, regulators allow a certain fraction $c + dX_0$ of the initial biomass to be taken each season. Thus the harvest target during any season t will be $Q_t = c + dX_{0,t}$ and we know that the end of the season biomass level will be $X_{T,t} = X_{0,t} - (c + dX_{0,t})$. As Fig. 2 shows, a linear exploitation rate leads the biomass to a long run equilibrium level X_{safe}. When the biomass is below X_{safe}, the linear rule results in a harvest level below the biological growth and the biomass approaches X_{safe} and similarly when the biomass is above X_{safe}. Along any path to a steady state, biomass dynamics may be expressed in terms of initial biomass levels, biological parameters a and b, and the exploitation rate parameters c and d, or

$$X_{0,t+1} = [(1 - d)X_{0,t} - c] + aX_{0,t} - bX_{0,t}^2. \tag{9}$$

In a long run steady state equilibrium we have

$$X_{0,t+1} - X_{0,t} = 0 = [(1 - d)X_{0,t} - c] + aX_{0,t} - bX_{0,t}^2 - X_{0,t}, \tag{10}$$

and if we let X_{safe} be the equilibrium value of the beginning of season biomass, we have, by substituting into (10) and solving, the expression

$$X_{safe} = \frac{a - d \pm \sqrt{(a - d)^2 - 4bc}}{2b}. \tag{11}$$

As expected, this is quadratic, representing the two intersections of the quota rule with the yield curve depicted in Fig. 2. We will assume $(a - d)$ is positive and that the desired steady state is the larger of the two roots, which would be considered the least vulnerable of the two stock levels.

Note that since the equilibrium biomass is a function of the regulatory parameters c and d, the notion of a safe biomass level is equivalent to a choice of the exploitation rule parameters c and d in the steady state. That is, we can view the regulatory goal as ensuring that a certain minimal long run biomass is maintained in the steady state, or alternatively, that the exploitation rule associated with the parameters is the one which achieves this goal.

---

[10] Alternatively, we could assume that biomass dynamics are governed by the alternative relationship $X_{0,t+1} = X_{T,t} + F(X_{T,t})$. This formulation would be appropriate if reproductive activities took place after the fishing season, while $X_{0,t+1} = X_{T,t} + G(X_{0,t})$ would reflect reproduction just prior to the season opening. The alternative relationship would change the corresponding expression for biomass to

$$X_{safe} = \frac{(1 + a + 2bc)(1 - d) - 1 \pm \sqrt{[(1 + a + 2bc)(1 - d) - 1]^2 - 4bc(1 - d)^2(1 + a + bc)}}{2b(1 - d)^2}.$$

Other expressions that embed biomass would change accordingly. Since halibut reproduce in the winter, just prior to the spring fishing season, the formulation used in the paper reflects halibut biomass dynamics more accurately than the alternative.

3. ANALYSIS OF REGULATED OPEN ACCESS BEHAVIOR

The three components described above characterize a regulated open access fishery. The industry is assumed to commit capacity E each season until rents are dissipated. Regulators are assumed to set a harvest target quota Q each season and then choose a season length T which ensures that the quota is achieved. A regulated open access equilibrium is achieved by the interaction of the industry and regulators each season. Biomass evolves between seasons according to whether the corresponding harvest is greater than, equal to, or less than biological growth. A long run steady state is achieved when the biomass is in equilibrium and when industry and regulatory behavior are constant.

In a steady state equilibrium, both beginning and ending biomass levels are given and constant from season and their levels depend upon c, d, a, and b. Since the harvest quota target is simply the difference between beginning and ending biomass levels, Q is predetermined also once c and d are chosen. Then a regulated open access equilibrium within the season is achieved when the industry commits a level of capacity E and the regulators choose a season length T such that

$$PX_0(1 - e^{-qET}) - vET - fE = 0$$

$$T = \frac{1}{qE}\ln\left[\frac{X_0}{X_T}\right] = \frac{1}{qE}\ln\left[\frac{X_0}{(1-d)X_0 - c}\right], \tag{12}$$

where X_0 satisfies (11) above. A regulated open access fishery is thus the joint outcome of the behavior of both the industry and regulatory agency, coupled with biological dynamics.

What happens to this equilibrium as economic, regulatory, and biological parameters change? This can be addressed by performing comparative statics computations on the equilibrium. Note first that this system is recursive in that the steady state biomass level is determined only by the biological parameters, a and b, and the regulatory parameters, c and d. The regulatory agency's instrument choice T depends only upon the exploitation rule parameters and the level of capacity chosen by the industry. Thus the two equations in (12) determine the equilibrium pair E, T which satisfy the open access regulatory equilibrium. For any arbitrary X_0, this pair of equations yields a temporary equilibrium $[E(T, X_0), T(E, X_0)]$ dependent upon X_0, and when X_0 satisfies (11), the system is in a full long run equilibrium with X_0 constant.

Comparative statics properties of the long run equilibrium are given in Table I and Appendix B. These are all plausible results. When prices P or harvest efficiency q rise (or costs v or f fall) the rent dissipating level of E rises. But if the biomass and quota are in long run equilibrium, this increase in potential fishing capacity must be stifled by corresponding reductions in season length. Since the biological system is recursive, price/cost parameters do not affect the steady state level of biomass which is determined only by biological and regulatory parameters (a, b, c, and d). On the other hand, changes in a, b, c, and d affect the levels T and E in potentially complicated ways via their effects on X and Q. For example, if the intrinsic growth rate a rises, then (*ceteris paribus*; holding the exploitation rule constant) both the allowable long run biomass and quota will rise. As it turns

REGULATED OPEN ACCESS RESOURCE USE 11

TABLE I
Comparative Statics

	Variable			
	E	T	X	Q
P	+	−	0	0
v	−	+	0	0
f	−	+	0	0
q	+	−	0	0
a	+	−	+	+
b	−	+	−	−
c	?	?	−	?
d	?	?	−	?

out, this higher quota is achieved with an increase in the equilibrium level of fishing capacity and a decrease in the regulated season length.

The only ambiguous results are those associated with the effects of changes in the regulatory parameters c and d on E and T. Recall that increasing the quota rule parameter d reduces the steady state biomass level while it increases the fraction of initial biomass targeted for harvest. Thus whether or not an increase in d will attract more entry depends upon the interplay between two forces in their effects on revenues. On the one hand, a lower steady state biomass level will reduce E *ceteris paribus* because there is a stock effect via the production function. On the other hand, increasing the exploitation rate increases the fraction of initial biomass allowed to be taken. If d is such that the steady state biomass is to the right of X_{msy}, then increasing d has opposing effects: a reduction in revenues via the stock effect of a reduced biomass level on the harvest function, and an increase in revenues via the fact that lower biomass levels actually increase the amount of growth and hence steady state yield. If d is relatively large, however, so that the equilibrium is to the left of X_{msy}, increases in d cause a negative stock effect and a reduction in allowable harvest, both of which lead to lower E and longer T. The bottom line is that for large enough values of d, the derivative of E with respect to d is negative and the derivative of T with respect to d is positive, whereas for smaller values these signs are indeterminate.

Note that the qualitative properties of the regulated open access model are different from those predicted by the pure open access Gordon model in fundamental ways. In the Gordon model, rents generate excess capacity, which in turn results in excessive harvest levels. These harvest levels, coupled with biological dynamics, determine an approach to a bioeconomic equilibrium. With higher prices and/or lower costs, the bioeconomic equilibrium will occur at lower, more vulnerable biomass levels. In the model presented here, the existence of the regulatory structure decouples the effects of economic parameters from impacts on the biomass. To the extent that the regulatory structure is effective, the biomass level will approach its safe level. Harvest quota targets ensure this, but the manner in which they are implemented depends upon the interplay between biological and economic factors. For example, when prices are high, potential rents attract more capacity which is stifled on the grounds with short seasons so that the targeted harvest quota is not exceeded. In the long run the regulated open access fishery

12   HOMANS AND WILEN

will be characterized by higher biomass levels and generally even higher levels of inefficient input use than the Gordon model would predict.

4. AN APPLICATION — THE NORTH PACIFIC HALIBUT FISHERY

The North Pacific Halibut fishery provides a good case study with which to estimate and test some of the hypotheses generated from the model of regulated open access described above. In the first place, the fishery has a long history of regulation dating back to the 1930s. Over this whole period, an extensive data collection effort has been maintained in order to regulate the fishery. Long time series exist on measures of fishing capacity, total fishing effort expended, biomass estimates, quota targets, season lengths, and other economic variables. Second, the halibut fishery has been conducted with relatively simple fishing technology over the whole period, a longline gear with fairly standardized design and under relatively constant practices. Third, the halibut fishery has been managed over this whole period under a structure much like that described above. In particular, harvest quotas have been set by regulators annually and season lengths have been used to constrain the application of fishing pressure in order to achieve the desired harvest target.

We assembled data from sources published by the International Pacific Halibut Commission (IPHC) over the 1935–1977 period. Two separate series were constructed as a consistency check of the models, one for each of two management areas referred to as Area 2 (off British Columbia and up to Cape Spencer in Southeastern Alaska) and Area 3 north of Cape Spencer off Alaska but excluding waters in the Western Aleutians and Bering Sea).[11] The data used include biomass estimates from Quinn *et al.* [16] which were computed from logbook entries over the entire halibut program history. These are computed *ex post*, of course, but we assume they are unbiased representations of estimates used by regulators over the period for seasonal regulation decisions, as well as by the industry for capacity commitment decisions. Our measure of fishing capacity is also derived from logbook information. These data were used to calculate the total number of standard skate soaks (units of longline gear, approximately 1800 feet long with 100 hooks attached at 18 foot intervals, set in the water for 12 hours) in each area, which we divided by season length to estimate average fishing capacity. This measure thus assumes that effort intensity does not vary over the sample and that each unit of standardized capacity has costs proportional by their conversion factors to actual costs. Harvest and price data are also derived from published IPHC sources and prices are deflated by a wholesale price index with base year 1982. Quotas for each of the areas are published in annual reports by the IPHC. Season lengths were derived from published annual reports by the IPHC and from a summary in Skud [20]. These are expressed in days of season length and are essentially continuous over the period examined.

---

[11] The period of estimation was truncated at 1977 for several reasons. First, the jurisdiction extension by both Canada and the U.S. led to a separation of the waters of Area 2 into separate Canadian and U.S. waters, each with new data collection procedures and management methods. In addition, Canada instituted a limited entry program in 1979 which would have added estimation complications associated with expected change in industry and perhaps regulatory behavior.

The econometric model consists of four structural equations: the biomass growth function (8), the quota rule (5), the entry/exit equation (4), and the season length determination function (7). Econometric error terms are appended to each of these equations, so that the model becomes

$$X_t + H_{t-1} = (1 + a)X_{t-1} - bX_{t-1}^2 + \epsilon_{1t} \tag{13}$$

$$Q_t = c + dX_t + \epsilon_{2t} \tag{14}$$

$$T_t = \frac{1}{qE_t} \ln\left[\frac{X_t}{X_t - Q_t}\right] + \epsilon_{3t} \tag{15}$$

$$0 = P_t X_t[1 - e^{-qE_tT_t}] - vE_tT_t - fE_t - f_wE_tD_{\text{WAR}} + \epsilon_{4t}. \tag{16}$$

Since the biomass level is predetermined each year, and the quota is set based upon the predetermined biomass level, Eqs. (13) and (14) are estimated individually using ordinary least-squares. The capacity level and the season length are determined simultaneously in equations (15) and (16) once the biomass and quota are known. We allow for a nonzero covariance between ϵ_{3t} and ϵ_{4t} and impose the cross equation restriction that the catchability coefficient q is the same in both equations. Because of the simultaneity, the correlation in the error terms, and the cross-equation restriction, Eqs. (15) and (16) are estimated jointly using nonlinear three-stage least-squares. All the predetermined variables in the system (P, Q, X, and D_{WAR}) as well as the regulatory variable ($\ln[X/(X - Q)]$) and lagged values for E, T, and X were used as instruments.

The entry/exit equation (16) is modified slightly by including a dummy variable for the war years. Since our estimation period spans the disruptive World War II years, we modify the fixed cost coefficient by including a dummy variable (D_{WAR}) for the years 1943–1945 when the war was active in the Pacific. Management reports suggest that fishermen were explicitly warned against fishing in the Bering Sea during the war due to dangers posed by Japanese submarines.

Results are presented in Table II. The biological dynamics and quota rule results are also shown in Fig. 4. The quota rule equation was corrected for autocorrelation, while the biological growth function was not.[12] The parameter estimates for these equations are all reasonable: the signs are as expected and all coefficients are significant at the 5% level, with most significant at the 1% level.

The implies maximum biomass levels are 318 and 416 million pounds for Areas 2 and 3, respectively. Maximum sustainable physical yields occur at half of these

---

[12] Durbin's h tests indicate that autocorrelation is present in equation (13). Our suspicion is that the positive test for autocorrelation, along with the recognition that the age of recruitment is 7–8 years, is an indication that a more complicated lag structure is in order. However, since our principal aim was to focus on the model of regulatory and effort choice rather than on biomass dynamics, we chose to keep our estimation (and our theoretical model) simple. Among population biologists who have paid serious attention to these modeling/estimation issues (see, for instance, Ludwig and Walters [15]), consensus seems to be that more accurate representations of biological dynamics do not necessarily lead to more accurate estimates of variables of interest such as optimum fishing effort or surplus yield. For example, simple autoregressive specifications often provide more accurate predictions of catch than more complicated age structured models, which demand more information out of limited data sets. Our corrections for higher order autocorrelation did not change parameter estimates appreciably, indicating bias associated with autocorrelation in the presence of a simple lagged dependent variable on the right-hand side is not too severe. Our estimates for the yield equations are also in accord with results from other studies, including Criddle [6], Criddle and Havenner [7], Deriso [9], and Capalbo [3].

14   HOMANS AND WILEN

TABLE II
Parameter Estimates

Parameter	Eq. (13)[a] Area 2	Area 3	Eq. (14)[a] Area 2	Area 3	Eqs. (16) and (15)[b] Area 2	Area 3
$1 + a$	1.3786^c (24.01)	1.3119^c (31.51)	—	—	—	—
b	0.00119^c (2.472)	0.00075^c (3.212)	—	—	—	
c	—	—	12.329^c (2.990)	16.417^c (2.807)	—	—
d	—	—	0.0897^c (2.943)	0.0575^d (1.703)	—	—
q	—	—	—	—	0.00114^c (24.984)	0.000975^c (30.895)
v	—	—	—	—	0.0555^c (8.4263)	0.07848^c (6.9705)
f	—	—	—	—	1.0318^d (2.5569)	2.0993 (1.8104)
f_W	—	—	—	—	0.79697 (1.2154)	9.6556^d (3.4284)
ρ	—	—	0.96	0.92	—	—
Durbin's h	6.21	5.76	—	—	—	—
D.W.	—	—	1.33	2.01	—	—
$\bar{R}^2$	0.95	0.97	0.94	0.84	—	—
Hansen's J	—	—	—	—	1.33	1.49
method	OLS		OLS		NL3SLS	

[a] t-statistics in parentheses.
[b] Asymptotic t-statistics in parentheses.
[c] Significant at the 1% level.
[d] Significant at the 5% level.

biomass levels and the implied levels are 159 and 208 million pounds. The four parameter estimates can also be used to compute the safe biomass level implied by regulators' choices of quota levels over this period. Using Eq. (11) above, the levels of X_{safe} implied by actual choices of harvest quota levels over the 1935–1977 period are 187 and 252 million pounds, respectively. Interestingly, the targeted or safe biomass levels implied by these equations are greater than the maximum sustainable yield biomass levels in both Areas 2 and 3, implying a conservative concern for stock safety.

Results from the estimation of the industry/regulator behavioral equations are also given in Table II. Parameter estimates are all of the correct sign and generally significant. Hansen's J test of overidentifying restrictions [11] indicates that the model is not misspecified.[13] With respect to magnitudes, the parameter estimates are also reasonable and similar across regions. We would expect higher costs in

[13] Hansen's test is for estimators in the class of Generalized Method of Moments estimators, of which non-linear three stage least-squares is one. It is calculated as the value of the criterion function multiplied by the number of observations, and has a Chi-square distribution where the degrees of freedom are the number of instruments (8) less the number of free parameters (4). Since Eq. (16) is an implicit equation where the dependent variable is zero, there is no adequate goodness of fit measure available.

REGULATED OPEN ACCESS RESOURCE USE 15

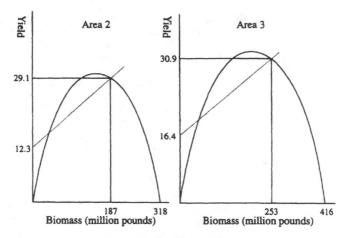

FIG. 4. Quota rule targets.

Area 3 due to its remoteness and this is the case for fixed costs. Fixed cost estimates suggest a twofold difference, with gear up cost estimates about $1,030 per unit of gear in Area 2 and $2,100 in Area 3. The War also had the expected effect of increasing the implicit opportunity cost of participating in the fishery. The implicit opportunity costs added by the hazards of war amount to an extra $790 per unit of gear in Area 2 and almost $9,700 in the Bering Sea. Variable costs are about $56 and $78 per gear set (skate soak) in Areas 2 and 3, respectively. This suggests that operating costs are lower in the more accessible region.

5. IMPLICATIONS AND ANALYSIS

The predictions generated from this model of regulated open access are significantly different from those from the standard open access paradigm. In view of the fact that most fisheries are regulated rather than pure open access, we should also expect that they are closer to what we observe in real fisheries. In this section we compare the predictions of our new model with those that would be generated using the basic H. S. Gordon model of open access equilibrium. The issue to be examined is: what are the implications of (correctly, we would argue) including the regulatory sector in a model of a fishery? To examine this question we calculate the steady state for two alternatives, using parameter estimates described above.

As discussed in the Introduction, the dominant paradigm in the renewable resource literature is the Gordon model of pure open access. We take this model to be represented by a zero rent condition together with a biological dynamics equation which describes how the species evolves. We thus assume instantaneous rent dissipation or fast dynamics for the entry/exit of fishing capacity, and slow dynamics for the biology. Since we have also brought into the analysis the concept of a season length, a question arises as to what to assume about $T(t)$. We assume that the industry ends up fishing for a period equal to T_{max}, which is the rent dissipating season length also. In Fig. 1, the industry reaches a temporary equilibrium which depends upon the biomass level each year, in which $E(t) = E_{max}$ and $T(t) = T_{max}$. Then, depending upon the corresponding level of harvest, the biomass will either rise or fall in an approach to the open access equilibrium.

The Gordon open access model thus consists of three equations

$$P(t)X(t)[1 - e^{-qE(t)T(t)}] - vE(t)T(t) - fE(t) = 0$$

$$X(t + 1) = (1 + a)X(t) - bX(t)^2 - X(t)[1 - e^{qE(t)T(t)}]$$

$$T(t) = T_{max} = \left[\frac{1}{qE(t)}\right]\ln\left[\frac{PqX(t)}{v}\right]$$

in three unknowns (E, T, X). In the long run steady state, $X(t)$ is constant and the system is in a full open access equilibrium.

The regulated open access model also includes the rent dissipation equation and the biological dynamics equation as in the Gordon model above. The principal difference is that the season length is fixed by regulators in the second stage of the regulatory process, at a length dictated by a quota rule tied to the biomass level. Thus the regulated open access model consists of four equations

$$P(t)X(t)[1 - e^{-qE(t)T(t)}] - vE(t)T(t) - fE(t) = 0$$

$$X(t + 1) = (1 + a)X(t) - bX(t)^2 - X(t)[1 - e^{qE(t)T(t)}]$$

$$T(t) = \left[\frac{1}{qE(t)}\right]\ln\left[\frac{X(t)}{X(t) - Q(t)}\right]$$

$$Q(t) = c + dX(t)$$

which determine four unknowns (E, T, X, Q).

We simulated these two systems using parameter estimates from Table II. We fixed the value of the exvessel price at its value in 1977. Table III shows the equilibrium values predicted by the alternative models for each of Areas 2 and 3. The main differences in predictions from the two models are significant. The regulated model predicts higher biomass and harvest levels, a higher level of capacity and a shorter fishing season. The reason for these differences are, of course, associated with the explicit inclusion of the regulatory sector. To the extent that regulations are successful, they hold the biomass at larger levels, which *ceteris paribus* generate higher rents and larger levels of capacity, which in turn must be

TABLE III

Regulated vs Unregulated Open Access: Steady State Predictions

	Area 2		Area 3	
	Gordon model	Regulated open access model	Gordon model	Regulated open access model
Capacity	4.5	45.9	2.1	23.6
Season length	79.6	3.3	153.1	5.7
Biomass	37.7	183	56.8	253
Quota/yield	12.6	28.7	15.3	30.9

stifled in order to hold the harvest at the targeted quota. For example, in Area 2 the regulated open access equilibrium has ten times the capacity of the unregulated open access equilibrium. However, instead of fishing over 80 days as in the unregulated case, the regulated season length is only 3 days. The actual values predicted by the regulated open access model are also relatively close to what has been observed in the actual fishery recently. For example, total exploitable biomass in Areas 2 and 3 combined reached about 275 million pounds in the late 1980s. Area 2 and 3 total harvest levels during the same period averaged about 24 and 55 million pounds, respectively. Season lengths off Alaska in both Areas 2 and 3 have fallen to about 3–5 days.[14]

The predictions from the regulated open access model point to and explain several interesting facts about modern fisheries. First, regulated fisheries are likely to attract even more redundant capital than was predicted by Gordon's unregulated open access model. The degree to which this is true depends, in fact, directly on the degree to which regulations are successful in holding biomass close to a safe level. At higher biomass levels, more potential rents exist, generating more entry pressure which must be controlled with efficiency decreasing policies. Second, as real prices rise (due to population growth), the industry equilibrium curve in Fig. 3 shifts upward, causing the equilibrium to slide along the regulatory equilibrium hyperbola. Thus it is almost inevitable that over the long run, more capacity is attracted and regulations are tightened. This explains the process we observe whereby in many fisheries, seasons have been reduced to a few weeks, days, and even hours. Finally, this model suggests some important but sometimes counterintuitive links between regulations, the output market, and rent dissipation. In particular, as seasons and other regulations are tightened, a likely market consequence is that exvessel prices are lower than they would otherwise be. For example, short seasons result in poorly delivered and handled product, and high storage costs. As prices are affected by regulatory actions, rent dissipating pressures are actually mitigated somewhat. Thus the direct effects of the regulatory process are to draw in more capacity than the Gordon model would suggest, but dampened somewhat due to indirect effects of regulations on the market. This in turn explains recent observations about fisheries that have become rationalized with individual transferable quota (ITQ) programs. In many of these, a surprising outcome has been that most of the immediate rent gains have seemed to emerge on the revenue rather than the cost side.[15] But this is what we would expect: as the system is released from its restrictive regulatory structure which reduces revenues, the first easy gains come from new marketing opportunities engendered both the unconstrained system and also by the new incentives generated under property rights based regulations.

APPENDIX A

Rent Dissipating Capacity Function

This appendix discusses the shape of the equation depicting industry equilibrium in a regulated open access setting. Industry behavior is assumed to be driven by

[14] Cf. U.S. Department of Commerce [23].
[15] Cf. Wilen and Homans [25], Wilen and Homans [26].

rent dissipation in a manner similar to that first described by Gordon in 1954. In particular, with a Schaefer production function and with biomass dynamics within the season governed by harvesting mortality, undiscounted rents for any season are simply

$$\text{Rents} = PX_0(1 - e^{-qET}) - vET - fE. \tag{A1}$$

This equation cannot be solved explicitly for capacity E as a function of other variables. Let $J(E, T; X_0, P, q, v, f)$ describe the function obtained by setting rents in (A1) above equal to zero. Then we can use the implicit function theorem to get

$$\frac{dE}{dT} = -\frac{J_T}{J_E} = \frac{E\left[v - PqX_0E^{-qET}\right]}{T\left[PqX_0e^{-qET} - v\right] - f}. \tag{A2}$$

The denominator of this expression is the net value of marginal physical product of an additional unit of capacity. If capacity enters until rents are dissipated, the net value of average physical product will be zero. Since marginal product is less than average product, and since average product is zero, the denominator must be negative.

The numerator of (A2) is the (negative of the) value of marginal physical product of an additional unit of season length. The sign of the numerator depends upon where in the domain of $E(T)$ we are operating. As it turns out, in the relevant region, the value of marginal product is positive and hence the numerator is negative. This can be seen as follows. Let $\Pi(ET)$ be variable profits associated with arbitrary levels of ET before fixed costs fE are subtracted, or

$$\Pi = PX_0(1 - e^{-qET}) - vET. \tag{A3}$$

Note first that Π is strictly concave in E, T, and ET. In addition, the average variable profit function has a finite limit as E approaches zero, namely

$$\lim_{E \to 0} (\Pi/E) = PqX_0T - vT. \tag{A4}$$

This leads to:

PROPOSITION 1. *A necessary and sufficient condition for positive levels of capacity is that the season length allowed by regulators T be larger than some $T_{\min}$ where*

$$T_{\min} = \frac{f}{PX_0q - v}. \tag{A5}$$

This can easily be seen by noting that capacity will enter only if average variable profits cover fixed entry costs. Since average variable profits are decreasing in E, there will be entry only if the first unit finds it profitable or if

$$\lim_{E \to 0} (\Pi/E) = PqX_0T - vT > f. \tag{A6}$$

But this is true only if T is greater than $T_{\min}$ defined in (A5). This defines a minimum season length below which no positive capacity is feasible.

There is also a maximum season length $T_{\max}$, beyond which the industry would not voluntarily choose to fish. This can be seen by noting that the numerator of

REGULATED OPEN ACCESS RESOURCE USE   19

(A2) approaches zero at some season length defined by $PqX_0e^{-qET} - v = 0$. Solving this simultaneously with the rent equation (A1) set equal to zero for T defines what we refer to as T_{max} and we have:

PROPOSITION 2. *The longest season length that the industry would voluntarily operate over is defined by*

$$T_{max} = \frac{f \ln[(PqX_0)/v]}{PqX_0 - v(1 + \ln[(PqX_0)/v])}. \tag{A7}$$

At season lengths longer than T_{max}, the industry would actually be incurring negative variable profits. This can be seen by noting that X_0e^{-qET} is equal to the terminal stock level X_T. Moreover, since the harvest rate is $h = qEX(t)$, $PqX_T - v$ is the average daily variable profit associated with the last unit of capacity utilized on the last day of the season. As the season progresses, average variable profits per day decline from their initial maximum of of $PqX_0 - v$ to their minimum of $PqX_T - v$ on the last day. Thus Proposition 2 ensures nonnegative variable profits throughout the season. Season lengths larger than T_{max} would be forcing the industry to fish the ending biomass down to levels that generate variable profit losses at season end and the industry would always choose not to incur such losses by truncating fishing at T_{max}. Hence season lengths where regulations are binding are limited to those between T_{min} and T_{max}.

If $T_{min} < T < T_{max}$, then the numerator of (A2) will be negative and the sign of the derivative will be positive. The second derivative of the entry function with respect to T can be shown to be

$$\frac{d^2E}{dT^2} = E\left\{2\left[\frac{v - PqX_0e^{-qET}}{T(PqX_0e^{-qET} - v) - f}\right]^2 + \frac{f^2Pq^2EX_0e^{-qET}}{[T(PqX_0e^{-qET} - v) - f]^3}\right\}. \tag{A8}$$

At T_{max}, the $(PqX_0e^{-qET} - v)$ terms vanish and hence the second derivative is negative. Hence, over the relevant range, as T increases E also increases monotonically, reaching a maximum at $E(T_{max})$ as shown in Fig. 1.

APPENDIX B

Comparative Statics and Existence of a Regulated Equilibrium

Comparative statics properties of the regulated open access model are tedious but easily derived. The basic model is as denoted in the text in equation system (12) along with the quota rule (5) and the biomass dynamics equation (8) with the biological equation assumed to be in a steady state. As discussed above, the system is recursive and the parameters a, b, c, and d determine equilibrium values for the biomass level X and the quota Q. Comparative statics properties of the equilibrium levels of E and T in the industry/regulator block can be computed most simply by determining the direct effects of parameters and the indirect effects of parameters operating on the biomass/quota block. (Comparative statics calculations are available upon request from the authors.)

20   HOMANS AND WILEN

The comparative statics properties summarized in Table I assume an "interior" equilibrium in the sense that the fishery is characterized by a joint equilibrium where regulations are binding. Circumstances where active regulation is necessary exist where there is an equilibrium intersection of the two equations representing the industry and regulator behavior. The circumstances which will lead to a regulated open access equilibrium can be described as follows. Begin with the relationship which describes the level of cumulative effort ET_{max} defined as

$$ET_{max} = \frac{1}{q} \ln\left[\frac{PqX_0}{v} \right] \tag{B1}$$

for some arbitrary level of biomass X_0. Note that this defines a rectangular hyperbola in E, T space and hence can be compared with the regulatory agency's choice which is also defined by a rectangular hyperbola, namely

$$ET_{reg} = \frac{1}{q} \ln\left[\frac{X_0}{(1-d)X_0 - c} \right]. \tag{B2}$$

A joint or binding temporary regulatory equlibrium occurs whenever the maximum level of effort which dissipates unregulated rents or ET_{max} is greater than the amount of total fishing effort desired by the regulatory agency ET_{reg}. Thus the condition for existence of a temporary regulatory equilibrium is

$$ET_{max} > ET_{reg} \Rightarrow \frac{PqX_0}{v} > \frac{X_0}{(1-d)X_0 - c}. \tag{B3}$$

Thus a regulated equilibrium is likely to occur when P, q, or X_0 are large, or when v is small, or whenever d or c are small. Note that d small implies that regulators are particularly concerned about high exploitation rates.

The above refers to the (temporary equilibrium) case where X_0 is some arbitrary value which could of course be its long run open access steady state value. In general, however, a full long run equilibrium must not only have the industry earning zero rents, but the biomass must also be in equilibrium at that state. The long run steady state biomass level that would evolve in an unregulated (Gordon) system is

$$X_G = \frac{a - 1 + \sqrt{(1-a)^2 - (4bv/Pq)}}{2b}, \tag{B4}$$

while the biomass level that would emerge in a regulated open access system is given by Eq. (11) in the text. The complete characterization of necessary conditions for a binding long run regulated open access equilibrium is given when these are inserted into Eq. (B3) above, or

$$\frac{PqX_G}{v} > \frac{X_{safe}}{(1-d)X_{safe} - c}. \tag{B5}$$

REFERENCES

1. J. Bhagwati, Directly unproductive profit seeking activities, *J. Polit. Econom.* **90**, 988–1002 (1982).
2. J. Buchanan, R. Tollison, and G. Tullock (Eds.), "Toward a Theory of a Rent Seeking Society," Texas A & M Univ. Press, College Station (1980).
3. S. Capalbo, "Bioeconomic Supply and Imperfect Competition: The Case of the North Pacific Halibut Industry," Ph.D. dissertation, University of California at Davis (1982).
4. B. Cook, "Optimal Harvest Levels for Canada's Pacific Halibut Fishery," Ph.D. dissertation, Simon Fraser University (1983).
5. J. Conklin and W. Kolberg, Chaos for the halibut?, *Mar. Resour. Econom.* **9**, 159–182 (1994).
6. K. Criddle, "Modeling Dynamic Nonlinear Systems," Ph.D. dissertation, University of California at Davis (1989).
7. K. Criddle and A. Havenner, Forecasting halibut biomass using systems theoretic time series, *Amer. J. Agr. Econom.* **71**, 422–431 (1989).
8. J. Crutchfield and A. Zellner, "Economic Aspects of the Pacific Halibut Fishery," U.S. Department of the Interior, Fishery Industrial Research, Washington, DC (1962).
9. R. Deriso, Risk averse harvesting strategies, *in* "Resource Management: Proceedings of the Second Ralf Yorque Workshop, Lecture Notes in Biomathematics" (M. Mangel, Ed.), Springer-Verlag, Berlin (1985).
10. H. S. Gordon, The economic theory of a common property resource: The fishery, *J. Polit. Econom.* **62**, 124–142 (1954).
11. L. P. Hansen, Large sample properties of generalized method of moments estimators, *Econometrica* **50**, 1029–1054 (1982).
12. R. Hilborn and C. J. Walters, "Quantitative Fisheries Stock Assessment: Choice, Dynamics, and Uncertainty," Chapman and Hall, New York (1992).
13. F. R. Homans, "Modeling Regulated Open Access Resource Use," Ph.D. dissertation, University of California at Davis, (1993).
14. J. Karpoff, Suboptimal controls in common resource management: The case of the fishery, *J. Polit. Econom.* **95**, 179–194 (1987).
15. D. Ludwig and C. J. Walters, A robust method for parameter estimation from catch and effort data, *Canad. J. Fish. Aquat. Sci.* **46**, 137–144 (1989).
16. T. Quinn, R. Deriso, and S. Hoag, "Methods of Population Assessment of Pacific Halibut," International Pacific Halibut Commission Scientific Report No. 72 (1985).
17. C. Rowley, R. Tollison, and G. Tullock (Eds.), "The Political Economy of Rent Seeking," Kluwer Academic, Boston (1988).
18. A. Scott, The fishery: The objectives of sole ownership, *J. Polit. Econom.* **63**, 116–124 (1955).
19. "SHAZAM User's Reference Manual Version 7.0," McGraw–Hill, New York (1993).
20. B. Skud, "Revised Estimates of Halibut Abundance and the Thompson–Burkenroad Debate," International Pacific Halibut Commission Scientific Report No. 63 (1975).
21. K. Stollery, A short-run model of capital stuffing in the Pacific Halibut Fishery, *Mar. Resour. Econom.* **9**, 137–153 (1986).
22. R. Turvey, and J. Wiseman (Eds.), "The Economics of Fisheries," FAO, Rome (1957).
23. U.S. Department of Commerce, National Ocean and Atmospheric Administration, Draft Environmental Impact Statement for Proposed Individual Quota Management Alternatives for the Halibut Fisheries in the Gulf of Alaska and Bering Sea/Aleutian Islands, July (1991).
24. J. E. Wilen, Towards a theory of the regulated fishery, *Mar. Resour. Econom.* **1**, 369–388 (1985).
25. J. E. Wilen and F. R. Homans, Marketing losses in regulated open access fisheries, *in* "Fisheries Economics and Trade: Proceedings of the Sixth Conference" (J. Catanzano, Ed.) IFREMER-SEM (1994).
26. J. E. Wilen and F. R. Homans, Unraveling rent losses in modern fisheries: Production, market, or regulatory inefficiencies? "Fisheries Management—Global Trends" (D. Huppert, E. Pikitch, and M. Sissenwine, Eds.) Univ. of Washington Press, Seattle, 1997.

Part VI
Applications

[41]

PROPERTY RIGHTS AND EFFICIENCY IN THE OYSTER INDUSTRY*

RICHARD J. AGNELLO and LAWRENCE P. DONNELLEY
University of Delaware

INTRODUCTION

OWNERSHIP rights to resources and product are important determinants of economic efficiency. The effects of a communal or common property right system for resource allocation have received the principal attention of contributors to the literature pertaining to fishery economics. In the context of this literature, common property means that any member of the community has the right to harvest the fish stock. However, the right in practice may be constrained by government enforced conservation policies such as bag limits and size limitations. Consensus among the discussants is that common property leads to over-exploitation of fish stocks and perhaps extinction of the species.[1] The property rights literature draws similar conclusions.[2] To date, the literature contains few examples of empirical investigations of the effects of property right systems and, in particular, applications to the fisheries.[3]

In this paper the significance of property rights for one aspect of economic efficiency, labor productivity, is empirically tested using data from the U.S.

* Research for this paper was aided by various grants received from the Sea Grant Program of the National Oceanic and Atmospheric Administration, U.S. Department of Commerce.

[1] The seminal articles in the literature are H. S. Gordon, The Economic Theory of a Common Property Resource: The Fishery, 62 J. Pol. Econ. 124 (1954); and Anthony Scott, The Fishery: The Objectives of Sole Ownership, 63 J. Pol. Econ. 116 (1955). Recent contributions in addition to modifying some conclusions dealing with regulatory policy have noted the potential for extinction of a species; see Vernon L. Smith, On Models of Commercial Fishing, 77 J. Pol. Econ. 181 (1969); J. R. Gould, Extinction of a Fishery by Commercial Exploitation: A Note, 80 J. Pol. Econ. 1031 (1972); Colin W. Clark, Profit Maximization and the Extinction of Animal Species, 81 J. Pol. Econ. 950 (1973).

[2] For examples, see Harold Demsetz, Toward a Theory of Property Rights, 57 Am. Econ. Rev. pt. 2, at 347 (Papers & Proceedings, May 1967); and Steven N. S. Cheung, The Structure of a Contract and the Theory of a Non-Exclusive Resource, 13 J. Law & Econ. 49 (1970).

[3] Frederick W. Bell, Technological Externalities and Common Property Resources: An Empirical Study of the U.S. Lobster Industry, 80 J. Pol. Econ. 148 (1972) estimates the redundant effort employed in the northern lobster fishery. He concludes that approximately 50% of the current level of fishing effort is required to achieve economic efficiency.

East and Gulf coast oyster industry. We first establish institutional characteristics of the oyster industry pertaining to the property rights issue. A
discussion follows as to how various property rights systems are expected to
affect labor productivity. Finally, a regression model employing a continuous
measure of property rights across coastal states is estimated and analyzed.
The evidence suggests that communal property rights have a significant
adverse impact on labor productivity in comparison with private rights.

THE U.S. OYSTER FISHERY

The American Eastern oyster represents the resource base of both the Gulf
and Atlantic coast oyster industries.[4] Following a brief larva stage (spat), the
oyster connects permanently to a firm subaqueous material such as rock or
shell deposits (cultch). Its habitat is the intermediate salinity waters of the
seacoast's intertidal zone and of inland rivers and bays. Water current,
temperature and biological productivity, in addition to salinity, are determinants of the resource productivity of a given parcel of subaqueous land.

Two property right structures characterizing the oyster industry in each of
the Atlantic and Gulf states can be identified, a private right structure based
mainly on leaseholds and a common right system. The courts have long
recognized rights to subaqueous land for the people of each state for their
common use. State legislatures have confirmed and modified these rights.
Although, the federal government claims jurisidiction over a three mile
coastal zone, Congress has ceded jurisdiction over land and resource use
rights within this zone back to the states.[5] The states have exercised their
jurisdiction over the oyster resource in similar ways. In general, natural
oyster beds have been set aside as a common fishery for state residents
whereas other submerged land parcels are available for private leasing.

The distinction made by each state between natural oyster beds and other
land permits the development of an empirical measure of a state's property
right variable used in subsequent sections of this paper. Oysters found on
subaqueous land classified as a "natural oyster bed" are in general an open
access (common property) fishery for state residents. Other areas may be
leased and used exclusively by the lessee for the cultivation of oysters. Regardless of how a state arrives at the determination of which lands are
natural oyster beds and which are available for leasing, each state makes a
clear demarcation of its common from leasable land. However, great variation among the states exists in the proportion of area and quality of land set

---

[4] This paper omits consideration of the Pacific coast oyster fishery since data classifying
harvests by private and common property origin are unavailable for Pacific coast states.

[5] See, Garrett Power, *More About Oysters Than You Wanted to Know*, 30 Md. L. Rev. 199,
216-23 (1970) for a detailed review of court decisions involving rights to submerged land.

PROPERTY RIGHTS IN THE OYSTER INDUSTRY 523

aside for public or common use versus private use depending on how broadly administrators define the term "natural oyster bed." An examination of the proportion of oyster catch by weight on private grounds to total catch by state reveals ratios ranging from a maximum of 1 to almost zero during the 1950's and 1960's.

The determination of which grounds are natural oyster beds uses criteria that are imposed either by state legislators, the courts, or administrative agencies. Variation in the value of the property rights variable across states is strongly influenced by the criteria used by decision makers to define natural beds.[6]

Some variation between states also occurs in the particular rights which the states attach to both private and common property. In the case of private property, leaseholds predominate. Though riparian rights are specified in several legislative codes, such rights are typically not sufficiently extensive to include productive oyster waters because of problems of siltation and/or inadequate biological productivity of riparian waters.[7]

Terms of the leasehold also vary across states. The legislative codes of most states specify who may obtain a leasehold, the duration of the lease, minimum fee and maximum and/or minimum acreage that is leaseable. In general, states attempt to restrict the leasing privilege to state residents including any corporation chartered in the state.[8] All states encourage long term leases.

Other minor differences between states arise in specifying minimum size limits, shell replenishment provisions, culling restrictions, and harvest seasons.

Most states also set certain gear restrictions. The oyster harvesting industry is largely comprised of many small firms. Harvesting technique is a crucial determinant of the size of the firm. In the case of the tonging technique, a single individual using a small boat is the typical firm, whereas the dredging technique requires a larger boat with a multiple-size crew. The choice of technique is usually constrained. Most states impose restrictions on harvesting technique in order to conserve the resource stock. As a general rule, states prohibit dredging on common property. Thus, where common

---

[6] Maryland, for example, classifies a natural bed as one such that the natural growth of oysters ". . . is of such abundance that during the preceding five years the public has resorted to them for livelihood." *Id.* at 214. Courts view one individual declaration of one day's work in a five year period as sufficient evidence of the existence of a natural bed. Most states employ a less restrictive definition for a natural bed.

[7] When specified in the legislative codes, riparian rights usually do not include natural oyster beds.

[8] Important exceptions are Massachusetts which specifies that only residents of coastal towns may lease and Maryland, which does not allow corporations to lease. Garrett Power, *supra* note 5, contains a discussion of the constitutionality of the non-resident and corporate distinction.

property rights prevail, the size of the firm is relatively small and capital intensity is low.[9]

COMMON PROPERTY AND ECONOMIC EFFICIENCY

In a world where transactions costs are significant and resources are scarce common property right systems result in less efficient resource allocation than private right systems. This occurs because communal rights do not ensure that the costs of an individual harvester's actions in exploiting the resource are borne fully by him. In attempting to maximize the value of his common right, the individual can be expected to over-exploit the resource leading to depletion of the stock, and in the extreme case to extinction of even replenishable resources. Private property internalizes the costs of the harvester's actions. In a similar way, a private right enables the producer to capture a greater proportion of the benefit of his activity in comparison to a communal right.

The type of externality manifested by communal rights to oysters depends on the stage at which the production process is being examined. In general, an oysterman can separate production into two stages: establishment of suitable conditions for oysters to mature, and second, the act of harvesting oysters for market.

The most important requirement under the control of an individual oysterman for ensuring maturation of oysters is that a clutch be established.[10] Other things equal, an individual could transform barren subaqueous land into productive grounds by depositing rock or shells on to it. He could then plant seed oysters on the material to hasten maturity. It should be obvious, however, that this kind of activity is minimal unless the planter is given the exclusive right of harvesting oysters from that land parcel. Under a communal right system, such a guarantee is nonexistent.

Three problems arise at the harvesting stage. First, the harvesting process typically removes cultch. Unless replaced, cultch removal implies disinvestment. On communal property once again, incentive to replenish this material is minimal. In practice, most states stipulate that some proportion of shells be redeposited on communal oyster grounds. In many states, however, processors rather than harvesters bear this cost.[11] Frequently, states

---

[9] Fundamental differences in technique such as the oyster raft culture, which is not commercially prevalent along the U.S. Gulf and Atlantic coasts, do not distinguish private from common grounds.

[10] Water quality and salinity are assumed not to be subject to control by the firm.

[11] Georgia, Mississippi, New Jersey, North and South Carolina and Virginia require processors to replant or turn over to the state varying proportions of their processed shells. Since an important part of the retail market consists of oysters in the half shell, this requirement may be constraining. Only New Jersey and Mississippi permit cash payments to be made to the state in lieu of shells.

also subsidize the communal system by undertaking cultch rehabilitation programs. Second, overcrowding of vessels on particularly rich water areas leads to congestion. Finally, immature oysters are caught indiscriminantly with the mature. Consequently, the growth behavior of the stock is affected. Most states set a minimum size restriction and require that culling of under-sized oysters be done immediately upon being caught.[12] Culling restrictions are also accompanied by stipulations that any cultch that was removed by the act of harvesting also be replaced immediately.

Given the variety of external costs associated with communal rights to oysters, it remains an empirical consideration as to whether the enforcement of private rights is economically feasible. It has been noted that in the case of migratory fish the policing problem may make exclusive use rights unman-ageable.[13] In the case of non-migratory species, however, property rights seem practical. In the case of sessile species such as oysters, private property rights are likely to incur minimal enforcement costs. Indeed, private rights have been instituted and enforced in all states to some degree.

THE MODEL

In order to test the hypothesis that common property is a less efficient ownership form than private property, a testable model relating property rights to measures of economic efficiency is developed. Measures of effi-ciency in oystering for which limited data are available include labor and capital productivities. However, due to difficulties in obtaining data that accurately measure capital used in the production of oysters, primary focus is placed on labor productivity as a resource efficiency measure. The hypothesis to be tested is that private property ownership has a positive relationship with labor productivity.

Several related arguments support this hypothesis. First, as noted above, exclusive user rights provide incentive for firms to pursue a policy of invest-ing in cultch and maintaining it at a desired level as influenced by market conditions. Second, congestion is not likely to occur. Third, in general a direct positive effect of private property results from an optimal schedule of harvesting the resource over time. It is well documented in the fishery litera-ture cited in the introduction that the externality effects of common property cause an over-utilization of the resource which eventually tends to deplete the resource, causing output and productivity to fall. It is likely that deple-

[12] Vernon L. Smith, *supra* note 1, has referred to the second harvesting externality as the crowding externality and to the third as the mesh externality. It should be noted that oyster harvesting cannot vary mesh size to effectively select particular sizes. The clusters can only be separated by hand. However, by mandating that the separation occur immediately rather than onshore, crew sizes and costs are increased.

[13] Steven N. S. Cheung, *supra* note 2, at 52-54.

tion will have already set in since the industry is well established for most states although the degree of exhaustion is unlikely to be uniform across states for a multitude of biological and economic factors. Finally, a communal property structure tends to lower productivity by requiring the use of inefficient technology as a conservation device. In oystering, inefficient technology often takes the form of obsolete capital regulated into use by state codes. In general, states relying on common property right structures tend to impose greater restrictions on the use of capital than private property states.[14] The intent of the legislation clearly is to prevent resource depletion over time by mandating old technologies. Although capital restrictions tend to mitigate the depletion tendency with common property rights, inefficiencies due to lack of investment incentives, congestion and the old technologies themselves remain.

The labor productivity model estimated and tested relates average physical product of labor (APL) to the extent of private property rights (R), a measure of capital intensity (K), biological factors (B), and a market wage rate (W).[15] A wage rate variable (outside the oyster industry) is included to measure the opportunity costs of oyster fishermen. If their alternative wage, for example, is greater in state A than state B, then the market adjustments will lead oyster firms to employ fewer workers in A than B, which in turn will lead to higher observed labor productivity in A than B, holding constant the structure of property rights and other relevant variables. The equation to be estimated and tested on cross sectional market data is

$$\overset{+}{APL} = \alpha_0 + \overset{+}{\alpha_1 R} + \overset{+}{\alpha_2 K} + \overset{-}{\alpha_3 B} + \overset{+}{\alpha_4 W}$$

where + and − signs refer to the expected signs of the parameter associated with each variable.

Other factors are doubtless important in explaining labor product variations across observations in the oyster industry. Unfortunately, little systematic and reliable information is available on state management policies, labor quality, hours worked and environmental factors such as weather, pollution, and siltation. Management of oyster beds by states is likely an important factor in explaining labor productivity. Replenishment of oyster

[14] For example, in the predominantly common property right state of Maryland power dredging is prohibited in the harvest of oysters. Consequently, dredging takes place through the use of sail powered craft called skipjacks, the newest of which is over fifty years old.

[15] An average physical product of labor function can be derived from a given production function specification. The production function that we have in mind relates physical output to capital, labor, and biological factors. Data availability limits us to a single negative biological factor which is discussed later in the paper. This approach is somewhat oversimplified since property rights have cumulative dynamic effects on the production function. Over time property rights will certainly impact the resource stock (a biological factor) and capital intensity.

cultch, pollution regulation, and direct subsidization through the provision of seed oysters are examples of how governmental actions can have a positive effect on labor productivity in the industry. Systematic information on state management activities and legislation affecting the oyster industry, although extremely important, is not readily available as a control in our empirical analysis on property rights.[16]

It is clear that the exogenous variables, private property and capital, are not entirely independent of each other due to state mandated gear restrictions often associated with common property. Thus property right structure affects labor productivity through both the R and K variables in the model. It is an empirical question, however, whether this potential multicollinearity problem is severe. From theory, in fact, one may argue that capital differences quite unrelated to private property rights may occur in data temporally or spatially observed. Different natural environments for harvesting the resource may account for variations in capital across observations. In oystering environmental differences affecting capital include oyster bed densities, water depth, and tidal conditions. For these reasons it is necessary to include both property rights and capital intensity measures in explaining differences in labor productivity across states. To the extent that R and K are related the effect that private property rights has on efficiency is imbedded in the coefficients associated with both R and K.

EMPIRICAL FINDINGS

Both cross section and time series data aggregated at the state level are used to test the efficiency hypothesis on private property rights using a linear regression framework. Cross section data are available for the 16 Atlantic and Gulf coastal states from Massachusetts to Texas for each year from 1950 to 1969. These annual data are averaged for the twenty year period for each state since individual years are not likely in long-run equilibrium with respect to omitted factors for which data are not readily available. Omitted factors that probably affect productivities more in the short-run than in the long-run include weather variation, biological factors, and industry expansion or contraction.

In the time series analyses annual observations from 1945 to 1969 for Maryland and Virginia are used. We focus on these states because of continuous data availability, their historical dominance of the U.S. oyster industry, and also since the major portion of the oyster harvest for each state is

---

[16] With respect to direct monetary subsidization to the oyster industry, it is generally the case that state expenditures are imbedded within a larger departmental or agency budget, thus making it difficult to isolate the subsidization. Also, when citations of state activities positively affecting the industry can be found, it is sometimes a recognition and response to past neglect rather than an enduring program or commitment of state support.

obtained from a common body of water, the Chesapeake Bay. The temporal
variation in private property rights for Maryland and Virginia is accen-
tuated when differences in productivity are related to differences in the
property right variable (R) for these states. This latter formulation is preferred in time series analyses since only limited variation exists in an individual state's property rights structure over time.

All variables except for a measure of wages are taken or derived from data
in the *Fishery Statistics of the United States* compiled annually by the Fish
and Wildlife Service of the U.S. Department of the Interior. Average physical product of labor (APL) data are derived from data on total harvest (meat
weight) and the total labor force in the oyster industry for each coastal state.
APL is measured as pounds per man year. The private property right variable (R) is obtained from data classifying catch by leased or common property origin (that is, private or public oyster beds).[17] A continuous variable
measuring the fraction of a state's total oyster catch harvested from privately
leased subaqueous grounds is constructed as the measure of private property.[18] The variable R encompasses legislative, judicial, and administrative
variation across coastal states or for a state through time, and provides a
continuous measure to contrast the effects of private property rights with
those of common property. Since data classifying labor and capital as harvesting on public versus private grounds are unavailable, it is not possible to
compare directly labor productivity on public grounds with private grounds
within a state. Also since consistent data over time and across states on the
size of areas devoted to common versus private ownership are unavailable,
an acreage measure of R is not possible.[19]

Since a total capital measure does not exist due to the nonadditivity of
various physical forms of capital and unavailability of reliable current val-

[17] This classification is based on harvest origin information provided the National Marine
Fishery Service by state fishery agencies which regulate common property grounds, grant
leaseholds, and collect fees. State fishery agencies in most states have relatively complete
statistical gathering and reporting systems. Where state data are not considered adequate, the
National Marine Fishery Service conducts its own surveys and visits wholesale dealers and
processors of fishery products to obtain information on purchases. More information on statistical survey procedures can be found in U.S. Dep't of Interior, Fish and Wildlife Service, Fishery
Statistics of the United States 1969, at 435-46.

[18] The 1950-1969 average value of R for each coastal state (in parentheses) indicates considerable cross sectional property right variation: Massachusetts (.67), Rhode Island (.86), Connec-
ticut (.98), New York (.96), New Jersey (.98), Delaware (.98), Maryland (.17), Virginia (.74),
North Carolina (.11), South Carolina (.89), Georgia (.99), Florida (.09), Alabama (.05), Mississippi (.04), Louisiana (.76), and Texas (.04).

[19] We note that an acreage specification of R has econometric merit since it eliminates any
potential positive simultaneous equations bias in the estimated coefficient associated with R
when R is measured as an output ratio. However, since it appears that most states allocate
highest quality oyster breeding grounds to common use, one could argue that the effect private
property rights have on labor productivity is likely underestimated by the statistical analysis.

PROPERTY RIGHTS IN THE OYSTER INDUSTRY 529

ues of capital, a proxy measure is employed. The capital variable measures capital intensity by focusing on the extent of dredging activity since gear for dredging is generally the most capital intensive method of oyster harvest. The ratio of the tonnage of dredge vessels to tonnage of all oyster boats and vessels for a given state is the empirical measure of K used.[20]

A biological variable (B) representing the MSX disease in oysters is constructed as a dummy variable from descriptive information on its prevalence contained in the *Fishery Statistics of the United States*. This protozoan oyster parasite struck oyster populations in New Jersey, Delaware, and Virginia in the 1960's. The MSX variable is used in both cross section and time series analyses. In cross section analyses MSX takes on a unit value in states affected and a zero value for unaffected states. In time series analyses MSX takes a unit value for Virginia from 1960 to 1969 and a zero value from 1945 to 1959. For Maryland MSX is zero since the state was unaffected by the MSX disease.[21]

Two wage variables are used to control for labor opportunity costs across states. W measures the average hourly earnings in dollars of production workers on manufacturing payrolls and comes from the *Handbook of Labor Statistics*.[22] W* represents the average annual income in dollars earned by a state's fishermen excluding those engaged in oyster harvesting. W* is calculated from data contained in the *Fishery Statistics of the United States*.

The results of testing the labor productivity model using ordinary least squares regression are found in Tables 1 and 2. We report the results first for the impact of R and B on APL and then controlling for variation in W and K across states and time. From the cross section results in Table 1 private property rights (R) is observed to positively influence labor productivity and is statistically significant in all cases. R is consistently the most significant single determinant of average product of labor, and its influence on APL remains remarkably stable when additional controlling factors are included in the model.

From Table 1 it can be seen that the MSX disease variable (B) has the expected negative impact on labor productivity although not significant for regressions on twenty year averages since the disease was not prevalent

---

[20] One tong boat is assumed to equal one ton. Alternative measures of capital were used in regression analyses yielding similar results to those reported. The ratio of the number of dredge vessels to total vessels was one such measure.

[21] Data limitations restrict us to one biological factor. A variable measuring average ocean water temperature was used as a control in some preliminary cross section analyses, but was not found to be significant since the effect of quicker maturation of oysters in warmer climates is apparently offset by a lower meat to shell ratio. The variable is omitted from the reported results. Also as noted previously no data are available on siltation and pollution levels for the coastal states through time which doubtless have had a negative impact on productivity in industrial regions.

[22] U.S. Dep't of Labor, Bureau of Labor Statistics, Handbook of Labor Statistics (1969).

TABLE 1
AVERAGE PHYSICAL PRODUCT CROSS SECTION REGRESSIONS

Exogenous Variables	Coefficients of Alternative Models					
	I	II	III	IV	V	VI
Constant	2441	−7447	1839	420	−4226	393
R	5790	4758	5967	5222	4803	5231
	(2.24)**	(1.85)*	(2.14) *	(2.23)**	(1.95)*	(2.06)*
K				4955	4035	4950
				(2.03) *	(1.43)	(1.93)*
B	−1436	−2013	−1619	−2542	−2630	−2550
	(−0.54)	(−0.79)	(−0.57)	(−1.05)	(−1.06)	(−0.98)
W		5391			2737	
		(1.48)			(0.69)	
W*			.1960			.0095
			(0.24)			(0.01)
R^2	.28	.39	.29	.47	.49	.47

Regressions performed on averages from 1950-1969 of annual observations from the 16 coastal states from Massachusetts to Texas. Number in parentheses are t values, and R^2 is the unadjusted coefficient of determination.
* Significant at the .05 level, one-tailed test.
** Significant at the .05 level, two tailed test.

during the 1950's. For unreported regressions on five year averages during the 1960's the MSX disease was found to be highly significant. The statistical results for the market wage variables are somewhat surprising, and indicate that controlling for wage differentials across states makes little empirical difference. Although important theoretically, wages lack explanatory power empirically primarily due to a small sample size and relatively little variation across observations. In addition, the wage variables for which data are available may be poor proxies for the true opportunity cost of oyster workers. Controlling for wages has a mixed effect on R, lowering R's impact in the case of W and increasing it for W*.

Private property rights are in general more significant than capital intensity in explaining labor productivity. Although not revealed in Table 1, R and K are somewhat collinear with simple correlations from between .2 to .3. Thus when capital intensity is added, the impact and significance of the private property rights variable is reduced in most cases. In determining the total effect private property has on the efficient use of resources, the equation specification omitting the capital variable may be preferable. We recall that common property rights adversely affects efficiency by discouraging investment, causing congestion and depletion, and necessitating technological restrictions as conservation devices. The equation specifications excluding the K variable pick up all property right effects through the R variable whereas including the capital variable attributes inefficient technologies to capital rather than a partial effect of common property rights. Thus the total

PROPERTY RIGHTS IN THE OYSTER INDUSTRY 531

impact of private property rights on labor productivity is revealed more clearly when K is omitted from the regression if we assume that differences in K across states are related only to environmental factors and are not systematically related to R.

Lastly we note from Table 1 that the availability of only 16 observations for the Atlantic and Gulf Coastal states reduces the sharpness of the findings. Less than half of the variation in average physical product is explained by the exogenous variables included. The low R^2 is caused by the many omitted variables especially environmental factors such as siltation and pollution and state management and subsidization activities for which no systematic data are available. Nevertheless the continuous private rights variable R remains highly significant with all of the data's shortcomings.[23]

Table 2 presents time series regression results over the 1945 to 1969 period for the states of Maryland and Virginia. In these regressions wages are not explicitly controlled for since we assume opportunity costs to be equal in the two contiguous states. Property rights again exhibit the expected parameter sign, and are highly significant with the exception of the Maryland regressions where R shows little temporal variation. The MSX disease which affected only the more saline waters of Virginia, and capital intensity both yield expected parameter signs. Autocorrelation is a potential problem in the time series analyses with Durbin-Watson statistic values often indicating positive autocorrelation or residing in the inconclusive range. It is clear that the addition of the capital variable lowers the significance of the private property rights variable in explaining labor productivities. The simple correlation between R and K is between .5 and .6 in the time series regressions. Again due to multicollinearity between R and K, the effect of private property rights on efficiency is likely understated when K is included in the regression.

An additional use can be made of the Maryland and Virginia time series data by comparing their long run average labor productivities in oystering in order to obtain a measure of relative efficiency. We find the 25 year APL average to be 59 per cent greater in Virginia than in Maryland.[24] Since they

[23] Although regressions for five year averages are not reported, we note that short run results are in general agreement with the twenty year averages reported. There is some variation in the explanatory power of the model between five year periods where omitted factors cause some dramatic changes in the APL variable for some states without corresponding changes in exogenous variables. We note in particular several northern industrial states for which labor productivity falls possibly due to a decrease in water quality, and some less industrial Southern states where productivity rises. Also in Florida where common property prevails and the oyster industry is comparatively young, some of the adverse effects on efficiency of common property rights may have not set in yet.

[24] In comparing the efficiencies of Australia's private and public airlines, David G. Davies, The Efficiency of Public versus Private Firms: The Case of Australia's Two Airlines, 14 J. Law & Econ. 149, 163-64 (1971), constructed a similar ratio of long run productivity averages.

TABLE 2
AVERAGE PHYSICAL PRODUCT TIME SERIES REGRESSIONS

Endogenous Variable	Exogenous Variables				Statistics	
APL	Constant	R	B	K	R^2	D-W
Maryland	2776	508 (0.19)			.001	0.39
Maryland	2342	−3518 (−1.17)		4865 (2.39)**	.21	0.40
Virginia	2560	3785 (3.78)**	−2048 (−8.06)**		.87	2.00
Virginia	2748	3006 (2.15)**	−2145 (−7.58)**	1175 (0.80)	.87	1.99
Va.—Md.	−2511	8310 (4.40)**	−1349 (−3.65)**		.69	1.01
Va.—Md.	−1649	6641 (2.95)**	−1775 (−3.64)**	2269 (1.31)	.71	0.95

Regressions performed on annual data from 1945 to 1969. In the Va.—Md. regressions all variables are measured as differences between the Virginia and Maryland values. Numbers in parentheses refer to t values. R^2 and D-W refer to the unadjusted coefficient of determination, and the Durbin-Watson statistic for autocorrelation respectively.

** Significant at the .05 level, two-tailed test.

share common bodies of water, we attribute much of the difference in labor productivity to the difference in the private property rights variable (R). During the 1945 to 1969 period the percentage of oyster harvests coming from privately leased beds were 74 per cent and 17 per cent for Virginia and Maryland respectively.[25]

CONCLUSIONS

The empirical findings suggest that private property rights do in general make a significant difference in a state's average labor productivity in oyster harvesting. Common property rights are associated with low labor productivity resulting from disinvestment, congestion, over exploitation and government restrictions. Regulation in common property rights states aims at conserving the oyster resource by mandating labor intensive technologies. If labor opportunity costs are roughly equal across states, it is likely that labor intensive methods in common property states are inefficient, and that social benefit could be increased by encouraging private leasing of oyster beds as an alternative to the common property structure utilized by many states.

The regression results provide an indication of the magnitude of the welfare loss due to common property rights. Based on R's coefficient of $5790 in

[25] Although Maryland and Virginia obtain most of their oyster harvest from the Chesapeake Bay, we note that differences exist in biological factors such as water temperature and salinity levels which have not been controlled for.

Model I, the point elasticity of labor productivity with respect to private property rights evaluated at the sample mean is .608. Thus a 10 per cent increase in private property rights across states could be expected to increase average physical product of labor by 6.08 per cent. Using 1969 average product and exvessel prices across states, a 10 per cent increase in private property rights would increase average physical product by 338 pounds per man and average income by $179. Furthermore if all coastal states had relied entirely on private property in oyster harvesting in 1969, R would have increased by over 70 per cent implying an increase in average oyster-men's incomes of around $1300 or almost fifty per cent of 1969 average income. Since costs of enforcing private rights do not seem to be a serious problem for sessile species in intertidal coastal waters, one can conclude that considerations other than economic efficiency are used by states relying on common property for the oyster industry.

[42]

Scand. J. of Economics 89(2), 145–164, 1987

Production Economics and Optimal Stock Size in a North Atlantic Fishery

Trond Bjørndal

Norwegian School of Economics and Business Administration, Bergen, Norway

Abstract

This paper contains an analysis of a fishery production function and an empirical analysis of the North Sea herring fishery. An intertemporal profit function is defined by introducing stock dynamics and the concept of a sole resource manager. Some new results with respect to the relationship between the optimal stock level and production technology are derived. Estimates of the optimal stock level for herring are also presented.

I. Introduction

In fisheries economics the production function is generally treated in a restrictive manner. The Schaefer (1957) function, which is linear in both effort and stock size, is commonly used. Although this model may be suitable for some fisheries, theoretical evidence suggests that it is inappropriate for modelling schooling species such as herring. In this paper, we specify a more general production function to represent the North Sea herring fishery. A Cobb–Douglas harvest supply function is estimated and the parameters of the underlying production function are derived. Interesting results from this analysis include a low stock output elasticity, while a hypothesis of increasing returns in all factors is generally accepted. In addition, the Schaefer function is found to be inappropriate for representing the herring fishery.

The static analysis is extended by introducing stock dynamics and sole-owner management. An intertemporal profit function is maxi-

*This paper is based on my Ph.D. dissertation; Bjørndal (1984). An earlier version of the paper was presented in 1984 at the meeting of the Canadian Economics Association and the European meeting of the Econometric Society. I would like to thank C. W. Clark, G. R. Munro, P. A. Neher, W. E. Schworm, R. S. Uhler and D. V. Gordon for helpful comments. This research has been supported by a grant from the Norwegian Fisheries Research Council.

146   *T. Bjørndal*

mized and an expression for the optimal stock level is derived. We analyze the relationship between the optimal stock level and production technology and derive some new analytical results. Optimal stock levels are estimated for the case of a sole resource manager. It is noteworthy that the quantitative results show that costs have a considerable effect on the optimal stock level, even if the stock output elasticity is low. Moreover, the optimal stock level is found to be rather insensitive to changes in the discount rate.

A production function for an open-access fishery is developed in Section II. A harvest supply function is estimated for North Sea herring. The concept of a sole owner and intertemporal profit maximization are introduced in Section III. Section IV contains estimates of the optimal stock level and corresponding harvests. The study is summarized in Section V.

II. A Fishery Production Function

Much of the literature in fisheries economics is based on the Schaefer (1957) production function:

$H = qES$, q constant.

Harvest (H) is linear in both effort (E) and stock size (S) and thus homogeneous of degree two. The conditions under which this model holds are stringent.[1] Our approach is to specify a general production function:

$$H_t = H(\mathbf{E}_t; S_t, \mathbf{K}_t). \tag{1}$$

Here, $\mathbf{E}_t$ is an n-dimensional vector[2] of variable inputs, while fixed inputs consists of stock size (S_t) and an m-dimensional vector of other factors $(\mathbf{K}_t)$. The latter typically describes technological aspects of the fleet. S and $\mathbf{K}$ are fixed in any given time period, but can change over time. The production function is assumed to be concave in $\mathbf{E}$ and increasing[3] in both variable and fixed inputs; see Lau (1978).

---

[1] The assumptions underlying the model include (i) catch per unit of effort (H/E) is proportional to stock size at all levels of E and S, and (ii) the distribution of fish is uniform. Both of these assumptions are unrealistic for fisheries of schooling species; see Clark (1976, p. 235).
[2] Bold type denotes a vector.
[3] Weak inequality is implied; similarly for other properties of the production and profit functions.

Scand. J. of Economics 1987

Economics and stock size in North Atlantic fishery　147

Profits can be written as revenues minus variable costs:

$$P_t = p_t H(\mathbf{E}_{it}; S_t, \mathbf{K}_t) - \sum_i c'_{it} E_{it} = p_t \left[H(\mathbf{E}_{it}; S_t, \mathbf{K}_t) - \sum_i c_{it} E_{it} \right], \tag{2}$$

where p_t is (nominal) price per unit of output, c'_{it}, $i = 1, \ldots, n$, is (nominal) price per unit of input i and $c_{it} = c'_{it}/p_t$ is the normalized price per unit of input i. Maximizing profits with respect to variable inputs gives the optimal levels of inputs as functions of prices and fixed factors:

$$E_i^* = f_i(\mathbf{c}_t; S_t, \mathbf{K}_t), \quad i = 1, \ldots, n. \tag{3}$$

Substituting (3) into (2) and dividing by output price yields the normalized profit function

$$\Pi_t = H(\mathbf{E}_{it}; S_t, \mathbf{K}_t) - \sum_i c_{it} E_{it} = G(\mathbf{c}_t; S_t, \mathbf{K}_t). \tag{4}$$

This function gives maximized profits for a given set of values $\{\mathbf{c}_t; S_t, \mathbf{K}_t\}$. The normalized profit function is (i) decreasing and convex in normalized prices of variable inputs, (ii) increasing in the nominal price of output, nominal input prices constant, (iii) increasing in fixed factors, and (iv) bounded, given S_t and $\mathbf{K}_t$; see Lau (1978).

In this model formulation, stock size enters into the profit function as a fixed factor, while stock dynamics are disregarded. This is because the fishery in question was unregulated and characterized by open access. Thus, stock size is considered a constant in any given year, available for exploitation. Furthermore, assuming no (static) externalities among boats in turn implies myopic profit-maximizing behaviour.

For the North Sea herring fishery, aggregate time-series data on prices, harvest quantities and fixed factors are available for the period 1963–77 (Appendix 1). These data can be used to estimate a harvest supply function. The limited number of observations calls for a functional form that is parsimonious in parameters. Thus a Cobb–Douglas function is chosen, of which the Schaefer function is a special case.

The production function in logarithmic form is:

$$\ln H_t = \ln A + \sum_{i=1}^n \alpha_i \ln E_{it} + \beta_0 \ln S_t + \sum_{j=1}^m \beta_j \ln K_{jt}. \tag{5}$$

Assuming profit-maximizing and price-taking behavior, the dual profit function can be derived; see Lau and Yotopoulos (1972):

148 *T. Bjørndal*

$$\ln \Pi_t = \ln A^* + \sum_{i=1}^{n} \alpha_i^* \ln c_{it} + \beta_0^* \ln S_t + \sum_{j=1}^{m} \beta_j^* \ln K_{jt}. \tag{6}$$

The harvest supply function is given by

$$\ln H_t = \ln A_0 + \sum_{i=1}^{n} \alpha_i^* \ln c_{it} + \beta_0^* \ln S_t + \sum_{j=1}^{m} \beta_j^* \ln K_{jt}. \tag{7}$$

Since the production and profit functions are self-dual, the parameters of one can be derived exactly from the parameters of the other.[4]

A very important behavioural characteristic of herring is their schooling behaviour. Schooling takes place to reduce the effectiveness of predators; see Partridge (1982). Moreover, schooling fish contract their feeding and spawning range as the stock is reduced, while the size of schools often remains unchanged. Schooling behaviour has permitted the development of very effective means of harvesting, epecially the purse seine. With modern fish-finding equipment, harvesting can be profitable even at low stock levels. For these reasons, changes in stock size may have little effect on harvest; see Ulltang (1980) and Clark (1982).

The herring fishery takes place in the central and northern North Sea. Data for the Norwegian purse seine fleet are used in the empirical work.[5] The fishery, utilizing this technology, started in 1963. Open access to the fishery caused severe stock depletion and the fishery was closed at the end of 1977. Regulations have been in effect ever since so as to allow the stock to recover (cf. Appendix 1, Table A1).

The following factors should be considered for the production function:

1. E_t: number of boat-days.
2. S_t: stock size.
3. K_{1t}: number of boats in the fishery. (This factor is important because we are dealing with a search fishery, where information about locations of fish tends to be shared. This phenomenon may represent a positive externality. On the other hand,

---

[4] The restrictions on the parameters are: (1) $A^* = f(A, \alpha)$. (2) $A_0 = g(A, \alpha)$. (3) $\mu = \Sigma_i \alpha_i < 1$. (4) $\alpha_i^* = -\alpha_i/(1-\mu), (i = 1, ..., n)$. (5) $\beta_j^* = \beta_j/(1-\mu), (j = 0, 1, ..., m)$.
[5] The purse seiners are 120–180 feet in length with crew size of 9 to 12, powerblock, modern fish-finding and communications equipment.

Economics and stock size in North Atlantic fishery 149

too many boats may cause overcrowding on the fishing grounds, a negative externality.)

4. K_{2t}: size of boats, measured in gross registered tonnes.[6] (This variable is considered because catch per trip may depend on boat size.)

Only one variable input has been specified. Cross-sectional analyses indicate that various inputs are commonly combined in fixed proportions; see Bjørndal (1984). This permits the use of a single, variable input. A harvest supply function is estimated for two specifications of the production function:

Model A1: $\quad H_t = A E_t^{\alpha} S_t^{\beta_0} K_{1t}^{\beta_1} K_{2t}^{\beta_2} e^{\varepsilon_t}$

Model A2: $\quad H_t = A E_t^{\alpha} S_t^{\beta_0} K_{1t}^{\beta_1} K_{2t}^{\beta_2} e^{-\beta_3 D_t + \varepsilon_t}$,

where

$$D_t = \begin{cases} 1 \text{ for 1963 and 1974--77} \\ 0 \text{ otherwise} \end{cases}$$

and by assumption $\varepsilon_t \sim N(0, \sigma^2)$. The model is specified with and without a dummy variable. The dummy accounts for trip quotas that were in effect during the years 1974--77.[7] The specification with the dummy is *a priori* preferred to the one without.

The cost-of-effort variable should be represented by its oppportunity cost. In this connection it should be noted that the herring fishery is a marginal one, as purse seiners participate in up to seven different seasonal fisheries. Thus, the herring fishery may be considered minor as compared to the others and investment decisions depend mainly on the prospects in the major fisheries. Moreover, no special equipment is required for this fishery. Once investments have been made, a decision to participate in this fishery is based on (expected) marginal revenues and costs. A cost-of-effort index is estimated in Appendix 1.

Two types of externalities are associated with open-access fisheries. Static externalities refer to interactions among boats. The stock externality refers to the effect of this period's harvest on next period's stock level. The production technology is estimated on the assumption that

---

[6] In the regressions, K_{2t} is represented by average boat size per year.
[7] This regulation set an upper limit on catch per trip. However, the number of trips per boat was not regulated.

150   *T. Bjørndal*

the static externality cannot be controlled. In the herring fishery, negative externalities are not believed to be serious.[8] The stock externality,
on the other hand, can be controlled by a sole owner (Section IV).

The supply function is estimated with the number of participating
boats (K_{1t}) as an endogenous variable, which is facilitated by two-stage
least squares (2SLS) estimation. This procedure requires extra instrumental variables in order to ensure consistent estimates. These are the
dependent, endogenous and exogenous variables lagged once and the
cost/price ratio in the mackerel fishery. Fishing mackerel may be
regarded as an alternative to herring, so that changes in the relative
prices in the herring and mackerel fisheries should be taken into
account when determining fleet participation. The results are shown in
Table 1, with derived production function parameters in Table 2. The
method for estimating the variances of the derived parameters was
given by Kmenta (1971).

The following results are derived from the supply function:

1. The hypothesis of increasing returns in all factors is accepted. The
    imputed degree of homogeneity is given in Table 2.
2. The supply elasticity of effort, α^*, is insignificant in model A1,
    while highly significant in model A2.

Table 1. *Estimated harvest supply function (Cobb–Douglas) for North
Sea herring. Time-series data, 1963–77*[1]

Model	ln A_0	α^*	β_0^*	β_1^*	β_2^*	β_3^*	Method[2]	r^2
A1	−20.77**	−0.62	1.01**	1.62**	1.38	—	2SLS	0.96
	(−1.96)	(−0.99)	(2.11)	(4.03)	(1.00)			
A2[3]	−21.38**	−1.26**	0.78**	1.57**	1.38*	−0.98**	2SLS + AUTO	0.98
	(−3.21)	(−2.46)	(2.45)	(5.43)	(1.65)	(−3.45)		

[1] t statistics in parentheses. * denotes significant at the 90% level. ** denotes significant at
the 95% level.
[2] AUTO means corrected for first-order autocorrelation. 2SLS involves the loss of one
degree of freedom (the first observation) due to the inclusion of lagged variables as extra
instrumental variables.
[3] In this model, the instrumental variable K_{2t} lagged once was dropped because of nearly
perfect collinearity with K_{2t}.

---

[8] This is because crowding and gear collisions are not very common in an ocean fishery. In
addition, the positive externality through sharing of information about locations of fish is
regarded as more important than any crowding externality.

Scand. J. of Economics 1987

Economics and stock size in North Atlantic fishery 151

Table 2. *Derived production function parameters (Cobb–Douglas) for North Sea herring*[1]

Model	A	α	β_0	β_1	β_2	β_3	$\alpha + \sum_{j=0}^{2} \beta_j$
A1	0.4×10^{-5}	0.38 (0.24)	0.62 (0.44)	1.00 (0.20)	0.85 (0.99)	—	2.85
A2	0.0001	0.56 (0.10)	0.34 (0.19)	0.69 (0.08)	0.61 (0.42)	−0.43 (0.20)	2.20

[1] Standard errors in parentheses. The variance of A has not been estimated.

3. The supply elasticity of stock size, β_0^*, is not significantly different from one in both model specifications. The estimated supply elasticity of fleet size, β_1^*, is significantly larger than one. Point estimates of β_0^*, β_1^* and β_2^* are fairly stable.
4. The dummy variable is significant. Judging from t statistics, specifications with the dummy variable appear to perform somewhat better than those without, as was expected.

The derived parameters of the production function (Table 2) reveal point estimates of the output elasticity of effort significantly less than one. Thus, utilizing the fleet more intensively on a given stock is met with decreasing returns, which appears reasonable. Another interesting result is the low stock output elasticity (model A1 has the higher point estimate, but with a higher standard error). Accordingly, for a given effort level and stock size, harvest is not very sensitive to changes in stock size. This must be attributed to the schooling nature of the species. Other results indicate constant or decreasing returns to fleet size and decreasing returns to boat size. Altogether, the results from estimating the production function appear to be robust.

The effects on harvest when more than one input changes at the same time may also be considered. The production function is shown to exhibit increasing returns to scale. However, it is important to note that stock growth is limited by nature. If the number of boats in the fishery (K_1) is increased marginally, while the total number of boatdays (E) is kept constant, the results indicate that there will be an increase in harvest. If this corresponds to a situation where more boats fish herring at the same time in a marginally shorter season, the results indicate a positive externality. This must be due to the search effect in the fishery.

Scand. J. of Economics 1987

152 *T. Bjørndal*

The empirical literature contains few studies of fisheries production functions. In a study of Norwegian spring-spawning herring, Ulltang (1976) found a stock output elasticity close to zero. Hannesson (1983) and Schrank *et al.* (1984) estimated Cobb–Douglas production functions for various demersal fisheries. Both studies found constant returns to effort, but with point estimates of the stock output elasticity less than one, although in some cases not significantly different from one.

As noted above, the Schaefer function *a priori* imposes output elasticities equal to one for both effort and stock size. While this function may be appropriate for demersal fisheries, our results show that the model is inappropriate for a fishery of a schooling species such as herring.

III. The Intertemporal Profit Function

We now undertake a dynamic optimization and derive an expression for the optimal stock level. It is assumed that the fish stock is managed by a sole owner whose objective is to maximize the present value of net revenues from the fishery. Normalized net revenues can be restated as

$$\Pi_t = H(E_t; S_t, \mathbf{K}_t) - c_t E_t. \tag{8}$$

The following model of population dynamics, defined by a delay-difference equation, is used to explain changes in stock size over time:

$$S_{t+1} = (S_t - H_t)\, e^{\delta(S_t)} + G(S_{t-\gamma}), \tag{9}$$

where $\delta(S_t)$ = net instantaneous natural growth rate (natural growth − natural mortality), depending on stock size, with $\delta(\bar{S}) \le 0$, where $\bar{S}$ is the carrying capacity of the stock, and $G(S_{t-\gamma})$ = recruitment to the stock, taking place with a time lag of γ periods. This model is formulated in terms of the spawning stock which, for North Sea herring, coincides with the harvestable population. Harvesting is assumed to occur at the beginning of the period. The escapement, $S_t - H_t$, is left to grow at the net growth rate δ, which depends on stock size. The model is formulated and estimated for North Sea herring by Bjørndal (1984). For this species, $\gamma = 2$.

In this formulation we have modelled stock changes over time, but not capital dynamics. It has been argued that the fishery under consideration is relatively minor and complementary to other fisheries, so that investment decisions are determined mainly by the prospects in

Economics and stock size in North Atlantic fishery 153

the major fisheries. This fleet is available during the herring season as it may then have no alternative employment opportunity. Consequently, the opportunity cost for capital is low. Under such circumstances, capital will always be available and its dynamics may be disregarded.

Let us suppose the following circumstances. The resource manager is faced with a cost/price ratio (c_t) and a given fleet with certain attributes (the K-vector). He must then optimize the use of variable effort (E_t) over time, subject to stock dynamics and fleet size. This gives a discrete time, dynamic bioeconomic model with E_t and S_t as control and state variables, respectively. The methods of Lagrange multipliers can be used to derive the equilibrium conditions. We define the Lagrangean

$$L= \sum_{t=0}^{\infty} \{d' \Pi(E_t; S_t, K_t) - q_t [S_{t+1} - (S_t - H_t) e^{\delta(S_t)} - G(S_{t-\gamma})]\}, \qquad (10)$$

where $d = 1/(1 + r)$ is the discount factor, r the discount rate and q_t the discounted value of the shadow price of the resource. Assuming the cost/price ratio to be constant and equal to c, an implicit expression for the optimal steady-state stock level S^* is derived (Appendix 2):

$$e^{\delta(S^*)} \frac{c\Pi_S}{\Pi_E} + \partial S_{t+1}/\partial St = 1 + r - d^{\gamma} G'(S^*). \qquad (11)$$

The equation can be interpreted in terms of stock adjustment. The *left-hand side* represents marginal benefits. The term $c\Pi_S/\Pi_H$ is the Marginal Stock Effect (MSE),[9] which measures the impact of stock density on marginal sustainable resource rent. In the model, harvesting occurs at the beginning of the time period; hence $e^{\delta(S^*)}$ represents what is foregone in growth due to harvesting. At the optimum, this factor is applied multiplicatively to the MSE. The second term, $\partial S_{t+1}/\partial S_t$, is understood in terms of stock productivity: the effect of a marginal change in stock size on next period's stock. Together, the two terms represent net benefits from a marginal stock adjustment. The *right-*

---

[9] The MSE may be illustrated by reference to the equilibrium condition in the Clark–Munro (1975) model: $F'(S^*) - c'(S^*) F(S^*)/[p - c(S^*)] = r$. $F(S)$ is the natural production function and $c(S)$ the unit cost of harvesting. At the optimum, the own rate of return on the resource — consisting of (i) the marginal physical product $F'(S^*)$ and (ii) $c'(S^*) F(S^*)/[p - c(S^*)]$, the MSE — is equal to the discount rate. The MSE represents the impact of stock density on harvesting costs and is analogous to the wealth effect in modern capital theory; see Kurz (1968).

154 T. Bjørndal

hand side represents the marginal cost of stock adjustment. Stock adjustment causes a marginal change in recruitment of $G'(S^*)$. Since this occurs with a time lag, the discounted value is $d^\gamma G'(S^*)$. If $G'(S^*) > 0$, this term is a net benefit (capital "appreciation"), i.e., $1 + r - d^\gamma G'(S^*)$ is the "net" (user) cost of stock (capital) adjustment. On the other hand, if $G'(S^*) < 0$, $-d^\gamma G'(S^*) > 0$, and the term enters as an additional cost (capital "depreciation"). Therefore, equation (11) states that at the margin, the net benefits from the resource should equal the net cost of stock adjustment.

If there is no discounting $(r = 0)$, equation (11) reduces to

$$c\Pi_S/\Pi_E + H'(S^*) = 0. \tag{12}$$

Assuming $c\Pi_S/\Pi_E > 0$, this implies $S^* > S_{msy}$, where S_{msy} is the stock level corresponding to Maximum Sustainable Yield (MSY).

As noted above, the MSE[10] represents the impact of stock size on marginal sustainable resource rent. Accordingly, economic and technological parameters — other than the discount rate — influence the optimal stock level through the MSE (cf. equation 11). It is of interest to study the MSE because the production technology is nonlinear as opposed to the commonly used Schaefer function. In the latter model, the MSE is a function of the cost/price ratio and stock size only; hence, the only prediction of the model is that S^* is increasing in the cost/price ratio (cf. footnote 9). The present approach may enhance our understanding of the relationship between the production technology and S^*. In particular, in the context of the Schaefer model,[11] it has been argued that a stock output elasticity close to zero implies that the cost/price ratio has little influence on S^*; see Clark and Munro (1980). This conclusion need not hold for a more general description of the production process. Moreover, the effects of changes in various parameter values on the optimal stock level may be predicted.

For the Cobb–Douglas net revenue function, the MSE is defined as

$$\text{MSE} = c\Pi_S/\Pi_E = \frac{c\beta_0 A E^\alpha S^{\beta_0-1} K}{\alpha A E^{\alpha-1} S^{\beta_0} K - c} = m(c, \alpha, \beta_0, S, K), \quad K = \pi_j K_j^{\beta_j}. \tag{13}$$

---

[10] All analytical results are derived in Appendix 2.
[11] This represents a modified version of the Schaefer function. $\alpha = 1$, but β_0 may take on values different from one; see Clark and Munro (1975).

Economics and stock size in North Atlantic fishery 155

First, it can be ascertained that the optimal stock level is increasing in the MSE. Second, the MSE is increasing in the cost/price ratio and vanishes in the zero-cost case. Furthermore, the MSE is decreasing in the output elasticity of effort and fixed factors. The intuition behind the latter result is that profits are increasing in both variables. Productivity improvements over time, reflected by increases in the value of α, imply diminishing importance of costs and hence a reduction in S^*. Moreover, the optimal level of effort is decreasing in K [12], which implies costs are also decreasing in K.

We now turn to the stock output elasticity.

$\beta_0 = 0 \rightarrow \Pi_S = \text{MSE} = 0$. In this case, the fish stock is not an argument in the production function and harvesting costs are independent of stock size. If the fishery is viable, it is established by equation (12) that

$$r = 0 \rightarrow S^* = S_{msy}, \quad H^* = H(S_{msy}) = \text{MSY}.$$

Moreover, Clark (1982) has shown that

$$r \rightarrow \infty \rightarrow S^* = 0, \quad H^* = 0.$$

Accordingly, [13] for $0 < r < \infty$,

$$0 < S^* < S_{msy} \text{ and } 0 < H^* < \text{MSY}.$$

This result may be modified somewhat, as S^* may go to zero for finite discount rates. This case is of particular relevance for fisheries of schooling species. The myopic behaviour characteristic of an open-access fishery can be shown to correspond to an infinite discount rate. Combined with a zero stock output elasticity, this implies that open access will cause stock extinction. If the stock output elasticity is positive but low, open access may cause severe stock depletion; see Bjørndal and Conrad (1987).

$\beta_0 > 0$. For positive β_0 values, two cases may emerge. In the first, assuming $\Pi_E > 0$, MSE will first be increasing and then decreasing in β_0. This is illustrated in Figure 1, with corresponding stock levels given in Figure 2. Such a relationship could emerge, for example, in the model

---

[12] The intuition may be understood by considering two effects: (i) The search effect. The area that can be searched increases with the number of boats. Modern communications equipment permit information about locations of fish to be shared, which reduces variable effort. (ii) Boat size. The larger the boat, assuming locations of fish to be known, the fewer the trips needed for a certain harvest.

[13] Here, S^* is determined by the condition $F'(S^*) = r$ (cf. note 9).

156 *T. Bjørndal*

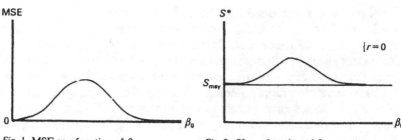

Fig. 1. MSE as a function of β_0.　　　　*Fig. 2.* S^* as a function of β_0.

of Clark and Munro (1975). In the second case, the MSE and thus S^* are decreasing functions of β_0. This occurs when the fishery is unprofitable at a zero stock output elasticity, i.e., $\Pi_E < 0$ given $\beta_0 = 0$. The consequence is that profits are increasing in the stock output elasticity. An increase in β_0 may then increase profits, which may reduce the optimal stock level and thereby increase the harvest.

As β_0 tends to infinity, the MSE goes to zero. The intuition is that increases in the stock output elasticity cause the "contribution" of the stock to its own harvest to increase. In a sense, the productivity of the stock improves. Thus, as β_0 increases, less variable effort is required to obtain a given harvest.

IV. The Optimal Stock Level

We now present estimates of the optimal stock level and the corresponding harvest for model A2 of the production function.[14] It is assumed that the sole-owner technology is the same as the open-access one. However, the sole owner can internalize the stock externality. As equation (11) cannot be solved explicitly for the optical stock level (S^*), a solution is found by a numerical computer routine. Results for the base case[15] are given in Table 3. Results are also given for the zero-cost case, which may serve as a point of reference.

---

[14]The dummy is set equal to zero because profit-maximizing behaviour is assumed (cf. Section II). Fleet size (K_1) has been set at 300, while boat size (K_2) has been set at the 1977 level.

[15] For the sole owner, costs include insurance and maintenance in addition to variable costs. In the open access fishery (Section II), only variable costs were considered.

Economics and stock size in North Atlantic fishery 157

Table 3. *Optimal stock level (S*) and corresponding harvest (H*)*

| Discount rate | Zero-Cost Case | | Base Case | | |
	S* Million tonnes	H* Million tonnes	S* Million tonnes	H* Million tonnes	Percentage increase in S*
0	1.57	0.61	2.26	0.54	44%
6%	1.40	0.60	2.21	0.55	58%
12%	1.23	0.59	2.17	0.55	76%
18%	1.06	0.56	2.13	0.56	101%

A number of conclusions can be drawn from the results in Table 3:

1. The inclusion of costs in the intertemporal profit function causes a considerable increase in S^*. This is noteworthy as the stock output elasticity is low. The assertion made by e.g. Clark and Munro (1980), that a low stock output elasticity implies that costs have a negligible effect on the optimal stock level is therefore not necessarily true.
2. S^* is not very sensitive to changes in the discount rate and less so in the base case than in the zero-cost case. This shows that costs have a stabilizing influence on the optimal stock level.
3. H^* is quite insensitive to changes in the optimal stock level and also to changes in the discount rate. This is because the growth curve is fairly flat over a wide range of stock values; see Bjørndal (1984).
4. In the zero cost case, $S^* = S_{msy}$ and $H^* = MSY$ for $r = 0$. Moreover, numerical results show that stock extinction is "optimal" for a discount rate of about 52 per cent.

The optimal stock level is increasing in the cost/price ratio. Therefore, a secular downward trend in the cost/price ratio would gradually diminish its influence on S^*. The sensitivity of S^* to a change of plus/minus 20 per cent in the cost/price ratio was investigated. For reasonable discount rates ($0 < r < 0.12$), the corresponding relative change in S^* is only 7–10 per cent. Therefore, the optimal stock level would be only marginally affected by moderate changes in the cost/price ratio.

In a study of a sardine fishery in the Gulf of Valencia, Gallastegui (1983) estimated the sole-owner stock level for reasonable discount rates to be more than twice the size of S_{msy}. Moreover, S^* is not very sensitive to changes in the discount rate. The consequences of dis-

158 *T. Bjørndal*

counting in the case of baleen whales were illustrated by Clark
(1976a). An increase in the discount rate from zero to 5 per cent
reduces the optimal stock level by almost 50 per cent. This is caused by
the low growth rate of whales.

Having analysed costs, we turn to the relationship between S^* and
the other arguments of the MSE. It was noted in Section III that the
optimal stock level is decreasing in the effort output elasticity. Sensitivity analyses show that S^* is fairly sensitive to changes in the effort
output elasticity (Table 4). In particular, a decrease in α causes a fairly
large increase in the optimal stock level. Moreover, a sufficiently low
effort output elasticity would render the fishery unprofitable. On the
other hand, the results tend towards the zero-cost case with increasing
α.

The optimal stock level for alternative values of the stock output
elasticity are given in Table 5. S^* is found to be sensitive to changes in
β_0, particularly for downward changes. Figure 3 illustrates the estimated relationship between the optimal stock level and the stock out-

Table 4. *Sensitivity to change in the effort output elasticity (α)* [1]

	$\alpha = 0.51$		$\alpha = 0.61$	
Discount rate	S^* Million tonnes	H^* Million tonnes	S^* Million tonnes	H^* Million tonnes
0	2.99	0.31	1.80	0.60
6%	2.98	0.32	1.68	0.60
12%	2.98	0.32	1.56	0.61

[1] In the base case, $\alpha = 0.56$.

Table 5. *Sensitivity to changes in the stock output elasticity (β_0)* [1]

	$\beta_0 = 0.29$		$\beta_0 = 0.39$	
Discount rate	S^* Million tonnes	H^* Million tonnes	S^* Million tonnes	H^* Million tonnes
0	2.86	0.37	1.88	0.59
6%	2.86	0.37	1.79	0.60
12%	2.85	0.37	1.69	0.60

[1] In the base case, $\beta_0 = 0.34$.

Scand. J. of Economics 1987

Economics and stock size in North Atlantic fishery   159

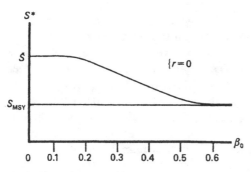

Fig. 3. S^* as a function of β_0.

put elasticity for a zero discount rate. The result is quite interesting, i.e., S^* is a decreasing function of β_0. In other words, the second type of relationship between the MSE and β_0 has been obtained (cf. Section III). As β_0 decreases, the profitability of the fishery also decreases, causing an increase in S^*. For sufficiently low values of β_0, the fishery becomes unviable so that $S^* = \bar{S}$, the carrying capacity of the stock.

The base case results refer to a fleet size of 300 boats and boat size as of 1977. Estimations for fleet sizes of 250 and 350, respectively, with boat size unchanged, show that the relative change in S^* is 6–10 per cent ($0 < r < 0.12$), i.e., less than the relative change in fleet size. Moreover, S^* is found to be decreasing in the fixed factors.

V. Summary

This paper set out to estimate the production function and optimal stock level for North Sea herring. Some new results in regard to the relationship between the optimal stock level and production technology were derived. In particular, the optimal stock level was found to be decreasing in the output elasticity of effort and the amount of fixed factors in the industry. The relationship between the optimal stock level and the stock output elasticity was shown to be ambiguous. Conditions under which the optimal stock level would be decreasing in the stock output elasticity were derived. The results from the empirical analysis can be summarized as follows:

1. The stock effect in the production function appears to be fairly low. However, the hypothesis of increasing returns in all inputs is

160   *T. Bjørndal*

generally accepted. The commonly used Schaefer production function was found to be inappropriate for the herring fishery.

2. Costs matter for the optimal stock level, even if the stock output
    elasticity is low. This result is contrary to what has previously been
    asserted on theoretical terms; see Clark and Munro (1980). In addi-
    tion, costs have a stabilizing influence on the optimal stock level.

3. The optimal stock level is not very sensitive to changes in the dis-
    count rate. Furthermore, the optimal harvest is quite insensitive to
    changes in both the discount rate and the optimal stock level.

4. The optimal stock level is found to be decreasing in the stock output
    elasticity.

5. The optimal stock level is fairly sensitive to changes in the output
    elasticities of effort and stock size, but not very sensitive to changes
    in the cost/price ratio and fixed factors.

6. For sufficiently large but finite values of the fixed factors, the output
    elasticities of effort and stock size, the MSE vanishes and the
    resource is managed as in the zero-cost case.

Appendix 1. Data for Estimation of the Harvest Supply Function

This Appendix contains the data for estimation of the harvest supply
function.

In the harvest supply function, the cost/price ratio enters as a variable. While data on the output price are available, this is not the case
for data on the cost of effort. Therefore, a time series has to be constructed for this variable. The cost per boatday, c'_t, includes labour, fuel
and material costs. In other words, only variable costs are considered.
Price indices for wages, fuel and materials are readily available. The
index for c'_t will then be a weighted average of these indices, which
raises the question of the appropriate set of weights. The Norwegian
Budget Committee for the Fishing Industry collects cost data for fishing vessels. Their data on purse seiners has been used to establish cost
shares for wages, fuel and materials. These shares have been used as
weights in construction of the cost-of-effort index.

The particular wage index chosen is average earnings in the manufacturing industry. We have used this rather than e.g. *ex post* seasonal
returns in the fishing industry. This is because the manufacturing industry represents alternative employment opportunities for fishermen.
Thus, over time, changes in wages in the fishing industry will have to
correspond to those in the manufacturing industry.

Economics and stock size in North Atlantic fishery   161

Table A1. *Spawning stock size and total catch, 1963–77*

Year	Spawning stock per January 1 [1] (1,000 tonnes)	Total catch [2] (1,000 tonnes)
1963	2,325	716
1964	2,529	871
1965	2,348	1,169
1966	1,871	896
1967	1,434	696
1968	1,056	718
1969	696	547
1970	717	564
1971	501	520
1972	509	498
1973	521	484
1974	345	275
1975	259	313
1976	276	175
1977	166	46

[1] Stock size is estimated by Virtual Population Analysis.
[2] Catch figures include juvenile herring.
Source: Bjørndal (1984).

Table A2. *Norwegian participation in the herring fishery*

Year	Norwegian harvest (tonnes)	Number of participating purse seiners	Index of average size of purse seiners [1]
1963	3,454	16	1.000
1964	147,933	195	1.060
1965	586,318	284	1.124
1966	448,511	334	1.192
1967	334,449	326	1.264
1968	286,198	352	1.340
1969	134,886	253	1.420
1970	220,854	201	1.505
1971	210,733	230	1.677
1972	136,969	203	1.900
1973	135,338	153	2.015
1974	66,236	165	2.115
1975	34,221	102	2.173
1976	33,057	92	2.331
1977	3,911	24	2.406

[1] In the regressions, actual boat size was used rather than the index listed here. It is immaterial which of these series is chosen.
Source: Bjørndal (1984).

162 *T. Bjørndal*

Table A3. *Cost-of-effort index and price data*

Year	Cost-of-effort index	Average price of herring (NOK/tonne)
1963	100.0	387
1964	105.7	339
1965	113.4	342
1966	120.0	357
1967	128.5	235
1968	137.2	213
1969	148.2	309
1970	164.7	437
1971	184.3	408
1972	198.7	358
1973	218.7	640
1974	256.5	830
1975	302.9	1,226
1976	349.7	1,422
1977	386.2	2,359

Source: Bjørndal (1984).

Appendix 2. Technical Derivations

The maximization problem is stated as equation (10). First-order necessary conditions for an optimum are: (i) $\partial L/\partial E_t = 0$, $t = 0, 1, \ldots$ and (ii) $\partial L/\partial S_t = 0$, $t = 1, 2, \ldots$; see Clark (1976). The shadow price, $q_t = d^t \Pi_E/(H_E e^{\delta(S_t)})$, is obtained from the first necessary condition. The steady state harvest is:

$$H = H(S) = S - (S - G(S))/e^{\delta(S)}. \tag{A1}$$

An implicit expression for the optimal stock level S^* is derived:

$$e^{\delta(S^*)}\left[\frac{c\Pi_S}{\Pi_E} + 1\right] + \delta'(S^*)[S^* - G(S^*)] + d^\gamma G'(S^*) = 1 + r. \tag{A2}$$

Equation (11) in the text is obtained from (A2) by using (9) and (A1).

Assume the Cobb–Douglas production function. The effort level required for a steady-state harvest is found from equation (A1):

$$E = (AK)^{-1/\alpha} S^{-\beta_0/\alpha}[S - (S - G(S))/e^{\delta(S)}]^{1/\alpha} \tag{A3}$$

(i) $\dfrac{\partial E}{\partial K} < 0$ and (ii) $\lim_{K \to \infty} E = 0$.

Economics and stock size in North Atlantic fishery   163

The Marginal Stock Effect (MSE) is defined in equation (13). For the Cobb–Douglas case with $A = \beta_0 = 1$, the change in the MSE with respect to the *effort-output elasticity* (α) is:

$$\frac{\partial \text{MSE}}{\partial \alpha} = -\frac{c^2 E^\alpha K \ln E + c E^{2\alpha-1} S K^2}{[\alpha E^{\alpha-1} S K - c]^2}. \tag{A4}$$

This derivative is negative provided (i) $c > 0$, (ii) $H_E \neq c$, i.e. $\Pi_E \neq 0$. If $E < 1$, a third condition needs to be satisfied; this special condition is disregarded. The limiting value of the MSE as $\alpha \to \infty$ is given by (for $E > 1$):

$$\lim_{\alpha \to \infty} \text{MSE} = \lim_{\alpha \to \infty} \frac{c E^\alpha K \ln E}{E^{\alpha-1} S K + \alpha E^{\alpha-1} S K \ln E} = \lim_{\alpha \to \infty} \frac{c K \ln E}{E^{-1} S K (1 - \alpha \ln E)} = 0.$$

Consider *fixed factors* (K). Assume that $A = \alpha = \beta_0 = 1$.

$$\frac{\partial \text{MSE}}{\partial K} = -\frac{c^2}{[SK - c]^2}. \tag{A5}$$

This derivative is negative provided (i) $c > 0$ and (ii) $\Pi_E \neq 0$.

$$\lim_{K \to \infty} \text{MSE} = \lim_{K \to \infty} \frac{cE}{S} = \frac{c}{S} \lim_{K \to \infty} E = 0.$$

Consider the *stock output elasticity* (β_0). Assume $A = \alpha = 1$.

(i) $$\frac{\partial \text{MSE}}{\partial \beta_0} = \frac{cE S^{\beta_0} K [S^{\beta_0} K - c \beta_0 \ln S - c]}{[S^{\beta_0} K - c]^2}.$$

Consider the derivative evaluated at $\beta_0 = 0$:

$$\left.\frac{\partial \text{MSE}}{\partial \beta_0}\right|_{\beta_0 = 0} = \frac{cEK}{K - c}.$$

A necessary and sufficient condition for this derivative to be negative is $K - c < 0$. However, note that $\Pi_E = K - c$ when $A = \alpha = 1$ and $\beta_0 = 0$. Hence, if the fishery is unprofitable given no stock effect (i.e., $\Pi_E < 0$ at $\beta_0 = 0$), then the MSE — at least locally — will be decreasing in β_0.

(ii) $$\lim_{\beta_0 \to \infty} \text{MSE} = \lim_{\beta_0 \to \infty} \frac{c \beta_0 E S^{\beta_0-1} K \ln S}{S^{\beta_0} K \ln S} = \lim_{\beta_0 \to \infty} \frac{c \beta_0 E}{S} = \frac{c}{S} \lim_{\beta_0 \to \infty} \beta_0 \lim_{\beta_0 \to \infty} E.$$

From (A3), it can be shown that this limit is zero provided $S > 1$.

164   *T. Bjørndal*

References

Bjørndal, T.: The optimal management of an ocean fishery. Ph.D. dissertati‿n, Department of Economics, University of British Columbia, 1984.
Bjørndal, T. & Conrad, J. M.: The dynamics of an open access fishery. *Canadian Journal of Economics 20*, 74–85, 1987.
Clark, C. W.: *Mathematical Bioeconomics*. Wiley, New York, 1976a.
Clark, C. W.: A delayed-recruitment model of population dynamics, with an application to Baleen whale populations. *Journal of Mathematical Biology 3*, 381–91, 1976b.
Clark, C. W.: Concentration profiles and the production and management of marine fisheries. In Eichorn, W., Henn, R., Neumann, K. & Shephard, R. W. (eds.), *Economic Theory of Natural Resources*, Physica Verlag, Würzburg, 1982.
Clark, C. W. & Munro, G. R.: The economics of fishing and modern capital theory: A simplified approach. *Journal of Environmental Economics and Management 2*, 92–106, 1975.
Clark, C. W. & Munro, G. R.: Fisheries and the processing sector: Some implications for management policy. *Bell Journal of Economics 11*, 603–16, 1980.
Gallastegui, C.: An economic analysis of sardine fishing in the Gulf of Valencia (Spain). *Journal of Environmental Economics and Management 10*, 138–50, 1983.
Hannesson, R.: The bionomic production function in fisheries: A theoretical and empirical analysis. *Canadian Journal of Fisheries and Aquatic Sciences 40*, 968–82, 1983.
Kmenta, J.: *Elements of Econometrics*. Macmillan, New York, 1971.
Kurz, M.: Optimal economic growth and wealth effects. *International Economic Review 9*, 348–57, 1968.
Lau, L. J.: Applications of profit functions. In Fuss, H. & McFadden, D. (eds.), *Production Economics: A Dual Approach to Theory and Applications*, North-Holland, Amsterdam, New York and Oxford, 1978.
Lau, L. J. & Yotopoulos, P. A.: Profit, supply and factor demand functions. *American Journal of Agricultural Economics 54*, 11–18, 1972.
Partridge, B. L.: The structure and function of fish schools. *Scientific American 246*, 90–9, 1982.
Schaefer, M. B.: Some considerations of population dynamics and economics in relation to the management of marine fishes. *Journal of the Fisheries Research Board of Canada 14*, 669–81, 1957.
Schrank, W. E., Tsoa, E. & Roy, N.: *An Econometric Model of the Newfoundland Groundfishery: Estimation and Simulation*. Department of Economics, Memorial University of Newfoundland, St. John's, 1984.
Ulltang, Ø.: Catch per unit effort in the Norwegian purse seine fishery for Atlanto-Scandian (Norwegian spring spawning) herring. *FAO Fisheries Technical Papers 155*, 91–101, 1976.
Ulltang, Ø.: Factors affecting the reaction of pelagic fish stocks to exploitation and requiring a new approach to assessment and management. *Rapports et Proces-verbaux des Réunions 177*, 489–504, 1980.

First version submitted October 1985;
final version received December 1986.

[43]

Optimal Timing of Harvest for the North Carolina Bay Scallop Fishery

Robert L. Kellogg, J. E. Easley, Jr., and Thomas Johnson

Substantial improvement is possible in economic returns for the North Carolina bay scallop fishery by delaying the opening of the season beyond its traditional date. A general bioeconomic harvesting model was developed for use in determining the optimal season opening/closing schedule for a seasonal fishery with the control specified as an on/off switch. One hundred and twenty separate scenarios were created by setting five exogenous variables to reasonable alternative values. The optimal season is contrasted with the unregulated case for each scenario. The optimal opening was typically two to three weeks later than the model of past practices.

Key words: bioeconomic models, fishery management.

State regulation of fisheries arises from the common property/open access nature of the resource. Without property rights, individual fishermen have no incentive to consider the opportunity cost of harvesting (i.e., the value of the productive capacity of the stock). Instead, fishermen tend to harvest as much and as fast as they can as long as expected net return is positive. Under these conditions, resource depletion can occur. The task facing the fishery manager is the promulgation of regulations designed to ensure continued harvests and, if possible, to enhance the value of those harvests. Most important, the fishery manager does not want to promulgate regulations that are inconsistent with optimal harvest strategies.

Bioeconomic models and optimal control theory can be used by fishery managers to help attain these goals. With a specified objective (such as maxi-

mizing net revenue) and a specified control variable, or regulatory device (such as the season opening date), optimal control theory can solve for the optimal resource harvest over time and the optimal regulation. The procedure requires a bioeconomic model of the fishery. Important components include the objective function, the price function, the fish production function, the cost function, and functions describing the population dynamics of the resource stock. A number of recent works have advanced theoretical optimal control models (for example, Clark; Huang, Vertinsky, and Wilimovsky; Strand and Hueth; Clark and Munro; Levhari, Michener, and Mirman; Conrad and Castro), but there are few examples of successful applications to fishery problems.

This article presents a general dynamic harvesting model to determine when to open the harvest season for a seasonal fishery. The model is applied to the North Carolina bay scallop fishery. Results show that even with data limitations, this analysis makes a strong case for significant returns to a change in traditional management of the fishery.

A Dynamic Seasonal Harvesting Model

A seasonal fishery has predictable fluctuations in yields throughout the year. In the common property/open access stiuation, individual fishermen are motivated to harvest seasonal fisheries early—when stocks are high and before they are depleted by other fishermen—even though the value of the catch may be much greater later in the season (Agnello

Robert L. Kellogg is an economic analyst with the U.S. government. He was a graduate student at North Carolina State University when the research reported in this paper was completed. J. E. Easley, Jr., is a professor of economics and business, and Thomas Johnson is a professor of economics and business and biomathematics and statistics, both at North Carolina State University.

The authors acknowledge partial support of this work by the Office of Sea Grant, NOAA, U.S. Department of Commerce, under grant no. NA83AA-D-00012, and the North Carolina Department of Administration.

Enthusiastic support for this work by the administrators of the University of North Carolina Sea Grant College is also acknowledged. Dennis Spitsbergen, North Carolina Division of Marine Fisheries, provided valuable insights into problems of the bay scallop fishery and a helpful review of an early draft of this paper. The authors also appreciate helpful reviews of earlier drafts by Charles Peterson, UNC Institute of Marine Sciences, Randall Rucker and Ann McDermed, both of NCSU's Economics and Business Department, and anonymous *Journal* referees.

and Donnelley). The harvesting model presented here determines the optimal season-opening (and season-closing) schedule for such a fishery.

The seasonal fisheries management model is formally stated as follows:

(1)   maximize with respect to $\Phi(t)$:

$$PV = \int_0^\infty [P(Q, t, w)g(z, t, y) - C(x, t, y)]e^{-\delta t}\,\Phi(t)\,dt$$

such that

$$\dot{x} = F(x, t, z) - M(x, t, z) - Q(x, t, y)\,\Phi(t),$$
$$x(t_0) \text{ and } t_0 \text{ given},$$

where $\phi(t)$ is the control variable, $P(Q,t,w)$ is the fish price function (price per pound), $g(z,t)$ is the size function (pounds per fish), $Q(x,t,y)$ is the fish production function (fishing mortality), $C(x,t,y)$ is the cost function, $e^{-\delta t}$ is the discount function, $F(x,t,z)$ is the population growth function, $M(x,t,z)$ is the natural mortality function, w is a vector of exogenous variables affecting market price, z is a vector of exogenous environmental variables, y is a vector of exogenous production inputs, $x(t)$ is population size in numbers, and t is time.[1]

This general model has been formulated with continuous time and an infinite time horizon and accommodates multiple openings and closings. The control variable, $\phi(t)$, takes only the values 0 and 1; $\phi(t) = 0$ implies a closed season and $\phi(t) = 1$ implies an open season. It is assumed that the decision to begin harvest is made prior to the potential harvest season and, once made, is irrevocable. Thus, the possibility of adaptive management is ignored. In practice, however, the season opening/closing schedule could be reassessed at any time in response to additional information.

The objective function [the integrand in equation (1)] is the part of the problem that is to be maximized by selection of an "optimal" control—$\phi(t)$. A discounting function, $e^{-\delta t}$, is required in the objective function to convert benefits from the fishery in future time periods to present values. Ideally, the objective function would represent the present value of net benefits to society. However, benefits and costs are inherently subjective and cannot be measured and aggregated over all members of society. The most common approach used in fisheries problems (and thus used here) is to substitute net revenue for society's net benefit function. The net revenue function represents the present value of net revenue summed over all future harvest periods. Net revenue is not a perfect measure of social value because it assumes that the marginal utility of money is the same for all individuals in society and at all points in time, making the model insensitive to income redistribution.

The price function, $P(Q,t,w)$, and the size function, $g(z,t)$, together define the market price of the fishery product in units of dollars per fish. Because only the population within the jurisdiction of the managing authority is modeled, the fishermen and the regulatory authority are price takers for most problems. Including water temperature in the size function is especially important. Fish are cold-blooded, and thus their growth and metabolic rates are determined predominantly by water temperature (Hall, Loucks and Sutcliffe).

The cost function, $C(x,t,y)$, defines the opportunity costs of producing, or harvesting, the fish. Fishing costs can be categorized into three groups: (a) fixed costs, such as investment cost of the boat, gear, and one-time seasonal maintenance costs; (b) daily non-labor costs; and (c) daily opportunity costs of the fisherman. Groups (b) and (c) represent variable costs. Because fixed costs are constant, they do not affect the optimal timing of the harvest and are therefore excluded from the analysis. For fuel and other inputs with alternative uses, the market price is a good estimate of the opportunity cost. Measuring the opportunity cost of labor is more difficult (Anderson).

The equation of motion,

$$\dot{x} = F(x, t, z) - M(x, t, z) - Q(x, t, y)\,\Phi(t),$$

comprises the biological sector of the fisheries management model. This equation defines how the state variable, $x(t)$, moves through time. In a fisheries problem, the state variable usually defines the population dynamics of the species involved. The presence of the population growth function, $F(x,t,z)$, in the equation of motion designates the resource as "renewable." A renewable resource (such as a fishery) cannot instantly replace the stock that is harvested. It takes time to replenish the population. This process usually depends on the absolute stock

---

[1] This model can be expanded to include multiple species when more than one species is vulnerable to capture by the harvesting operation. For example, several species of shrimp can be included, each with a different state variable, size function, recruitment function, natural mortality function, and even price function (see Kellogg, Easley and Johnson 1986 for an application to a multispecies fishery). Catch of incidental (nontarget) species, which usually have little or no direct commercial value at the time of collection but which may have commercial value at a later date, can be included by adding the appropriate equations of motion and assigning value to the "bycatch" on the basis of its eventual commercial value. (See Waters, Easley and Danielson for an economic analysis of the forgone value of bycatch of immature shrimp in relation to proposed restrictions on the timing of harvest.) Population dynamics of predator and prey species can be included in the equations of motion, as well.

size, water quality variables, habitat availability, food availability, predators, and other factors.[2]

Application to the North Carolina Bay Scallop Fishery

The bay scallop fishery in North Carolina is a winter fishery, traditionally opening in December and extending through early spring. Bay scallops spawn in their first year and most do not survive to spawn the next year. Harvesting is prohibited during the fall spawning period (September–November) to ensure continued harvests in subsequent years (Kellogg and Spitsbergen). The fishery is predominantly a small-boat fishery (under 25 feet) because bay scallops live in shallow water. Bay scallops are harvested primarily by use of a scallop drag (dredge or scrape).

The state regulatory agency (North Carolina Department of Natural Resources and Community Development, Division of Marine Fisheries) controls the season opening. In addition, fishing has been allowed only on 2–3 days per week during the first two months of the season. As the season progresses, fishing days are often increased to five per week. Other controls include quotas, closure of the commercial fishery on weekends, restriction of fishing to daylight hours, and certain gear restrictions to prevent destruction of the habitat.

Components of the Model

The general seasonal harvesting model presented above was modified for application to the North Carolina bay scallop fishery. Because the bay scallop in North Carolina essentially has a single cohort available for harvest each year and regulations prohibit harvest during the fall spawning months, $F(x,t,z)$ reduces to $F(t,z)$. Current harvesting decisions will not affect the potential for future harvests, and so the time horizon can be restricted to a single harvest season. A terminal time, $t = T$, replaces the infinite time horizon in the objective function, where T represents the natural end of the harvest period.[3] The harvesting model is now in

the form of a model for an exhaustible resource. The state variable, x (the number of fish), cannot increase during the time horizon of the control problem except according to a prescribed recruitment pattern—$F(t,z)$. It can decrease either by natural mortality or fishing mortality. In addition, the equation of motion was simplified to include only the harvest rate, fish production is represented by the catch-per-unit-effort production function,[4] and costs are represented by a cost-per-unit-effort function. The modified model is

(2)   maximize with respect to $\Phi(t)$

$$PV = \int_0^T [P(t,w)g(z,t)Eqx(t)$$

$$- cE]e - \delta t\ \Phi(t)\ dt,$$

such that

$$\dot{x} = -Eqx(t)\ \Phi(t),\ x(0)\ \text{given},$$
$$0 \le t \le T,\ \text{and}\ x(t) \ge 0,$$

where $\Phi(t)$ is the decision variable ($\Phi(t) = 0$ implies a closed season and $\Phi(t) = 1$ implies an open season);

$P(t, w)$ is $a_0 + a_1 INCOME_t + a_3 SEAP_t + a_4 SEAP_{t-1} + a_5 SEAP_{t-2} + a_7 CALQ_t + a_8 t + a_9 t^2 + a_{10} t^3$;

$g(z, t) = 0.002205[M_{max}(1 - e^{-B(C,t)}) + M_0 e^{-B(C,t)}]$;

$B(C, t) = b_1(t + 4, 3) + b_2 C + b_3 C^2/(t + 4.3)$;

$M_{max} = m_1 S_t^3$;

$S_t = S_{max}(1 - e^{-(t+4.3)+c_2 C}) + S_0 e^{-(c_2)t+4.3)+c_2 C)}$.

$C = 20.203(t + 4.3) - 1.012(t + 4.3)^2 + 0.027(t + 4.3)^3$;

t is time in weeks starting from 1 December, and

T is 31 March.

---

fluctuate significantly from year to year independent of harvest activities—because of changes in habitat availability or environmental conditions—can often be modeled with a finite time horizon, where the time interval is a single harvest season.

[4] A general functional form for the production function is $\phi(x)\ \psi(E)$, where $\psi(E)$ defines the effect of fishing effort on a stock (the mortality rate), and ϕ defines the total fishing mortality generated by ψ acting on x (Clark, Hannesson). A common representation of the production function in the fisheries literature (and the one used in this study) is where $\psi(E) = qE(t)$ and $\phi(x) = x(t)$, represented as

$$Q(x,t,y) = h(t) = qE(t)x(t).$$

The function $h(x)$ is the harvest rate and q is a "catchability" coefficient which is needed to transform $E(t)$ (measured in nominal terms, such as number of vessels or number of fishermen) into a fishing mortality rate. This is sometimes referred to as the "catch per unit effort hypothesis" (Clark). Important assumptions are associated with this function, including nonsaturation of fishing gear, no congestion of fishing vessels, and uniform distribution of the stock.

---

[2] If modeled fully, the equations of motion more aptly could be called "ecosystem constraints," since they represent how the ecosystem (or rather, a subset of the ecosystem) would respond to a prescribed harvest rate of one or more of the species involved.

[3] The essential difference between a finite and an infinite time horizon is that the latter leads typically to an optimal steady state, or long-run equilibrium solution. Fisheries with strong stock-recruitment relationships (where next year's harvest depends on this year's harvest) are best modeled with an infinite time horizon, whereas fish populations that

The potential harvest season spans from 1 December ($t = 0$) to 31 March ($t = T$). No quotas are imposed, and fishing is allowed Monday through Friday. The weekly discount rate, δ, is set equal to 0.001827 for this study, which is equivalent to an annual discount rate of 10%.[5] (A real rate is required because price and cost are in units of 1967 dollars.) Parameter estimates and values used for exogenous variables are summarized in table 1.

[5] An increase (decrease) in the discount rate, δ, leads to a faster (slower) depletion of the resource. The value chosen for δ is very important for infinite time horizon problems (Clark), but it is less important when the time horizon is less than a year.

Ex-vessel price equation. Three species of scallops are harvested in the United States—bay, sea, and calico scallops. The three species are probably close substitutes for some uses and perfect substitutes for others. Most scallops are frozen and stored for transportation to inland markets or for later consumption, creating an inventory demand in addition to consumption demand. Local demand for fresh scallops may also be an important factor. Thus, the price of bay scallops is a complex interplay of both consumption and inventory demand, supply of all three scallop species, and local (seasonal) demand for the fresh product. Histori-

Table 1. Estimates of Coefficients and Exogenous Variables Used for the North Carolina Bay Scallop Harvesting Problem

Parameter Estimates (standard errors in parentheses)

Price equation:

$$a_0 = -4.24904127 \ (.94401)$$
$$a_1 = 0.00473008 \ (.00119)$$
$$a_3 = -1.85448147 \ (.54878)$$
$$a_4 = 0.56224238 \ (.55072)$$
$$a_5 = 1.69072116 \ (.57156)$$
$$a_7 = -0.0000005476 \ (.00000013)$$
$$a_8 = 0.38661194 \ (.17669)$$
$$a_9 = -0.04133764 \ (.02381)$$
$$a_{10} = 0.00119521 \ (.00091)$$

Size equations:

$$M_0 = 2.522 \ (.09919)$$
$$b_1 = -0.4415 \ (.04480)$$
$$b_2 = 0.0969 \ (.00890)$$
$$b_3 = -0.0034 \ (.00037)$$
$$m_1 = 0.0270 \ (.00023)$$
$$S_{max} = 6.378 \ (.03510)$$
$$c_1 = -0.0298 \ (.00730)$$
$$c_2 = 0.0065 \ (.00082)$$

Exogenous Variables Were Assigned the Following Values:

$E = 540$ or 750 standard boat-days per week
$q = .0001, .0002,$ or $.0003$
$c = 42.55$ or 34.04 (1967 dollars)
$K = 0.001827$
$S_0 = 5.9$ centimeters
$x_0 = 13, 18, 23, 28,$ or 33 million scallops

Sets of Values for the Three Exogenous Demand Variables

Variable	High price (1980–81)	Low price (1981–82)
$INCOME_t$	882	887
$SEAP_t = SEAP_{t-1} = SEAP_{t-2}$	2.00	1.31
$CALQ_t$	531,369	1,084,457

54 *February 1988* *Amer. J. Agr. Econ.*

cally, the North Carolina bay scallop price has increased from early December through late January to mid-February. In part, this seasonal price movement is a result of shifts in supplies of northeastern U.S. bay scallops as that season draws to a close. It is beyond the scope of this study to model this system completely. Instead, a prediction equation for North Carolina ex-vessel price was estimated, which was used to forecast North Carolina price movements through the season.

Because sea scallops dominate the scallop market, the price of bay scallops largely is determined by factors that are important in the sea scallop market. The prediction equation was initially formulated as follows:

$$(3) \quad NCP_t = a_0 + a_1 INCOME_t + a_2 NCQ_t$$
$$+ a_3 SEAP_t + a_4 SEAP_{t-1}$$
$$+ a_5 SEAP_{t-2} + a_6 PSHRIMP_t$$
$$+ a_7 CALQ_t + a_8 WEEK$$
$$+ a_9 WEEK^2 + a_{10} WEEK^3$$
$$+ a_{11} INVENT_t,$$

where NCP_t is the ex-vessel price of North Carolina bay scallops in dollars per pound of meats, NCQ_t is the quantity of North Carolina bay scallops in pounds of meats, $SEAP_t$ is the ex-vessel price of sea scallops, $INCOME_t$ is total personal income for the U.S., $INVENT_t$ is the inventory of frozen stocks of scallops at the beginning of the period t, $PSHRIMP_t$ is a shrimp price index (published by the U.S. Department of Commerce in "Current Fisheries Statistics"), $CALQ_t$ is the landings of calico scallops, and time was incorporated as a cubic function in terms of weeks starting from 1 December. $SEAP_t$ and $PSHRIMP_t$ represent prices of substitutes, and together with $INCOME_t$ and NCP_t comprise the standard variables expected in a demand equation. Time was included to capture seasonal changes in demand, which can be important for seasonally available products. Time would also capture any seasonal price effects from increasing size of the meat as the season progresses. $INVENT_t$ represents the inventory demand response in the scallop market. The $CALQ_t$ variable is a demand shifter because high calico scallop landings have sometimes depressed the North Carolina bay scallop market when the harvests coincided (Dennis Spitsbergen, Division of Marine Fisheries, personal communication).

Ideally, the above equation would be estimated with weekly data. However, price data were only available in monthly aggregates. Thus, monthly data for the potential harvest season—December through March—were used to fit the model. Since

the price equation requires time in units of weeks, the midpoint of each month, measured in weeks, was used for the time variables. All prices and income were adjusted for inflation prior to the estimation by dividing by the consumer price index; thus, all prices are in units of 1967 dollars. The model was estimated using data from 1974–75 through 1982–83 (see Kellogg for a complete compilation of the data).

After fitting the above price equation, the variables NCQ_t, $PSHRIMP_t$, and $INVENT_t$ were jointly insignificant ($F_{3,23} = 0.308$) in a mean square error test to minimize the mean square error of prediction (Wallace).[6] All other variables were statistically significant $\alpha = 0.05$) (the three SEAP variables were tested jointly as were the three time variables). As expected, the North Carolina scallop industry is a price taker. Parameters of the equation were reestimated after omitting the three nonsignificant variables (see table 2).

For application of the price prediction equation in the optimal control problem, the averages of the independent variables ($INCOME_t$, $SEAP_t$, and $CALQ_t$) for two contrasting seasons (1980–81 and 1981–82) were used. The 1980–81 set of values produced a relatively high price path, whereas the 1981–82 set produced a relatively low price path.

Scallop size equation. The scallop size equation, $g(z,t)$, was developed by Kellogg and Spitsbergen.[7] The growth rate of the scallop meat was modeled as a function of meat size and the growth rate of the shell, which were in turn determined by water temperature. The basic model was a Brody-Bertalanffy growth equation with a temperature-dependent growth coefficient, as follows:

$$(4) \quad M_t = M_{max}(1 - ((M_{max} - M_0)/M_{max})e^{-B(C, t)},$$

where

$$B(C, t) = b_1 t + b_2 C + b_3 C^2/t,$$

M_t is meat size in grams at time t, M_{max} is maximum attainable meat size, M_0 is initial meat size at $t = 0$ (1 November), t is time in weeks from 1 No-

---

[6] The mean square error criterion is appropriate here because the purpose is price forecasting, not testing a theory on the structure of the scallop industry.

[7] Two unit changes were necessary before the scallop size equation could be compatible with the seasonal harvesting model. First, meat size is predicted in grams, whereas the harvesting model requires meat size in pounds. Thus, the meat size equation was multiplied by 0.002205 to convert grams to pounds. Second, the time unit for the meat size equation is weeks starting from 1 November, whereas the seasonal harvesting model requires time in weeks from 1 December. This was reconciled by replacing t with $t + 4.3$ (4.3 is the number of weeks in November) for every t in the size equation. Consequently, $t = 0$ would correspond to 1 December and $t = 17.3$ to 31 March, as required.

Table 2. Statistical Results from Estimating Two Models for Ex-Vessel Price of North Carolina Bay Scallops

Full Model
Analysis of Variance Table

Source	DF	Sum of Squares	Mean Square	F-Value	R^2
Model	11	3.54716445	0.32246950	5.06	0.70758
Error	23	1.46592456	0.06373585		
Corrected total	34	5.01308901		$(PR > F = 0.0005)$	

Parameter Estimates

Variable Name	Parameter Estimate	Standard Error	t-Value	$PR > |T|$
$INTERCEPT$	−3.62847863	1.64382944	−2.21	0.0375
$INCOME_t$	0.00379481	0.00269188	1.41	0.1720
NCQ_t	0.00000011	0.00000185	0.61	0.5454
$SEAP_t$	−1.97299849	0.62789450	−3.14	0.0046
$SEAP_{t-1}$	0.68808097	0.59989948	1.15	0.2632
$SEAP_{t-2}$	1.77670043	0.61348777	2.90	0.0082
$PSHRIMP_t$	−0.05640425	0.17662204	−0.32	0.7523
$CALQ_t$	−0.00000057	0.00000017	−3.42	0.0024
$WEEK$	0.34991078	0.19991462	1.75	0.0934
$WEEK^2$	−0.03577499	0.02714856	−1.32	0.2006
$WEEK^3$	0.00098686	0.00104226	0.95	0.3536
$INVENT$	0.00000003	0.00000006	0.47	0.6426

Restricted Model
Analysis of Variance Table

Source	DF	Sum of Squares	Mean Square	F-Value	R^2
Model	8	3.48834848	0.43604356	7.44	0.695848
Error	26	1.52474053	0.05864387		
Corrected total	34	5.01308901		$(PR > F = 0.0001)$	

Parameter Estimates

Variable Name	Parameter Estimate	Standard Error	t-Value	$PR > |T|$
$INTERCEPT$	−4.24904127	0.94400997	−4.50	0.0001
$INCOME_t$	0.00473008	0.00119327	3.96	0.0005
$SEAP_t$	−1.85448147	0.54878495	−3.38	0.0023
$SEAP_{t-1}$	0.56224238	0.55072151	1.02	0.3167
$SEAP_{t-2}$	1.69072116	0.57155936	2.96	0.0065
$CALQ_t$	−0.00000055	0.00000013	−4.15	0.0003
$WEEK$	0.38661194	0.17668992	2.19	0.0378
$WEEK^2$	−0.04133764	0.02380936	−1.74	0.0944
$WEEK^3$	0.00119521	0.00090949	1.31	0.2003

vember, and C is cumulative water temperature in degrees centigrade (degree-weeks) from 1 November to time t. The model was refined further by substituting an expression for M_{max}:

$$M_{max} = m_1 S_t^3$$

where S_t is the length of the shell. This relationship is used commonly in fisheries to relate weight to length. Shell size was then also modeled as a function of cumulative temperature in the same manner as the meat model:

$$(5) \qquad S_t = S_{max}[1 - ((S_{max} - S_0)/S_{max}) e^{-B_S(C, t)}],$$

where

$$B_S(C, t) = c_1 t + c_2 C,$$

S_t is shell length in centimeters at time t, S_{max} is maximum attainable shell size, and S_0 is initial shell length at $t = 0$ (a variable). Coefficients for the two models were estimated by Kellogg and Spitsbergen and are presented in table 1.

Using the above models, scallop meat size can be predicted for any week in the potential harvest season. Information needed to estimate the size equation includes an initial measure (or estimate) of shell size on about 1 November and projections

56   February 1988   Amer. J. Agr. Econ.

of cumulative water temperature. The initial value for shell size can be estimated by sampling, and an expected water temperature curve could be based on regional long-term weather predictions and pre-season temperatures. For the present study, the variable S_0 was set equal to 5.9 centimeters, which is the average shell size for November (Kellogg and Spitsbergen). Using a seven-year database, Kellogg and Spitsbergen estimated cumulative water temperature (degrees centigrade) for a "normal" winter as a function of time during the potential harvest season:

$$(6)\quad C = 20.203t - 1.012t^2 + 0.027t^3,$$

where C is cumulative temperature in degree-weeks from 1 November and t is time in weeks from 1 November. The resulting size function and price-per-scallop functions are shown in figures 1 and 2.

Fishing mortality equation. Fishing mortality is composed of two parts: (a) a standard measure of fishing effort (E) and (b) the catchability coefficient (q). A difficult aspect of any fishery management problem is the definition and subsequent measurement of a standard unit of effort. Any unit of effort can be used if the associated catchability coefficient is known or can be measured and if all other nominal units of effort in the fishery can be expressed in terms of the standard unit. Because no information is available for either E or q for this fishery, only ad hoc estimates of these variables are possible.

The number of fishermen engaged in scallop fishing at any moment during the harvest season depends on expected profit (and thus expected catch, price, and fishing costs) and profitability of alter-

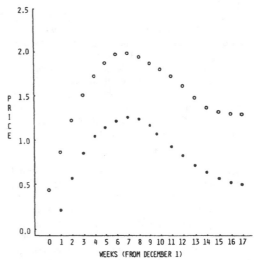

Note: The price equation using 1980–81 values for exogenous variables is indicated by "ooo"; the equation using 1981 values is indicated by "•••."

Figure 2.   Ex-vessel price per scallop, $P(w,t)g(z,t)$, for North Carolina bay scallops in 1967 cents

native fisheries or employment opportunities. Consequently, fishing effort should be modeled as an endogenous variable. Because quantitative data on fishing effort are not available for this fishery, fishing effort is taken as a constant average level throughout the season.[8] Kellogg, Easley, and Johnson (1985) estimated that 540 boat-days per week for E was generally consistent with the recent historical record. Because the effort level likely would be greater if the fishery maximized returns to the harvesting sector, the optimal control problem was also solved with a value of 750 boat-days per week for E.

With effort (E) in units of boat-days per week, the catchability coefficient (q) is the fraction of the scallop population that is harvested by one boat-day. It represents an average value over the entire harvest period and over all vessels. In actuality, the catchability coefficient varies from vessel to vessel and even from day to day for the same vessel. Kellogg, Easley, and Johnson (1985) estimated that fishing mortality has ranged historically from about 5% to 10% per week. For the standard unit of effort defined here, this is equivalent to q values ranging from about 0.0001 to 0.0003. Three q val-

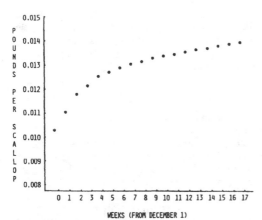

Figure 1.   The bay scallop meat size function, $g(z,t)$, in pounds per scallop

[8] A large project more costly than the present study is required to determine the supply response of effort to the North Carolina fisheries including the bay scallop fishery. Such a study has not yet been funded.

ues were used in this study to calculate optimal harvesting solutions: 0.0001, 0.0002, and 0.0003. These q values correspond to fishing mortalities ranging from 5% to 16% per week for $E = 540$ boat-days per week and 7% to 22% per week for $E = 750$. Without more specific information, these Eq values likely bracket expected fishing mortality for this fishery.

Cost equation. The cost function was modeled as a cost-per-unit-effort function,

$$(7) \qquad \text{cost per week} = cE,$$

where E is the number of boat-days per week and c is a constant cost coefficient. The cost coefficient, c, represents the average variable cost per standard unit of effort, or the average cost per boat day. Whereas the cost coefficient is modeled here as a constant, there is good reason to suspect that both daily fuel costs and oppotunity costs of the average fisherman vary over the harvest season because of employment, competition, and depletion factors. Thus, daily fuel costs may be a function of stock size rather than effort. However, because the relationship between stock density and fuel costs is not known, daily fuel costs were modeled on a per-effort basis.

The opportunity cost of scallop fishermen and the daily operating costs were estimated indirectly. Individuals employed in the processing sector were reported to make about $40 per day hand-shucking scallops in January 1982 (Raleigh, North Carolina *News and Observer*, 31 Jan. 1982). This wage was taken as an estimate of the opportunity cost of a scallop fisherman. That is, the fisherman could remain in port and shuck scallops if his income from fishing was less than $40 per day. Fricke cites a division of income from a day's catch as giving one-third of the gross to the boat to cover operating expenses (mostly fuel costs), one-third to the owner-captain (operator), and one-third to the crew member. (The standard scallop operation is one boat, captain, and one crew member.) If the opportunity cost of a fisherman is $40 per day, the total minimum daily opportunity cost for the two fishermen would be $80 and daily operating expenses would be $40. The total daily cost per standard unit of effort would be $120, which is equivalent to $42.55 after conversion to 1967 dollars (determined by dividing the nominal amount by the consumer price index for Jan. 1982).

The opportunity costs of scallop fishermen can vary markedly from season to season and from individual to individual because of other employment alternatives. Since this cost estimate affects the optimal season opening/closing schedule, a sec-

ond (lower) estimate of the cost coefficient was also calculated. Assuming an opportunity cost of $3.50 per hour (rather than the $5.00 per hour used above) and retaining the $40 per day operating cost, an alternative cost per standard unit of effort is $34.04 per day in 1967 dollars. The harvesting problem was solved for both of these cost estimates.

Equation of motion. Two features of bay scallop biology that simplify the equation of motion are (a) no recruitment during the potential harvest season and (b) very low natural mortality during the potential harvest season. Whereas natural mortality of bay scallops is high during the spring and summer, Division of Marine Fisheries biologists believe that the natural mortality rate is low during the winter months when harvesting occurs. Many of the important predators of bay scallops favor warmer temperatures, and bay scallops seem to thrive at the cooler winter temperatures (Kellogg and Spitsbergen). The natural mortality rate is not always near zero, however, because of extremely cold temperatures and low salinities. But without estimates of the natural mortality rate, the assumption of zero mortality during the potential harvest season is reasonable.

With zero natural mortality and no recruitment during the harvest season, the recruitment function, $F(z,t)$, collapses to a single initial value for population size, x_0, at $t = 0$. The change in population number—the equation of motion—can be modeled as equal to the harvest rate, as follows:

$$(8) \qquad \dot{x} = -Eqx(t)\,\Phi\,(t),$$
$$X(0) = x_0.$$

For this analysis, five values of x_0 that span the range of probable values were selected—13, 18, 23, 28, and 33 million scallops (Kellogg, Easley, and Johnson 1985).

The Optimal Harvesting Period

The maximum principle is used to solve for the optimal opening/closing schedule, $\Phi(t)$ (Johnson; Kamien and Schwartz). The hamiltonian for this problem is

$$(9) \quad H(t) = \{[P(t,w)g(z,t)Eqx(t) - cE]e^{-\delta t}$$
$$- \lambda(t)Eqx(t)\}\,\Phi(t),$$

which leads to the following switching function:

58   *February 1988*   *Amer. J. Agr. Econ.*

$$
(10) \quad \Phi(t) = \begin{cases} 1 \text{ if } \lambda(t) < [P(t, w)g(z, t) \\ \qquad - c/qx(t)]e^{-\delta t} \\ \\ 0 \text{ if } \lambda(t) \geq [P(t, w)g(z, t) \\ \qquad - c/qx(t)]e^{-\delta t}. \end{cases}
$$

The term on the right-hand side of the inequality is the discounted net revenue per scallop harvested. The switching function indicates that the season should remain closed as long as the user cost per scallop, λ, is greater than the discounted net revenue per scallop harvested. The season should open at the point where λ equals the discounted net revenue per scallop and remain open as long as the marginal discounted net revenue is greater than zero (shown graphically in figure 3).

The system of differential equations is as follows:

$$
(11) \quad \Phi = 0 \Longrightarrow \dot{x} = 0 \text{ and } \dot{\lambda} = 0,
$$

$$
\Phi = 1 \Longrightarrow \dot{x} = -Eqx(t), \quad x_0 \text{ given,}
$$

$$
\dot{\lambda} = \lambda(t)Eq - \frac{P(t, w)g(z,t)}{Eqe^{-\delta t}}.
$$

Sufficient conditions for a solution cannot be derived because boundary conditions for the adjoint equation are not specified. (That is, only the general solution of the adjoint equation can be obtained; the particular solution requires that λ be defined at a specific point in time.) The optimal solution is obtained numerically by varying $\lambda(0)$ (designated λ_0) until the $\Phi(t)$ corresponding to the maximum net present value of the season harvest

is identified. For a given estimate of λ_0, the procedure gives an optimal opening date conditional on this initial shadow price per scallop. Figure 3 shows one such solution with λ_0 equal to 0.007. With this initial shadow price, it is optimal to open the season four weeks from 1 December (i.e., at the beginning of the fifth week), when the present value of the net revenue per scallop first exceeds the user cost per scallop. Similarly, it is optimal to discontinue fishing after about eleven weeks from 1 December, when the present value of the net revenue per scallop falls below zero and it is thus no longer profitable to harvest fish. For the example shown in figure 3, a numerical search over possible values for the initial shadow price, λ_0, showed that the maximum cumulative net present value of the harvest would be obtained for λ_0 between 0.008 and 0.009 (see figure 4).

The optimal opening/closing schedule was determined to the nearest week. Further precision is not warranted in view of the many assumptions and approximations that were made in estimating the functions. The switching function is solved at the beginning of each week to see if the season should

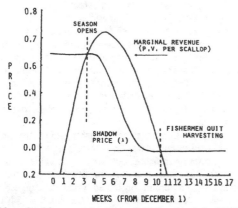

Note: In this example, $\lambda_0 = 0.007$; price is in 1967 cents.

Figure 3.   A graphical illustration of the procedure for solving for the optimal harvest period (see text). Price is in 1967 cents

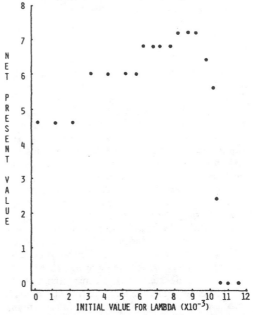

Note: The cumulative net present value of the harvest (in tens of thousands of 1967 dollars) is plotted against values for λ_0. In this example, the objective function is maximized when λ_0 is between 0.008 and 0.009.

Figure 4.   The value of the objective function

open or not. If the switching function is negative, the program skips to the beginning of the next week and repeats the check. If it is positive, the two differential equations are solved for the new x and λ before advancing to the beginning of the next week.

Obtaining the optimal λ_0 also results in an optimal season closing time. This is important in establishing the optimal present value of the harvest. However, it would not be regulated by the regulatory agency. In practice, fishermen will stop fishing when it is no longer profitable to harvest scallops, which will occur—under the assumptions of the model—as long as $P(t,w)g(z,t) > c/qx(t)$. The focus here is on finding a season opening time that is consistent with profit maximization and economic efficiency.

To estimate the gains from regulation, a comparison between the optimal solution (regulated case) and the unregulated case was made using the same set of assumptions. The "unregulated" case was determined by setting $\lambda(t) = 0$. The resulting solution delineates a period when it is profitable to harvest scallops under the assumptions of the model, but maintains the common property/open access condition that user costs are disregarded.

Results

The problem was solved using two price estimates, two cost estimates, five population size estimates, three q estimates, and two effort levels (table 1), resulting in 120 separate solutions. The range of values for each variable brackets the values most likely to occur. Consequently, the 120 optimal solutions provide general guidelines for opening the scallop season when better information on the exogenous variables is not available.

Table 3 contains optimal solutions for both the regulated and unregulated fishery for these combinations of population size, catchability coefficient, price, and cost with effort equal to 540 boat-days per week. The season length and net revenue entries in table 3 were generated by varying λ_0 as discussed above and illustrated in figures 3 and 4. Under the table heading of "fishing mortality," the results for the unregulated case can be compared directly to the results for the optimal solution, or regulated case.

Our most important result is that net revenue from the fishery can be increased by delaying the season opening beyond the traditional opening in early December. Four possibilities for the optimal time to open the season resulted from the 120 simulations, ranging from about the first week in January ($t = 4$) to about the first week in February ($t = 7$).[9] In contrast, the fishery has historically opened in the first or second week of December. In a move toward more optimal season openings, the North Carolina Division of Marine Fisheries this past winter opened the season temporarily for four days in December 1986 and delayed the permanent season opening to 12 January 1987 (personal communication, Division of Marine Fisheries personnel).

A useful comparison is between the regulated fishery and the unregulated fishery because assumptions about population size, market price, harvesting costs, and catchability are held constant for both cases. The season opening for the unregulated case typically ranged from one to three weeks after 1 December, depending primarily on the size of the scallop population. Optimal harvesting solutions typically delayed the opening of the season an additional two to three weeks. The difference between the net revenue for the regulated case and the unregulated case is a measure of the gains to the fishery from regulation. From the results shown in table 3, it is clear that delaying the season opening beyond the time at which fishing first became profitable—under the assumptions of the model—substantially increased the present value of the harvest for nearly all comparisons.

The most important determinant of the season opening was population size (x_0). At low fishing mortality levels with all factors except population size constant, solutions ranged from opening at the fifth week (high x_0) to not opening at all (low x_0). The effects of x_0 on the season opening were less pronounced at the higher fishing mortality values.

Price and cost also had an effect on the season opening, though not as dramatic as population size. For a given population size and fishing mortality, the season opening was generally one week earlier at the high price level than at the low price level. The two cost levels had a similar effect (the lower the cost, the earlier the season opening). As expected, higher prices or lower costs also extend the time before fishing becomes unprofitable to the fisherman.

The predominant effect of fishing mortality on the season opening was in determining whether it was profitable to fish or not. At low Eq values (less than 0.08), it was generally profitable to fish only at the higher population levels. Over the range of Eq values from 0.118 ($q = 0.0002$ and $E = 540$)

---

[9] The technique of stochastic dominance can be applied to reduce the number of "optimal" options facing the fishery manager when there is uncertainty about values for exogenous variables. See Kellogg, Easley, and Johnson (1985) for an application to the bay scallop harvesting problem.

60   *February 1988*   *Amer. J. Agr. Econ.*

Table 3. Summary of Harvesting Solutions for the North Carolina Scallop Fishery for Sixty Combinations of Exogenous Variables

| | Low Fishing Mortality (q = 0.0001) | | | | Medium Fishing Mortality (q = 0.0002) | | | | High Fishing Mortality (q = 0.0003) | | | |
| | unregulated[a] | | regulated | | unregulated[a] | | regulated | | unregulated[a] | | regulated | |
x_0^b	season[c]	net revenue	optimal season[c]	net revenue	season[c]	net revenue	optimal season[c]	net revenue	season[c]	net revenue	optimal season[c]	net. revenue
					Price = low,[d]	Cost = 42.55						
13	none		none		none		none		5	-68[e]	7	1,573
18	none		none		6	-38[e]	none		3-5	746	6-8	17,418
23	none		none		4-6	4,052	6-8	10,239	3-7	23,122	6-9	42,861
28	none		none		3-7	12,453	5-8	28,110	2-7	29,195	5-9	73,701
33	none		none		3-8	36,233	5-9	51,870	2-8	58,883	5-10	107,737
					Price = low,	Cost = 34.04						
13	none		none		none		none		4-6	2,214	6-7	8,304
18	none		none		4-6	1,983	6-8	6,829	3-7	16,115	6-9	31,967
23	none		none		3-7	12,669	6-9	25,276	2-7	26,911	5-9	62,970
28	6	-107	7	118	3-8	37,361	5-9	49,510	2-8	57,689	5-10	98,006
33	4-7	1,643	6-8	6,084	2-9	47,036	5-10	77,237	2-9	90,917	5-10	134,934
					Price = high,	Cost = 42.55						
13	none		none		4-5	-403[e]	7-8	3,818	2-5	4,144	6-8	29,457
18	none		none		2-6	7,025	6-9	31,065	1-6	21,990	5-9	78,080
23	5,7[f]	614[e]	7	932	2-8	45,999	5-10	71,919	1-8	67,819	5-10	136,033
28	4-8	10,408	6-9	13,838	1-9	66,814	5-11	119,718	1-9	121,070	5-11	199,263
33	3-9	28,925	5-10	36,439	1-10	115,532	5-12	171,734	0-9	120,886	5-12	265,578
					Price = high,	Cost = 34.04						
13	none		none		3-6	6,376	6-8	15,474	2-7	22,837	5-9	47,569
18	6	115	7	335	2-8	33,227	5-10	53,920	1-7	51,240	5-10	104,096
23	3-8	5,549	6-9	13,309	1-9	59,274	5-11	101,736	1-9	103,842	5-11	167,077
28	2-9	23,489	5-10	37,511	1-10	109,056	4-12	154,969	0-9	113,619	5-12	234,114
33	2-11	55,442	4-11	67,695	1-11	162,232	4-13	212,422	0-10	165,852	5-13	303,550

Note: Present value of net revenue is in 1967 dollars. Results for an effort level of 540 boat-days per week are presented.
[a]The unregulated solution was determined by setting λ equal to zero, which is equivalent to setting the user cost equal to zero, as occurs in the unregulated open access fishery.
[b]Millions of scallops.
[c]The season is in weeks where the first week is the first seven days in December. Numbering begins with zero. For example, an optimal season of "5–12" denotes that the season opens at the beginning of the sixth week (delaying the opening five weeks past 1 December) and remains open through the thirteenth week.
[d]The "high" price was produced by solving the price equation using the 1980–81 average of the exogenous variables (sea scallop price, calico scallop landings, and income). The "low" price was produced using the 1981–82 average of these variables.
[e]Negative values resulted because there was a positive net return at the beginning of the week which was offset by losses in the latter part of the week. Recall that the decision to harvest was made on a weekly basis.
[f]The fishery was not profitable during the second harvest week, but returned to being profitable in the third week because of growth in value.

to 0.225 (q = 0.0003 and E = 750), the optimal season opening varied by a maximum of one week (all other inputs constant) for about one-half of the input combinations and did not change for the remaining input combinations.

Summary and Conclusions

A bioeconomic optimal control model was constructed for the bay scallop fishery to determine the optimal season opening/closing schedule. Quotas were not imposed in the model, nor were re-

strictions on the number of fishing days allowed per week. Other regulations in current practice were maintained. A total of 120 separate scenarios were created using two price estimates, two cost estimates, five population size estimates, three estimates of the catchability coefficient, and two effort levels.

Results of this analysis clearly suggest gains can be obtained from delaying the season opening for bay scallops beyond the traditional opening date in December. The size of the gain depends on prices, costs, population size, and other variables.

Gains also come from eliminating the quota and

daily fishing restrictions. The basic principle behind optimal harvesting of a resource is to delay harvesting until the increase in resource value no longer exceeds the return from harvesting the resource and investing the proceeds elsewhere. For an annual fishery such as the North Carolina bay scallop fishery, the optimal harvest strategy is to devote all available fishing effort to the fishery (as long as each unit of effort remains profitable) once the optimal harvest time has arrived. Restrictions on catch and effort are inherently inconsistent with this harvesting strategy.

While this analysis provides useful insight into the problem of when to open the bay scallop season, several aspects of the model could be further developed. The assumption of a constant effort level throughout the entire harvest season is perhaps the most implausible aspect of the model.[10] Fishing effort is a function of expected profit, which in turn depends upon costs, market price, and the density of the scallop beds. In addition, effort in the first few weeks of the season is typically greater because of participation by part-timers, many of whom stop fishing when the weather gets colder and the scallop population becomes less dense. A supply function for effort is required to reflect these factors (see Kellogg, Easley, and Johnson 1986 for an example).

Another model simplification is the assumption of zero natural mortality during the harvest season. A nonzero natural mortality would yield an earlier optimal season opening than predicted here. While biologists currently believe the natural mortality rate during the winter is low, it surely is not zero.[11]

Bioeconomic optimal control models are not the only input that should be used by the fishery manager in promulgating regulations. Some aspects of a fishery are not easily incorporated into a model, such as income redistribution, political realities, dynamics of ecosystems, and catastrophic weather events. But management models such as the one presented here provide important insights.

[Received December 1985; final revision received June 1987.]

References

Agnello, Richard J., and Lawrence P. Donnelley. "Some Aspects of Optimal Timing of Intraseasonal Catch." *Economic Impacts of Extended Fisheries Jurisdiction*, ed. Lee G. Anderson, pp. 209–21. Ann Arbor MI: Ann Arbor Science Publishers, 1977.

Anderson, Lee G. *The Economics of Fisheries Management*. Baltimore MD: Johns Hopkins University Press, 1977.

Clark, Colin W. *Mathematical Bioeconomics: The Optimal Management of Renewable Resources*. New York: John Wiley & Sons, 1976.

Clark, Colin W., and Gordon R. Munro. "Fisheries and the Processing Sector: Some Implications for Management Policy." *Bell J. Econ.* 11(1980):603–16.

Conrad, Jon M., and Jose M. Castro. *Bioeconomics and the Harvest of Two Competing Species*. Agr. Econ. Staff Pap. No. 83-8, Cornell University, 1983.

Fricke, Peter H. *Socioeconomic Aspects of the Bay Scallop Fishery in Carteret County, North Carolina*. Sea Grant College Program Work Pap. 81-12, University of North Carolina, 1981.

Hall, Darwin C. "A Note on Natural Production Functions." *J. Environ. Econ. and Manage.* 4(1977):258–64.

Hannesson, Rognvaldur. "Bioeconomic Production Function in Fisheries: Theoretical and Empirical Analysis." *Can. J. Fisheries and Aquatic Sci.* 40(1983):968–82.

Huang, C. C., I. B. Vertinsky, and N. J. Wilimovsky. "Optimal Controls for a Single Species Fishery and the Economic Value of Research." *J. Fisheries Res. Board of Canada* 33(1976):793–809.

Johnson, Thomas. *Growth and Harvest without Cultivation: An Introduction to Dynamic Optimization*. Dep. Econ. and Bus., Econ. Res. Rep. No. 48, North Carolina State University, 1985.

Kamien, Morton I., and Nancy L. Schwartz. *Dynamic Optimization: The Calculus of Variations and Optimal Control in Economics and Management*. New York: Elsevier-North-Holland Publishing Co., 1981.

Kellogg, R. L. "A Bioeconomic Model for Determining the Optimal Timing of Harvest with Application to Two North Carolina Fisheries." Ph.D. thesis. North Carolina State University, 1986.

Kellogg, R. L., J. E. Easley, Jr., and Thomas Johnson. *A Bioeconomic Model for Determining the Optimal Timing of Harvest for the North Carolina Bay Scallop Fishery*. Sea Grant College Program Publ. UNC-SG-85-25, North Carolina State University, 1985.

———. *Application of a Seasonal Harvesting Model to Two North Carolina Shrimp Fisheries*. Sea Grant College Program Pub. UNC-SG-86-03, North Carolina State University, 1986.

Kellogg, R. L., and Dennis Spitsbergen. *Predictive Growth Model for the Meat Weight (Adductor Muscle) of Bay Scallops in North Carolina*. Sea Grant College Program Pub. UNC SG-83-6, North Carolina State University, 1983.

Levhari, David, Ron Michener, and Leonard J. Mirman. "Dynamic Programming Models of Fishing: Competition." *Amer. Econ. Rev.* 71(1981):649–61.

Loucks, R. H., and W. H. Sutcliffe, Jr. "A Simple Fish-Pop-

---

[10] The assumption of constant fishing effort is more reasonable when the season is short. For the 120 combinations of exogenous variables, the model determined that optimal fishing periods ranged from 1 to 10 weeks in length (table 3), with most less than 6 weeks.

[11] A research project has been funded to estimate the natural mortality rate of North Carolina bay scallops.

62 *February 1988*

Amer. J. Agr. Econ.

ulation Model Including Environmental Influence, for Two Western Atlantic Shelf Stocks.'' *J. Fisheries Res. Board of Canada* 35(1978):279–85.

Strand, I. E., and D. L. Hueth. "A Management Model for a Multispecies Fishery." *Economic Impacts of Extended Fisheries Jurisdiction*, ed. Lee G. Anderson, pp. 331–48.

Ann Arbor MI: Ann Arbor Science Publishers, 1977.

Wallace, T. D. "Pretest Estimation in Regression: A Survey." *Amer. J. Agr. Econ.* 59(1977):431–43.

Waters, James R., J. E. Easley, Jr., and Leon E. Danielson. "Economic Trade-Offs and the North Carolina Shrimp Fishery." *Amer. J. Agr. Econ.* 62(1980):124–29.

[44]

Bulletin of Mathematical Biology Vol. 54, No. 2/3, pp. 219–239, 1992.
Printed in Great Britain.

0092–8240/92$5.00 + 0.00
Pergamon Press plc
© 1991 Society for Mathematical Biology

A BIOECONOMIC MODEL OF THE PACIFIC WHITING*

■ JON M. CONRAD
  Department of Agricultural Economics,
  Cornell University,
  Ithaca, NY 14850, U.S.A.

The Pacific whiting (*Merluccius productus*) is a highly migratory fish occupying the continental shelf and slope off the west coast of North America. The species spawns in January off southern California and northern Mexico. During spring and summer the older and larger fish will migrate as far north as central Vancouver Island. Recruitment is highly variable, with strong year classes often supporting the commercial fishery during several years of low recruitment. The level of recruitment appears to be independent of the size of the spawning population.

A simple bioeconomic model of the Pacific whiting is constructed with independent recruitment. Fishery production functions are estimated from data on U.S. catch, average annual biomass and the number of vessels in the U.S. fleet. A stochastic optimization problem, seeking to maximize the expected value of industry profit, is formulated. Its solution would require a joint distribution on future recruitment and other bioeconomic parameters. Such a distribution is problematic. As an alternative, the certainty-equivalent problem is solved yielding solution values for the stochastic equilibrium and an approximately-optimal rule that sets allowable catch based on an estimate of current-year biomass.

Adaptive management can result in large changes in fleet size and allowable catch from year to year. The whiting fishery might be characterized as an opportunistic fishery, requiring a generalist fleet to expand or contract as bioeconomic conditions warrant. It is possible that long-run conditions would not support a profitable fishery, but that short-run fishing is profitable based on previous years of strong recruitment. The situation is not dissimilar to that facing the owner of a marginal gold mine that opens or closes depending on the price of gold. In the case of the whiting fishery, the optimal level of short-run fishing will depend not only on price, but on current biomass, the annual cost of fishing, the discount rate and vessel productivity. A simple interactive program is provided for would-be managers.

1. Introduction. With the development of a joint-venture fishery, the Pacific whiting (*Merluccius productus*) has become a commercially valuable species. Trawlers from California, Oregon, Washington and the province of British Columbia harvest whiting (also called hake) and then off-load the cod-end of their nets to a foreign factory vessel where the whiting is quickly processed to preserve freshness and texture. In 1989, the U.S. fleet delivered approximately 204 000 metric tons of whiting to foreign processing vessels, earning about $21 million in revenues.

The Pacific whiting is a highly migratory species, spawning off the coasts of southern California and northern Mexico in January (Bailey *et al.*, 1982).

* This paper was written in July 1990 while the author was a Summer Faculty Fellow at the Southwest Fisheries Center, La Jolla, California. The author gratefully acknowledges the support of the U.S. National Marine Fisheries Service.

During the spring and summer the population migrates northward, with the older and larger fish crossing into Canadian waters in August. Joint-venture arrangements have proved profitable to both U.S. and Canadian trawlers, and the distribution of allowable catch between the U.S. and Canada has taken on greater importance. While there is no formal treaty, fisheries managers from both countries have met to work out a long-term plan for binational allocation.

Recruitment in the whiting fishery appears to be independent of spawning biomass, but is positively correlated to surface temperature during spawning. Temperature in the month of January is affected by Eckmann transport, a process where warmer nearshore surface waters are pushed offshore, followed by an upwelling of deeper, cooler water (Bailey, 1981).

By August the whiting stock is distributed along the coast by age. While the location of a cohort in a particular year will depend on temperature, cohorts aged two to six are likely to be found off northern California and Oregon, while cohorts 7 to 14 are likely to be found off the coasts of Washington and British Columbia. In September and October whiting begin their southward migration from the feeding grounds to the spawning areas, and the cycle repeats itself.

The age structure of the resource and its reasonably stable migratory pattern have lead previous researchers to develop cohort models with population dynamics, migration and trophic interactions (Francis, 1983), stochastic recruitment (Swartzman *et al.*, 1983), and a game-theoretic approach to U.S.–Canadian management (Swartzman *et al.*, 1987). Dorn and Methot (1989) also employ a cohort model with recruitment randomly generated by iterative resampling from estimates of recruitment for the period 1959–1986. Constant and variable effort strategies are examined by averaging yields from 10 replicate, 1000-year simulations. Estimates of average yield ranged from 178 000 to 244 000 tons for the constant-effort strategy and from 205 000 to 251 000 tons for the variable-effort strategy. They recommend that total allowable catch be split 80 and 20% for the U.S. and Canada, respectively.

A simpler approach is taken in this paper. All the numerical results can be derived from the nine observations on catch, mean annual biomass and effort (vessel numbers) in Table 1, and by using the 20-line program (in BASIC) listed in Table 3. Analytical expressions for stochastic equilibrium and the approximately-optimal policy rule for adaptive management require some calculus and a fair amount of tedious algebra.

While the model is simple, it incorporates economic elements which have been absent in all the previous modeling of the Pacific whiting. Specifically, the program in Table 3 will employ estimates of a vessel productivity parameter, natural mortality, annual cost per vessel, dockside (or exvessel) price, the real rate of discount (interest) and long-run average recruitment to calculate what has been called the stochastic equilibrium. More relevant to short term management is the adaptive-management rule which, given an updated set of bioeconomic parameters and an estimate of current-year biomass, will suggest levels for allowable catch and fleet size. The issue of distributing allowable catch between the U.S. and Canada is left for resolution by managers from both countries.

The rest of this paper is organized as follows. In the next section we constuct a bioeconomic model and derive equations defining stochastic equilibrium and the adaptive-management rule. In the third section we estimate production functions for the Pacific whiting fishery and calibrate the model for price and cost *circa* 1988. Section 4 examines stochastic equilibria and the performance of the adaptive-policy rule for allowable catch under a range of values for the bioeconomic parameters. The paper concludes with a discussion of the implications and limitations of the model.

2. Bioeconomics: Stochastic Equilibrium and Adaptive Management.   Let X_t denote the average biomass of Pacific whiting in year t, E_t the level of fishing effort in year t and Y_t the level of harvest or catch. We assume there exits a production function relating annual catch to biomass and effort and write $Y_t = F(X_t, E_t)$, where the partial derivatives of $F(X_t, E_t)$ are denoted with subscripts and assumed to have the following signs: $F_X > 0$, $F_E > 0$, $F_{X,E} = F_{E,X} > 0$, $F_{X,X} \leqslant 0$ and $F_{E,E} \leqslant 0$. If p denotes the exvessel price per unit of catch (say, \$/metric ton) and c the cost of effort (say, cost/vessel/year), then we may write net revenue or profit in year t as:

$$\pi_t = pF(X_t, E_t) - cE_t. \tag{1}$$

Average annual biomass is assumed to change according to the following first-order difference equation:

$$X_{t+1} = (1 - M)[X_t - F(X_t, E_t)] + R_t \tag{2}$$

where M is annual natural mortality and R_t is a random variable denoting recruitment in year t. Maximization of the present value of expected profits subject to the dynamics of mean annual biomass may be stated mathematically as:

$$\text{Maximize } E\left\{ \sum_{t=0}^{\infty} \rho^t [pF(X_t, E_t) - cE_t] \right\}$$

$$\text{Subject to } X_{t+1} = (1 - M)[X_t - F(X_t, E_t)] + R_t$$

where $\rho = 1/(1 + \delta)$ is a discount factor and δ is the real rate of discount (or real annual interest rate).

This stochastic optimization problem might be solved by dynamic programming if a distribution for future recruitment were known. If other bioeconomic parameters are also random variables then one would need a joint distribution over all random variables. Such a distribution is problematic. As an alternative we consider what is called the "certainty-equivalent problem". The name is a bit of a misnomer, because the solution to the certainty-equivalent problem will not be the same as the solution to the stochastic dynamic programming problem (when the necessary distribution is known). The actual degree of suboptimality associated with the solution to the certainty-equivalent problem will depend on the specifics of the problem, the functional forms, the presence of irreversibilities, and the degree to which initial conditions differ from the long-run "stochastic equilibrium".* Before discussing the issue of suboptimality further, it may be useful to pose and solve the certainty-equivalent problem.

Let the expected value of R_t be denoted by R. The certainty-equivalent problem is the deterministic problem obtained by substituting the expected value for its random variable. This results in a problem with a Lagrangian expression that may be written as:

$$L = \sum_{t=0}^{\infty} \rho^t \{ pF(X_t, E_t) - cE_t$$
$$+ \rho\lambda_{t+1}[(1-M)[X_t - F(X_t, E_t)] + R - X_{t+1}] \} \qquad (3)$$

where λ_{t+1} is the Lagrange multiplier associated with biomass in period $t+1$, and may be interpreted as the marginal value of an additional unit (say, metric ton) of fish in the water in year $t+1$. The Lagrange multiplier is also called the "shadow-price" of the fish stock. Note that R becomes a parameter in the certainty-equivalent problem.

In the Appendix we derive the first-order necessary conditions for this problem. They can be evaluated in steady state and are shown to imply the following two equations:

$$\frac{c(1-M)F_X}{(pF_E - c)} = \delta + M \qquad (4)$$

$$R = MX + (1-M)F(X, E). \qquad (5)$$

Equation (4) is a special case of what has been called the "fundamental equation of renewable resources" (see Conrad and Clark, 1987, p. 34). With independent recruitment the first derivative of the net growth function vanishes

---

* Perhaps a more accurate name would be "certainty-equivalence equilibrium" since the concept of stochastic equilibrium is usually associated with a stationary probability distribution.

and we are left equating the "marginal stock effect" to the sum of the rate of discount and natural mortality. The marginal stock effect measures the incremental cost savings from larger biomass relative to the immediate benefit if that increment in biomass were harvested this year.

Equation (5) requires that expected (or long-run average) recruitment offset the reduction in biomass from natural mortality plus that portion of biomass that would have survived had it not been harvested. Equations (4) and (5) collectively define what Burt (1967) refers to as the stochastic equilibrium. Burt was concerned with the optimal management of a groundwater stock when recharge (form rain or melting snow) was stochastic. He notes that the stochastic equilibrium is "always approached, but rarely experienced".

The stochastic equilibrium for our problem is portrayed in Fig. 1. From the implicit function theorem, equation (4) will define a curve in $X-E$ space. Totally differentiating equation (4) and making use of the partials of $F(X, E)$, we can show that along this curve dE/dX is positive. Depending on the form of $F(X, E)$ it may be possible to solve for an explicit relationship, $E = E(X)$, that is positively sloped.*

Equation (5) also implies a curve in $X-E$ space. Total differentiation and the signs for F_X and F_E will imply that along this curve $dE/dX < 0$. If an explicit relationship, $E = R(X)$, can be obtained from equation (5), it will be negatively sloped. Thus, the partials of $F(X, E)$ imply that a nonzero stochastic equilibrium, (X_R, E_R) in Fig. 1, will be unique.

While the stochastic equilibrium may be of interest in determining the long-run effects of changes in the bioeconomic parameters, it is not very useful for short-term management. When fish biomass is not at its long-run equilibrium we would need to solve the deterministic certainty-equivalent problem, or a finite-horizon stochastic dynamic programming problem to determine the first step along an "approach path". With $F(X_t, E_t)$ nonlinear, this is not a trivial problem.

Instead of taking this tack we make use of an "approximately-optimal" technique proposed by Burt (1964, 1967) for groundwater management and more recently examined by Kolberg (1990) for management of a fishery. This approach makes use of equation (4) by noting that it can be regarded as defining a relationship between X_t and E_t in the vicinity of long-run equilibrium. Could we use this relationship for short-run management? If we do, how inferior would the resulting decisions be, relative to the solution

---

* The curve $E(X)$ has nothing to do with the expectation operator. It is a smooth, positively-sloped curve obtained from equation (4) and will be used to identify the approximately optimal level of effort given an estimate of biomass, X. The curve implied by equation (5) will be denoted $E = R(E)$, since this curve will depend on parameters of the production function, the annual mortality rate and the long-run, expected level of recruitment, R. This latter curve is used in defining the stochastic or certainty-equivalence equilibrium, but only $E(X)$ is needed for short-run, adaptive management.

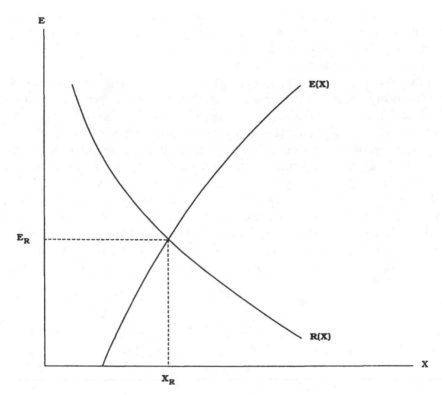

Figure 1. The long-run stochastic equilibrium.

obtained for a stochastic dynamic programming problem (with a known distribution for recruitment)? We will take these two questions in order.

The procedure for using equation (4) as an adaptive management rule is shown in Fig. 2. In the northeast quadrant we have redrawn the $E(X)$ curve from Fig. 1. Its position depends on all of the bioeconomic parameters except R, expected recruitment, which only appeared in equation (5). Suppose that biologists, using data from a series of scientific trawls or through a cohort model taking into account last year's total (U.S. plus Canadian) harvest, could provide managers with an estimate of biomass for the forthcoming year. With an estimate of X we could project up to the $E(X)$ curve to determine the recommended level of effort. The estimate of current biomass will also imply a specific production function in $E - Y$ space drawn in the northwest quadrant. Projecting E over to the appropriate production function results in a catch rate which might be used as allowable catch for the forthcoming year. Because recruitment is stochastic and because fishermen may exceed or fail to harvest allowable catch in a particular year, the subsequent estimates of X may bounce around. From Fig. 2 we can get a qualitative feel for how recommended effort

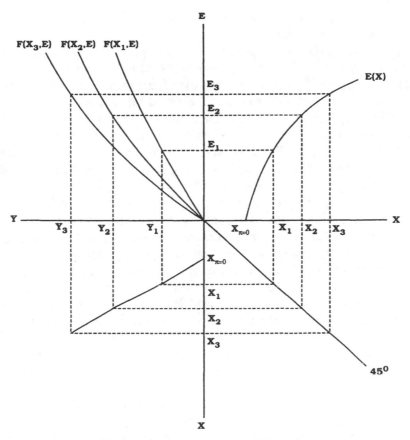

Figure 2. A depiction of the approximately-optimal feedback policy.

and allowable catch vary with X. First, note that there is likely to be an intercept of the $E(X)$ curve on the X axis. This has a straightforward interpretation. For a given set of bioeconomic parameters (per unit price, annual cost, the rate of discount, natural mortality, and perhaps catchability) there is likely to be some stock level below which fishing today would reduce present value. This is denoted by $X_{\pi=0}$. As X increases we see a less than proportional increase in E. The resulting change in Y is less easy to assess qualitatively because the production function shifts upward with increases in X. With a particular form for $F(X, E)$, and given estimates of the bioeconomic parameters, we could numerically examine the change in Y for a change in X. (We will do this for the whiting fishery in Section 4.) If we wish, we could collect the (X, Y) pairs by constructing a 45° transfer line in the southeast quadrant, project X downward, across, and then pair it with the corresponding Y projected downward from the Y-axis of the northwest quadrant. This is done in the southwest quadrant and the four "dots" have been (arbitrarily) "connected" by a series of line segments.

Burt compared the level of groundwater pumping recommended by such a procedure to the level of pumping recommended when using stochastic dynamic programming, taking the estimate of current X as an initial condition, with all other parameters the same. In his study he found that the pumping rates differed by less than 2% when the current groundwater stock was within 42% of the stochastic equilibrium. As the current stock got closer to the stochastic optimum, the difference went to zero. On the basis of this relatively small departure from the optimal pumping rate, Burt dubbed this rule "the approximately-optimal" pumping rule.*

Burt and Cummings (1977), in considering this rule for other renewable resources, found that the difference between the approximately-optimal harvest rate and the optimal rate obtained via stochastic dynamic programming was likely to exhibit a consistent and perhaps attractive bias. When $X < X_R$ the harvest rate from the approximate rule was likely to be less than the harvest rate from the optimal rule. When $X > X_R$ harvest was likely to be slightly more than optimal. This would lead to a more rapid approach to equilibrium in a deterministic model. The slightly lower levels for recommended harvest when the resource stock was less than its stochastic equilibrium caused Burt and Cummings to regard the approximate rule as also being "conservative". Managers may find this built in conservatism (when stock is low) to be attractive.

In a recent study and application to the anchovy fishery in northern California, Kolberg (1990) analyzed the above approximate procedure and compared it to the optimal solution (obtained via dynamic programming) and two other approximate solutions obtained from first- and second-order Taylor approximations to the value function at the steady state optimum. Burt's original approximate rule [equation (4) in this paper] is equivalent to the first-order approximation of the value function. Kolberg finds that both first- and second-order rules result in harvesting decisions that produce a stream of discounted profits within 1% of the maximum.

---

* The approximate optimality of the $E(X)$ curve as employed in Fig. 2 can be derived from the first-order conditions in the Appendix. Suppose that the optimal approach from some stock level X_t involves values for the Lagrange multipliers of $\lambda_t = \lambda + \varepsilon_t$ and $\lambda_{t+1} = \lambda + \varepsilon_{t+1}$, where λ is the value of the multiplier at the certainty-equivalence equilibrium. Then the first and second of the first-order conditions can be shown to imply:

$$\frac{(pF_E - c)[\delta + M + (1 - M)F_X]}{(1 - M)F_E} = pF_X + \varepsilon_{t+1} + \varepsilon_t.$$

If the sum of ε_{t+1} and ε_t is small (which seems especially likely if convergence is oscillatory), then the above equation will be closely approximated by equation (4) in the text, and hopefully not too far from the solution to the underlying stochastic optimization problem.

It is difficult to make such comparisons in the whiting fishery. Without a joint distribution for recruitment and other bioeconomic parameters we do not have the necessary ingredients for the appropriate stochastic problem. As we will see in the next section, many of the estimates for equilibrium stock are around 1.0 million metric tons. This is within 30% of the 1989 estimate of 1.3 million metric tons for mean annual biomass (Dorn and Methot, 1989; Table 12). It would appear, at least superficially, that the approximately-optimal decision procedure described above can be appropriately applied to the Pacific whiting fishery.

3. Calibration of the Model for the Pacific Whiting Fishery.   In the general model of the preceding section the production function, $Y_t = F(X_t, E_t)$, took on central importance in defining the stochastic equilibrium and the adaptive-management rule. When one attempts to specify and estimate such a function, one encounters at least two problems. First, where does one obtain a time series of estimates for average annual biomass, and second, how should one define effort?

In calibrating the model to the Pacific whiting fishery the author was fortunate to have estimates of average annual biomass from a stock-synthesis model developed by Dorn and Methot (1989). This time series seemed the best available and would also allow a comparison of yield levels from two otherwise disparate modeling perspectives.

The definition of effort has always proven difficult. Ideally, one would like as precise a measure as possible of the actual volume of water "strained" per unit time. The closest practical measure might be the number of hours that a vessel had net in the water fishing. In a bioeconomic model, the analyst is further removed from the ideal measure because of the need to estimate the unit cost of effort. The measure adopted here is the number of vessels in the fishery. This measure is open to criticism because it may not correspond to the volume of water strained during a season, but it is a measure for which we have some data on unit annual cost.

Table 1 contains data on catch by U.S. vessels, estimates of mean annual biomass, and the number of vessels in the U.S. whiting fleet from 1981 to 1989. From 1985 onward the fleet has increased, with a jump from 42 vessels in 1988 to 65 in 1989. The estimate of mean annual biomass has declined from 2.225 million metric tons in 1986 to 1.315 million metric tons in 1989. Dorn and Methot believe that this reflects the "mining" of the strong 1980 and 1984 year classes that were recruited into the fishery in 1982 and 1986, respectively. (Note the jump in average annual biomass in those years.)

Table 2 contains the regression results when the data in Table 1 were used to estimate Cobb–Douglas and exponential production functions. The Cobb–Douglas function takes the form $Y = qX^\alpha E^\beta$, and is linear logs. It contains, as a

Table 1. Data on catch, mean annual biomass and effort in the
U.S. Pacific whiting fishery

Year (t)	Catch (Y_t)*	Biomass (X_t)†	Effort (E_t)‡
1981	44 395	1 384 000	21
1982	68 488	2 000 000	17
1983	73 150	1 805 000	19
1984	81 610	1 742 000	21
1985	35 586	1 685 000	17
1986	85 103	2 225 000	25
1987	110 792	2 012 000	31
1988	142 657	1 688 000	42
1989	204 038	1 315 000	65

*U.S. catch is measured in metric tons and is the sum of joint-venture and domestic catch from Table 1 of Dorn and Methot (1989).

†Biomass is measured in metric tons and is the mean annual estimate of biomass from Table 12 of Dorn and Methot (1989).

‡Effort is measured as the number of vessels in the U.S. Pacific whiting fleet as listed in the fax from D. E. Squires, NMFS, Southwest Fisheries Center, La Jolla, California, 18 June 1990.

special case, the standard catch-per-unit-effort production form (when $\alpha = \beta = 1$).

The exponential function takes the form $Y = X(1 - e^{-\alpha E})$. With this form it is impossible to catch more than current biomass, a logical characteristic, unfortunately not exhibited by the Cobb–Douglas production function. (Note: With the Cobb–Douglas form, as effort goes to infinity, so does catch.)

The exponential function may be estimated by regressing the natural log of the fraction of surviving biomass on effort. Ideally one would like to obtain an intercept not significantly different from zero and a significantly negative coefficient on effort. Alternatively, one can force the regression through the origin by suppressing the intercept.

The regression results for Cobb–Douglas and exponential production functions are given in Table 2. For the Cobb–Douglas form the coefficient for the natural log of q is not significantly different from zero, implying that q is not significantly different from one. The estimates of α and β are significant at the 5% level for a one-tail test. The adjusted R^2 is 0.8369. The Durbin–Watson statistic would appear to be in the inclusive range. The small sample size, however, makes determination of autocorrelation difficult.

In the least squares regression for the exponential function, the estimates of both the constant and the effort coefficient were significant. The adjusted R^2 was 0.9568. Because the Durbin–Watson seemed at the lower end of the inconclusive range the regression was run correcting first for first-order autocorrelation, then for second-order autocorrelation. Neither $AR(1)$ nor

Table 2. Regression results for the Cobb–Douglas and exponential production functions

A. Cobb–Douglas: $\ln Y_t = \ln q + \alpha \ln X_t + \beta \ln E_t + \varepsilon_t$
Least squares

Variable	Coefficient	Standard error	t-statistic	Two-tail significance
$\ln q$	−7.4226	7.4350	−0.9983	0.357
$\ln X$	1.0274	0.4979	2.0631	0.085
$\ln E$	1.2240	0.1878	6.5169	0.001
$R^2 = 0.8776$,	Adjusted $R^2 = 0.8369$,	$F = 21.5285$,	$D - W = 1.4724$	

B. Exponential: $\ln(1 - Y_t/X_t) = \gamma + \alpha E_t + \varepsilon_t$
Least squares

Variable	Coefficient	Standard error	t-statistic	Two-tail significance
γ	0.0215	0.0068	3.1732	0.016
E	−0.0028	0.00021	−13.35	0.000
$R^2 = 0.9622$,	Adjusted $R^2 = 0.9568$,	$F = 178.232$,	$D - W = 1.1146$	

Least squares, $AR(1)$

Variable	Coefficient	Standard error	t-statistic	Two-tail significance
γ	0.0272	0.0187	1.4575	0.205
E	−0.0030	0.00043	−7.0358	0.001
$AR(1)$	0.5962	0.5030	1.1852	0.289
$R^2 = 0.9708$,	Adjusted $R^2 = 0.9591$,	$F = 83.209$,	$D - W = 0.9544$	

Least squares, $AR(2)$

Variable	Coefficient	Standard error	t-statistic	Two-tail significance
γ	0.0225	0.0121	1.8498	0.138
E	−0.0028	0.00036	−7.7592	0.001
$AR(2)$	−0.0783	0.6942	−0.1128	0.916
$R^2 = 0.9651$,	Adjusted $R^2 = 0.9477$,	$F = 55.368$,	$D - W = 0.9676$	

$AR(2)$ were significant and in fact the Durbin–Watson statistics became smaller. The constant term became insignificant in both the $AR(1)$ and $AR(2)$ regressions but the estimate of α was essentially unchanged and remained significant.

The exponential form, with $\alpha > 0$ implies that production is strictly concave in effort, while the Cobb–Douglas function with $\beta > 1$ is not concave and would cause the stochastic equilibrium to be locally unstable. For this reason, and others noted above, we adopt the exponential form and run sensitivity analysis on α over the interval [2.0E-2, 3.0E-2].

In an analysis of the tax returns of 13 vessels participating in the whiting fishery in 1988, Squires (1990) estimates annual variable costs per vessel to be approximately \$150 000. More difficult to estimate is the portion of fixed costs that should also be included when estimating annual operating costs. Squires calculates annual fixed costs by adding the costs of insurance, rent, association dues, professional services and 7% of vessel acquisition costs (for vessels bought in 1978–1986, inclusive), for a total of approximately $237 000 in 1988. The sum of annual variable and fixed cost payments comes to $387 000.

It is difficult to argue that all the costs filed (*ex post*) on a tax form are relevant when a fisherman chooses to fish whiting, as opposed to some other species. One also suspects that there is an incentive to report as high a cost as possible (to reduce taxable income). In the numerical analysis of the next section we restrict our estimate of c to the interval ($200 000, $300 000).

Francis (1983), in fitting a cohort model to survey data, concluded that annual mortality was likely to be age-dependent, with rates varying from 0.195 for 5-year-old fish, to 0.757 for 11-year-old fish. Dorn and Methot use a constant rate of 0.20 for all cohorts. An average annual mortality rate of 0.25 is used in the Base-Case, with values of $M = 0.20$ and $M = 0.30$ also examined.

The price per metric ton for whiting has fallen since the early 1980s, when it peaked at slightly over $151 in 1982. From 1986 to 1989 the price has been relatively stable between $106 and $110 per metric ton. Stochastic equilibria and adaptive management are examined for prices of $100, $110 and $120 per metric ton.

Modeling by Dorn and Methot also provided estimates of recruitment, measured as billions of age two fish entering the fishery. They construct a time series from 1958 to 1988. There is a large range, from a low of 0.017 in 1987 to a high of 5.16 in 1963. The average over this 31-year period was 0.991 billion fish. An average 2-year-old whiting will weigh about 250 grams, transforming the 0.991 billion fish into an average recruitment of approximately 250 000 metric tons per year.

Though imprecise, we set $R = 250 000$ metric tons in the Base-Case parameter set. It is important to emphasize that while recruitment is highly variable, mean annual biomass is much less variable. Adaptive management does not depend directly on recruitment, only on an estimate of mean annual biomass. This has ranged from a high of 3.695 million metric tons in 1965, to low of 1.315 in 1989; with most year to year changes being less than 15%.

The final parameter required for both stochastic equilibrium and adaptive management is an estimate of the real (inflation-free) rate of discount. There has been a long standing debate among economists as to the appropriate rate of discount to employ when evaluating public investments or managing publicly held resources. There appears to be no simple answer. It depends on where the funds are coming from (whether they are displacing private investment or consumption) and whether the beneficiaries of the project derive a significant portion of their income from the investment or resource.

The question is perhaps more easily answered when managing a fishery resource. If there are a large group of fishermen, or if the species being managed constitutes only a small portion of the total income derived from fishing, then the discount rate should be risk-free as well. Discount rates of 2, 4 and 6% will be evaluated.

4. Results. Restricting our analysis to the exponential production function, $Y = X(1 - e^{-\alpha E})$, we note that $F_X = (1 - e^{-\alpha E})$, and that $F_E = \alpha X e^{-\alpha E}$. Substitution into equation (4) results in:

$$\frac{c(1 - M)(1 - e^{-\alpha E})}{(p\alpha X e^{-\alpha E} - c)} = \delta + M. \tag{6}$$

Equation (5) takes the form:

$$R = MX + (1 - M)X(1 - e^{-\alpha E}). \tag{7}$$

It is possible to solve equation (6) for an explicit expression for E, yielding:

$$E = -\ln\left[\frac{c(1 + \delta)}{[p\alpha(\delta + M)X + c(1 - M)]}\right]\bigg/\alpha. \tag{8}$$

This is our $E = E(X)$ curve in Figs 1 and 2. It will be used in the adaptive-management program.

Using equation (7) it is possible to eliminate E from equation (6) and obtain a quadratic expression in X. The positive root gives an expression for the optimal (stochastic) equilibrium stock. This expression is tedious to derive but some careful algebra should reveal:

$$X_R = (-B + \sqrt{B^2 - 4N})/2 \tag{9}$$

where:

$$B = -\frac{[p\alpha R(\delta + M) + c(1 - M)\delta]}{[p\alpha(\delta + M)]} \tag{10}$$

and

$$N = -\frac{c(1 - M)R}{[p\alpha(\delta + M)]}. \tag{11}$$

With X_R we can calculate long-run optimal effort as:

$$E_R = -\ln\{(X_R - R)/[(1 - M)X_R]\}/\alpha. \tag{12}$$

From the production function we know $Y_R = X_R(1 - e^{-\alpha E_R})$.

In the program in Table 3 we define and read the parameters α, c, δ, M, p and R and then calculate the stochastic equilibrium X_R, E_R and Y_R. You are then asked if you would like to adaptively manage. If you answer yes, you are asked for an estimate of current-year biomass. Using this as the value of X in equation (8), and the same bioeconomic parameters as specified in line 10, the program calculates the approximately-optimal E (lines 140–150), then catch (line 160), and finally prints the results.

Table 3. A listing of the BASIC program to calculate the stochastic equilibrium and to adaptively manage based on estimates of current biomass

```
 10  DATA 0.25E−2,250000,0.04,0.25,110,250000
 20  READ A,C,D,M,P.R
 30  B=−(P*A*R*(D+M)+C*(1−M)*D)/(P*A*(D+M))
 40  N=−C*(1−M)*R (P*A*(D+M))
 50  XR=(−B+SQR(B 2−4*N))/2
 60  ER=−LOG((XR−R)/((1−M)*XR))/A
 70  YR=XR*(1−EXP(−A*ER))
 80  PRINT:PRINT "Long-Run Average Biomass=";XR
 90  PRINT:PRINT "Long-Run Average Effort=";ER
100  PRINT:PRINT "Long-Run Average Catch=";YR
110  PRINT:INPUT "Do you want to Adaptively Manage? Yes=1, No=0.";W
120  IF W=0 GOTO 200
130  PRINT:INPUT "Current Biomass=";X
140  NUM=C*(1+D):DEN=P*A*(D+M)*X+C*(1−M)
150  E=−LOG(NUM DEN)/A
160  Y=X*(1−EXP(−A*E))
170  PRINT:PRINT "Current Biomass=";X
180  PRINT:PRINT "Recommended Effort=";E
190  PRINT:PRINT "Recommended Catch=";Y
200  END
```

The Base-Case parameter set is $\alpha = 0.25\text{E-}2$, $c = \$250\,000$, $\delta = 0.04$, $M = 0.25$, $p = \$110$ and $R = 250\,000$. Table 4 reports the calculated values for stochastic equilibrium and the approximately-optimal values for effort and allowable catch when the current biomass is 1.0E6, 1.5E6 and 2.0E6 metric tons. Each parameter (with the exception of R) is varied above and below its Base-Case value to determine its effect on the stochastic equilibrium and the adaptively-managed levels of effort and allowable catch. The results are presented in the 10 subcases (B through K) also contained in Table 4. Increases in long-run expected recruitment, R, will increase the equilibrium levels for biomass, effort and yield, but given the imprecise estimate of this value and the previously noted fact that the stochastic equilibrium is "seldom experienced", we do not present these results.

For the Base-Case parameter set the stochastic equilibrium occurs at a mean annual biomass of 957 748 metric tons, supporting a fleet of six vessels

Table 4. Stochastic equilibria and adaptive management

A. Base-Case parameter set

$\alpha = 0.25E\text{-}2$	$c = \$250\ 000$	$\delta = 0.04$
$M = 0.25$	$p = \$110$	$R = 250\ 000$ mt

$X_R = 957\ 748$ mt		$E_R = 6$ vessels	$Y_R = 14\ 083$ mt
When	$X = 1.0E6$	$E = 11$	$Y = 27\ 128$
	$X = 1.5E6$	$E = 67$	$Y = 230\ 159$
	$X = 2.0E6$	$E = 115$	$Y = 501\ 441$

B. $\alpha = 0.20E\text{-}2$

No commercial fishery in the long run. Vessels not sufficiently productive.

When	$X = 1.0E6$	No commercial fishery	Stock too low
	$X = 1.5E6$	$E = 43$	$Y = 122\ 881$
	$X = 2.0E6$	$E = 96$	$Y = 349\ 730$

C. $\alpha = 0.30E\text{-}2$

$X_R = 883\ 051$ mt		$E_R = 15$ vessels	$Y_R = 38\ 983$ mt
When	$X = 1.0E6$	$E = 28$	$Y = 81\ 921$
	$X = 1.5E6$	$E = 81$	$Y = 321\ 930$
	$X = 2.0E6$	$E = 126$	$Y = 627\ 606$

D. $c = \$200\ 000$

$X_R = 867\ 361$ mt		$E_R = 21$ vessels	$Y_R = 44\ 213$ mt
When	$X = 1.0E6$	$E = 40$	$Y = 94\ 668$
	$X = 1.5E6$	$E = 104$	$Y = 342\ 837$
	$X = 2.0E6$	$E = 159$	$Y = 655\ 897$

E. $c = \$300\ 000$

No commercial fishery in the long run. Fishing too costly.

When	$X = 1.0E6$	No commercial fishery	Stock too low
	$X = 1.5E6$	$E = 40$	$Y = 142\ 002$
	$X = 2.0E6$	$E = 84$	$Y = 377\ 113$

F. $\delta = 0.02$

$X_R = 958\ 887$ mt		$E_R = 6$ vessels	$Y_R = 13\ 704$ mt
When	$X = 1.0E6$	$E = 10$	$Y = 25\ 788$
	$X = 1.5E6$	$E = 64$	$Y = 220\ 201$
	$X = 2.0E6$	$E = 110$	$Y = 482\ 143$

G. $\delta = 0.06$

$X_R = 956\ 702$ mt		$E_R = 6$ vessels	$Y_R = 14\ 432$ mt
When	$X = 1.0E6$	$E = 12$	$Y = 28\ 414$
	$X = 1.5E6$	$E = 70$	$Y = 239\ 956$
	$X = 2.0E6$	$E = 120$	$Y = 519\ 553$

Table 4—*continued*

H. $M = 0.20$

$X_R = 1\ 075\ 564$ mt		$E_R = 17$ vessels	$Y_R = 43\ 609$ mt
When	$X = 1.0E6$	$E = 9$	$Y = 22\ 556$
	$X = 1.5E6$	$E = 56$	$Y = 195\ 652$
	$X = 2.0E6$	$E = 98$	$Y = 433\ 735$

I. $M = 0.30$

No commercial fishery in the long run. Natural mortality too high.

When	$X = 1.0E6$	$E = 13$	$Y = 31\ 657$
	$X = 1.5E6$	$E = 77$	$Y = 262\ 887$
	$X = 2.0E6$	$E = 132$	$Y = 563\ 536$

J. $p = \$100$ mt^{-1}

No commercial fishery in long run. Price too low.

When	$X = 1.0E6$	$E = 0$	$Y = 0$
	$X = 1.5E6$	$E = 52$	$Y = 183\ 544$
	$X = 2.0E6$	$E = 98$	$Y = 436\ 090$

K. $p = \$120$ mt^{-1}

$X_R = 921\ 131$ mt		$E_R = 12$ vessels	$Y_R = 26\ 289$ mt
When	$X = 1.0E6$	$E = 22$	$Y = 52\ 823$
	$X = 1.5E6$	$E = 81$	$Y = 273\ 585$
	$X = 2.0E6$	$E = 132$	$Y = 561\ 549$

Vessel numbers are rounded to nearest whole vessel. Catch is rounded to nearest whole metric ton. Catch is calculated before rounding effort. Thus, fractional effort less than 0.5 vessels may give rise to slightly different catch for same biomass (Subcase H to J when $X = 2.0E6$).

harvesting 14 083 metric tons per year. These values are significantly below those observed in the previous decade (see Table 1).

When the current biomass increases from 1.0 to 2.0 million metric tons the adaptive rule recommends that fleet size increase from 11 to 115 vessels and that catch be allowed to increase from 27 128 to 501 441 metric tons. When current biomass is 1.5 million metric tons, a recommended fleet of 67 vessels would harvest 230 159 metric tons. These latter values are very similar to the "observed" values for catch, biomass and effort in 1989 from Table 1.

From this single piece of analysis we might hazard a characterization of the whiting fishery. It is a fishery that will be strongly influenced by current bioeconomic conditions. It should be managed opportunistically. When stochastic recruitment "deals a full house", maximization of expected present value says the fleet should significantly expand to harvest the windfall. The downside, of course, is that when recruitment deals nothing, the fleet must "fold 'em" and walk. To quote the Kenny Rodgers song, fisheries managers have to "know when to hold 'em and know when to fold 'em". In the U.S. and elsewhere, unfortunately, managers and fishermen have been slow to walk, trying to stay in the game when bioeconomic conditions indicate one should leave (at least temporarily).

The program in Table 3 indicates when fishing would reduce present value by returning a negative value for effort and catch. This can occur in the long-run stochastic equilibrium or in the short-run under adaptive management. In fact, for a given set of bioeconomic parameters a fishery that is unprofitable in the long-run may continue to be fished if strong recruitment or favorable prices prevail. Conversely, a fishery which is profitable in the long-run (stochastic equilibrium) may be shut down in the short-run because biomass has declined below a level that would support positive effort and catch along the optimal approach path. Recall the interpretation of $X_{\pi=0}$ in Figs 1 and 2.

The first situation is shown in Subcase B where, when vessel productivity declines from $\alpha = 0.25\text{E-2}$ to $\alpha = 0.20\text{E-2}$, there is no fishing in the stochastic equilibrium. If a run of strong recruitment (or a temporary moratorium) pushes biomass up to 1.5E6 metric tons a fleet of 43 would be allowed to harvest 122 881 metric tons. In Subcase C, where $a = 0.30\text{E-2}$, the stochastic equilibrium has a biomass of 883 051 metric tons supporting 15 vessels and an annual yield of 38 983 metric tons. If recruitment pushes biomass up to 1.5E6 metric tons, adaptive managers would send out 81 vessels to harvest 321 930 metric tons.

In Subcase E, with an annual vessel cost of $300 000 there would be no fishing in the stochastic equilibrium. A biomass level of 1.0E6 is still below $X_{\pi=0}$. At a biomass of 1.5E6 a fleet of 40 vessels is allowed to harvest 142 002 metric tons.

The value for $X_{\pi=0}$ when $p = \$100$ is precisely 1.0E6 metric tons (see Subcase J). The fishery is not profitable in the long run at this price, but short-run biomass levels of 1.5E6 and 2.0E6 would support fleets of 52 and 98 vessels.

A systematic analysis of the results in Table 4 will reveal:

(i) an increase in α will reduce equilibrium biomass while increasing fleet size and catch (Subcase A to C);

(ii) an increase in cost, c, will increase equilibrium biomass, reducing effort and catch (Subcase D to A);

(iii) an increase in the discount rate has relatively little impact, reducing equilibrium biomass slightly, causing a fractional increase in effort and a slight increase in catch (Subcase F to G);

(iv) an increase in natural mortality might shut down the fishery in the long run and has the effect (similar to an increase in the discount rate) of increasing effort and catch in the short run (before fish die of natural causes; see Subcase H to I), finally;

(v) an increase in price may make the fishery tenable in the long run and will increase effort and yield when adaptively managed at the same level of biomass (Subcase J to K).

It is a bit difficult to compare the results of Table 4 to the results of Francis (1983), Swartzman *et al.* (1983, 1987) and Dorn and Methot (1989). All of the models are cohort models and none are bioeconomic, in the sense of maximizing a present value measure. Perhaps the only common denominator is average yield. This is difficult to calculate in anything but a naive way because the cohort models are frequently run with constant fishing mortality or with constraints that prevent the biomass from declining below some bound. With that caveat in mind, we note that a simple average of yields listed in the first row of Table 3 from Swartzman *et al.* (1983) is 193 666 metric tons. The average yield from Table 2 of Swartzman *et al.* (1987) is 184 000 metric tons. From Dorn and Methot (1989) we have previously noted that average yield ranges from 178 000 to 244 000 metric tons for low risk runs and from 205 000 to 251 000 metric tons for high risk runs. If one averages the 44 yields (including zero yield when the fishery is shut down in the long or short run) from Table 4 in this paper one obtains 195 552 metric tons. While the models are very different in their biological and economic details, from the perspective of average yield they would appear to be in the same ballpark.

5. Conclusions.   The Pacific whiting has become an important commercial species for both the U.S. and Canada. Both countries participate in joint-venture fisheries, where domestic trawlers capture whiting and off-load onto foreign processing vessels. Several papers published in the 1980s have examined population dynamics within age-structured models. Recruitment is thought to be independent of spawning biomass, and has been treated as a random variable. Because older and larger fish migrate further north, the age-structure of the resource can influence the availability of fish in Canadian waters.

While these models have been rich in biological detail, they have not adequately incorporated the economic factors which affect the commercial value of the resource, nor have they tried to determine optimal fleet size. The biological detail present in these models necessitates numerical analysis, such as Monte Carlo simulation, to determine the properties of the model and to develop average yields that might be used in making recommendations for allowable catch.

In this paper we have traded-off the biological detail of a cohort model in order to incorporate some of the economic factors thought to be important in the Pacific whiting fishery. The simple bioeconomic model of Section 2 permitted us to: (1) pose a stochastic optimization problem that sought to maximize the present value of expected net revenue; (2) solve the certainty-equivalent problem for the stochastic equilibrium and an approximately-optimal rule for adaptive management; and (3) portray the equilibrium (Fig. 1) and show how the adaptive-management rule would operate (Fig. 2).

Data on catch, mean annual biomass and vessel numbers allowed for the direct estimation of a fishery production function. Cobb–Douglas and an exponential function both gave reasonable fits. The exponential form makes more sense biologically, gave a slightly better fit and for the parameter estimate was strictly concave in effort. This form was used and a range of values for the other bioeconomic parameters was obtained from previous biological and economic research.

In the bioeconomic model, long-run (stochastic) equilibrium depended on the production parameter, annual vessel cost, the discount rate, natural mortality, exvessel price and long-run average recruitment. In the short-run, using the adaptive-management rule, fleet size and allowable catch depended on the first five parameters and current biomass (instead of recruitment). Recommendations for short-run fleet size and allowable catch could fluctuate widely depending on the bioeconomic parameters, especially current biomass. From the Base-Case parameter set we observed that a current biomass of 1.0E6 metric tons would commend a fleet of only 11 vessels harvesting 27 128 metric tons. If current biomass were 1.5E6 metric tons, 67 vessels could harvest 230 159 metric tons and, if biomass increased to 2.0E6 metric tons (perhaps in the vicinity of "pristine equilibrium"), 115 vessels could harvest a 501 441 metric tons.

Such results characterize what might be called an opportunistic fishery, requiring a flexible fleet of generalist vessels able to respond to windfall recruitment and to shift to other fisheries when bioeconomic conditions are no longer favorable. Such flexibility has not been present in the U.S. or Canadian fishing industry, where effort seems quick to expand, but slow to contract. Managers and the fishing industry need to explore ways of increasing flexibility.

To use the adaptive-management rule we need an estimate of current-year biomass. The cohort models, especially the stock-synthesis model of Dorn and Methot (1989), can provide such an estimate. The age-structured models also have the advantage of being able to project changes in the abundance of particular cohorts. Such information might be important in determining spawning potential and the availability of whiting in Canadian waters.

This model should not be viewed as a replacement or even as a competitor for the niche occupied by the more complex biological models within the current "management landscape". Rather, it should be used to complement the analysis of such models in seeking the economically efficient and equitable distribution of the Pacific whiting resource.

APPENDIX

The Lagrangian for the certainty-equivalent problem has first-order conditions requiring:

$$\frac{\partial L}{\partial E_t} = \rho^t \{pF_E - c - \rho\lambda_{t+1}(1-M)F_E\} = 0$$

$$\frac{\partial L}{\partial X_t} = \rho^t \{pF_X + \rho\lambda_{t+1}(1-M)[1-F_X]\} - \rho^t\lambda_t = 0$$

$$\frac{\partial L}{\partial[\rho\lambda_{t+1}]} = \rho^t[(1-M)[X_t - F(X_t, E_t)] + R - X_{t+1}] = 0.$$

In steady state these conditions imply:

$$\rho\lambda = (pF_E - c)/[(1-M)F_E]$$

$$\rho\lambda[(1-M)[1-F_X] - (1+\delta)] = -pF_X$$

$$R = MX + (1-M)F(X, E).$$

The second steady state equation can be further simplified to:

$$-\rho\lambda[(\delta+M) + (1-M)F_X] = -pF_X.$$

Multiplying through by -1 and substituting the first steady-state expression for $\rho\lambda$ yields:

$$(pF_E - c)[(\delta+M) + (1-M)F_X] = pF_X[(1-M)F_E].$$

This last expression can be further simplified to:

$$\frac{c(1-M)F_X}{(pF_E - c)} = \delta + M$$

which is given as equation (4) in the text. Equation (5) in the text is the third of the steady-state equations listed above.

LITERATURE

Bailey, K. M. 1981. Larval transport and recruitment of Pacific hake *Merluccius productus*. *Mar. Ecol. Prog. Ser.* **6**, 1–9.

Bailey, K. M., R. C. Francis and P. R. Stevens. 1982. The life history and fishery of Pacific whiting, *Merluccius productus*. *Calif. Coop. Oceanic Fish. Invest.* **23**, 81–98.

Burt, O. R. 1964. Optimal resource use over time with an application to groundwater. *Mgmt Sci.* **11**, 80–93.

Burt, O. R. 1967. Temporal allocation of groundwater. *Water Resources Res.* **3**, 45–56.

Burt, O. R. and R. G. Cummings. 1977. Natural resource management, the steady state and approximately optimal decision rules. *Land Econ.* **53**, 1–22.

Conrad, J. M. and C. W. Clark. 1987. *Natural Resource Economics: Notes and Problems*. Cambridge, U.K.: Cambridge University Press.

Dorn, M. W. and R. D. Methot. 1989. Status of the Pacific whiting resource in 1989 and recommendations to management in 1990. In *Status of the Pacific Coast Groundfish Fishery*

Through 1989 *and Recommended Acceptable Biological Catches for* 1990, pp. A.1–A.61. Portland, Oregon: Pacific Fishery Management Council.

Francis, R. C. 1983. Population and trophic dynamics of Pacific hake (*Merluccius productus*). *Can. J. Fish. Aquatic Sci.* **40**, 1925–1943.

Kolberg, W. C. 1990. *Approach Paths to the Steady State: A Performance Test of Current Period Decision Rule Solution Methods for Models of Renewable Resource Management*. Ithaca, New York: Working Paper, Department of Economics, Ithaca College.

Squires, D. E. 1990. Notes on the economics of the Pacific whiting fishery. Fax transmission. Southwest Fisheries Center, 18 June 1990.

Swartzman, G. L., W. M. Getz and R. C. Francis. 1983. A management analysis of the Pacific whiting (*Merluccius productus*) fishery using an age-structured stochastic recruitment model. *Can. J. Fish. Aquatic Sci.* **40**, 524–539.

Swartzman, G. L., W. M. Getz and R. C. Francis. 1987. Binational management of Pacific hake (*Merluccius productus*): a stochastic modeling approach. *Can. J. Fish. Aquatic Sci.* **44**. 1053–1063.

Received 24 September 1990
Revised 14 January 1991

[45]

ASSESSING EFFICIENCY GAINS FROM INDIVIDUAL TRANSFERABLE QUOTAS: AN APPLICATION TO THE MID-ATLANTIC SURF CLAM AND OCEAN QUAHOG FISHERY

QUINN WENINGER

Delayed fishing fleet restructuring complicates the assessment of efficiency gains from individual transferable quota (ITQ) fisheries management programs. This article presents a methodology to estimate harvest sector efficiency gains in lieu of incomplete fleet restructuring. The methodology is applied to assess the efficiency gains in the Mid-Atlantic surf clam and ocean quahog fishery ITQ program. While roughly 128 vessels harvested clams under the previous management regime, the analysis suggests that 21–25 vessels will remain under ITQs. The efficiency gains are estimated to be between $11.1 million and $12.8 million annually (1990 dollars).

Key words: efficiency gains, fleet restructuring, individual transferable quotas.

An important benefit of individual transferable quota (ITQ) fisheries management programs is the efficiency gains that may emerge under enhanced property rights. Quota rights provide a mechanism to eliminate redundant capital that may have accumulated under the pre-ITQ management regime and encourage cost-efficient production once industry restructuring is complete. Benefits emerge as retired capital is employed in other more productive uses, and as remaining fishers exploit production economies under the ITQ operating rules. For example, the elimination of input controls and harvest time restrictions can improve (input) allocative efficiency and vessel capacity utilization on fishing vessels that remain active under the ITQ management regime.

In initially overcapitalized fisheries, industry restructuring will be a key determinant of the total efficiency benefits that emerge under the ITQ program. Because restructuring can take time, possibly years, to complete, ITQ program benefits emerge over the longer term. Nonetheless, the appropriate benchmark

for assessing the performance of ITQ programs is the efficiency gains that are generated after all economies are realized. Benefits generated during the restructuring phase can underestimate the full program benefits and can bias against ITQ management reform. Moreover, the transition phase benefits are arbitrary because they depend on the extent to which restructuring is complete or the extent to which all economies available under the ITQ program have been captured.

This article presents a methodology to analyze harvest sector efficiency gains from ITQ management reform in the presence of incomplete fleet restructuring. The approach is to exploit the economic incentives implicit in the ITQ system to predict the fleet structure and individual vessel output levels expected to prevail under ITQs. The anticipated efficiency gains in the harvest sector are then estimated from the predicted fleet structure. Specifically, harvest sector efficiency gains are calculated as the reduction in total harvesting costs expected under the ITQ-regime fleet structure. The methodology is applied to the Mid-Atlantic surf clam and ocean quahog fishery (hereafter, the MA clam fishery), which switched from limited entry (LE) management to ITQs in October 1990. During the LE regime (1977–90) roughly 128 fishing vessels actively harvested surf clams and ocean qua-

Quinn Weninger is assistant professor in the Department of Economics at Utah State University.

The author would like to acknowledge valuable comments from Richard E. Just, Dale Squires and two *Journal* reviewers. This research was supported in part by the Utah Agricultural Experiment Station, Utah State University, Logan, Utah 84322-4810, project number UTA 025. Approved as journal paper 7034.

hogs. Many of these vessels operated well below an efficient scale of production due to the stringent harvest time restrictions that were imposed under LE. The analysis in this article reveals that the ITQ-regime fleet will consist of 21–25 vessels operating at a cost-efficient output scale. Hence, the elimination of redundant harvesting capital and the realization of scale economies is an important source of efficiency gain in the MA clam fishery.

A second source of efficiency gain in the MA clam fishery is associated with returns to specialization or single-species production. Strict surf clam harvest time restrictions under the LE-regime induced fishers to diversify into the production of ocean quahogs to take advantage of otherwise idled vessel capital (Strand, Kirkley, and McConnell; Lipton and Strand). The analysis of this article suggests that the clam harvesting technology exhibits scope diseconomies (cost anti-complementarities) under the ITQ-regime operating rules. As a consequence, clam fishers should eventually return to specialized production (single clam species) under the ITQ program, resulting in additional efficiency gains.

The total fleet harvesting cost incurred under LE is estimated to be $28.4 million in 1990. All values are reported in 1990 dollars. Total cost incurred to harvest the same total allowable catch (TAC) under ITQs, conditional on the predicted ITQ-regime fleet structure, is estimated to be $15.6 million annually. Harvest sector efficiency gains in the range of $12.8 million annually are thus possible under the ITQ management program.

The analysis of the MA clam fishery underscores the importance of controlling for long-run fleet adjustments when assessing the performance of ITQ programs. Roughly fifty vessels remained active in the MA clam fishery at the end of 1994, four years after the ITQ program was introduced. Of these fifty vessels, many continued to operate below cost efficient output levels, and fourteen vessels continued to harvest multiple clam species. Evidence suggests that the 1994 fleet was still in a transition phase and 25–29 vessels would eventually exit the fishery. Vessels that remained would adjust quota holdings to facilitate single-species production. Total harvest cost incurred by the 1994 fleet is estimated to be $19.9 million. Accordingly, the efficiency gain estimates based on the 1994 fleet structure would underestimate the post-transition phase gains by $5.8 million annually.

The following section presents a model of equilibrium fleet structure under an ITQ program. Then we briefly review the regulatory history and industry background in the MA clam fishery. The analysis of the harvest cost technology that is used to identify the ITQ-regime fleet structure is then presented, followed by estimates of harvest sector efficiency gains. The final section summarizes the main results and discusses implications for future assessment of ITQ programs.

Model of Equilibrium Fleet Structure in the ITQ Fishery

It is well known that ITQs provide economic incentives that promote efficient resource use (Montgomery). The reason is that the residual return to quota ownership, or quota rent, is maximized on a cost-efficient harvesting operation. Transferability implies that quota will eventually gravitate into the hands of those who are able to generate the largest residual return from ownership. Any quota owner that earns less than the maximal rent can be made better off by trading the quota asset (Weninger and Just). This incentive structure leads to an equilibrium quota distribution and fleet structure.

Consider a fishery that generates a surplus harvest of two fish species in each period. Assume that the resource manager distributes quota in an amount that corresponds to the desired TAC of each species.[1] Let Q_i, $i = 1$, 2 denote the TAC and total available quota for species i. The ITQ regulation requires that fishers own or lease quota in an amount that corresponds to their output level; thus, output and quota will be used synonymously. To facilitate the identification of the ITQ-regime fleet structure, the unit of analysis will be an individual (representative) vessel operation.[2]

Assume that all fishers have access to the same harvest technology. Denote the cost function for a single-product vessel as $C(q_i, v, S_i)$, $i = 1, 2$, where q_i is the quantity of output i, v is a vector of strictly positive factor input prices and S_i is an index for the ith fish stock. A "tilde" is used to distinguish mul-

---

[1] See Clark for an analysis of the social rent maximizing TAC in a multispecies fishery.

[2] The focus on a representative vessel operation is simplistic but facilitates presentation of the basic intuition. The section "ITQ Regime Fleet Structure and Returns to ITQ Management Reform" briefly discusses implications of heterogeneity in the ITQ-regime fleet structure.

752   *November 1998*   *Amer. J. Agr. Econ.*

tioutput harvest operations. Let $\tilde{C}(\bar{q}, v, S)$ denote the multiproduct cost function, where $\bar{q}$ = $(\bar{q}_1, \bar{q}_2)$, $\bar{q}_i > 0$, $i = 1, 2$, is the output vector and $S = (S_1, S_2)$ is the vector of stock indices. Assume that the single- and multioutput cost functions are twice differentiable, increasing and convex over a bounded output space, increasing and concave in v, and nonincreasing in S. Hereafter S and the TAC for each product are assumed constant.

In a multiple output fishery, three possible fleet structures must be considered. First, individual vessel operations may harvest a single output leading to a specialized fleet structure. Second, all vessels may produce multiple outputs in which case the fleet will be fully diversified. Finally, MacDonald and Slavinsky identify a mixed-product market structure under which multiproduct (diversified) and single product (specialized) firms operate simultaneously. Similar conditions apply in a multiproduct ITQ fishery leading to a third, mixed ITQ-regime fleet structure comprised of both single- and multioutput operations. Conditions under which a specialized, diversified and mixed fleet structure will emerge under ITQs are discussed in the following sections.

Specialized Fleet Structure

Under single-output production, the residual return to the marginal unit of output is the difference between the output price and the marginal harvest cost:

$$(1) \quad p_i - \frac{\partial}{\partial q_i}\{C(q_i, v, S_i)\}, \qquad i = 1, 2$$

where p_i is the ex-vessel price for output i. The residual return in equation (1) attains a maximum at the average cost-minimizing output level given by

$$(2) \quad q_i^* = \arg\min_{q_i \leq \bar{q}_i} \left\{\frac{C(q_i, v, S_i)}{q_i}\right\},$$

$$i = 1, 2$$

where $\ddot{q}_i$ is the upper bound for output space i.

In the presence of a well-functioning ITQ trading market, a quota lease rate, L_i, will emerge in the fishery. Any single-product operator who cannot match the fully efficient technology or production plan will earn a re-

sidual return that is less than the market lease rate and is better off trading the ITQ asset (Weninger and Just). In the ITQ equilibrium, specialized operators will harvest q_i^* units and earn the maximal per unit (per period) return, $L_i^* = p_i - [C(q_i^*, v, S_i)/q_i^*]$.

The number of operations comprising the specialized ITQ-regime fleet is determined by

$$(3) \quad N_i^* q_i^* \approx Q_i, \qquad i = 1, 2$$

where N_i^* denotes the number of vessels harvesting species i. Note that Q_i/q_i^* may not be an integer and thus equation (3) will hold only approximately ($\approx$). The number of single-species vessels that prevail will be an integer that lies closest below or above Q_i/q_i^*. In equilibrium, N_i^* is expected to equal the integer at which the corresponding per vessel harvest level, Q_i/N_i^*, attains the lowest average harvest cost, and thus the largest residual return to quota ownership. When Q_i/q_i^* is large, the difference between Q_i/N_i^* and q_i^* will be small.

Observe that if $N_i^* - 1$ fishers are active, they must produce in excess of q_i^* and, by definition, earn a residual return less than L_i^*. A fisher could bid sufficient quota away from the $N_i^* - 1$ active fishers and profitably enter the fishery. Alternatively, if $N_i^* - 1$ fishers are active, the corresponding per vessel output must be less than q_i^*, and the residual return to the quota asset would be less than L_i^*. An active fisher would find it profitable to sell their quota allocation and exit the fishery.

The total cost of harvesting the TAC under the specialized fleet structure, denoted TC_S, will be

$$(4) \quad TC_s = \sum_{i=1}^{2} N_i^* C(q_i^*, v, S_i).$$

Multiproduct Fleet Structure

Multiproduct operations are viable if the residual return to the ITQ asset(s) under multioutput production exceeds or is at least equal to the residual earned by a specialized operator. This will be the case only if the harvest technology exhibits economies of joint production or economies of scope. A sufficient condition for scope economies is cost complementarity among outputs (Baumol, Panzar, and Willig).[3] Cost complementarity implies that the marginal cost of harvesting output i is reduced if the operator also harvests output j ($i \neq j$).

Multiproduct cost concepts can be used to identify the output bundle that generates the greatest return to the quota asset(s). Let $\tilde{\mathbf{q}}^*$ denote a reference output vector that minimizes the multiproduct ray average cost (RAC). The vector $\tilde{\mathbf{q}}^*$ fully exhausts all cost advantages from adjusting the scale of production as well as the product mix. Formally,

$$\tilde{\mathbf{q}}^* = (\bar{q}_1^*, \bar{q}_2^*) = \arg\min_{\tilde{q} \leq q, \alpha} \left\{ \frac{\tilde{C}(\tilde{\mathbf{q}}, v, S)}{\alpha \cdot \tilde{\mathbf{q}}} \right\},$$

where $\ddot{\mathbf{q}}$ is the upper bound for the multiproduct output space, and α is a proportional weight vector (Panzar). The equilibrium quota lease rate evaluated at $\tilde{\mathbf{q}}^*$ is

$$(5) \quad \bar{L}_i^* = p_i - \frac{\partial}{\partial \tilde{q}_i} \{ C(\tilde{\mathbf{q}}^*, v, S) \},$$

$$i = 1, 2.$$

Under the cost complementarity assumption, a specialized operator will be unable to profitably bid quota away from a diversified operator because the residual earning will be less than $\bar{L}_i^*$. If all fishers produce $\tilde{\mathbf{q}}^*$, and the TAC of each species is exactly harvested, the equilibrium number of vessel operations, $\tilde{N}^*$, will be determined by

$$(6) \quad \tilde{N}^* \tilde{\mathbf{q}}^* \approx Q.$$

The integer problem applies in the diversified fleet. The actual number of vessels in the ITQ-regime fleet will be determined by similar logic as in the specialized fleet case. The total cost under the fully diversified fleet structure, TC_D, is

$$(7) \quad TC_D = \tilde{N}^* \tilde{C}(\tilde{\mathbf{q}}^*, v, S).$$

Mixed Production Fleet Structure

The conditions under which the ITQ regime fleet is fully diversified are somewhat unique. In particular, $\bar{q}_1^*/\bar{q}_2^*$ must coincide with the ratio of TAC quantities Q_1/Q_2 set by the management authority. A more likely scenario is that $Q_1/Q_2 \neq \bar{q}_1^*/\bar{q}_2^*$, in which case the ITQ regime fleet may be mixed.

For concreteness, suppose the ratio of Q_1/Q_2 is greater than $\bar{q}_1^*/\bar{q}_2^*$. The analysis of Q_1/Q_2 less than $\bar{q}_1^*/\bar{q}_2^*$ follows symmetrically. A fully diversified fleet cannot maintain the output mix implied by $\tilde{\mathbf{q}}^*$, and simultaneously harvest the entire Q_1. With surplus Q_1, the value of species 1 quota will fall providing the opportunity for a specialized operator to enter the fishery. The resulting quota distribution and fleet composition will depend on the nature of the cost complementarity. First, if the lease rate for species 1 quota is low, diversified operators will expand production of $\tilde{q}_1$ in order to maximize their profit. Let $\tilde{\mathbf{q}}^\circ$ denote an output vector that is proportional to Q_1/Q_2, and maximizes the residual return for each output:

$$(8) \quad \tilde{\mathbf{q}}^\circ = \arg\max_{\tilde{q} \leq q} \left\{ \bar{L}_i^\circ = p_i - \frac{\partial}{\partial \tilde{q}_i} \tilde{C}(\tilde{\mathbf{q}}, v, S) \right\}$$

$$\text{s.t. } \bar{q}_1^\circ/\bar{q}_2^\circ = Q_1/Q_2, \qquad i = 1, 2.$$

If the residual return defined in equation (8) exceeds the residual earned on a specialized operation, as defined in equation (2), then the ITQ-regime fleet structure will remain fully diversified. The situation is one where the economies of scope are large so that the marginal harvesting cost on a diversified operation is less than the marginal harvesting cost for single-product operators. The residual return at $\bar{q}_1^\circ$ remains above a specialized operator's maximum offer price, and specialized operators cannot profitably enter the fishery. In this case, the number of vessels is given by $\tilde{N}^\circ \tilde{\mathbf{q}}^\circ \approx Q$ and fleet costs are $\tilde{N}^\circ \tilde{C}(\bar{q}^\circ, v, S)$.

If the scope economies are less pronounced, or if Q_1/Q_2 is significantly larger than $\bar{q}_1^\circ/\bar{q}_2^\circ$, then $\bar{L}_i^\circ$ will fall below the residual earned under specialized production, and the ITQ-regime fleet will be mixed. Recall that in equilibrium all gains from quota trading must be exhausted. In a mixed fleet, single- and multiple-output producers will adjust their species 1 quota to satisfy

$$(9) \quad \left[p_1 - \frac{\partial}{\partial \tilde{q}_1} \tilde{C}(\tilde{\mathbf{q}}^*, v, S) \right]$$

$$= \left[p_1 - \frac{\partial}{\partial q_1} C(q_1^*, v, S_1) \right].$$

---

[3] Scope economies can arise from product-specific fixed costs (Gorman). The analysis here focuses on scope economies that arise from cost complementarity.

754   *November 1998*   *Amer. J. Agr. Econ.*

With $\bar{q}_1^*$ satisfying equation (9), diversified operators will adjust $\bar{q}_2$ to satisfy

$$(10) \quad \bar{q}_2^* = \underset{\bar{q}_2 \leq \bar{q}_2}{\arg \max}$$

$$\left\{ p_2 - \frac{\partial}{\partial \bar{q}_2} \bar{C}(\bar{q}_1^*, v, S) \right\}.$$

The equilibrium lease rates under the mixed fleet structure are given by

$$\bar{L}_1^* = \left[p_1 - \frac{\partial}{\partial \bar{q}_1} \bar{C}(\bar{\mathbf{q}}^*, v, S) \right]$$

$$= \left[p_1 - \frac{\partial}{\partial q_1} C(q_1^*, v, S_1) \right]$$

for species 1 quota and

$$\bar{L}_2^* = \left[p_2 - \frac{\partial}{\partial \bar{q}_2} \bar{C}(\bar{\mathbf{q}}^*, v, S) \right]$$

for species 2 quota. The number of multioutput operations in the mixed fleet is determined by

$$(11) \quad \bar{N}^* \approx Q_2 / \bar{q}_2^*$$

and the number of single-product operations in the mixed fleet is determined by[4]

$$(12) \quad N_1^* \approx (Q_1 - \bar{N}^* \bar{q}_1^*)/q_1^*, \quad N_2^* = 0.$$

Finally, the total harvesting cost under a mixed fleet, TC_M, is given by

$$(13) \quad TC_M = \bar{N}^* \bar{C}(\bar{\mathbf{q}}^*, v, S) + N_1^* C(q_1^*, v, S_1).$$

Harvest sector rents under the ITQs are the total revenues generated from the TAC, less the minimum of equations (4), (7), and (13);

$$(14) \quad \sum_{i=1}^{2} p_i Q_i - \min\{TC_s, TC_D, TC_M\}.$$

It follows that the harvest cost technology must be investigated to determine cost-efficient output levels and product mixes. The ITQ-regime fleet structure and harvest sector rents in equation (14) may then be compared

---

[4] The mixed fleet structure will include diversified vessels and specialized vessels of both types only if species 2 specialists earn exactly the same return as a diversified operator at $\bar{q}^*$.

to pre-ITQ harvesting costs to obtain an estimate of the anticipated efficiency gains. The remainder of the article discusses the application of this methodology to the MA clam fishery.

Industry Background

This section discusses features of the regulatory history and industry background that influence the subsequent empirical analysis. Additional information may be obtained from the Mid-Atlantic Fisheries Management Council 1988, 1996; Lipton and Strand; McCay and Creed; Strand, Kirkley, and McConnell; and Wang.

Regulatory History

The Mid-Atlantic Surf Clam and Ocean Quahog Fisheries Management Plan (FMP) was approved in November 1977. Limited entry permits were issued to 184 surf clam and/or ocean quahog vessels, and an entry moratorium was imposed. Vessel replacement for surf clam permit holders was prohibited unless the vessel left the fishery involuntarily (e.g., sinking or fire). The FMP set quarterly quotas for surf clams and ocean quahogs. To maintain quarterly quotas, limits were placed on the days that fishing was allowed and on the total hours of fishing per week. During the LE regime, increasingly stringent harvest time restrictions were placed on surf clam fishers even though quarterly quotas did not dramatically change throughout the period. In response to severe harvest time restrictions, some vessels expanded ocean quahog production to take advantage of otherwise idle vessel capital (Strand, Kirkley, and McConnell).

Amendment #8 to the FMP was adopted on 25 October 1989, and was approved by the National Marine Fisheries Service (NMFS) on 23 March 1990. The amendment changed the management system to ITQs on October 1, 1990. An initial distribution of species-specific quota rights was distributed gratis to 161 vessel owners: 154 received surf clam quota and 117 received ocean quahog quota. The number of initial allocations exceeded the number of active vessels because some vessels received quota for both clam species. Quota was delineated as percentage shares of the TAC, based on a formula of historical catch rates (80%) and vessel size (20%). All harvest time restrictions were lifted under ITQs.

Empirical Considerations

Surf clams and ocean quahogs are harvested with hydraulic dredges towed by the vessel. Clam fishers travel from port to a chosen site and dredge the sea bottom. Most trips are completed in a single day, although two-day ocean quahog trips occur. Fishers may return to a lucrative site until its yield declines, at which point the fisher may search for an alternate site. Catch uncertainty is small relative to other fisheries and will be ignored.

Surf clams and ocean quahogs are sold directly to processing firms that conduct value-adding activities before selling the final consumable product in downstream markets. The perishable nature of the clams, scheduling of processing activities, and the need to coordinate with downstream buyers requires tight vertical coordination between fishers and processors. Processors may place orders with a vessel captain weeks in advance of actual harvest (Wallace, personal communication). Fishers are assumed to minimize the cost of delivering the surf clam and ocean quahog orders placed by processors, subject to exogenous prices, technology, and clam stock levels.

Multispecies clam fishing from the same vessel is feasible. Ocean quahogs are smaller and require adjusting the dredge knife spacing and the angle at which the dredge is set relative to surf clams (almost flat for ocean quahogs to approximately two degrees for surf clams). Ocean quahogs are located farther from shore, and quahog fishing is subject to more severe weather conditions. Slightly larger vessels and additional crew may be preferred on quahog trips. Larger vessels miss fewer days at sea because of poor weather, and the extra crew facilitates the dredging operation during bad weather.

The number of clam cages that can be carried on the vessel and the number of trips per calendar period impose a short-run constraint on total harvest capacity. Fishing trips involve one to two days at sea, plus off-loading and replenishment of fuel and supplies. Harvest costs may rise sharply (become infinite) beyond a particular harvest capacity. Logbook data sources and industry participants were consulted to determine the maximum harvest capacity for each clam species under a variety of vessel size classes as measured by the vessel gross registered tonnage (GRT).

Empirical Specification and Estimation of the Harvest Cost Technology

Harvesting costs are separated into fixed and variable costs within each three-month or quarterly production period. Within each quarter, the maximum flow of vessel capital services is treated as a fixed operating variable. The GRT of the fishing vessel is used as a proxy for the flow of vessel capital services available for production.

Discussions with industry participants indicate that the number of crew on board is rarely adjusted; labor services are used in proportion to the flow of vessel services. Furthermore, if the crew is paid under a revenue share system, labor costs are analogous to an ad valorem tax on revenue. In this setting, it is not clear that the labor input is adjusted at its cost margin. For these reasons, the flow of labor services used in harvesting activities is approximated by the GRT variable.[5]

Variable inputs include (*i*) fuel, engine oil, and lubricants; and (*ii*) gear, supplies, and repairs (hereafter the "gear" input). The gear input includes food consumed by the captain and crew, and maintenance costs for the clam dredge. Dredge maintenance is the largest component of the gear input cost (Wallace, personal communication). Denote the variable harvest cost function as

$$(15) \quad c(v_1, v_2, q_1, q_2 \,|\, z, S_1, S_2)$$

where v_1 and v_2 are the respective prices of fuel and gear, q_1 and q_2 are the respective quantities of surf clams and ocean quahogs, z is the vessel GRT, and S_1 and S_2 are, respectively, surf clam and ocean quahog stock indices. Denote the vector of arguments in $c(\cdot)$ as $\mathbf{Z} = (v_1, v_2, q_1, q_2, z, S_1, S_2)$ with individual element Z_j, $j = 1, \ldots, 7$. The translog variable cost function is specified for $c(\cdot)$:

---

[5] The GRT variable will overestimate the actual vessel and labor services that are used in production when the vessel harvest capacity is not fully utilized. Previous studies have attempted to control for this problem by including days-at-sea in developing the proxy for capital services (Squires). The product of trips-taken and GRT was highly collinear with the remaining endogenous variables in the model and could not be used as an indicator of capital and labor.

Amer. J. Agr. Econ.

$$(16) \quad \ln c(\mathbf{Z}, \boldsymbol{\Gamma}) = a_0 + \sum_{j=1}^{7} a_j \ln Z_j$$

$$+ \frac{1}{2} \sum_{j=1}^{7} \sum_{k=1}^{7} b_{jk} \ln Z_j \ln Z_k$$

where $b_{jk} = b_{kj}$ for $j, k = 1, \ldots, 7$, and $\boldsymbol{\Gamma} = \{a_0, a_j, b_{jk}\}$ denotes the parameter vector. Necessary and sufficient parameter restrictions to ensure linear homogeneity in input prices are $a_1 + a_2 = 1$, $\Sigma_k \, b_{jk} = 0$, $j = 1, 2$, and $b_{1j} + b_{2j} = 0$, $j = 1, \ldots, 7$. Shepherd's lemma can be used to recover factor share equations for the fuel and gear input,

$$(17) \quad \frac{\partial \ln c(\cdot)}{\partial \ln Z_j} = \frac{v_j x_j(\cdot)}{c(\cdot)}$$

$$= a_j + \sum_{k=1}^{7} b_{jk} \ln Z_k \qquad j = 1, 2$$

where $x_j(\cdot)$ is the demand for input j. Equation (16) and (17) form a system of structural equations that may be used to estimate $\boldsymbol{\Gamma}$. The factor share equations and the cost function are linearly dependent and thus the gear equation is dropped from the econometric estimation.

Data

Data are from several sources: (*i*) the NMFS logbook reporting system, (*ii*) the FMP, (*iii*) the Mid-Atlantic Fisheries Management Council, (*iv*) a survey of industry participants conducted by McCay and Creed, and (*v*) discussions with industry participants. Detailed vessel harvest information is recorded as part of the NMFS logbook reporting system. Vessel-specific cost information from expenditure reports was not available. Instead, per vessel costs were estimated from logbook data and estimates of hourly and per trip costs as reported in the FMP. Quarterly fuel consumption was estimated from an hourly fuel consumption function.[6] Fuel prices for #2 diesel were obtained from the Energy Information Administration (U.S. Department of Energy). Fuel consumption and price were combined

to estimate the total quarterly fuel costs for each vessel.

The FMP provides estimates of per trip gear costs for various vessel size classes. These cost estimates were used in conjunction with the fishing trip information from logbook sources to obtain an estimate of the quarterly gear expenditures for each vessel.[7] A detailed gear price was not available and is instead approximated by the gross national product implicit price deflator. All remaining prices are deflated using the gross national product implicit price deflator. Based on calculations obtained from the above procedure, fuel expenses were estimated to be 24% of variable harvesting costs. Gear expenses were estimated to be 76% of variable harvesting costs.

Total exploitable stock estimates are from the *Report of the 19th Northeast Regional Stock Assessment Workshop* (NOAA/National Marine Fisheries Service, available on a yearly basis only). It is assumed that stock abundance does not appreciably change within a given year. Ocean quahog harvest is considered small relative to the exploitable biomass. Moreover, no appreciable variation in ocean quahog stocks occurred during the study period. For these reasons, the quahog stock index was dropped from the empirical specification.

Some ports of departure may be located farther from lucrative clam beds, and thus require additional travel time and fuel consumption.[8] To allow for this possibility, three dummy variables for the two northernmost ports (Point Pleasant and Atlantic City, NJ) and the southernmost port (Cape Charles, VA) are included in the cost equation. The base case (75% of the sample) included ports located in the central region (Chincoteague, VA, Cape May and Wildwood, NJ, and Ocean City, MD).

Vessels that take fewer than 50% of the maximum feasible number of trips per quarter were dropped from the analysis. For these observations, the capital and labor services used in production are likely to be overestimated by the GRT proxy. Vessels exiting the fishery before the end of 1994 were dropped. It is likely that exiting vessels sold quota because they were relatively cost-inefficient. Exiting

---

[6] Fuel used when the vessel is docked before and after the trip is calculated as engine horse power $\times$ 6 hrs. $\times$ 0.02. Fuel used steaming to and from the site is calculated as engine horse power $\times$ hrs. steaming $\times$ 0.04. Fuel used while fishing is calculated as engine plus dredge pump horse power $\times$ hrs. fishing $\times$ 0.05.

[7] Gear cost estimates range from $526 to $1,044 per trip on surf clam vessels and $526 to $1,699 per trip on ocean quahog vessels.

[8] Vessel operators develop trading relationships with land-based processing firms and rarely change their port of departure. It is assumed that all trips depart from the same port in each quarter.

vessel observations may provide inaccurate information about the cost technology that will prevail under ITQs. Finally, quota redistribution and other adjustments to the ITQ-regime operating rules occurred throughout the data period. It is reasonable to believe observations from later stages provide more information about the ITQ-regime cost structure. To account for this in the analysis, each observation is weighted by a factor $\tau \cdot [\bar{\tau}]^{-1}$, where τ denotes the cumulative quarter since the introduction of the ITQ program, and $\bar{\tau}$ is the sample average.[9]

An error term is added to each observation. The error term is assumed contemporaneously correlated across equations but independent across time and vessels. Each vessel/quarter observation is interpreted as an independent observation on the average harvest technology.

Empirical Results: The Structure of the Harvest Technology

The data included 501 observations.[10] The parameter vector is estimated by the iterative seemingly unrelated regression technique using Gauss software. Parameters found to be statistically different from zero at the 95% confidence level are denoted with an asterisk (table 1). The adjusted R-squared statistic for the cost function equation is 0.99 indicating a very good fit to the data. The fuel share equation does not contain a free intercept term; thus, the related R-square statistic is invalid. To further assess the reliability of the results, the empirical cost function was checked for monotonicity and concavity in input prices. Both requirements were satisfied at all data points.[11] A Wald test of the null hypotheses that the variable cost function is linearly homogeneous in input prices was not rejected at the 95% confidence level. A Wald test of the null hypotheses that the harvest technology (*i*) exhibits constant returns to scale and (*ii*) is separable in outputs were both rejected at the 95% confidence level. Non-joint-in-inputs implies the existence of output-specific variable cost functions (Denny and Pinto). The null hypothesis that the harvest technology is nonjoint-in-inputs could not be rejected at the 95% confidence level.[12]

The parameters of flexible functions are difficult to interpret. All economic effects are recoverable from the parameter estimates. For the purposes of identifying the ITQ-regime fleet structure, scale and scope efficiency (cost complementarity or anti-complementarity) measures are required. Kim reports multi-product and product-specific measures of returns to scale for the translog cost function as well as convenient measures of cost complementarity. Space limitations do not permit a complete presentation of these measures. See Kim (pp. 186–93) for further details. Measures of scale economies and cost complementarity are functions of the data. The measures that follow were calculated for a 175 GRT vessel harvesting 35,000 bushels of surf clams and 35,000 bushels of ocean quahogs per quarter. Prices and stock levels were set at 1990 levels. The sign of all reported measures were robust to the evaluation point.

The cost elasticities for surf clams and ocean quahogs $[\partial \ln c(\cdot)/\partial \ln q_j]$ equaled 0.31 and 0.29, respectively. All else equal, a 1% increase in surf clam (ocean quahog) harvest results in a 0.31% (0.29%) increase in variable costs. The derivative of the surf clam cost elasticity with respect to additional surf clam harvest is -0.32; the cost elasticity declines as surf clam harvest is increased. The derivative of the ocean quahog cost elasticity with respect to additional quahog harvest is -0.45. The degree of overall scale economies for a multiproduct operation (obtained as the inverse of the sum of the cost elasticities) is 1.66, indicating overall increasing returns to scale. Product-specific scale economies are measured as the ratio of average incremental cost and marginal cost for each output. Based on the necessary calculations, product-specific scale economies were estimated to be 1.04 for surf clams and 1.51 for ocean quahogs. While the surf clam measure is very close to 1 at the point of evaluation, both measures suggest product-specific increasing re-

---

[9] The study period begins in 1990:4 ($\tau = 1$) and ends in 1994:4 ($\tau = 22$).

[10] A small value (1,000 bushels) is inserted when the quantity of surf clams or ocean quahogs harvest is zero. The results were not sensitive to the value that is inserted for zero output levels.

[11] A necessary and sufficient condition for concavity is that the Hessian matrix of the variable cost function be negative semidefinite. This requirement was satisfied at all data points. The estimated input shares were positive at all data points indicating monotonicity.

[12] The null hypothesis of constant returns and output separability were both rejected at the 95% confidence level. The constant returns to scale, chi-square statistic was 14.28 with critical level 9.49. The output separability, chi-square statistic was 81.24 with critical value 12.59. The nonjoint-in-inputs hypothesis was tested by comparing the likelihood value under the restriction $b_{34} + a_3 \cdot a_4 = 0$. The chi-square statistic was 0.57 with 95% confidence level, critical value 3.84.

758   *November 1998*   *Amer. J. Agr. Econ.*

Table 1. Parameter Estimates for Translog Variable Cost Function

Parameter	Variable	Estimate	Standard Error	Parameter	Variable	Estimate	Standard Error
a_0	const.	-0.021	0.025	b_{33}	$\ln q_1 \ln q_1$	-0.114[a]	0.017
a_1	$\ln v_1$	0.018	0.014	b_{34}	$\ln q_1 \ln q_2$	-0.078[a]	0.013
a_2	$\ln v_2$	0.982[a]	0.036	b_{35}	$\ln q_1 \ln z$	0.007	0.031
a_3	$\ln q_1$	-1.599	1.649	b_{36}	$\ln q_1 \ln S_1$	0.340[a]	0.368
a_4	$\ln q_2$	0.421	1.499	b_{44}	$\ln q_2 \ln q_2$	0.075[a]	0.016
a_5	$\ln z$	4.325	6.196	b_{45}	$\ln q_2 \ln z$	-0.132[a]	0.030
a_6	$\ln S_1$	1.092	8.117	b_{46}	$\ln q_2 \ln S_1$	0.132	0.338
$b_{11} = -b_{12}$	$\ln v_1 \ln v_1$	-0.088[a]	0.038	b_{55}	$\ln z \ln z$	-1.185[a]	0.238
$b_{22} = -b_{12}$	$\ln v_2 \ln v_2$	-0.088[a]	0.044	b_{56}	$\ln z \ln S_1$,	0.654	1.393
$b_{13} = -b_{23}$	$\ln v_1 \ln q_1$	0.017[a]	0.004	b_{66}	$\ln S_1 \ln S_1$	-1.986	3.628
$b_{14} = -b_{24}$	$\ln v_1 \ln q_2$	0.023[a]	0.004	D_1	PP	-0.060	0.040
$b_{15} = -b_{25}$	$\ln v_1 \ln z$	-0.014[a]	0.017	D_2	AC	-0.014	0.029
$b_{16} = -b_{26}$	$\ln v_1 \ln S_1$	0.028	0.017	D_3	CCH	0.124[a]	0.033

Source: Estimated.

Note: D_1 *(PP)* is the dummy variable for home port at Point Pleasant, NJ; D_2 *(AC)* is the dummy variable for home port at Atlantic City, NJ; and D_3 *(CCH)* is the dummy variable for home port at Cape Charles, VA.

[a] The parameter is different from zero at the 95% confidence level.

turns to scale. Note that increasing returns to scale over observed output levels is not unexpected. If fleet adjustments are incomplete, some fishers may be quota constrained and forced to continue to operate in regions of increasing returns. Earlier it was suggested that operating in regions of decreasing returns is equally unlikely.

Cost complementarity or anticomplementarity can be determined by examining the cross derivatives of the cost elasticities, $(\partial / \partial q_j)\{\partial \ln c(\cdot)/\partial \ln q_i\}$, $i = 1, 2, i \neq j$. If $(\partial / \partial q_j)\{\partial \ln c(\cdot)/\partial \ln q_i\} < 0, (>0)$, an increase in q_j reduces (increases) the cost elasticity for q_i, indicating cost complementarity (anticomplementarity) among outputs. The cross derivative of the surf clam (ocean quahog) cost elasticity with respect to ocean quahog (surf clam) production is 0.05 (0.05). Hence, the estimated variable cost function exhibits scope diseconomies at the output vector of 35,000 bushels of each clam species. To ensure that the cost anticomplementarity finding was robust, several output vectors were evaluated. The results were similar in both sign and magnitude.

Following Panzar, a measure of the degree of scope economies is obtained by calculating the percentage change in cost under multioutput versus single-output production. Harvesting 70,000 bushels of surf clams and 70,000 bushels of ocean quahogs on two specialized vessels rather than two diversified vessels results in a $12,378 (6%) variable cost savings. While the degree of scope diseconomies does

not appear large, it must be considered in identifying the ITQ-regime fleet structure.

All remaining economic effects, except for the surf clam stock effect, conform to expectations. The cost elasticities with respect to the price of fuel and gear are 0.28 and 0.72, respectively. The elasticity of substitution between fuel and gear is 0.28, indicating nonzero input substitution possibilities. The derivative of the cost elasticity with respect to GRT indicates that a 1% increase in GRT results in a 0.28% decline in variable cost. The derivative of the surf clam (ocean quahog) cost elasticity with respect to GRT is -0.26 (-0.70). As expected, both elasticity estimates decline with larger GRT.

The cost elasticity with respect to the surf clam stock level is 0.63, which is not the expected sign. A Wald test of the null hypothesis that the surf clam stock effect is zero could not be rejected at the 95% confidence level. The inability to identify cost-reducing stock effects may be due to lack of stock variability over the four-year study period. Dummy variables for home-port cost effects, D_1–D_3 in table 1, indicate a significant variable cost increase for vessels originating from Cape Charles, Virginia, and no appreciable cost differences for the remaining ports.

In summary, the variable cost technology exhibits overall and product-specific increasing returns to scale, cost anticomplementarity, and is nonjoint-in-inputs. Only the latter finding is supported by a statistical test. The variable cost technology favors specialized pro-

Table 2. Hull Values, Maximum Harvest Capacity and Fixed Costs

| | | | | Maximum Harvest Capacity | | | | Fixed Costs ($) | | | |
| | | Labor | | Max. Trips/ Qtr. | | Max. Catch/Qtr. | | Total FC (Per Qtr.) | | Min. AFC | |
GRT	Hull Value ($)	SC	OQ	SC	OQ	SC	OQ	SC	OQ	SC	OQ
100	129,333	3	4	35	30	29,120	24,960	37,045	43,939	1.29	1.76
125	213,416	4	5	40	35	40,960	35,840	48,705	54,695	1.19	1.53
150	401,503	4	5	45	40	57,600	51,200	57,958	63,478	1.01	1.24
175	529,687	4	5	50	45	76,800	69,120	64,516	69,566	0.84	1.01
200	563,744	4	5	50	50	96,000	96,000	66,838	71,418	0.70	0.74

Source: Logbook data sources and industry participants.
Notes: SC denotes surf clam, OQ denotes ocean quahog, FC denotes fixed cost, Min. AFC denotes minimum attainable average fixed cost (at maximum capacity). Catch is in bushels.

duction. In the absence of product-specific fixed costs, a specialized clam fleet should be expected to emerge under the ITQ management regime. A final step in identifying the number of vessels under ITQs is to combine variable and fixed cost information to determine the long-run cost-efficient output level.

Fixed Costs

Remaining costs are associated with the vessel capital and labor services. Vessel maintenance and capital cost estimates are obtained from the FMP. Maintenance costs include (*i*) costs associated with haul-out and maintenance, required, on average, every 1.5 years at $22,600, $33,900, and $44,200 for class one (<50 GRT), class two (50–100 GRT), and class three (>100 GRT) vessels, respectively; (*ii*) administrative expenses, approximated at 2% of gross revenues; (*iii*) professional services such as legal and accounting costs: $5,650; (*iv*) docking fees: $2700 annually; and (*iv*) miscellaneous expenses: $3,390 per year. Vessel capital costs include (*i*) hull insurance at 5.5% of hull value, (*ii*) interest payments, and (*iii*) depreciation costs. It is assumed that capital markets are efficient, and interest payments reflect the foregone value of vessel capital. An interest rate of 10% is used. Following the FMP, depreciation is approximated linearly over a thirty-year life of the vessel.

Average crew sizes were obtained from the Mid-Atlantic Fisheries Management Council. Labor remuneration was adjusted in the transition to ITQs (McCay and Creed). The previous share system (wherein 33% of vessel revenues were paid to the captain and crew) no longer reflected the market wage under ITQs because vessel catch rates and revenues increased. It is reasonable to assume that adjustments to the wage will proceed until an equilibrium wage or share emerges to reflect the outside earning opportunities for crew labor. An average salary ($22,716) for Mid-Atlantic States was obtained from the U.S. Department of Labor, *Employment and Earnings* statistics. Personal and indemnity insurance, $4,520 per crew, was added to obtain an estimate of the annual wage rate (FMP).

Estimates of maximum per period harvest capacity and average hull values were obtained from industry members and the Fisheries Management Council. Table 2 reports crew size, maximum capacity and fixed cost estimates for representative vessel size classes. All values reported in table 2 were validated from discussions with industry participants.

Quarterly fixed costs were divided by the maximum harvest capacity to obtain an estimate of the minimum attainable average fixed cost (AFC). Note that smaller vessels have lower fixed costs but significantly smaller maximum harvesting capacities. Larger vessels are able to spread fixed costs over larger outputs, and as a result attain lower AFC. Surf clam average fixed cost is $1.32 per bushel for a 100-GRT vessel operating at maximum capacity. Minimum attainable AFC declines to $0.70 per bushel on a 200-GRT vessel. Because an additional crew member is used on board ocean quahog vessels, the minimum attainable AFC is $1.83 on a 100-GRT vessel and declines to $0.75 on a 200-GRT vessel. Given the finding of increasing returns to scale for the variable cost function and declining AFC, it is expected that larger vessels will attain greater total cost efficiency under ITQs.

ITQ Regime Fleet Structure and Returns to ITQ Management Reform

Identifying the ITQ-regime fleet structure requires estimates of average total cost (ATC) minimizing output levels. For a specialized fleet, the number of vessels is then obtained by calculating the smallest fleet size that remains capable of harvesting the TAC in the fishery. The quarterly TAC is set at 712,500 bushels of surf clams and 1,325,000 bushels of ocean quahogs (FMP).

The quarterly TAC for each species was divided by different fleet sizes to obtain an estimate of the average per vessel harvest. ATC is calculated as the sum of fixed and variable costs (from table 2 and the fitted variable cost function, respectively) divided by the implied harvest level. This procedure was repeated for 150, 175 and 200 GRT vessel classes.[13] Average cost estimates on smaller vessels (<150 GRT) exceed the cost estimates for the 150 to 200 GRT vessel classes and are not reported.

Per vessel output, average variable cost (AVC), ATC, and total fleet cost estimates for various fleet structures are reported in table 3. To show the efficiency benefits from quota consolidation, harvest levels and costs are reported for incrementally smaller fleet sizes, ending with the smallest feasible fleet size. For example, a fleet of eight 200-GRT vessels is the smallest that is capable of harvesting the surf clam TAC and thus smaller surf clam specialized fleets were not considered.

Surf Clam Vessels

At a fleet size of twenty-five boats, the per vessel allocation of the TAC is 28,500 bushels per quarter. At this output level, vessels are operating below efficient scale so ATC remains high. Because of increasing returns to scale, AVC and ATC decline as fleet size is decreased. For example, the ATC for a 200-GRT vessel is $4.11 at a harvest level of 28,500 bushels per quarter (output corresponding to a twenty-five-vessel fleet). ATC for a 200-GRT vessel declines to $1.94 per bushel at a harvest level of 89,063 bushels per quarter (the output level corresponding to an eight-vessel fleet).

The estimated quarterly fleet costs indicate the cost savings from smaller fleet sizes. A fleet consisting of twenty-five, 150-GRT boats can harvest the TAC at a quarterly cost of $2.6 million ($2.9 for 175- and 200-GRT vessels). Fleet costs decline as the number of vessels in the fleet is reduced and the total harvest is consolidated onto fewer boats. At a fleet size of ten vessels, for example, each vessel harvests 71,250 bushels per quarter, and ATC is $2.15 and $2.21 for 175- and 200-GRT vessels, respectively. (Note that 71,250 bushels per quarter exceeds the maximum output level for a 150-GRT vessel and thus a ten-boat 150-GRT fleet is not feasible.) The cost analysis reveals that the surf clam portion of the ITQ-regime fleet will consist merely of eight 200-GRT vessels that harvest 89,063 bushels of surf clams per quarter. The corresponding fleet cost is $1.4 million per quarter.

Ocean Quahog Vessels

Similar results are indicated for ocean quahog vessels. ATC declines as the number of vessels is reduced and per vessel harvest is increased. ATC attains a minimum of $2.04 per bushel when harvested by thirteen 200-GRT vessels. The quarterly fleet cost for the ocean quahog specialized vessels is $2.5 million. Combining surf clam and ocean quahog boats, the ITQ-regime fleet will be comprised of eight 200-GRT surf clam vessels and thirteen 200-GRT ocean quahog vessels. The total harvest cost incurred is $1.4 + $2.5 = $3.9 million per quarter or $15.6 million annually.

Diversified Fleet Structure

For comparison, table 3 reports per vessel and fleet cost for a fully diversified fleet structure.[14] As with specialized vessels, ATC declines as per vessel output increases. However, a diver-

---

[13] The 150-, 175-, and 200-GRT vessel classes are representative of the clam fleet active in 1994. Of the fifty active vessels in 1994, six were 87.5-GRT or less, five were 100-GRT (includes 87.5- to 112.5-GRT), fourteen were 125-GRT, 9 were 150-GRT, twelve were 175-GRT, and four were 200-GRT.

[14] The output mix for diversified vessels is obtained by minimizing total fleet cost conditional on a fully diversified fleet structure. Note that a mixed fleet structure would incur a lower total harvest cost. The cost estimates in table 3 are presented to illustrate the additional savings from specialized production.

Table 3. Harvesting Costs Estimates: ITQ-Regime Clam Harvesting Fleet

Surf Clam Portion of Specialized ITQ-Regime

| | Per Vessel Catch (Bushels/Qtr.) | | Per Vessel Costs | | | | | | Quarterly Fleet Cost ($ Millions) | | |
| | | | GRT = 150 | | GRT = 175 | | GRT = 200 | | GRT = 150 | GRT = 175 | GRT = 200 |
Boats	SC	OQ	AVC	ATC	AVC	ATC	AVC	ATC			
25	28,500	0	1.69	3.71	1.79	4.01	1.84	4.11	2.6	2.9	2.9
20	35,625	0	1.54	3.15	1.63	3.41	1.68	3.50	2.2	2.4	2.5
15	47,500	0	1.38	2.59	1.46	2.79	1.50	2.86	1.8	2.0	2.0
10	71,250	0	nf	nf	1.27	2.15	1.30	2.21	nf	1.5	1.6
8	89,063	0	nf	nf	nf	nf	1.21	1.94	nf	nf	1.4

Ocean Quahog Portion of Specialized ITQ-Regime Fleet

| | Per Vessel Catch (Bushels/Qtr.) | | Per Vessel Costs | | | | | | Quarterly Fleet Cost ($ Millions) | | |
| | | | GRT = 150 | | GRT = 175 | | GRT = 200 | | GRT = 150 | GRT = 175 | GRT = 200 |
Boats	SC	OQ	AVC	ATC	AVC	ATC	AVC	ATC			
30	0	44,167	1.85	3.35	1.81	3.44	1.73	3.40	4.4	4.6	4.5
25	0	53,000	1.73	2.98	1.68	3.04	1.61	2.99	4.0	4.0	4.0
20	0	66,250	1.59	2.59	1.54	2.63	1.47	2.58	3.4	3.5	3.4
15	0	88,333	nf	nf	1.39	2.20	1.32	2.15	nf	2.9	2.8
13	0	94,643	nf	nf	nf	nf	1.28	2.06	nf	nf	2.5

Fully Diversified ITQ-Regime Fleet

| | Per Vessel Catch (Bushels/Qtr.) | | Per Vessel Costs | | | | | | Quarterly Fleet Cost ($ Millions) | | |
| | | | GRT = 150 | | GRT = 175 | | GRT = 200 | | GRT = 150 | GRT = 175 | GRT = 200 |
Boats	SC	OQ	AVC	ATC	AVC	ATC	AVC	ATC			
40	17,812	33,125	1.68	2.81	1.66	2.90	1.60	3.00	6.0	6.1	5.6
35	20,357	37,857	1.60	2.58	1.57	2.65	1.51	2.74	5.5	5.6	5.6
30	23,750	44,166	nf	nf	1.47	2.40	1.42	2.47	nf	5.1	5.0
25	28,500	53,000	nf	nf	nf	nf	1.31	2.19	nf	nf	4.5
22	32,386	60,227	nf	nf	nf	nf	1.25	2.02	nf	nf	4.1

Source: Estimated.

Notes: "nf" indicates nonfeasible output level. AVC indicates average variable cost. ATC indicates average total cost. GRT indicates vessel gross registered tonnage.

sified operation will be unable to fully exploit product-specific returns to scale, and thus cannot match the cost efficiency of a specialized operation. For example, the estimated (short-run) marginal harvesting cost on a specialized surf clam vessel (200-GRT vessel and an output level of 89,063 bushels per quarter) is $0.79. The estimated surf clam marginal harvesting cost on diversified vessel operation (200-GRT and an output level of 32,386 bushels of surf clams and 60,227 bushels of ocean quahogs per quarter) is $0.98. A diversified fleet consisting of 22 200-GRT vessels is estimated to incur a quarterly fleet cost of $4.1 million. This is $0.9 million more per year than under the specialized fleet structure.

Efficiency Gains under ITQs

Fleet cost estimates from table 3 can be compared to costs incurred under LE. McCay and Creed report the total labor force under LE to be 207 individuals in 1990. Using this estimate of total labor, per vessel harvest costs were estimated following the above procedure and summed over active LE-regime vessels. The cost incurred by the LE-regime fleet is estimated to be $28.4 million in 1989.[15] A comparison to the ITQ-regime cost estimate

___
[15] The last complete year of LE management was 1989. Fleet costs were estimated to be $22.9 million in 1987 and $27.6 million in 1988.

762   November 1998   Amer. J. Agr. Econ.

indicates a harvest sector cost savings of $12.8 million annually. While these gains may seem remarkable, it should be remembered that 128 vessels actively harvested clams in 1989. Apparently the fleet was roughly six times larger under LE than is expected under the ITQ management regime.

Sensitivity Analysis

Assessment of the ITQ fleet structure and the efficiency gains relies on an accurate estimate of the harvest cost structure. Empirical errors can occur at various stages of the analysis. In this study, cost data were measured indirectly from information on fishing times and fishing trips per calendar period. Systematic bias in these data could result in misspecification of the ITQ-regime cost and fleet structure and bias estimates of efficiency gains. Other sources of error include econometric misspecification and poor estimates of stock abundance. It is difficult to determine the net impact of error on the final efficiency gain estimates. Thus, results reported here must be viewed accordingly.

One source of error is the estimate of per vessel maximum harvest capacity. If per vessel capacity is overestimated, the ITQ-regime fleet may consist of more that eight 200-GRT surf clam and thirteen 200-GRT ocean quahog vessels. In this case, the harvesting cost incurred by the ITQ-regime fleet will be underestimated. If, instead, ten surf clam and fifteen ocean quahog vessels are required to harvest the TAC, the quarterly costs for the fleet would rise to roughly $4.3 million per quarter and $17.3 million annually (table 3). The efficiency gains under ITQs are reduced to $11.1 million per year.

A more encouraging observation is that it may not be necessary to identify all aspects of the ITQ-regime fleet structure to obtain a reasonable estimate of the efficiency gains under ITQs. Further examination of table 3 indicates that ATC costs decline sharply as per vessel output is increased. In contrast, vessel GRT impacts ATC estimates to a lesser degree. The ATC at 71,250 bushels per quarter is $2.20 for a 175-GRT vessel and $2.15 for a 200-GRT vessel. Identifying the ITQ-regime vessel size appears to be less critical than identifying the efficient scale of operation and the corresponding minimum ATC. Any vessel that can match the minimum ATC is viable under ITQs. For example, differences in fishing skill, which were not modeled in this article, may lead to comparable cost efficiency on different-sized vessels or at different output levels. The ITQ-regime fleet structure may be heterogenous except with respect to the residual return from owning the ITQ asset. See Anderson for further discussion of fisher heterogeneity under an ITQ management program.

A second encouraging observation is that the findings in this article appear consistent with ancillary information and observed trends in the MA clam fishery. First, logbook data indicates that eighteen of the fifty (36%) active vessels in 1994 took fewer than half of the maximum number of trips per quarter. This suggests that the harvesting capacity of a fifty-vessel fleet exceeds the TAC. This evidence is consistent with the prediction that 21–25 vessels will make up the ITQ-regime fleet. Second, the Overview of the Surf Clam and Ocean Quahog Fisheries and Quota Recommendations for 1997 and 1998 reports that clam fishers continue to adjust quota holdings to focus on single-species production. This further supports the prediction of a specialized ITQ-regime fleet structure.

Conclusion

The post-restructuring efficiency gains provide the appropriate benchmark for assessing ITQ fisheries management policy. We present a methodology to estimate harvest sector efficiency gains from ITQs, in lieu of delayed fleet restructuring. A model of equilibrium fleet structure is presented to identify the number of vessels and individual output that is expected to emerge under ITQs. Harvest-sector costs based on the anticipated ITQ-regime fleet are then estimated to provide a post-restructuring assessment of the efficiency gains. The methodology is applied to the MA clam fishery.

While data limitations may have influenced the final estimates of the efficiency gains in this study, the results illustrate the importance of considering long-run fleet adjustments when assessing efficiency gains from ITQs. A long history of capital replacement restrictions in the MA clam fishery led to a fleet structure that bore little resemblance to that expected under the ITQ program. The analysis of this article suggests that roughly 21–25 vessels will remain active under the ITQ-regime as compared to 128 vessels under the previous LE program. The total harvest cost savings under the ITQ program is estimated to be between $11.1 million and $12.8 million annually.

In this study, estimates of efficiency gains were obtained after the ITQ management scheme was implemented. Clearly, *ex ante* estimates would be invaluable to resource managers who are considering ITQ programs. The methodology used in this article may be applied, but the analysis should proceed with caution. Harvesting activities and costs may be strongly influenced by existing regulations. For example, Lipton and Strand estimated harvesting costs and equilibrium fleet structure in the MA clam fishery during the LE regime. Their analysis suggested an equilibrium fleet structure consisting of 149 multispecies vessels in which each vessel harvested 14,933 of surf clams and 56,134 bushels of ocean quahogs annually (Lipton and Strand, p. 205). The stark difference in fleet structures can be explained by vastly different management systems. Future research that focuses on methods to extrapolate ITQ-regime cost and fleet structures from pre-ITQ regime data sources could provide robust estimates of the efficiency gains expected from an ITQ management program.

Additional policy questions emerge from the analysis in this article. Indications are that prolonged fleet restructuring in the MA clam fishery continued to dissipate available rents even four years after the ITQ program was implemented. Further policy actions designed to expedite fleet restructuring may be warranted. For example, an initial quota auction rather than a gratis initial allocation to active fishery participants could hasten industry restructuring. Furthermore, our analysis suggests that production activities during the pre-ITQ management regime may hold little resemblance to the production activities expected under ITQs. Basing initial quota allocations on pre-ITQ harvest levels may in fact prolong the transition to the equilibrium quota distribution and the equilibrium ITQ-regime fleet structure and reduce the possible benefits of ITQ programs.

[Received June 1997;
accepted March 1998.]

References

Anderson, L.G. "Conceptual Constructs for Practical ITQ Management Policies." *Rights Based Fishing.* P.A. Neher, R. Arnason, and N. Mollett, eds., pp. 191–209. Boston: Kluwer Academic Publishers, 1989.

Baumol, W., J. Panzar, and R. Willig. *Contestable Markets and the Theory of Industry Structure.* San Diego CA: Harcourt Brace Jovanovich, 1982.

Clark, C.W. *Mathematical Bioeconomics: The Optimal Management of Renewable Resources.* New York: John Wiley and Sons, 1990.

Denny, M., and C. Pinto "An Aggregate Model With Multi-Product Technologies." *Production Economics: A Dual Approach to Theory and Applications,* M. Fuss and D. McFadden, eds., pp. 249–67. Amsterdam: North Holland, 1978.

Gorman, I.E. "Conditions for Economies of Scope in the Presence of Fixed Costs." *Rand J. Econ.* 16(Autumn 1985):431–36.

Kim, H.Y. "Economies of Scale in Multi-product Firms: an Empirical Analysis." *Economica* 54(May 1987):185–206.

Lipton, D.W., and I.E. Strand "Effect of Stock Size and Regulations in Fishing Industry Cost and Structure: The Surf Clam Industry." *Amer. J. Agr. Econ.* 74(February 1992): 197–207.

MacDonald, G., and A. Slivinski "A Simple Analysis of Competitive Equilibrium with Multiproduct Firms." *Amer. Econ. Rev.* 77(December 1987):941–53.

McCay, B.J., and C.F. Creed "Social Impacts of ITQ's in the Sea Clam Fisheries." Unpublished, February 1993.

Mid-Atlantic Fisheries Management Council. *Amendment # 8, Fisheries Management Plan for the Atlantic Surf Clam and Ocean Quahog Fishery.* Unpublished, July 1988.

Mid-Atlantic Fisheries Management Council. *Overview of the Surf Clam and Ocean Quahog Fisheries and Quota Recommendations for 1997 and 1998.* Unpublished, September 1996.

Montgomery, D.W. "Markets in Licenses and Efficient Pollution Control Programs." *J. Econ. Theory* 5(December 1972):395–418.

NOAA/National Marine Fisheries Service. *Report of the 19th Northeast Regional Stock Assessment Workshop.* Unpublished, Northeast Fisheries Science Center, March 1995.

Panzar, J.C. "Technological Determinants of Firm and Industry Structure." *Handbook of Industrial Organization,* vol. 1. R. Schmalensee and R.D. Willig, eds., pp. 3–59. Amsterdam: Elsevier Science Publishing, 1989.

Squires, D. "Production Technology, Costs and

764 *November 1998* *Amer. J. Agr. Econ.*

Multiproduct Industry Structure: An Application of the Long-Run Profit Function to the New England Fishing Industry." *Can. J. Econ.* 21(May 1988):359–78.

Strand, I.E., J.E. Kirkely, and K.E. McConnell. "Economic Analysis and the Management of Atlantic Surf Clams." *Economic Analysis for Fisheries Management Plans.* L.G. Anderson, ed., pp. 245–66. Ann Arbor MI: Ann Arbor Science, 1981.

U.S. Department of Energy, Energy Information Administration. *Monthly Energy Review.*

Washington DC, December 1992 and December 1994 issues.

U.S. Department of Labor, Bureau of Labor Statistics. *Employment and Earnings,* vol. 37, no. 12, 1990.

Wang, S.D. "The Surf Clam ITQ Management: An Evaluation." *Marine Resour. Econ.* 10(Spring 1995):93–98.

Weninger, Q., and R.E. Just "An Analysis of Transition from Limited Entry to Transferable Quota: Non-Marshallian Principles for Fisheries Management." *Natural Resour. Model.* 10(Winter 1997):53–83.

[46]

Marine Resource Economics, Volume 13, 51–74
Printed in the U.S.A. All rights reserved

0738-1360/98 $3.00 + .00
Copyright © 1998 Marine Resources Foundation

Bioeconomic Analysis of Alternative Selection Patterns in the United States Atlantic Silver Hake Fishery

E.M. THUNBERG
National Marine Fisheries Service
T.E. HELSER
West Virginia University
R.K. MAYO
National Marine Fisheries Service

Abstract   *In this paper a bioeconomic simulation of the U.S. fisheries for silver hake, Merluccius bilinearis, is presented. The model design combines elements of age-structured population and harvest yield models with economics of the silver hake fishery. The analysis evaluates both biological and economic effects of interest to managers, such as future yields or rebuilding of parental stock, as well as future revenues and net returns to vessels. The bioeconomic model is used to evaluate the economic implications of tradeoffs between alternative selection patterns in the U.S. Atlantic silver hake fishery. Throughout the study, a selection pattern is defined as the suite of age-specific selection coefficients that are applied to a fish population over time. The results indicate that shifting fishing pressure to younger age classes could result in short-run gains in economic value that may not be sustainable due to longer run declines in biomass, hence lowered fishery yield and value. By contrast, strategies to delay age at first capture may improve economic value over current levels with only modest reductions in short-run fishery yield.*

Key words   Bioeconomic model, fishery management, selection pattern, silver hake, simulation.

Introduction

Bioeconomic models provide an integrated approach to evaluation of alternative fishery management strategies. Such models have generally been used to examine economic performance or rent dissipation in a fishery either by consideration of resource and fleet interactions or through market effects. Much of the existing work in bioeconomic modeling has concentrated on the former interaction with major themes being developed around resource extraction rates and issues of fleet size or

E.M. Thunberg is an economist in the Social Sciences Branch of the Northeast Fisheries Science Center, National Marine Fisheries Service, 166 Water Street, Woods Hole, MA 02543, USA; e-mail: ethunber@whsun1.wh.whoi.edu. T.E. Helser is assistant professor in the Department of Wildlife and Fisheries in the College of Agriculture and Forestry, West Virginia University, P.O. Box 6125, Morgantown, WV 26505, USA. R.K. Mayo is a fisheries research biologist in the Population Dynamics Branch of the Northeast Fisheries Science Center, National Marine Fisheries Service.

   We thank Fred Serchuk, Terry Smith, Steve Edwards, and Phil Logan of the Northeast Fisheries Science Center and two anonymous reviewers for constructive comments on previous drafts of the manuscript.

capacity. Considerably less attention has been focused on economic losses associated with the failure to consider the market implications of fishery management strategies.

Gates (1974) was among the first to point out the implications of fish size and prices for ex-vessel demand analysis and, more importantly, for fishery management policy. Conrad (1982) examined optimal resource management strategies when output prices are based on size. Similarly, Anderson (1989) developed a set of analytical models to demonstrate how resource management strategies might differ in cases where output prices are based on individual size. Work by Christensen and Vestergaard (1993) and Overholtz, Edwards, and Brodziak (1993) also incorporated size dependent price effects to maximize the value of harvested fish. Taking a somewhat different approach, Wilen and Homans (1992) evaluated the marketing losses associated with regulatory approaches that tend to result in poor timing of harvest or compromised product quality.

Many of the most important food-fish and shellfish harvested in the northeastern United States are sold in several different market categories where the price paid is frequently based upon the size of the fish. Price premiums paid for size may be attributable to a variety of factors including texture, product recovery rates, and interactions between product form and preferences in final markets. However, the relationship between fish size and price may differ markedly depending upon the species, suggesting that optimal management strategies may also differ. Several different possible price/size relationships are illustrated in figure 1. Hard clam prices (the subject of Conrad's investigation) exhibit a negative relationship between individual size and unit price. In cases such as these, optimal resource management may be to

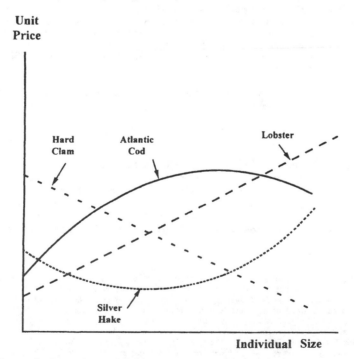

Figure 1. Hypothetical Price Structure for Northeastern U.S. Species

utilize the larger size classes exclusively as brood stock to increase the harvestable quantities of the smaller and more valuable size classes. Lobster prices exhibit the opposite pattern with higher unit prices being paid for larger, older individuals. Studies of the American lobster fishery have generally concluded that optimal harvest would involve a mixed strategy with greater emphasis placed on delaying the size at first capture to take advantage of the higher prices paid for larger individuals while assuring sufficient escapement of the larger individuals as brood stock (see Botsford, Wilen, and Richardson 1986; Richardson and Gates 1986; Wang and Kellogg 1988). Market prices for Atlantic cod first increase with size then decrease for the largest market categories. A combined strategy may be desirable in this case where delaying the size at first capture may be advantageous while reserving the largest size categories to serve as brood stock to increase harvestable quantities of the smaller size classes. Market prices for silver hake, or whiting (*Merlucius bilinearis*) exhibit a convex price pattern with relatively higher prices paid for the smallest and largest size categories as compared to intermediate sizes.

The U.S. Atlantic silver hake fishery is prosecuted in Federal waters and is subject to management as prescribed in the former Magnuson Act (now the Magnuson-Stevens Act or the Sustainable Fisheries Act or SFA). Silver hake is one of thirteen species regulated under the Northeast Multispecies Fishery Management Plan (hereafter referred to as the Multispecies Plan) which includes the complex of principal groundfish species harvested in the region. Initially, management concern over silver hake was prompted by two developments. The first was passage of Amendments 5 and 7 to the Multispecies Plan. The Amendments place limits on the number of days individual vessels may fish for groundfish (principally cod, haddock, and yellowtail flounder). Although included in the Multispecies Plan, silver hake is designated as a "small mesh" fishery, and time spent fishing for silver hake does not count toward a vessel's allocated fishing time. Additionally, the majority of the silver hake fishery is prosecuted in southern New England and mid-Atlantic waters; areas that have relatively low bycatch of the principal groundfish species. Thus, from a multiproduct perspective, silver hake can be characterized as a substitute species with respect to other species in the Multispecies Plan. For these reasons, resource managers and industry participants who have traditionally targeted silver hake are concerned that displaced effort from other groundfish fisheries will be redirected to silver hake stocks that are already considered to be fully exploited.[1]

The second event was the development of a European export market for juvenile fish. Since 1993, U.S. exports of juveniles have been filling a marketing void left by the prohibition on landings of small hake (less than 19 cm) in European waters. Historic at-sea observations indicate that discards of small silver hake are significant in U.S. fisheries (Helser and Mayo 1994). At least a portion of the export demand might be met with increased retention of fish that have traditionally not been marketed. However, there is substantial concern that vessels may instead simply change their targeting behavior to concentrate on aggregations of juvenile fish.

The passage of the amended SFA has added increased impetus to addressing silver hake management. One of the two Atlantic Silver hake stocks was designated by the National Marine Fisheries Service (NMFS) as one of eighty-six species nationwide determined to be overfished (NMFS 1997) and the other stock was determined to be approaching an overfished status. Under the SFA, this designation requires that the New England Fishery Management Council (NEFMC) develop a plan to address the overfished status of the resource within one year's time. Further, the Act requires

---

[1] Of course, even without Amendments 5 and 7, reduced groundfish landings have prompted a search for alternative fisheries such as silver hake.

that fishery resources be managed to achieve optimum yield subject to maintaining a stock size no less than that which would be required to allow harvest at maximum sustainable yield (MSY).

In this paper we develop an age-structured bioeconomic model to evaluate the economic implications of tradeoffs between alternative selection patterns in the U.S. Atlantic silver hake fishery. The present study is based upon previous work where combinations of fishing mortality rates and fishery selection were systematically varied to generate a response surface for management strategies for silver hake fisheries in the Northeast (see Helser, Thunberg, and Mayo 1996). While not ignoring the importance of reducing fishing mortality rates in this study, we focus on the economic and resource implications of alternative fishery selection patterns. Throughout the study, a selection pattern is defined as the suite of age-specific selection coefficients that are applied to a fish population over time. The selection pattern reflects both the mesh sizes of trawls used in the fishery, as well as biological factors such as recruitment and distribution of fish. As such, our definition is related, but should not be confused with, selection at age from trawl mesh experiments.

The present study differs from the earlier work in that the economic component has been expanded by adding vessel harvesting costs, different fleets by vessel size are introduced, and adjustments to vessel productivity are made based on changes in stock sizes. The biological model has also been revised by incorporating a parametric stock-recruitment function. Of particular interest is exploration of whether tradeoffs among segments of a fish population can result in increased resource value. This question is explored within the context of whether a limited juvenile fishery can co-exist with traditional fisheries consistent with management objectives for silver hake stocks and if so, under what conditions might such a fishery operate.

History of the Fishery and Management

Silver hake is a widely distributed species whose range extends from Newfoundland to South Carolina (Bigelow and Schroeder 1953), but is most abundant between Nova Scotia and the New York Bight. Silver hake was estimated by Edwards (1968) to comprise the largest biomass of any demersal species inhabiting the Northeast continental shelf between Nova Scotia and the New York Bight during the mid 1960s and based on current resource assessments has the greatest long-term potential yield among Northeast demersal species (National Marine Fisheries Service 1996).

The U.S. silver hake fishery began in the middle 1800s with the principal center of activity ranging from the Middle Atlantic Bight to Georges Bank (Fritz 1960). The commercial fishery expanded rapidly during the 1930s and 1940s due to the introduction of otter trawling vessels and technical advances in quick-freezing and automatic scaling. During this time, silver hake were processed largely for human consumption in fresh and frozen form and marketed as a lesser quality product for animal food (Anderson, Lux, and Almeida 1980). Landings continued to increase in the 1950s and 1960s with the development of a specialized "industrial" fishery for reduction and industrial purposes (Anderson, Lux, and Almeida 1980; Hennemuth and Rockwell 1987).

Exploitation intensified between 1961 and 1965 with the arrival of distant water fleets (DWF, principally the former Union of Soviet Socialist Republic). Total domestic and foreign fleet landings increased markedly to a peak of over 351,000 t in 1965, followed by a sharp decline to 55,000 t (40% U.S.) in 1970 (Helser and Mayo 1994). During this period, spawning stock biomass (SSB) and recruitment declined by more than 50% (Helser, Almeida, and Waldron 1995). Limits on total allowable catch (TAC) were introduced in 1973 by the International Commission for the Northwest Atlantic Fisheries (ICNAF) followed by areal/spatial fishing "windows" in 1976 which restricted the DWF to a relatively narrow portion of the outer continental shelf of Georges Bank. After a brief increase during 1971–75, landings subsequently declined as the DWF left the fishing grounds.

The silver hake fishery has been managed under exclusive control of the U.S. since passage of the Fisheries Conservation and Management Act in 1976, but the U.S. domestic fishery was not regulated until 1988 when silver hake was added to the Multispecies Plan by the New England Fishery Management Council (NEFMC). Under U.S. management, silver hake resources have been assessed as two separate stocks: a northern stock extending from the Gulf of Maine to the northern flank of Georges Bank and a southern stock extending from southern Georges Bank into the Middle Atlantic region. Due to its lower value relative to alternative groundfish species, landings from both stocks have remained at low, but stable levels compared to the earlier years of the fishery, averaging about 17,500 t over the last decade (Helser and Mayo 1994).

There are currently no regulations on the fisheries exploiting the southern stock, while a regulated small mesh fishery, restricted seasonally and spatially since 1988, has been conducted over an area of northern Georges Bank. At this time, the NEFMC is exploring options to deal with the silver hake resource including, but not limited to, implementing a minimum size limit on silver hake that, in some cases, would eliminate the juvenile market category. Further discussion of the history of the silver hake fisheries and their management is provided in Helser, Almeida, and Waldron (1995).

Bioeconomic Model

Figure 2 provides a schematic view of the biological and economic components of the model. Rectangles denote endogenous components, ovals denote exogenous factors, and arrows indicate the directional flows. Management controls are limited to selection of a fishing mortality rate and a fishery selection pattern. The population dynamics are modeled using an age-structured Leslie matrix population model (Leslie 1945). Recruitment to the fishery is modeled using a Ricker curve (Ricker 1954). Catch-per-day fished (CDF) is determined by the amount of fishing capital, and stock abundance and total landings are determined by a modified harvest yield model (Thompson and Bell 1934). Historical fleet shares are used to allocate total landings among vessel size classes. Prices are determined using a price model holding exogenous factors constant. Fishery profitability in the aggregate and by fleet are determined through a set of accounting relationships. The model is used to examine the economic yield (in terms of vessel profitability) under alternative fishery selection patterns.

Fishery Population Dynamics

The biological component of the model incorporates both silver hake stocks, although a stock-specific index has been dropped in the following for clarity. A more detailed description of the biological component can be found in Helser, Thunberg, and Mayo (1996). A glossary of symbols is reported in table 1, and the mathematical specification of the essential population dynamics is captured in equations (1) – (4) of table 2.

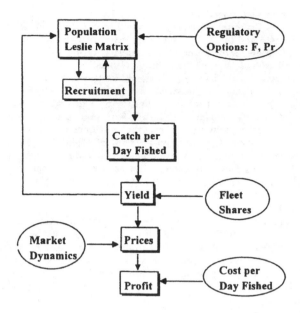

Figure 2. Schematic Representation of Silver Hake Model

Table 1
Glossary of Symbols

Variables		Coefficients	
A	Abundance index, number age 2+	e	Quarterly proportion of effort
C	Catch at age in numbers	F	Annual fishing mortality rate
CDF	Catch per day fished	f	Fraction of fishing mortality prior to spawning season
DF	Number of days fished	M	Annual natural mortality rate
K	Fishing capital	m	Fraction of natural mortality prior to spawning season
N	Numbers at age	q	Relative selection at age
P	Ex-vessel price	p	Fraction mature at age
SSB	Spawning stock biomass	a,b	Ricker parameters
TR	Total revenue	w	Residuals from Ricker curve
VC	Variable cost per day fished	w	Average weight at age
W	Landed weight	α	Constant coefficient for price model
WJ	Landed weight of juveniles	β	Slope coefficient for price model
WR	Landed weight of round	γ	Fleet share of landed weight
WK	Landed weight of king	λ	Cost per day fish
P	Net return	δ	Marginal productivity of capital
Pr	Selection pattern	θ	Marginal productivity of abundance

Indices

c	Market category (c = 1, 2, 3)
i	Age in years (i = 1, 2, ..., a)
j	Quarter (j = 1, 2, 3, 4)
t	Year (t = 1, 2, ..., 10)
v	Fleet (v = 1, 2, ..., 6)

Table 2
Listing of Equations Used in the Mathematical Model

Dynamics Equations

Annual recruitment

$$N_i(t) = a \cdot SSB(t-1)e^{-h \cdot SSB(t-1)} \cdot e^w \quad \text{where } w \sim N(0, \sigma_w^2) \tag{1}$$

Quarterly numbers-at-age

$$N_{i+1,j+1,t+1} = N_{i,j,t} \cdot e^{-(M/4 + Pr_ie_jF_t)} \tag{2}$$

Annual spawning stock biomass

$$SSB_t = \sum_{i=1}^{a} \sum_{j=1}^{4} N_{i,j,t} \cdot e^{-(mM/4 + fPr_ie_jF_t)p_iw_i} \tag{3}$$

Quarterly total catch-at-age

$$C_{ijt} = \left(\frac{Pr_ie_jF_t}{M/4 + Pr_ie_jF_t} \right)\left[1 - e^{-(M/4 + Pr_ie_jF_t)}\right]N_{ijt} \tag{4}$$

Accounting Equations

Quarterly total landings by market category

$$WJ_{jt} = \sum_{i=0}^{2} C_{ijt}w_i; \quad WR_{jt} = \sum_{i=3}^{5} C_{ijt}w_i; \quad WK_{jt} = \sum_{i=6}^{a} C_{ijt}w_i \tag{5}$$

Predicted quarterly ex-vessel unit price by category

$$P_{cjt} = \alpha_{cjt} + \beta_{cjt}W_{cjt} \tag{6}$$

Quarterly total landings by fleet and category

$$W_{cjtv} = \gamma_{cjtv}W_{cjt} \tag{7}$$

Quarterly catch per day fished by fleet

$$CDF_{jtv} = \sigma_{jtv}K_{jtv}^{\delta}A_t^{\theta} \tag{8}$$

Quarterly total days fished by fleet

$$DF_{jtv} = \frac{\sum_{i=1}^{3} W_{cjtv}}{CDF_{jtv}} \tag{9}$$

Quarterly total revenues by fleet

$$TR_{jtv} = \sum_{i=1}^{3} P_{cjt}W_{cjtv} \tag{10}$$

Quarterly total variable cost by fleet

$$VC_{jtv} = \lambda_t DF_{jtv} \tag{11}$$

Quarterly total net returns by fleet

$$\Pi_{jtv} = TR_{jtv} - VC_{jtv} \tag{12}$$

Stock-Recruitment Relationship. In general, stock and recruitment data are highly variable owing to intrinsic variability in factors governing survival and measurement error in the estimates of recruitment $[N_1(t)]$ and the spawning stock biomass that generated it $[SSB(t - 1)]$. It is usually desirable to characterize the uncertainty in recruitment predictions, but standard statistical approaches fail to provide the type of information desirable for such characterization. In our model, the stock-recruitment relationship ultimately defines the sustainable level of harvest and its expected variability over time. We evaluated the suitability of Beverton-Holt (1957), Ricker (1954), and Shepherd (1982) models to represent the relationship between spawning biomass and recruitment. We found that a Ricker curve with lognormal error provided the best fit to the stock recruitment data for both stocks. This curve provides a parametric model for simulating recruitment where survival to recruitment age is density dependent and subject to stochastic variation. The Ricker curve with lognormal error generates recruitment as:

$$N_1(t) = a \cdot SSB(t - 1)e^{-b \cdot SSB(t-1)} \cdot e^w$$

where $w \sim N(0, \sigma_w^2)$ and the stock-recruitment parameters a, b, and σ_w^2 were estimated with nonlinear regression using stock-recruitment data (Hilborn and Walters 1992). For the southern stock, the parameter estimates were $a = 9.4418$ (1.6441); $b = 0.0022$ (0.0007); $\sigma_w^2 = 0.591$ while for the northern stock the estimates were $a = 5.0572$ (1.0258); $b = 0.0021$ (0.001), $\sigma_w^2 = 0.6024$ where values in parentheses are standard errors. Within the range of observed values, recruitment decreases at high levels of SSB for the southern stock while for the northern stock recruitment increases somewhat asymptotically at higher SSB levels.

Stock Dynamics. Silver hake stocks are composed of a recognizable age classes, where N_i, $i = 1, ..., a$, is used to represent the numbers of individuals in the ith age-class. The numbers alive at the beginning of the age interval [table 2; equation (2)] are reduced quarterly throughout year t by the instantaneous natural mortality rate, $M/4$, and during the fishing season by a harvesting rate, $Pr_i e_j F_t$, where Pr_i is the relative selection coefficient for age-class i, e_j is the quarterly proportion of the total annual fishing effort (days fished), and F_t is the annual instantaneous fishing mortality rate. Initial age specific selectivity, Pr_i, was set equal to recent stock assessment estimates based on VPA (virtual population analysis) for the years 1989 to 1992 (Helser and Mayo 1994). For the time period of analysis the selection at age was 28.1% at age 1, 50% at age 2, and 100% at age 3 for the northern stock and 10% at age 1, 50% at age 2, and 100% at ages 3 and over for the southern stock. Alternative selection patterns were developed by "sliding" the baseline selection coefficients forward or backward as appropriate. For example, a selection pattern of 50% at age 1 would be simulated by setting selection at age 1 to 50% and selection at age for all other age classes to 100%. In cases where selection at age was delayed (50% at age 3 for example) the selectivity coefficients for younger age classes were halved from their baseline values. For the northern stock age specific selectivity would be adjusted as follows; age 3 = 50%, age 2 = 28.1%, and age 1 = 14.05%. Selectivity for the southern stock would be similarly adjusted.

New recruits enter the fishery at age one (one year after hatching) at the beginning of each year. The size of the spawning stock that produced recruits in year t [table 2; equation (3)] is determined by the fraction of population mature at age i, the average weight of an individual at age, and the fraction of the natural and fishing mortality rates that occur before the spawning season. The catch at age i in quarter j of year t was calculated using a slight modification of the usual catch equation [table 2; equation (4)] where the overall level of the yield depends on stock size, the fishing mortality rate, F_t, and the relative selection at age, Pr_i.

Fishery Economics

Silver hake is one of thirteen species that may be sought by otter trawl vessels regulated under the Multispecies FMP. For this reason, the fishery economics were modeled in a manner analogous to a partial budget based only on time spent engaged in the silver hake fishery. Time spent in the fishery is based on days fished. This approach was selected for several reasons. First, days fished is one of the primary management tools used by the NEFMC under both Amendments 5 and 7 to the Multispecies FMP.[2] Second, days fished provides a potential means through which explicit links between vessel activity and fishing mortality rates may be developed. Last, management debate has begun to turn to the potential for making days fished a tradeable commodity.

In the present study, vessel entry-exit behavior is not modeled, since landings are determined by the fishing mortality rate, not the vessel activity. Although the fishing mortality rate serves as a proxy for fishing effort, without explicit consideration of the behavioral response to changes in management or economic conditions there is no assurance that the assumed fishing effort will ever materialize. While the lack of an entry-exit model creates a recognized disconnect between vessel owner behavior and fishing mortality rates, an explicit model was not developed due to two considerations. From 1986 to 1993 (the time period over which much of the model structure is based), the maximum interannual change in number of trips upon which silver hake was landed did not exceed 13%. This suggests that while different vessels may be involved in the fishery in any given year, the aggregate supply of effort appears to have been relatively constant. Second, silver hake is prosecuted as one of several small-mesh fisheries in an annual round or as one species in a mixed-trawl fishery. Given this mixed-species mixed-fishery context it was determined that available data was not adequate to support a credible behavioral model. Without such a behavioral model, the silver hake fishery economics are represented by a series of accounting relationships that determine the level of fishing activity and allocate landings and revenues across several fleets. The procedures used to develop these relationships are described below.

Ex-Vessel Price. Although dockside pricing of whiting is based on multiple grades for size and quality, only three categories are recorded in the NMFS data. These categories are based on size and their estimated corresponding ages are; juveniles (1–2), round (3–5), and king (6+). Annual landings and average ex-vessel prices by market category are reported in table 3. Note that data for the juveniles are not reported. This is due to the fact that no known markets for this product existed prior to 1993. A simple price model was developed and used in our earlier study to produce forecasts of ex-vessel prices under alternative management scenarios. We retain this approach in the present study as well.[3] In the model, prices are assumed to adjust to quantities supplied. This is based on observations that harvested supplies tend to be affected largely by stock conditions, environmental factors, and regulation and not by price (Lin, Richards, and Terry 1988; and Burton 1992). Alternatively, a structural model of silver hake demand might have been estimated or prices could have been held constant. Since we do not include an entry-exit component, nor do we attempt to estimate consumer benefits, the former approach was deemed beyond the scope of the study. Although a simplifying assumption in many bioeconomic models, (see for example, Gilbert 1988; Clarke,

---

[2] Under certain circumstances time spent engaged in the silver hake fishery may be exempt from days fished controls.

[3] Due to the recent development of the juvenile export product, data for this market category did not exist prior to 1994. Therefore, insufficient observations were available to estimate a price model for the juvenile market category.

Table 3
Landings and Average Annual Ex-Vessel Price by Market Category

| Year | Round (Ages 2–5) | | King (Age 6+) | |
	Landings (1,000 lbs)	Ex-Vessel Price ($/lbs)	Landings (1,000 lbs)	Ex-Vessel Price ($/lbs)
1987	32,698.7	0.353 (0.244)	1,079.8	0.504 (0.259)
1988	33,911.7	0.263 (0.164)	835.1	0.486 (0.216)
1989	37,585.8	0.248 (0.168)	670.2	0.480 (0.243)
1990	42,196.8	0.277 (0.180)	1,090.1	0.498 (0.198)
1991	34,107.0	0.410 (0.227)	1,182.9	0.605 (0.262)
1992	32,903.6	0.408 (0.264)	1,185.6	0.629 (0.264)
1993	34,385.4	0.407 (0.222)	828.9	0.752 (0.246)
1994	30,501.4	0.424 (0.240)	896.8	0.781 (0.251)

Note: Values in parentheses are standard deviations.

Yoshimoto, and Pooley 1992; Christensen and Vestergaard 1993; and Campbell, Hand, and Smith 1993) constant price was also rejected because it was deemed too simplistic and ignores observed price-quantity relationships in silver hake ex-vessel markets.

The details of the price modeling are described in Helser, Thunberg, and Mayo (1996) and are only discussed in general terms here.[4] Variables included in the price models were own-supply, supplies of groundfish substitutes (cod, haddock, pollock, white hake, and redfish), and wholesale market prices. Seasonality in price was modeled by incorporating quarterly dummy variables. The estimated price models are incorporated in the bioeconomic model in a manner similar to that used by Resosudarmo (1995). Specifically, price forecasts were obtained by using equation (6) (table 2) where the constant α_{cj}, was obtained by setting all exogenous variables to their conditional means, multiplying by the appropriate coefficient estimate and summing. Landings by market category, W_{cj}, were determined by summing landings at age [table 2, equation (4)] over appropriate age classes and multiplying by average weight at age [table 2, equation (5)].

Fleet Shares. Omission of an entry-exit model is equivalent to assuming that current and future vessel behavior will mimic the past. This is clearly an oversimplification of vessel behavior; however, many vessel owners do tend to follow "traditional" patterns of fishing activity. Further, management actions (whether intended or otherwise) often preserve historical access and levels of participation. In this manner, landings shares tend to be perpetuated. The landings shares by stock and market category were calculated for vessels falling into three size classes based on gross registered tons (grt):< 50 grt (class 1), 51–150 grt (class 2), and > 151 grt (class 3). The landings shares for these three fleets are reported in table 4 for the period 1987 to 1993. The landings shares are relatively stable over time with fleet shares in the northern stock indicating somewhat greater variability as compared to the southern stock. Landings by quarter (j) and market category (c) were allocated among the different fleets (v) by multiplying total landings by the fleet share γ_{cjv} [table 2, equation (7)].

Catch-per-Day Fished (CDF). Catch may be assumed to be a constant proportion of biomass (Hilborn and Walters 1992). Cooke and Beddington (1984) discuss a variety

---

[4] Reprints are available from the authors upon request.

Table 4
Annual Landings Shares by Fleet, Stock, and Market Category

Landings Shares by Fleet/Stock

	Northern Stock			Southern Stock		
	Class 2	Class 3	Class 4	Class 2	Class 3	Class 4
1987	0.43	0.55	0.02	0.12	0.67	0.21
1988	0.26	0.52	0.22	0.09	0.70	0.21
1989	0.24	0.39	0.37	0.17	0.64	0.19
1990	0.30	0.38	0.32	0.17	0.57	0.27
1991	0.26	0.31	0.43	0.14	0.60	0.26
1992	0.25	0.36	0.39	0.05	0.53	0.42
1993	0.23	0.51	0.26	0.04	0.75	0.21
Mean	0.28	0.43	0.29	0.11	0.64	0.25
St. dev.	0.07	0.09	0.14	0.05	0.08	0.08

Landings Shares by Fleet/Stock/King

	Northern Stock			Southern Stock		
	Class 2	Class 3	Class 4	Class 2	Class 3	Class 4
1987	0.33	0.65	0.02	0.14	0.55	0.31
1988	0.32	0.55	0.13	0.09	0.60	0.31
1989	0.45	0.38	0.17	0.11	0.47	0.42
1990	0.42	0.48	0.10	0.12	0.43	0.45
1991	0.31	0.51	0.19	0.06	0.53	0.40
1992	0.34	0.39	0.27	0.05	0.50	0.45
1993	0.25	0.51	0.25	0.03	0.61	0.36
Mean	0.35	0.49	0.16	0.09	0.53	0.38
St. dev.	0.07	0.09	0.08	0.04	0.07	0.06

Landings Shares by Fleet/Stock/Round

	Northern Stock			Southern Stock		
	Class 2	Class 3	Class 4	Class 2	Class 3	Class 4
1987	0.43	0.55	0.02	0.12	0.68	0.21
1988	0.26	0.52	0.22	0.09	0.70	0.21
1989	0.23	0.39	0.37	0.17	0.64	0.19
1990	0.29	0.38	0.33	0.17	0.57	0.26
1991	0.26	0.30	0.44	0.14	0.60	0.25
1992	0.24	0.35	0.40	0.05	0.53	0.42
1993	0.23	0.51	0.26	0.04	0.75	0.21
Mean	0.28	0.43	0.29	0.11	0.64	0.25
St. dev.	0.07	0.10	0.14	0.05	0.08	0.08

of cases in which catch may not simply be a linear function of stock size. Although, not universal (see Resosudarmo 1995, for example) variants of the Cobb-Douglas production function seem relatively common [Bjørndal 1988; Kennedy 1992; Campbell, Hand, and Smith 1993; Squires and Huppert 1988). To estimate CDF [table 2, equation (8)], a study fleet was constructed based on Northeast Fisheries Science Center (NEFSC) landings data. The study fleet was constructed by first identifying all vessels that landed at least one pound of silver hake in every year from 1982–92. Data on landings and trip duration were then extracted for every recorded trip upon which silver hake was landed. In addition to landings data, the following data on vessel characteristics were recorded: length, grt, and main engine horsepower. CDF was estimated as a log-linear Cobb-Douglas production function where landed catch was hypothesized to be positively related to fishing vessel capital and stock abundance. Fishing vessel capital was measured as vessel horsepower, and stock abundance, A_t, was measured as the estimated biomass of age 2+ individuals (1982–92). Seasonality effects were measured by using quarterly dummies. Interannual differences in productivity due to availability or weather conditions were assumed to be reflected in a time variable. Last, a stock dummy was introduced to reflect historic patterns of fishery development where southern New England ports (Point Judith, Rhode Island, in particular) have been the primary developers of domestic silver hake markets and fisheries. Therefore, CDF for the southern stock was hypothesized to be higher than that for the northern stock.

After correcting for first-order serial correlation, the resulting parameter estimates are reported in table 5. The relatively low model R^2 is not particularly surprising since a number of factors that might affect interannual and intervessel variability in CDF (skill level, availability, technology changes, to name just a few) could not be controlled. Several alternative model specifications using different measures of fishing capital (*e.g.*, vessel length and grt) and stock abundance (*e.g.*, survey indices of catch per tow) were attempted but were deemed unsatisfactory.

Fishing Effort. Within the current model structure, fishing effort is measured by its proxy, F. However, the fishing mortality rate alone provides little information about the realized vessel or fleet effort that would be consistent with its proxy. In the current study, realized fishing effort is defined in terms of numbers of days fished. Given a fishing mortality rate, selection pattern, and stock size, an *ad hoc* measure of realized effort can be calculated by multiplying realized total landed catch by fleet landings shares and dividing by CDF [table 2, equation (9)]. The resulting quotient provides an estimate of the number of days fished by each fleet (DF_{jtv}) that is implied by the selected fishing mortality rate.

Fleet Costs and Returns. Days fished for silver hake represent only a portion of an otter trawl vessel's activity during a fishing year. This required development of cost

Table 5
Catch Per Day Fished Coefficient Estimates

Variable	Coefficient	Standard Error
Intercept	55.10*	7.69
Horsepower	0.20*	0.11
Age 2+ biomass	0.76*	0.15
Year	−13.19*	1.78
Southern stock	0.47*	0.11
Quarter 2	0.27*	0.14
Quarter 3	0.13	0.15
Quarter 4	0.52*	0.14

Notes: $N = 1.671$. $R^2 = 0.06$. time series = 1982–92.
* Denotes statistically significant at the 0.05 level or greater.

and returns on the basis of numbers of days fished. Gross returns are easily computed by multiplying predicted prices by total landings [table 2, equation (10)]. Costs per day fished were estimated using cost data from vessels participating in the vessel Capital Construction Fund (CCF). The CCF is a Federal program that offers certain tax advantages to fund planned new vessel construction or retrofitting of existing vessels. One of the participation requirements is provision of annual tax returns. Gautam and Kitts (1996) have developed the tax return data into a cost and earnings data base including variable and fixed cost categories, gross revenues, and vessel characteristics data. Additionally, vessel records have been matched with NEFSC landings and effort data. The resulting database makes it possible to relate vessel costs to vessel characteristics and vessel activity levels in terms of time spent fishing. To estimate costs per day fished in the silver hake fishery, a simple OLS model of vessel operating costs as a function of vessel characteristics was specified. An easier approach would have been to use the CCF data to compute mean variable costs by vessel class. Unfortunately, no observations were available for smaller (Class 2) vessels and while the problem of making out-of-sample forecasts for vessels that are not represented in the database is recognized, the regression approach was adopted to make inferences about vessels that were not CCF program participants.

Variable costs included (ice, water, food, fuel, oil, gear, supplies and lumping, auction, and packing fees). Since only vessel or fleet activity while engaged in the silver hake fishery was of interest, fixed costs were omitted. Crew costs were also omitted because crew remuneration is based on landings value and not on an hourly or daily wage. The results of the statistical cost equation are reported in table 6. As long as technology is static and operating costs are largely a function of vessel characteristics, then vessel costs per day fished can be treated as a constant. Thus, the coefficient, 1, [table 2, equation (11)] represents the mean estimated cost per day fished based on the statistical cost equation and mean vessel characteristics of the study fleets. Quarterly variable costs by fleet are determined by multiplying λ_v by DF_{jtv}.

Net returns above variable costs are calculated by subtracting gross revenues from total operating costs [table 2, equation (12)]. Positive returns indicate residual revenues that can be used to pay labor (crew and captain) and fixed costs. Negative returns suggest that vessels may not be able to cover their trip costs from the sale of silver hake alone. Even though a relatively high degree of specific targeting takes place for silver hake, bycatch of other species does occur and revenues from the sale of species other than silver hake may be an important source of trip income. Thus, negative returns to silver hake may not necessarily be indicative of an unprofitable fishing activity.

Table 6
Coefficients for Cost Per Day Fished

Variable	Coefficient	Standard Error
Intercept	619.372	506.533
Horsepower	5.551*	0.551
Gross registered tons	−18.756*	4.552

Notes: $N = 87$, $R^2 = 0.5527$, F = 54.13, time series = 1983–90. Predicted cost per day Fished: Class 2 = $1,323, Class 3 = $1,036, and Class 4 = $1,658. * Denotes statistically significant at the 0.01 level or greater.

Simulation Methods

Since there are two control variables, it is possible to systematically vary the selection pattern and the fishing mortality rate. This approach was taken in earlier work (Helser, Thunberg, and Mayo 1996) to produce a response surface which elucidated the tradeoffs between the two control variables. This was accomplished, however, under assumptions of constant recruitment and long-run equilibrium conditions. For the current study, recruitment was generated for each year for a ten year trajectory by using Monte Carlo methods. In each realization a random variate is drawn from a normal distribution with mean 0 and variance σ_w^2 from the estimated Ricker stock- recruitment model. The variate is entered as a multiplicative term to the expected value of recruitment conditional on SSB and the Ricker parameter values [table 2; equation (1)] (Hilborn and Walters 1992). The result is a simulated distribution of recruitment about mean recruitment at a given SSB level. Empirical means and standard deviations for biological and economic variables were estimated from 500 random realizations from the Ricker stock-recruitment relationship for each time period.

In all, three scenarios were evaluated. First a baseline condition was developed. The baseline mimics, to the extent possible, fishery conditions that existed prior to the development of a juvenile export market. Previous analysis indicated that the baseline selection pattern was approximately 50% at age 2 (Helser and Mayo 1994). Holding the selection pattern constant, the fishing mortality rate was systematically varied until stable (*i.e.*, the distribution around SSB for each projection year was relatively constant) SSB trajectories were produced. Thus the baseline condition was defined as being a selection pattern consistent with 50% selection at age 2 and fishing mortality rates of $F = 0.25$ and $F = 0.70$ for the northern and southern silver hake stocks respectively.

To examine the implications of a significant effort shift toward directed fishing on juveniles, a second scenario was developed in which the selection pattern was adjusted to 50% at age 1. Although short-run gains may be produced, questions regarding the biological sustainability of such a fishing strategy have been raised. The economic value of the fishery may also be compromised should increased juvenile takes result in lower availability of older, more valuable individuals. A similar question may be raised under the baseline scenario since population surveys already indicate a truncated age distribution. Thus, the question of whether increased fishery returns could be achieved by delaying age at first capture is examined by introducing a third scenario in which the selection pattern is set at 50% at age 3. In each of these scenarios fishing mortality rates were held constant at the baseline level. The starting values for the biological variables for the baseline and alternative scenarios are reported in table 7.

Results

For each of the three scenarios, means and standard deviations for annual projections of SSB, fishery yield, returns above operating costs, and days fished by stock are reported in table 8. In table 8 the scenarios are denoted as PR_i with the subscript denoting 50% selectivity at age. Note that recruitment is the sole source of variability in the simulation. This means that the simulation statistics reported in table 8 do not reflect the full extent of uncertainty associated with the model projections. Not captured in the estimated standard deviations are forecast errors associated with the estimated price, cost, and catch models. Similarly, the simulation statistics do not reflect a variety of biological uncertainties, such as, maturity-at-age, growth rates, natural mortality rates, or selectivity-at-age. The results reported below must be interpreted in light of these economic and biological uncertainties.

Table 7
Input Parameter Values Used in Model Simulations

Parameters[1]	Age-class						
	1	2	3	4	5	6	7+

Northern Stock							

Selection Pattern

50% at age 1	0.500	1.000	1.000	1.000	1.000	1.000	1.000
50% at age 2	0.281	0.500	1.000	1.000	1.000	1.000	1.000
50% at age 3	0.140	0.281	0.500	1.000	1.000	1.000	1.000

Fraction Mortality Pre-Spawning

Natural (m)	0.500	0.500	0.500	0.500	0.500	0.500	0.500
Fishing (f)	0.660	0.660	0.660	0.660	0.660	0.660	0.660
Landed catch weight (w_i)	0.078	0.144	0.218	0.301	0.394	0.559	0.674
Stock weight (s_i)	0.017	0.076	0.177	0.312	0.466	0.632	0.674
Nat. mortality (M)	0.400	0.400	0.400	0.400	0.400	0.400	0.400
Maturity (p_i)	0.080	0.700	0.990	1.000	1.000	1.000	1.000
Recruitment $(N_1)^2$	$a = 5.057$						
	$b = 0.0021$						
	$\sigma_w^2 = 0.6024$						

Southern Stock							

Selection Pattern

50% at age 1	0.500	1.000	1.000	1.000	1.000	1.000	1.000
50% at age 2	0.100	0.500	1.000	1.000	1.000	1.000	1.000
50% at age 3	0.050	0.250	0.500	1.000	1.000	1.000	1.000

Fraction Mortality Pre-Spawning

Natural (m)	0.500	0.500	0.500	0.500	0.500	0.500	0.500
Fishing (f)	0.500	0.500	0.500	0.500	0.500	0.500	0.500
Landed catch weight (w_i)	0.075	0.122	0.178	0.238	0.322	0.425	0.522
Stock weight (s_i)	0.012	0.091	0.216	0.348	0.442	0.522	0.551
Nat. mortality (M)	0.400	0.400	0.400	0.400	0.400	0.400	0.400
Maturity (p_i)	0.090	0.800	0.99	1.000	1.000	1.000	1.000
Recruitment $(N_1)^2$	$a = 9.4418$						
	$b = 0.0022$						
	$\sigma_w^2 = 0.5910$						

[1] Input parameters taken from Helser and Mayo (1994).
[2] Paramater estimates from Ricker function.

Given starting stock sizes and age structure, SSB for both stocks were projected to increase gradually over time under baseline exploitation rates and selectivity. For the northern stock, a change in selectivity favoring younger age classes resulted in a slight lowering of mean SSB over initial levels, but did not result in any appreciable decline in SSB. Altering the selection pattern to 50% at age 3 resulted in a more rapid buildup in SSB as compared to baseline conditions. By contrast, even though the selection patterns were applied equally to both stocks, the SSB trajectories show greater impacts on the southern stock. The greater impact is due to the fact that the

Table 8

Simulation Results for Annual Spawning Stock Biomass, Yield, Days Fished, and Returns Net of Operating Costs by Stock

Year	Stock Biomass (1,000 metric tons)			Potential Yield (1,000 metric tons)			Net Returns ($1,000,000)			Days Fished (1,000s)		
	PR_1	PR_2	PR_3	PR_1	PR_2	PR_3	PR_1	PR_2	PR_3	PR_1	PR_2	PR_3
						Northern Stock						
1	20.1 (0.2)	20.4 (0.2)	22.8 (0.2)	5.9 (1.0)	5.4 (0.6)	5.4 (0.3)	-0.5 (0.0)	-0.6 (0.0)	-0.6 (0.0)	3.8 (0.0)	3.8 (0.0)	4.0 (0.0)
2	18.9 (3.1)	19.8 (3.7)	22.6 (4.4)	6.3 (2.4)	5.1 (1.3)	4.8 (0.9)	-1.1 (0.0)	-1.2 (0.0)	-1.1 (0.0)	3.7 (0.0)	3.8 (0.0)	3.8 (0.0)
3	18.9 (6.1)	20.8 (7.3)	25.3 (9.7)	6.5 (2.7)	5.7 (2.5)	4.7 (1.6)	-0.8 (0.2)	-0.9 (0.2)	-0.4 (0.3)	4.0 (0.8)	4.2 (0.9)	3.5 (0.4)
4	18.3 (7.7)	21.2 (9.4)	27.1 (12.6)	6.6 (3.1)	6.1 (2.8)	5.6 (2.7)	-0.9 (0.3)	-1.0 (0.4)	-0.5 (0.5)	4.2 (0.8)	4.4 (0.9)	3.9 (0.6)
5	16.4 (9.0)	20.2 (11.4)	27.4 (15.6)	6.2 (3.4)	5.6 (3.3)	5.6 (3.1)	-1.0 (0.4)	-1.2 (0.5)	-0.7 (0.5)	3.7 (0.9)	4.1 (1.0)	3.6 (0.7)
6	18.1 (10.1)	23.1 (13.1)	32.4 (18.6)	6.5 (4.0)	6.5 (3.8)	6.5 (3.8)	-1.3 (0.7)	-1.3 (0.8)	-0.7 (0.9)	4.4 (0.8)	4.7 (0.9)	4.2 (0.6)
7	19.1 (11.6)	25.1 (15.2)	36.7 (23.1)	6.5 (4.6)	7.0 (4.6)	7.4 (4.5)	-0.9 (0.7)	-1.0 (0.8)	-0.3 (1.0)	4.4 (0.9)	4.8 (1.1)	4.3 (0.7)
8	18.3 (12.4)	25.3 (17.6)	38.6 (27.2)	6.4 (4.4)	7.1 (5.1)	7.9 (5.3)	-0.7 (0.8)	-0.7 (0.9)	-0.01 (1.1)	4.1 (0.9)	4.5 (1.0)	4.2 (0.8)
9	18.3 (12.6)	26.1 (18.5)	41.4 (29.1)	6.5 (4.5)	7.3 (5.1)	8.4 (6.1)	-0.7 (0.8)	-0.7 (0.9)	-0.01 (1.2)	4.0 (0.8)	4.5 (0.9)	4.2 (0.8)
10	18.3 (13.2)	27.3 (19.8)	45.0 (32.4)	6.4 (4.7)	7.6 (5.6)	9.0 (6.3)	-0.6 (0.8)	-0.6 (0.9)	0.1 (1.3)	4.1 (1.0)	4.6 (1.2)	4.3 (0.8)
						Southern Stock						
1	18.6 (0.2)	24.3 (0.3)	33.6 (0.3)	11.2 (2.3)	10.6 (0.6)	8.7 (0.3)	2.9 (0.0)	3.3 (0.0)	3.1 (0.0)	2.9 (0.0)	3.5 (0.0)	3.1 (0.0)
2	15.9 (5.5)	25.4 (10.5)	36.7 (13.1)	10.4 (4.7)	10.4 (3.8)	10.5 (2.4)	0.6 (0.0)	1.5 (0.0)	2.7 (0.0)	3.4 (0.0)	4.0 (0.0)	3.9 (0.0)
3	14.3 (6.8)	27.5 (14.5)	45.7 (24.7)	9.3 (4.7)	11.7 (6.2)	11.6 (4.9)	0.4 (0.9)	2.0 (0.2)	3.3 (1.9)	3.6 (0.7)	4.4 (0.9)	3.9 (0.4)
4	12.8 (7.4)	29.6 (18.6)	55.0 (32.3)	8.3 (5.2)	12.4 (7.4)	14.6 (8.2)	0.1 (0.8)	2.1 (0.2)	4.4 (2.9)	3.4 (0.6)	4.4 (0.8)	4.2 (0.6)
5	11.4 (7.5)	31.3 (22.0)	64.3 (42.8)	7.5 (4.9)	13.3 (9.2)	17.2 (10.2)	-0.1 (0.9)	2.2 (0.3)	5.0 (3.5)	3.3 (0.6)	4.4 (0.9)	4.3 (0.6)
6	10.5 (7.4)	34.5 (25.8)	77.6 (53.4)	6.9 (5.4)	14.4 (10.4)	20.4 (13.4)	-0.3 (0.9)	2.3 (0.3)	5.8 (4.4)	3.2 (0.6)	4.4 (0.9)	4.4 (0.7)
7	9.6 (8.3)	37.3 (33.4)	92.6 (72.9)	6.1 (5.6)	15.7 (13.4)	24.2 (17.1)	-0.4 (0.8)	2.6 (0.3)	6.9 (5.3)	3.2 (0.7)	4.6 (1.1)	4.7 (0.8)
8	8.4 (7.6)	38.2 (34.6)	103.1 (83.5)	5.4 (4.7)	16.1 (14.7)	27.9 (21.2)	-0.5 (0.9)	2.7 (0.4)	7.9 (6.5)	3.0 (0.6)	4.6 (1.1)	4.9 (0.9)
9	7.5 (6.7)	40.2 (34.3)	114.9 (84.0)	4.9 (4.4)	16.8 (14.0)	31.2 (23.2)	-0.6 (0.7)	2.7 (0.3)	8.4 (6.5)	2.9 (0.6)	4.6 (1.1)	5.0 (0.9)
10	6.8 (6.3)	42.2 (36.1)	129.4 (94.5)	4.4 (4.1)	17.7 (15.1)	34.6 (23.9)	-0.6 (0.6)	2.8 (0.4)	9.1 (6.7)	2.8 (0.7)	4.7 (1.2)	5.1 (0.9)

Note: Figures in parentheses are standard deviations.

initial southern stock population structure had proportionally greater biomass in younger age classes than the northern stock. This means that targeting of younger fish in the south removes proportionally more fish and leads to greater reductions in SSB over time as compared to the northern stock. Similarly, as targeting is shifted to older age classes, proportionally more fish become contributors to SSB resulting in a more rapid build-up of SSB in the southern stock.

The projected pattern of SSB over time indicates that neither the baseline nor the Pr_1 scenarios would be consistent with the requirements of the SFA. At this writing, the biological reference points for SSB levels consistent with MSY have not been determined hence, whether or not the Pr_3 SSB trajectories would meet the SFA requirements is not known. The current model configuration could be used to examine the bioeconomic implications of MSY biological reference points once they have been established.

Fishery yield follows the same general pattern as that of SSB with a gradual increase in yield coming from each stock over time under baseline conditions. Mean fishery yield from the northern stock is virtually constant as the fishery shifts to juvenile fish while yield increases at a slightly greater rate as catch is shifted to older fish. By contrast, yield from the southern stock was projected to suffer as effort shifted to younger age classes and expanded at a considerably faster rate as effort shifted to older age classes. The reader will note that fishery yield is denoted as "potential" yield. Current management proposals would potentially eliminate the juvenile market category. Thus, for purposes of analysis, landings of the juvenile market category were deducted from the calculation of net returns to labor and fixed costs.

After deducting juvenile revenues, projected net returns above operating costs were negative under all scenarios for the northern stock.[5] The baseline and 50% selectivity at age 1 scenarios resulted in roughly equivalent projections of mean net returns for the north. For both the northern and southern stocks the standard deviation for years one and two are zero. Symptomatic of the declines in landings and SSB with a selection pattern of 50% at age 1, fishery returns start out positive for the southern stock, but decline and turn negative by the midpoint of the ten year projection period. Returns under the baseline scenario remain positive and relatively constant for the southern stock, but increase steadily under a selection pattern favoring delayed capture at age. The increased returns are made possible by taking advantage of slight reductions in fishery yield in early years followed by increased yield from all market categories during the latter half of the projection time period.

Projections of mean annual revenues by market category and selection pattern are shown in figure 3. The value of forgone juvenile revenues provides a measure of the opportunity cost of management action that would prohibit landings of juveniles. The top portion of figure 3 illustrates the potential revenue streams that would be foregone in the event that a minimum size limit would eliminate the juvenile market category. Adding the juvenile revenues to net returns resulted in positive returns to labor and fixed costs under all scenarios. However, due to declines in revenues from all market categories from years 5–10, returns under a selection pattern of 50% at age 1 were positive but declined steadily. Compared to the baseline, a reduction in the age at first capture results in higher levels of juvenile revenues during the first five years of the simulation horizon. By that year, however, aggregate SSB is halved, and juvenile revenues lie below the baseline throughout the remainder of the simula-

---

[5] The reader will note that the standard deviation for net return in table 8 is zero for years one and two. Juveniles recruit into the next larger market category at age 2. Therefore, for the first two years of the simulation period the only source of variability in fishery yield is in juvenile catches. Once these catches are removed from fishery revenues, net returns are constants until new recruits grow into the round market category.

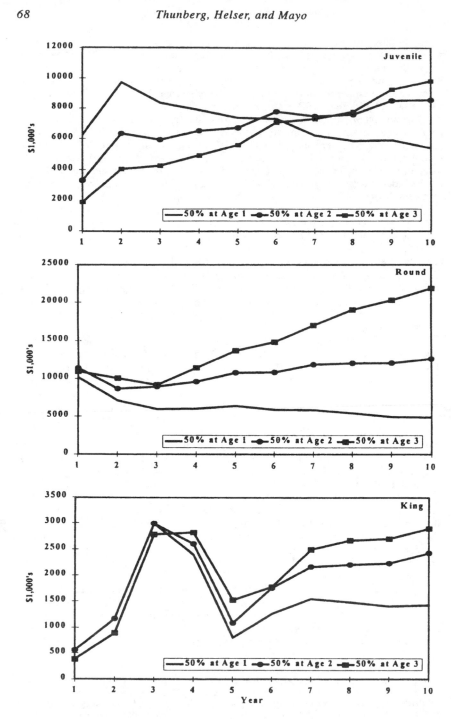

Figure 3. Mean Annual Revenues by Market Category and Selection Pattern

tion period. Although the full impact on the juvenile market may not be felt for several years, the impact of changing the fishery selection pattern is more immediate. As larger quantities of juveniles are taken, fewer individuals remain to enter the next larger market category. This phenomenon is illustrated in figure 3 where increased juvenile takes result in an immediate and persistent downward trend in revenues from the round market category and in later years in the king market category. By contrast, increasing the age at first capture leaves more individuals to enter the next larger market category. Consequently, projected revenues of round silver hake generally increase throughout the simulation horizon as targeting is shifted to older age classes. Projected landings of silver hake in the king market category follow the same general pattern as the round category with declining revenues as age at entry is reduced, and increasing revenues as the selection pattern is shifted to older fish. The reader will note that the revenue trajectories exhibit pronounced changes over the simulation horizon particularly in the projections for the king market category. These changes are due to the movement of the starting population age classes through each time step. For this reason, the first 5–6 years of the simulation horizon represents a transition period to the longer run conditions in the later years of the simulation.

The last three columns of table 8 report the estimated number of days fished that would be consistent with the combination of fishing mortality and selection patterns selected for the study. With no change in the fishery selection pattern, mean days fished remains virtually constant. As the selection pattern is altered to reflect targeting of younger age classes, reductions in fishery landings and revenues are reflected in a general downward trend in projected days fished, particularly for the southern stock. Conversely, projected days fished follows an increasing trend as the selection pattern favors older fish.

Projected annual returns (less juvenile revenues) above operating costs per day fished by vessel class are shown in figure 4. For Class 2 vessels insufficient revenues are earned from the sale of silver hake alone to cover total operating costs under any of the selection patterns evaluated for this study. However, without an estimate of revenues from the sale of other species it would be premature to conclude that these vessels are not profitable. Estimated net returns for Class 3 vessels indicate that vessels in this fleet would earn sufficient silver hake revenues to cover all operating costs under any of the simulated selection patterns. However, the results indicate a general downward trend in returns per day fished as effort is redirected to younger age classes. Among Class 3 vessels the opposite trend emerges as the selection pattern is shifted to older age classes. For larger, Class 4 vessels returns per day fished are close to break-even under the baseline, but decline over time with a selection pattern favoring younger silver hake. Projected returns per day fished are substantially improved, however, with a selection pattern favoring older age classes.

Under the baseline condition the present value of mean returns to labor and fixed costs was $11.0 million over the ten year simulation period exclusive of juvenile revenues.[6] Adjusting the selection pattern to 50% at age 1 resulted in a present value of –$4.1 million. Increasing age at entry to the fishery to 50% at age 3 resulted in a present value of $35.7 million. Including juvenile revenues narrows the difference between the three scenarios somewhat; $50.0, $60.7, and $78.8 million for the 50% at age 1, baseline, and 50% at age 3 scenarios respectively. Compared to the other scenarios, the 50% at age 1 selection pattern yielded higher revenue streams during the first 3–4 years of the simulation period. These revenue streams were not sustainable

---

[6] A discount rate of 7% was used. While debate over private and public discount rates is duly acknowledged, the U.S. Office of Management and Budget mandates that 7% be used for evaluating public investment projects.

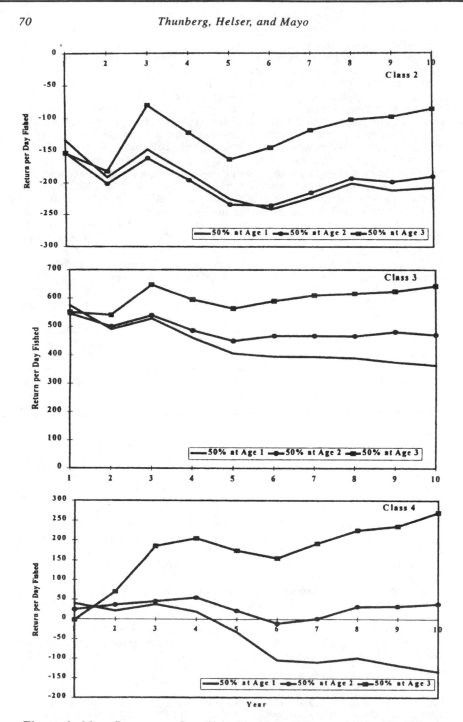

Figure 4. Mean Returns per Day Fished by Vessel Class and Selection Pattern

as increased takes of juveniles resulted in continuously declining biomass levels hence declining landings and revenues from all market categories over time.

Mean present value of returns to labor and fixed costs provides one basis for comparing management alternatives. Figure 5 shows the empirical cumulative probability distributions on present value of returns (exclusive of juvenile revenues) for each of the three scenarios evaluated for this study. Since the error structure of the stock-recruitment function was lognormal the probability distribution for present value of returns was also lognormal. For the baseline scenario, the probability that returns above operating costs will be zero or negative was 35%. By contrast, the probability of zero or negative returns for the 50% at age 1 selection pattern was 88.2%, and the probability of zero or negative returns for the 50% at age 3 selection pattern was less than 1%.

Conclusions

The bioeconomic simulation developed herein provides a framework within which expected biological and economic benefits to the U.S. Atlantic silver hake fishery can be evaluated. While a number of economic and biological uncertainties are not captured in the simulation itself, the principal bioeconomic relationships that characterize the silver hake fishery are embedded in the model. The simulation results indicate that a 50% at age 1 selection pattern, that may be expected from a targeted U.S. juvenile fishery, could reduce economic benefits and reduce long-term stock sizes. In contrast, simulation results indicate that a selection pattern delaying 50% selection to age 3 could result in increased long-term economic and biological benefits as compared to the baseline. Under such a strategy, longer term fishery yields are higher for all market categories including juveniles. This indicates that under certain conditions a targeted juvenile fishery may be able to co-exist with the "traditional" fishery targeting older age classes without compromising current management objectives for silver hake stocks.

The focus on the selection pattern as a management instrument raises the issue

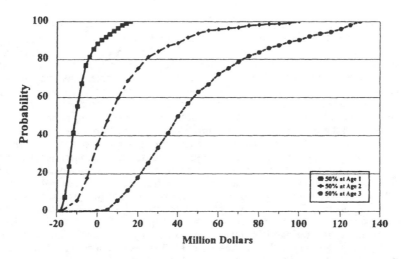

Figure 5. Cumulative Probability for Present Value of Net Returns

of how to convert the analytical findings to operational rules and regulations. As defined, a target selection pattern could be achieved in any number or combination of methods ranging from gear modifications to times and areas fished. For example, historical data on seasonal patterns of juvenile and adult distributions could be used to identify times and areas where adult aggregations would be targeted using larger mesh, while at other times or areas a targeted juvenile fishery might be prosecuted. Incorporation of NEFSC survey data to allow finer detail in time and space may improve the ability to refine the analysis to evaluate such strategies.

A desired selection pattern may also be achieved through design of incentive-compatible management regimes. The effectiveness of input and/or time/area controls may be difficult to predict, hard to monitor, and may create undesired economic incentives to engage in capital stuffing or input substitution. Various forms of property rights-based strategies (territorial use rights or individual transferable quotas, for example) may provide the incentive to maximize the value of the various components of the silver hake stocks.

Under current regulations, there is a moratorium on the issuance of new multispecies permits, but the moratorium does not affect targeting behaviors of individual vessels that are currently permitted. Thus, omission of an explicit model of entry and exit patterns in the silver hake fishery remains a limitation in the current model. At this time, projected silver hake landings are driven by an assigned fishing mortality rate; the implied assumption being that fishing effort will simply materialize or evaporate depending upon the fishing mortality rate. The result is a disconnect between management objectives as reflected by target fishing mortality rates and vessel owner behavior. This disconnect may significantly reduce the reliability of forecasted impacts. For example, improvements in landings and/or fishery revenues are often followed by increased fishing effort. This scenario is consistent with the projected increase in days fished under a selection pattern of 50% at age 3 (table 8). Unfortunately, fishing effort tends to persist even in the face of declining landings and revenues. This phenomenon is not captured in the simulation results as total days fished were projected to decline as landings and revenues fall under the 50% at age 1 scenario. The consequence of this forecast error may be that (*a*) effective fishing mortality rates may be higher than projected, (*b*) SSB levels may be depleted at a rate faster than projected, and (*c*) landings and revenue projections may be lower than projected.

An alternative formulation of the model would begin with specification of a behavioral entry-exit model to provide projections of effective effort from which a fishing mortality rate would then be back-calculated. Formulated in this manner, management actions could be evaluated within the context of their ability to effect behavioral change in the fishery to accomplish the desired management objective. This approach may be more important in cases where fish stocks are on the decline. For example, if fishing effort tends to persist in the face of stock declines, projections based on fishing mortality rates may consistently underestimate effective fishing effort.

In spite of the lack of an entry-exit model, the present formulation does provide a useful analytical tool to evaluate the implications of fishery management that may affect the age structure of silver hake stocks. Given the fact that price premiums are paid based on individual size, the age-structured bioeconomic model provides greater capability to identify management strategies that may enhance the value of U.S. Atlantic silver hake resources. Further, incorporation of the fleet component offers two advantages. First, distributional effects across fleets can be evaluated. Second, conversion of fishing mortality rates into equivalent days fished provides a mechanism for monitoring effort levels. The latter may be of particular utility since groundfish management in the northeastern U.S. has adopted a set of effort management tools that rely heavily on reductions in days fished.

References

Anderson, L.G. 1989. Optimal Intra- and Interseasonal Harvesting Strategies When Price Varies With Individual Size. *Marine Resource Economics* 6(2):145–62.

Anderson, E.D., F.E. Lux, and F.P. Almeida. 1980. The Silver Hake Stocks and Fishery Off the Northeastern United States. *Marine Fisheries Review* 4(1):12–20.

Beverton, R.J.H., and S.J. Holt. *On the Dynamics of Exploited Fish Populations*. Fisheries Investment Series 2, Vol. 19 U.K Ministry of Agriculture and Fisheries, London.

Bigelow, H.B., and W.C. Schroeder. 1953. Fishes of the Gulf of Maine. *United States Fishery Bulletin* 53(74):1–577.

Bjørndal, T. 1988. The Optimal Management of North Sea Herring. *Journal of Environmental Economics and Management* 15:9–29.

Botsford, L.W., J.E. Wilen, and E.J. Richardson. 1986. Biological and Economic Analysis of Lobster Fishery Policy in Maine. A Report Submitted to the Committee on Maine Resources.

Burton, M.P. 1992. The Demand for Wet Fish in Great Britain. *Marine Resource Economics* 7(2):57–66.

Campbell, H.F., A.J. Hand, and A.D.M. Smith. 1993. A Bioeconomic Model for Management of Orange Roughy Stocks. *Marine Resource Economics* 8(2):155–72.

Christensen, S., and N. Vestergaard. 1993. A Bioeconomic Analysis of the Greenland Shrimp Fishery in the Davis Strait. *Marine Resource Economics* 8(4):345–65.

Clarke, R.P., S.S. Yoshimoto, and S.G. Pooley. 1992. A Bioeconomic Analysis of the Northwestern Hawaiian Islands Lobster Fishery. *Marine Resource Economics* 7(3):115–40.

Conrad, J.M. 1982. Management of a Multiple Cohort Fishery: The Hard Clam in Great South Bay. *American Journal of Agricultural Economics* 64(3):461–74.

Cooke, J.G., and J.R. Beddington. 1984. The Relationship Between Catch Rates and Abundance in Fisheries. *IMA Journal of Mathematics Applied in Medicine and Biology* 1:391–405.

Edwards, R.L. 1968. Fishery Resources of the North Atlantic Area. In *The Future of the Fishing Industry of the United States*, D. Gilbert, ed. University of Washington Press.

Fritz, R.L. 1960. A Review of the Atlantic Coast Whiting Fishery. *Commercial Fisheries Review* 22(11):1–11.

Gates, J.M. 1974. Demand Price, Fish Size and the Price of Fish. *Canadian Journal of Agricultural Economics* 22(3):1–12.

Gautam, A.B., and A.D. Kitts. 1996. Data Description and Statistical Summary of the 1983–92 Cost-Earnings Data Base for Northeast U.S. Commercial Fishing Vessels. NOAA Technical Memorandum NMFS-NE-112. Northeast Fisheries Science Center, Woods Hole, MA.

Gilbert, D.J. 1988. Use of a Simple Age-Structured Bioeconomic Model to Estimate Optimal Long-Run Surpluses. *Marine Resource Economics* 5(1):23–42.

Helser, T.E., and R.K. Mayo. 1994. Estimation of Discards in the Silver Hake Fisheries and a Re-analysis of the Long-Term Yield from the Stocks. National Marine Fisheries Service, Northeast Fisheries Science Center Reference Document 94-01, Woods Hole, MA.

Helser, T.E., F.P. Almeida, and D.E. Waldron. 1995. Biology and Fisheries of Northwest Atlantic Hake (Silver Hake: *M. bilinearis*). In *Hake: Fisheries, Products and Markets*, J.A. Alheit and T. Pitcher, eds., pp. 203–37. London: Chapman and Hall.

Helser, T.E., E.M. Thunberg, and R.K. Mayo. 1996. An Age-Structured Bioeconomic Simulation of the United States Silver Hake Fisheries. *North American Journal of Fisheries Management* 16:783–94.

Hennemuth, R.C., and S. Rockwell. 1987. History of Fisheries Conservation and Management. In *Georges Bank*, R.H. Backus and D.W. Bourne, eds., pp. 431–66. Cambridge, MA: MIT Press.

Hilborn, R., and C.J. Walters. 1992. *Quantitative Fisheries Stock Assessment.* New York: Chapman and Hall.

Kennedy, J.O.S. 1992. Optimal Annual Changes in Harvests from Multicohort Fish Stocks: The Case of Western Mackerel. *Marine Resource Economics* 7:95–114.

Leslie, P.H. 1945. On the Use of Matrices in Certain Population Mathematics. *Biometrika* 33:183–212.

Lin, B.H., H.S. Richards, and J.M. Terry. 1988. An Analysis of Exvessel Demand for Pacific Halibut. *Marine Resource Economics* 4(4):305–14.

National Marine Fisheries Service. 1996. Our Living Oceans Report on the Status of U.S. Living Marine Resources, 1995. U.S. Department of Commerce, NOAA Technical Memorandum NMFS-F/SPO-19.

National Marine Fisheries Service. 1997. Status of Fisheries of the United States. First Annual Report to Congress. National Marine Fisheries Service, Silver Springs, MD.

Overholtz, W.J., S.F. Edwards, and J.K.T. Brodziak. 1993. Strategies for Rebuilding and Harvesting New England Groundfish Resources. Proceedings of the International Symposium on Management Strategies for Exploited Fish Populations. AK-SG-93-02. Alaska Sea Grant College Program.

Resosudarmo, B.P. 1995. The Construction of a Bioeconomic Model of the Indonesian Flying Fish Fishery. *Marine Resource Economics* 10:357–72.

Richardson, E.J., and J.M. Gates. 1986. Economic Benefits of American Lobster Fishery Management Regulations. *Marine Resource Economics* 2(4):353–82.

Ricker, W.E. 1954. Stock and Recruitment. *Journal of the Fisheries Research Board of Canada* 11:559–623.

Shepherd, J.G. 1982. A Versatile New Stock-Recruitment Relationship of Fisheries and Construction of Sustainable-Yield Curves. *Journal du Conseil*, Conseil International pour l'Exploration de la Mer. 40:67–75.

Squires, D., and D.D. Huppert. 1988. Measuring Harvest Capacity in the Pacific Coast Groundfish Fleet. Administrative Report LJ-88-24. Southwest Fisheries Science Center, La Jolla, CA.

Thompson, W.F., and F.H. Bell. 1934. Biological Statistics of the Pacific Halibut Fishery. Effects of Changes in Fishing Intensity Upon Total Yield and Yield per Unit of Gear. Report of the International Fishery (Pacific Halibut) Commission, 8.

Wang, S.D.H., and C.B. Kellogg. 1988. An Econometric Model for American Lobster. *Marine Resource Economics* 5(1):61–70.

Wilen, J.E., and F.R. Homans. 1992. Marketing Losses in Regulated Open Access Fisheries. Sixth International Conference of International Institute of Fisheries Economics and Trade (IIFET).

Name Index